Medicinal Chemistry: A Practical Approach

Medicinal Chemistry: A Practical Approach

Editor: Kayla Nicholson

STATES
ACADEMIC PRESS
www.statesacademicpress.com

Published by States Academic Press
109 South 5th Street,
Brooklyn, NY 11249, USA
www.statesacademicpress.com

Medicinal Chemistry: A Practical Approach
Edited by Kayla Nicholson

International Standard Book Number: 978-1-63989-347-8 (Hardback)

Cataloging-in-Publication Data

Medicinal chemistry : a practical approach / edited by Kayla Nicholson.
p. cm.
Includes bibliographical references and index.
ISBN 978-1-63989-347-8
1. Pharmaceutical chemistry. 2. Chemistry. I. Nicholson, Kayla.
RS403 .M43 2022
615.19--dc23

Contents

Preface

The intersection of chemistry and pharmacology is called medicinal chemistry. This discipline is concerned with the designing, development and chemical synthesis of pharmaceutical drugs. It focuses on the identification and synthetic alteration of new chemical entities in order to make them appropriate for therapeutic use. Medicinal chemistry also studies the existing drugs and agents and their structure-activity relationships. The discovery of pharmaceutical drugs involves the assay of compounds for carrying out any desired biological activity. This happens through repurposing existing pharmaceutical agents towards any new pathologic process. Structural analysis in medicinal chemistry focuses on assessing the structure of pharmaceuticals. It helps in predicting their stability, efficacy and accessibility. This book brings forth some of the most innovative concepts and elucidates the unexplored aspects of this discipline. Different approaches, evaluations, methodologies and advanced studies on medicinal chemistry have been included herein. In this book, using case studies and examples, constant effort has been made to make the understanding of the difficult concepts of this field, as easy and informative as possible, for the readers.

Various studies have approached the subject by analyzing it with a single perspective, but the present book provides diverse methodologies and techniques to address this field. This book contains theories and applications needed for understanding the subject from different perspectives. The aim is to keep the readers informed about the progresses in the field; therefore, the contributions were carefully examined to compile novel researches by specialists from across the globe.

Indeed, the job of the editor is the most crucial and challenging in compiling all chapters into a single book. In the end, I would extend my sincere thanks to the chapter authors for their profound work. I am also thankful for the support provided by my family and colleagues during the compilation of this book.

Editor

1

Synthesis and *In Vitro* AMPK Activation of Cycloalkyl/Alkarylbiguanides with Robust *In Vivo* Antihyperglycemic Action

Erika Gutierrez-Lara,[1,2] Carlos Martínez-Conde,[1] Edgar Rosales-Ortega,[1] Juan José Ramírez-Espinosa,[1] Julio C. Rivera-Leyva,[1] David Centurión,[2] Karla Carvajal,[3] Daniel Ortega-Cuellar,[3] Samuel Estrada-Soto,[1] and Gabriel Navarrete-Vázquez[1]

[1]*Facultad de Farmacia, Universidad Autónoma del Estado de Morelos, 62209 Cuernavaca, MOR, Mexico*
[2]*Departamento de Farmacobiología, Cinvestav-Coapa, 14330 Ciudad de México, Mexico*
[3]*Laboratorio de Nutrición Experimental, Instituto Nacional de Pediatría, 04150 Ciudad de México, Mexico*

Correspondence should be addressed to Samuel Estrada-Soto; enoch@uaem.mx
and Gabriel Navarrete-Vázquez; gabriel_navarrete@uaem.mx

Academic Editor: Teodorico C. Ramalho

This work describes the design, synthesis in one step, and the *in vitro, in vivo*, and in silico antidiabetic evaluation of a series of ten alicyclic and aromatic (alkyl +aryl: alkaryl)biguanides, analogues of metformin and phenformin. The design was conceived using isosteric replacement, chain-ring transformation, and lower and higher homologation strategies. All compounds were obtained as crystals and their structure was confirmed on the basis of their spectral data (NMR and mass spectra), and their purity was ascertained by microanalysis. Compounds were *in vitro* evaluated as activators of AMP-Activated Protein Kinase (AMPK). The results indicated that compounds **4**, **5**, and **6** showed similar or even better effect compared to metformin. Docking analysis was performed with regulatory subunit γ of AMPK, showing several interactions with nucleotide binding pocket. The *in vivo* evaluation of compounds **4**–**6** at a single dose of 50 mg/kg was performed in a murine experimental model of diabetes. The results showed an important and robust decrease of plasmatic glucose levels (−40%). Compound **6** was selected for an oral glucose tolerance test, showing an antihyperglycemic effect similar to metformin. The *in vivo* results indicated that compounds **4**–**6** may be effective in treating experimental T2DM.

1. Introduction

Type 2 diabetes mellitus (T2DM) is a long-lasting and progressive metabolic disease characterized by insulin resistance in several peripheral tissues such as liver, muscle, and adipose, as well as impaired insulin secretion by the pancreas [1]. Metformin (a biguanide, Figure 1) is the most extensively prescribed oral antidiabetic drug for the treatment of T2DM [2]. The main effect of metformin is to decrease hepatic glucose production, being a perfect agent for controlling fasting hyperglycemia. Several mechanisms of action have been proposed, but a previous study [3] reported that metformin increases phosphorylation of the AMP-Activated Protein Kinase (AMPK), with subsequent activation of AMPK activity in hepatocytes. AMPK is a heterotrimeric protein kinase comprised of three subunits: α catalytic subunit and the regulatory β and γ subunits [3]. AMPK is activated by phosphorylation of Thr172 residue on the α-subunit by kinases LKB1 and CaMKKb. Furthermore, AMP binds to nucleotide binding sites on γ-subunits to allosterically activate as well as facilitate Thr172 phosphorylation. AMPK plays a role of "fuel gauge" of the cell as it works to guarantee that ATP levels are maintained under situations of energetic stress such as exercise, hypoxia, and starvation [4]. Due to the central role played by AMPK in cellular energy homeostasis, it has emerged as an attractive drug target for the treatment of a

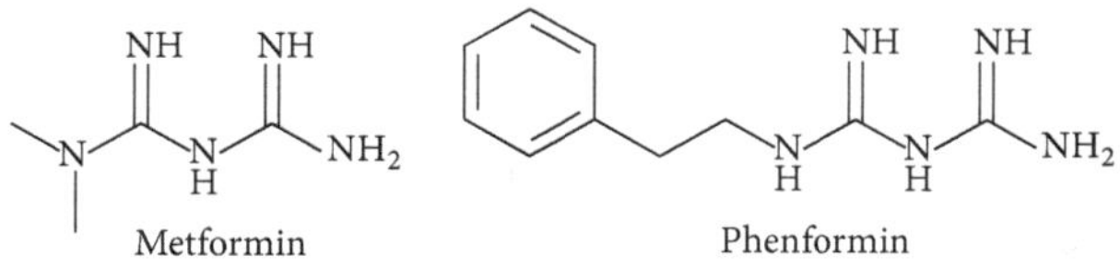

FIGURE 1: Biguanides that activate AMPK.

number of metabolic diseases such as type 2 diabetes mellitus (T2DM) [2].

Another biguanide, phenformin (Figure 1), also activates AMPK, but in 1978 it was retired from the market due to its toxicity, producing lactic acidosis as a major side effect that prompted the withdrawal of phenformin as a treatment for diabetes [5].

In our ongoing research on molecules with antidiabetic activity [6, 7], we report in this manuscript the preparation of ten alicyclic and aromatic biguanides (Table 1), as well as the *in vitro* activation of AMPK. We also describe the molecular docking of the most active compounds in the nucleotide binding site on γ-subunit of AMPK, and their *in vivo* antihyperglycemic effect using a streptozotocin–nicotinamide rat model of noninsulin dependent diabetes mellitus.

2. Materials and Methods

2.1. Chemicals and Analytical Methods. Starting materials and solvents were purchased from Sigma-Aldrich and were used without any further purification. Melting points were determined using an EZ-Melt MPA120 automated melting point apparatus from Stanford Research Systems and are uncorrected. Reactions were monitored by thin layer chromatography on 0.2 mm precoated silica gel 60 F_{254} Merck plates. ^{1}H NMR spectra were recorded on Varian Oxford (400 MHz) and ^{13}C NMR (100 MHz), as well as ^{1}H NMR (200 MHz) and ^{13}C NMR (50 MHz) instruments. Chemical shifts are given in ppm relative to tetramethylsilane (Me_4Si, $\delta = 0$) in DMSO-d_6 and $CDCl_3$; J values are given in Hz. The following abbreviations are used: s, singlet; d, doublet; q, quartet; dd, doublet of doublet; t, triplet; m, multiplet; bs, broad signal. MS were recorded on a JEOL JMS-700 spectrometer by electronic impact. Element analyses have been carried out on an Elementar Vario ELIII instrument.

2.2. General Procedure for the Synthesis of Compounds ***1–10***. To a solution of dicyandiamide **21** (0.5 g, 0.0060 mol) in toluene (5 mL) was added 1.1 mol equiv of corresponding alkylamines **11–15** or arylamines **16–20**. After the reaction mixture was stirred at room temperature for 15 min, a mixture 50 : 50 of HCl diluted in water (2.5 mL) was added dropwise. This mixture was stirred at reflux for 3 to 16 h. After that, the obtained residue was neutralized with a diluted solution of NH_4OH. Solvent was removed under vacuum, and the residues were washed with water. The crude solid products were then recrystallized from ethanol affording title compounds (Figure 2, Table 1).

*2.2.1. N,N-Diethylimidodicarbonimidic Diamide (**1**).* Yield 37%. White crystals obtained from ethanol. Mp 190°C (Dec). ^{1}H NMR (400 MHz, DMSO-d_6) δ: 1.16 (t, 6H, $CH_3 \times 2$), 2.89 (q, 4H, N-$CH_2 \times 2$), 6.60 (bs, 5H, -NH $\times$ 5) ppm. ^{13}C NMR (100 MHz, DMSO-*d6*) δ: 11.1 (-CH_3), 41.4 (CH_2), 162.9 (C=NH) ppm. MS/EI: m/z (% int. rel). 157 (M^+, 25%), 85 (100%). Anal. Calcd for $C_6H_{15}N_5$: C, 45.84; H, 9.62; N, 44.55. Found: C, 45.80; H, 9.58; N, 44.57.

*2.2.2. N-[Amino(imino)methyl]pyrrolidine-1-carboximidamide (**2**).* Yield 46%. White crystals obtained from ethanol. Mp 156.4–158.7°C. ^{1}H NMR (400 MHz, DMSO-d_6) δ: 1.82 (q, 4H, CH_2-CH_2), 3.07 (t, 4H, N-$CH_2 \times 2$), 6.60 (bs, 5H, -NH $\times$ 5) ppm. ^{13}C NMR (100 MHz, DMSO-*d6*) δ: 23.7 (CH_2-CH_2), 44.7 (CH_2-N), 162.9 (C=NH) ppm. MS/EI: m/z (% int. rel). 155 (M^+, 2%), 85 (100%). Anal. Calcd for $C_6H_{13}N_5$: C, 46.43; H, 8.44; N, 45.12. Found: C, 46.50; H, 8.48; N, 45.01.

*2.2.3. N-[Amino(imino)methyl]piperidine-1-carboximidamide (**3**).* Yield 63%. White crystals obtained from ethanol. Mp 116.5–118.0°C. ^{1}H NMR (400 MHz, DMSO-d_6) δ: 1.03 (m, 4H, CH_2-CH_2), 1.05 (m, 2H, CH_2) 3.40 (t, 4H, N-$CH_2 \times 2$), 6.58 (bs, 5H, -NH $\times$ 5) ppm. ^{13}C NMR (100 MHz, DMSO-*d6*) δ: 18.5 (CH_2-CH_2), 30.6 (CH_2), 56.1 (CH_2-N), 162.9 (C=NH) ppm. MS/EI: m/z (% int. rel). 169 (M^+, 40%), 85 (100%). Anal. Calcd for $C_7H_{15}N_5$: C, 49.68; H, 8.93; N, 41.38. Found: C, 49.62; H, 8.90; N, 41.40.

*2.2.4. N-[Amino(imino)methyl]morpholine-4-carboximidamide (**4**).* Yield 22%. White crystals obtained from ethanol. Mp 178.4–180.5°C. ^{1}H NMR (400 MHz, DMSO-*d6*) δ: 3.06 (t, 4H, CH_2-O-CH_2), 3.75 (t, 4H, N-$CH_2 \times 2$), 6.61 (bs, 5H, NH $\times$ 5) ppm. ^{13}C NMR (100 MHz, DMSO-*d6*) δ: 43.2 (CH_2-N), 63.7 (CH_2-O), 162.3 (C=NH) ppm. MS/EI: m/z (% int. rel). 171 (M^+, 15%), 85 (100%). Anal. Calcd for $C_6H_{13}N_5O$: C, 42.09; H, 7.68; N, 40.91. Found: C, 42.10; H, 7.65; N, 40.88.

*2.2.5. N-[Amino(imino)methyl]-4-methylpiperazine-1-carboximidamide (**5**).* Yield 54%. White crystals obtained from ethanol. Mp 148°C (Dec). ^{1}H NMR (400 MHz, DMSO-d_6) δ: 2.48 (s, 3H, CH_3), 2.73 (m, 4H, CH_2-N-CH_2), 3.38 (m, 4H, CH_2-NMe-CH_2), 6.71 (bs, 5H, NH $\times$ 5) ppm. ^{13}C NMR (100 MHz, DMSO-*d6*) δ: 42.5 (CH_2-N), 43.3 (CH_2-N), 49.6 (CH_3), 163.3 (C=NH) ppm. MS/EI: m/z (% int. rel). 184 (M^+, 2%), 85 (100%). Anal. Calcd for $C_7H_{16}N_6$: C, 45.63; H, 8.75; N, 45.61. Found: C, 45.70; H, 8.75; N, 45.58.

*2.2.6. N-Benzylimidodicarbonimidic Diamide (**6**).* Yield 56%. White crystals obtained from ethanol. Mp 184.2–186.8°C. ^{1}H NMR (400 MHz, DMSO-d_6) δ: 3.99 (d, 2H, CH_2), 7.30 (m, 2H, H-2, H-6), 7.39 (m, 2H, H-3, H-5), 7.50 (m, 1H, H-4), 8.60 (bs, 6H, NH $\times$ 6) ppm. ^{13}C NMR (100 MHz, DMSO-*d6*) δ: 42.6 (CH_2-N), 129.4 (C-4), 128.8 (C-2, C-6), 129.0 (C-3, C-5), 134.5 (C-1), 154.9 (C=NH), 156.1 (C=NH) ppm. MS/EI: m/z (% int. rel). 191 (M^+, 2%), 91 (100%), 85 (40%). Anal. Calcd for $C_9H_{13}N_5$: C, 56.53; H, 6.85; N, 36.62. Found: C, 56.50; H, 6.85; N, 36.61.

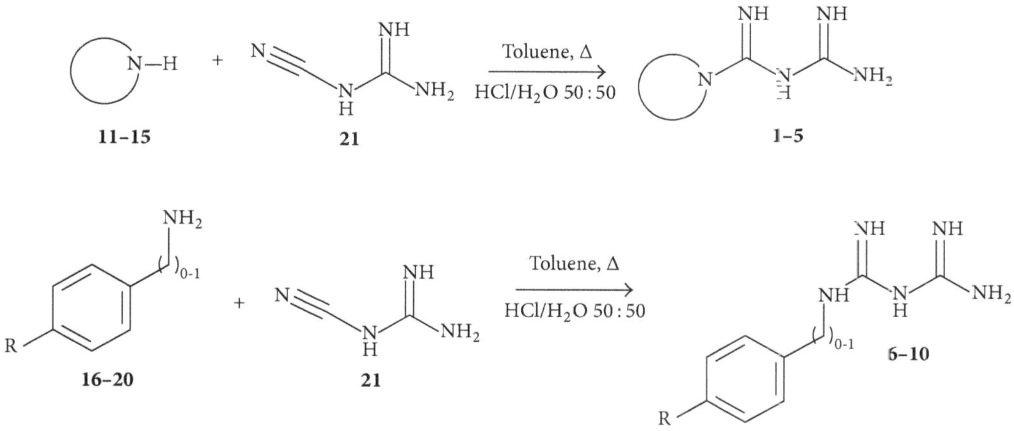

FIGURE 2: Scheme reaction to obtain compounds **1–10**.

TABLE 1: Physicochemical properties of biguanides **1–10**.

Compd.	R	Yield (%)	Mp (°C)	Mol. Weight (g/mol)	Reaction time (h)
1	Diethylamine	37	190 (dec)	157	5
2	Pyrrolidine	46	116.5–119	155	5
3	Piperidine	63	156.4–158.7	169	5
4	Morpholine	22	178.4–180.5	171	7
5	4-Methylpiperazine	54	148 (dec)	184	7
6	Benzylamine	56	184.2–186.8	191	7
7	Aniline	75	185.3–187.4	177	7
8	4-Nitroaniline	68	145.5–148.7	222	16
9	4-Chloroaniline	80	161.9–163.3	211	10
10	4-Fluoroaniline	85	207.5–209.2	195	4

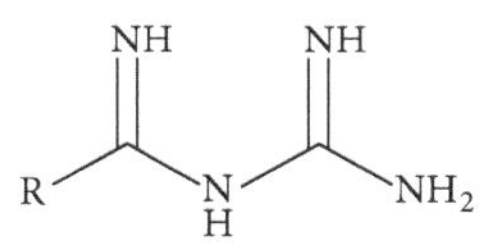

*2.2.7. N-Phenylimidodicarbonimidic Diamide (**7**).* Yield 75%. White crystals obtained from ethanol. Mp 185.3–187.4°C. ^{1}H NMR (400 MHz, DMSO-d_6) δ: 7.19 (m, 5H, H-2 to H-6), 9.91 (bs, 6H, NH × 6) ppm. ^{13}C NMR (100 MHz, DMSO-*d6*) δ: 121.4 (C-4), 123.7 (C-2, C-6), 129.1 (C-3, C-5), 139.1 (C-1), 155.7 (C=NH), 161.6 (C=NH) ppm. MS/EI: *m/z* (% int. rel). 177 (M$^+$, 70%), 119 (100%), 85 (20%). Anal. Calcd for $C_8H_{11}N_5$: C, 54.22; H, 6.26; N, 39.52. Found: C, 54.20; H, 6.21; N, 39.60.

*2.2.8. N-(4-Nitrophenyl)imidodicarbonimidic Diamide (**8**).* Yield 68%. White crystals obtained from ethanol. Mp 145.5–148.7°C. ^{1}H NMR (400 MHz, DMSO-d_6) δ: 7.68 (d, 2H, H-2, H-6, *Jo* = 9.2 Hz), 8.21 (d, 2H, H-3, H-5, *Jo* = 9.2 Hz), 11.07 (bs, 6H, NH × 6) ppm. ^{13}C NMR (100 MHz, DMSO-*d6*) δ: 119.4 (C-2, C-6), 125.5 (C-3, C-5), 142.9 (C-1), 144.5 (C-4), 151.6 (C=NH), 155.4 (C=NH) ppm. MS/EI: *m/z* (% int. rel). 222 (M$^+$, 70%), 165 (100%), 85 (5%). Anal. Calcd for $C_8H_{10}N_6O_2$: C, 43.24; H, 4.54; N, 37.82. Found: C, 43.31; H, 4.50; N, 37.79.

*2.2.9. N-(4-Chlorophenyl)imidodicarbonimidic Diamide (**9**).* Yield 80%. White crystals obtained from ethanol. Mp 161.9–163.3°C. ^{1}H NMR (200 MHz, DMSO-d_6) δ: 7.30 (d, 2H, H-2, H-6, *Jo* = 8.4 Hz), 7.39 (d, 2H, H-3, H-5, *Jo* = 8.8 Hz), 9.99 (bs, 6H, NH × 6) ppm. ^{13}C NMR (50 MHz, DMSO-*d6*) δ: 122.5 (C-2, C-6), 127.27 (C-3, C-5), 128.9 (C-4), 136.3 (C-1), 155.3 (C=NH), 161.8 (C=NH) ppm. MS/EI: *m/z* (% int. rel). 211 (M$^+$, 90%), 154 (100%), 85 (10%). Anal. Calcd for $C_8H_{10}N_5Cl$: C, 45.40; H, 4.76; N, 33.09. Found: C, 45.38; H, 4.78; N, 33.12.

*2.2.10. N-(4-Fluorophenyl)imidodicarbonimidic Diamide (**10**).* Yield 85%. White crystals obtained from ethanol. Mp 207.5–209.2°C. ^{1}H NMR (200 MHz, DMSO-d_6) δ: 7.18 (m, 2H, H-3, H-5, *Jo* = 9.2 Hz), 7.42 (m, 2H, H-2, H-6, *Jo* = 9.2 Hz), 9.91 (bs, 6H, NH × 6) ppm. ^{13}C NMR (50 MHz, DMSO-*d6*) δ: 117.8 (d, C-3, C-5, $^2J_{C\text{-}F}$ = 11.25 Hz), 125.6 (d, C-2, C-6, $^3J_{C\text{-}F}$ = 3.9 Hz), 137.6 (d, C-1, $^4J_{C\text{-}F}$ = 1.55 Hz), 160.9 (d, C-4, $^1J_{C\text{-}F}$ = 294.7 Hz), 159.8 (C=NH), 162.1 (C=NH) ppm. MS/EI: *m/z* (% int. rel). 195 (M$^+$, 80%), 111 (100%), 85 (70%). Anal. Calcd for $C_8H_{10}N_5F$: C, 49.22; H, 5.16; N, 35.88. Found: C, 49.21; H, 5.20; N, 35.90.

2.3. Biological Activity

2.3.1. In Vitro AMPK Activation. Primary rat hepatocytes were obtained by collagenase digestion as described by Berry and Friend [8]. Fresh isolated hepatocytes were equally distributed in collagen-coated dishes and incubated at 37°C for 2 h in medium α MEM (Gibco; cat 11900-024), 10% FBS, 100 U/mL penicillin, and 100 μg/mL streptomycin. The cultured dishes were then washed three times with PBS to remove unattached dead cells. Dishes were randomly assigned to receive fresh FBS-free medium without or with different concentrations of both metformin (10 mM) and ten of its analogues (1 and 10 mM) for 1 h. After this time, the plated cells were washed three times with cold PBS. The cells were then lysed using a buffer containing 50 mM HEPES (pH 7.5), 50 mM KCl, 1 mM EDTA, 5 mM EGTA, 1 mM glycerolphosphate, 0.1% (vol/vol) Triton X-100, 50 mM NaPPi, 1 nM orthovanadate, and 1 nM DTT, a standard complete protease inhibitor mixture (Roche). The lysates were then centrifuged (10,000g, 10 min, 4°C), and the supernatants were stored at −80°C. The activation of AMPK and its target protein, acetyl-CoA carboxylase (ACC), was determined by immunoblot detection with antibodies against phospho-Thr^{172} AMPK and phospho-Ser^{79} ACC1, as well as total AMPK, ACC, and actin as loading controls. All antibodies were purchased from Cell Signaling.

2.3.2. In Vivo Antidiabetic Activity. Male Wistar rats weighing an average of 300 g were used. They were maintained at 25°C in a 12 h a light/dark cycle and at 45–65% of humidity during experimentation time. All animal procedures were developed in accordance with the Mexican Federal Regulations for Animal Experimentation and Care, ratified by the Institutional Animal Care and Use Committee (UNAM) based on US National Institute of Health Publication #85-23 [9, 10].

2.3.3. Induction of Diabetes. Streptozotocin (STZ) was dissolved in citrate buffer (pH 4.5) and nicotinamide was dissolved in normal physiological saline solution. Non-insulin-dependent diabetes rat model was induced in overnight fasted rats by a single intraperitoneal injection of 100 mg/kg STZ, 15 min before the i.p. administration of 40 mg/kg of nicotinamide. Hyperglycemia was confirmed by elevated glucose concentration in plasma, determined after 2 weeks by strip-glucometer. The animals with blood glucose concentration higher than 200 mg/dL were used for the antidiabetic study [11, 12].

2.3.4. Non-Insulin-Dependent Diabetes Mellitus Rat Model. The diabetic animals were divided into three groups of five animals each ($n = 5$). Rats of experimental groups were given a solution of compounds **4–6** (50 mg/kg body weight, prepared in tween 80, 10%). Control group animals were also treated with saline plus tween 80, 10%. Metformin (50 mg/kg) was used as antihyperglycemic reference drug. Blood samples were collected from the caudal vein at 0, 1, 3, 5, and 7 h after vehicle, compound, and drug administration. Blood glucose concentration was estimated by enzymatic glucose oxidase method using a commercial glucometer [13]. The percentage variation of glycemia for each group was calculated in relation to initial (0 h) level, according to

$$\% \text{ variation of glycemia} = \left(\frac{\text{Glu}x - \text{Glu0}}{\text{Glu0}}\right) \times 100, \quad (1)$$

where Glu0 were initial glycemia values and Glu$x$ were the glycemia values at 1, 3, 5, and 7 h, respectively. All values were expressed as mean ± SEM. Statistical significance was estimated by analysis of variance (ANOVA); $p < 0.05$ implies significance.

2.3.5. Glucose Tolerance Test. Normoglycemic rats were divided into groups of five animals each ($n = 5$). Thirty min after administration of test compounds, a dose of 2 g/kg of glucose solution was orally administered to each animal. Compound **6** (50 mg/kg), metformin (50 mg/kg), and vehicle (tween 80, 10%) were administered to rats in the same volume of solution. Blood samples were collected from the tail tip at 0 (before oral administration), 1, 1.5, 2, 2.5, and 3 h after vehicle, positive control, or test compound administration.

2.4. Docking Studies. Discovery Studio version 3.5 and Pymol version 1.0 were used for visualization. The crystal structure of AMPK was retrieved from the PDB with the accession code: 2UV4. Docking calculations were conducted with AutoDock Vina. The program performs several runs in each docking experiment. Each run provides one predicted binding mode. All water molecules and also cocrystal ligand (5′-adenylic acid) were removed from the crystallographic structure. The AutoDock Vina plugin through Pymol program was used, where we generated the grid maps. Each grid was centered at the crystallographic coordinates of the cocrystal ligand. The grid dimensions were $20 \times 20 \times 20\,\text{Å}^3$ with points separated by 1.0 Å. Also, the protein file was selected as the rigid part, and the ligand file as the flexible one, allowing all its torsions to rotate during docking. AutoDock Vina uses default algorithms of searching and automatically prepares the files for use as it adds charges and polar hydrogens to the protein necessary to perform scoring calculations; it clusters showing only the main results. The number of docking runs was 10. After finishing, the poses were visualized on Pymol and compared against the cocrystalized ligand over the protein.

2.4.1. Docking Validation. The molecular docking protocol was validated by redocking of cocrystal ligand (5′-adenylic acid) into the active site of the structure of AMPK. The root mean square deviation between the cocrystal ligand and the docked structure was less than 2.5 Å. This value indicates that the parameters for docking simulations are good in reproducing orientation and conformation in the X-ray crystal structure of enzyme and receptors.

3. Results and Discussion

3.1. Chemistry. Compounds **1–10** were designed on the basis of the structure of metformin and phenformin (Figure 1,

Table 1), maintaining the biguanide group, removing both dimethyl and phenylethyl side chains, and substituting the proximal amino group with diethyl or cycloalkyl groups using a straightforward approach called chain-ring transformation, attaining conformational constraint connecting alkyl substituents to give the corresponding cyclic analogues. The homologation criteria were employed to pass from pyrrolidine to piperidine. A homologous series is a group of compounds that differ by a constant unit, generally a methylene group [14]. Morpholine and 4-methylpiperazine derivatives were selected as isosteric replacements of piperidine. Benzyl or 4-substituted phenyl groups were designed as lower homologues, with one or two methylene groups less than those presented by phenformin. Some physicochemical properties of compounds **1–10** are described in Table 1.

Compounds **1–10** were prepared in a single step starting from cyanoguanidine (**21**), which was condensed with several alkylamines **11–15** or aryl amines **16–20** under reflux conditions (Figure 2). Title compounds were recovered with 22–85% yields and purified by recrystallization with ethanol. Their chemical structures were confirmed by spectral data (NMR and mass spectra), and their purity was ascertained by elemental analysis.

3.2. In Vitro Biological Activity. To test the ability of each derivative to activate AMPK, an *in vitro* assay was performed on a primary culture of hepatocytes. Aliquots of stock solutions of the analogues (dissolved in DMSO) were diluted with the assay buffer, using metformin as positive control. The phosphorylation of AMPK and its target ACC were assessed by immunoblot analysis, where it is observed that aliphatic or alicyclic compounds **1–3** were not able to activate the enzyme (data not shown). Conversely, we found that AMPK phosphorylation (activation) was increased in a concentration-dependent manner with compounds **4**, **5**, and **6**, being more pronounced with compound **6**; in fact, the concentrations of three of the compounds required for activation of AMPK were significantly lower than those of metformin (Figure 3). On the other hand, aromatic biguanides **7–10** were unable to activate the AMPK. These results are in accordance with those reported in a parallel work performed with closely related biguanides [15].

In the immunoblot not only is the phosphorylated AMPK observed, but also the phosphorylated ACC (acetyl-CoA carboxylase) in serine 79 can be detected. This enzyme is one of the targets of AMPK. The phosphorylation of this enzyme causes its inactivation, and this leads to an increase in the oxidation of free fatty acids [16]. The phosphorylation of ACC indicates that analogues induce the phosphorylation (activation) of AMPK, which leads to modification of the different metabolic pathways.

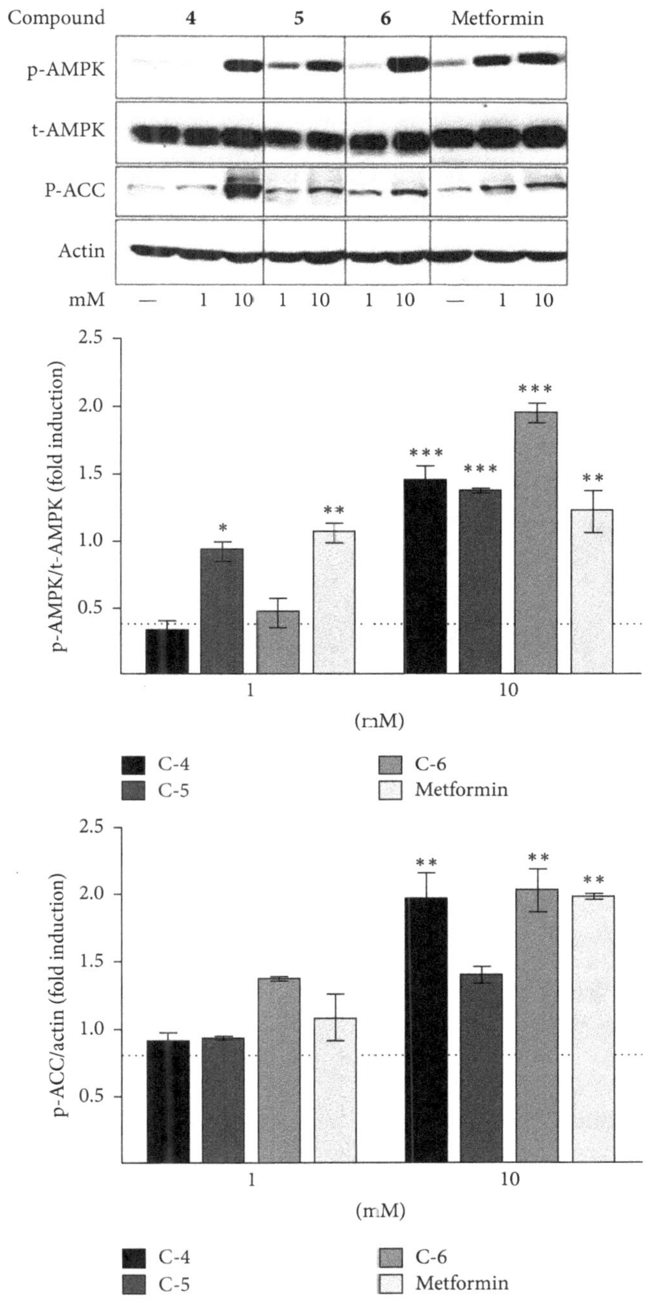

FIGURE 3: AMPK phosphorylation by compounds **4–6**. The analogues **4** and **5** are able to activate the enzyme in a similar way to metformin, whereas the analogue **6** appears to induce AMPK phosphorylation in a larger extent. $^{*}p < 0.05$, $^{**}p < 0.01$, and $^{***}p < 0.001$.

3.3. In Silico Studies

3.3.1. Docking Analysis. Based on the *in vitro* biological assay of AMPK activation, the most active compounds were selected to explain the experimental activities. On this basis, a preliminary molecular docking study was conducted to evaluate the putative binding mode of compounds **4–6** into the regulatory γ-subunit of AMPK. This molecular docking reveals that compounds **4** and **5** internalize into the nucleotide binding pocket of AMPK and interact by electrostatic and hydrogen bonds with Asp-317 and Thr-200, and both residues are essential for the activation of this enzyme (ΔG = 5.2 kcal/mol for both compounds). However, compound **6** (the most active *in vitro*) showed an

TABLE 2: Toxicity profiles predicted for compounds **4–6**, metformin, and phenformin.

Compd.	LD_{50} (mg/kg)				Probability of inhibition (IC_{50} or $Ki < 10\,\mu M$)			
	Mouse		Rat		CYP450 isoform			hERG
	i.p.	p.o.	i.p.	p.o.	3A4	2D6	1A2	
4	400	510	260	850	0.01	0.06	0.01	0.02
5	190	640	130	750	0.01	0.04	0.01	0.03
6	170	710	240	850	0.01	0.10	0.03	0.05
Metformin	247	810	220	960	0.01	0.02	0.01	0.01
Phenformin	160	720	240	890	0.02	0.12	0.04	0.07

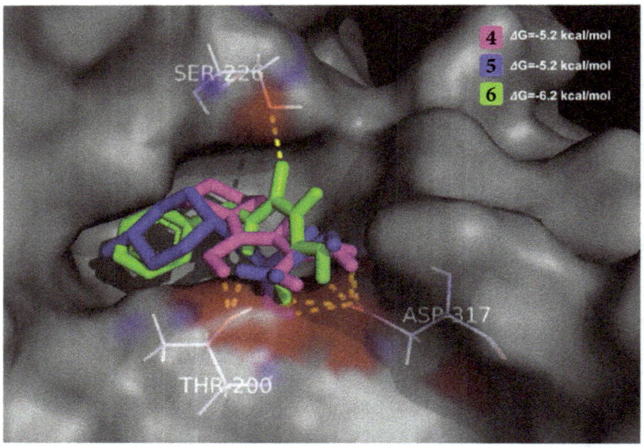

FIGURE 4: 3D binding model of compounds **4–6** into the nucleotide binding site of AMPK (γ subunit). Compounds **4** (cyan), **5** (blue), and **6** (green) are shown as stick models, whereas aminoacids are depicted as lines. A yellow dashed lines represent polar or electrostatic interactions.

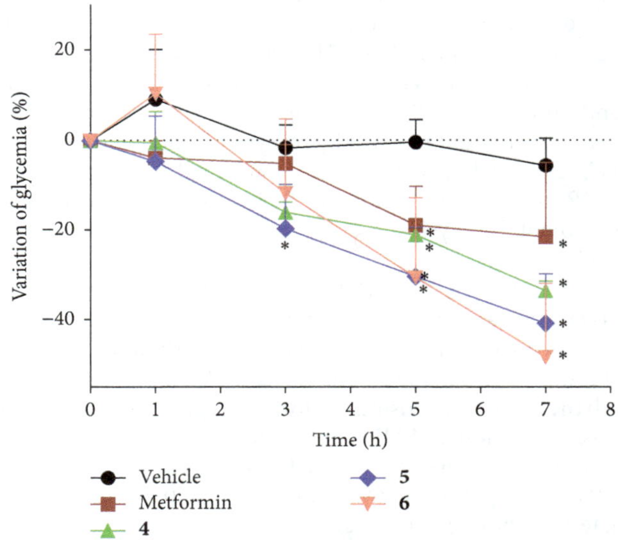

FIGURE 5: Effect of a single dose of compounds **4–6** and metformin (50 mg/Kg, intragastric, $n = 5$), or vehicle in streptozotocin-nicotinamide-induced diabetes rat model. $^{*}p < 0.001$ versus vehicle group.

additional interaction with Ser-226, increasing the docking energy to −6.2 kcal/mol. Figure 4 shows the binding mode of compounds **4–6** found by AutoDock showing an extensive hydrogen bonds network.

These results contribute to explaining at the molecular level the relevant activities of compounds **4–6** in the *in vitro* test.

3.3.2. In Silico Toxicity. With the aim of anticipating potential toxicity issues of compounds **4–6**, a computational prediction of safety profiles was performed. The toxicity parameters of **4–6**, metformin, and phenformin were calculated through the ACD/ToxSuite software, v. 2.95 (Table 2).

The in silico calculation of inhibition for the three main isoforms of CYP450 for compounds **4–6** was comparable to that of metformin at relevant clinical concentrations (<10 μM), showing low probabilities of drug-drug interactions and undesirable adverse effects [17]. Several basic nitrogen compounds are associated with cardiovascular risks due to human ether-a-go-go related gene (hERG) channel blockade [18–20]. Compounds **4–6** showed low prediction of hERG channel blockage at clinically relevant concentrations ($K_i < 10\,\mu M$), being considered as potentially noncardiotoxic compounds. In the calculation of acute toxicity, compounds **4–6** demonstrated similar predicted LD_{50} than metformin and phenformin by two different administration routes.

3.4. In Vivo Antidiabetic Effect of Compounds ***4–6****.* Compounds **4–6** were the most potent AMPK activators of the series, and they were selected in order to evaluate their *in vivo* antidiabetic activity using an STZ-nicotinamide non-insulin-dependent diabetes mellitus rat model. Metformin (50 mg/kg) was used as a positive control. The antidiabetic activity of compounds **4–6** was determined using a 50 mg/kg single dose, by intragastric route (Figure 5).

The antidiabetic assay shows that analogue **4** significantly reduced glucose levels compared to the vehicle and is as good as to the control group (metformin), having at 7 hours after administration a percentage of glucose decrease of 33.5%. Analogue **5** also decreased glucose levels compared to

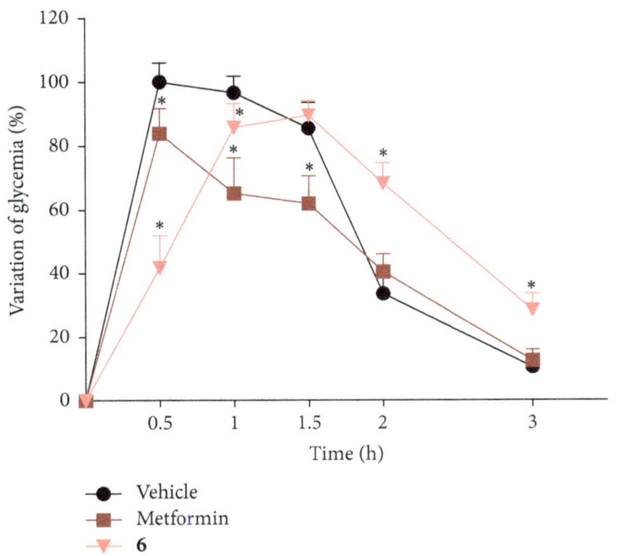

FIGURE 6: Oral glucose tolerance test. Effect of a single dose of compound **6** and metformin (50 mg/Kg, intragastric, $n = 5$), or vehicle in streptozotocin-nicotinamide-induced diabetes rat model. $^{*}p < 0.05$ versus vehicle.

vehicle. In addition, its effect was similar to that presented by metformin at the same dose. The activity was retained during the 7 hours of experimentation. At 7 hours after administration, a 40% decrease in blood glucose levels was observed. Analogue **6**, which was the compound that showed the best activity on AMPK activation, was also active in the *in vivo* assay, and its antihyperglycemic effect was retained throughout the assay.

In order to verify the plausible antihyperglycemic effect of compound **6**, glucose tolerance test curves in normoglycemic rats were obtained. Dose of 50 mg/kg for **6** and metformin was employed. As shown in Figure 6, compound **6** displayed a significant reduction of hyperglycemic peak which was attained at 0.5 h after glucose administration. In Figure 6, it can be seen that the animals treated with analogue **6** reached a lower hyperglycemic value than the animals treated with metformin and compared to the vehicle, 0.5 h after the treatment.

During the experiment, glucose levels did not decrease beyond baseline, indicating that the antidiabetic effect of compound **6** is due to an antihyperglycemic action rather than a hypoglycemic effect. Also, compounds **4**–**6** did not increase the lactic acid concentrations in plasma of rats tested (less than 1.9 mM), so any evidence of lactic acidosis was found. With these results, it can be concluded that the mechanism of action of the analogues **4**–**6** that confer their antidiabetic activity is similar to metformin, through the activation of AMPK and of some of the pathways that are regulated by this enzyme. Previously *in vitro* reports agree with the AMPK results obtained with compound **6** [15]. However, in our current study, we have demonstrated the robust *in vivo* effect produced by this compound after an oral administration. Further studies are being conducted by us in order to demonstrate the cardiovascular action of compound **6** in a murine model of fructose-induced insulin resistance [21].

4. Conclusion

In summary, ten alkarylbiguanides have been developed as promising compounds for the treatment of type 2 diabetes mellitus. Compounds **4**–**6** (a) exhibited AMPK activation similar to or greater than metformin, (b) demonstrated a robust reduction of glucose levels with marked *in vivo* anti-hyperglycemic efficacy, and (c) showed predicted low toxicity profiles and any experimental evidence of lactic acidosis. These compounds could be an alternative to metformin, the only biguanide currently available.

Disclosure

The paper is taken in part from the M. Pharm. thesis of E. Gutierrez-Lara.

Conflicts of Interest

The authors declare that there are no conflicts of interest regarding the publication of this paper.

Acknowledgments

This work was supported in part by the Consejo Nacional de Ciencia y Tecnología (CONACyT) under Grant no. 253814 (CB-2015). The authors are in debt with Abraham Gutierrez-Hernandez, M. Pharm. for technical assistance. This article is dedicated to all the people, mainly Pharmacist, who helped in the medication management and classification after the recent Mexico's earthquake.

References

[1] American Diabetes Association, "Diagnosis and classification of diabetes mellitus," *Diabetes Care*, vol. 32, supplement 1, pp. S62–S67, 2009.

[2] S. Meng, J. Cao, Q. He et al., "Metformin activates AMP-activated protein kinase by promoting formation of the $\alpha\beta\gamma$ heterotrimeric complex," *The Journal of Biological Chemistry*, vol. 290, no. 6, pp. 3393–3802, 2015.

[3] E. Moreno-Arriola, M. El Hafidi, D. Ortega-Cuéllar, and K. Carvajal, "AMP-activated protein kinase regulates oxidative metabolism in Caenorhabditis elegans through the NHR-49 and MDT-15 transcriptional regulators," *PLoS ONE*, vol. 11, no. 1, Article ID 0148089, 2016.

[4] K. O. Cameron and R. G. Kurumbail, "Recent progress in the identification of adenosine monophosphate-activated protein kinase (AMPK) activators," *Bioorganic & Medicinal Chemistry Letters*, vol. 26, no. 21, pp. 5139–5148, 2016.

[5] L. Yang, H. Sha, R. L. Davisson, and L. Qi, "Phenformin activates the unfolded protein response in an AMP-activated protein kinase (AMPK)-dependent manner," *The Journal of Biological Chemistry*, vol. 288, no. 19, pp. 13631–13638, 2013.

[6] S. Hidalgo-Figueroa, J. J. Ramírez-Espinosa, S. Estrada-Soto et al., "Discovery of thiazolidine-2,4-dione/biphenylcarbonitrile hybrid as dual ppar α/γ modulator with antidiabetic effect: in vitro, in silico and in vivo approaches," *Chemical Biology & Drug Design*, vol. 81, no. 4, pp. 474–483, 2013.

[7] S. Hidalgo-Figueroa, G. Navarrete-Vázquez, S. Estrada-Soto et al., "Discovery of new dual PPARγ-GPR40 agonists with robust antidiabetic activity: Design, synthesis and in combo drug evaluation," *Biomedicine & Pharmacotherapy*, vol. 90, pp. 53–61, 2017.

[8] M. N. Berry and D. S. Friend, "High-yield preparation of isolated rat liver parenchymal cells: a biochemical and fine structural study.," *The Journal of Cell Biology*, vol. 43, no. 3, pp. 506–520, 1969.

[9] U. Albus, "Guide for the Care and Use of Laboratory Animals (8th edn)by the National Research Council of the NationalAcademiesWashington, DC: National Academies Press, 2011," *Laboratory Animals*, vol. 46, no. 3, pp. 267-268, 2012.

[10] J. A. García-Díaz, G. Navarrete-Vázquez, S. García-Jiménez et al., "Antidiabetic, antihyperlipidemic and anti-inflammatory effects of tilianin in streptozotocin-nicotinamide diabetic rats," *Biomedicine & Pharmacotherapy*, vol. 83, pp. 667–675, 2016.

[11] E. J. Verspohl, "Recommended testing in diabetes research," *Planta Medica*, vol. 68, no. 7, pp. 581–590, 2002.

[12] S. N. Goyal, N. M. Reddy, K. R. Patil et al., "Challenges and issues with streptozotocin-induced diabetes-a clinically relevant animal model to understand the diabetes pathogenesis and evaluate therapeutics," *Chemico-Biological Interactions*, vol. 244, pp. 49–63, 2016.

[13] R. R. Ortiz-Andrade, J. C. Sánchez-Salgado, G. Navarrete-Vázquez et al., "Antidiabetic and toxicological evaluations of naringenin in normoglycaemic and NIDDM rat models and its implications on extra-pancreatic glucose regulation," *Diabetes, Obesity & Metabolism*, vol. 10, no. 11, pp. 1097–1104, 2008.

[14] R. B. Silverman and M. W. Holladay, *The organic chemistry of drug design and drug action*, Academic Press, San Diego, Calif, USA, 3rd edition, 2014.

[15] H. R. Bridges, V. A. Sirviö, A.-N. A. Agip, and J. Hirst, "Molecular features of biguanides required for targeting of mitochondrial respiratory complex I and activation of AMP-kinase," *BMC Biology*, vol. 14, no. 1, article 65, 2016.

[16] O. Scudiero, E. Nigro, M. L. Monaco et al., "New synthetic AICAR derivatives with enhanced AMPK and ACC activation," *Journal of Enzyme Inhibition and Medicinal Chemistry*, vol. 31, no. 5, pp. 748–753, 2016.

[17] L. Xu, Y. Chen, Y. Pan, G. L. Skiles, and M. Shou, "Prediction of human drug-drug interactions from time-dependent inactivation of CYP3A4 in primary hepatocytes using a population-based simulator," *Drug Metabolism and Disposition*, vol. 37, no. 12, pp. 2330–2339, 2009.

[18] O. Taboureau and F. S. Jørgensen, "In silico predictions of hERG channel blockers in drug discovery: from ligand-based and target-based approaches to systems chemical biology," *Combinatorial Chemistry & High Throughput Screening*, vol. 14, no. 5, pp. 375–387, 2011.

[19] G. Navarrete-Vázquez, H. Torres-Gómez, S. Hidalgo-Figueroa et al., "Synthesis, in vitro and in silico studies of a PPARγ and GLUT-4 modulator with hypoglycemic effect," *Bioorganic & Medicinal Chemistry Letters*, vol. 24, no. 18, pp. 4575–4579, 2014.

[20] G. Navarrete-Vázquez, A. Austrich-Olivares, B. Godínez-Chaparro et al., "Discovery of 2-(3,4-dichlorophenoxy)-N-(2-morpholin-4-ylethyl)acetamide: A selective σ1 receptor ligand with antinociceptive effect," *Biomedicine & Pharmacotherapy*, vol. 79, pp. 284–293, 2016.

[21] E. J. Gutiérrez-Lara, G. Navarrete-Vázquez, A. Sánchez-López, and D. Centurión, "Pharmacological evaluation of metformin and N- benzylbiguanide, a novel analogue of metformin, on the vasopressor responses to adrenergic system stimulation in pithed rats with fructose-induced insulin resistance," *European Journal of Pharmacology*, vol. 814, pp. 313–323, 2017.

2

Acetylation Modification Improves Immunoregulatory Effect of Polysaccharide from Seeds of *Plantago asiatica* L.

Le-Ming Jiang, Shao-Ping Nie, Dan-Fei Huang, Zhi-Hong Fu, and Ming-Yong Xie

State Key Laboratory of Food Science and Technology, Nanchang University, Nanchang 330047, China

Correspondence should be addressed to Shao-Ping Nie; spnie@ncu.edu.cn and Ming-Yong Xie; myxie@ncu.edu.cn

Academic Editor: Ji Kang

The current study was conducted to investigate effects of acetylated *Plantago asiatica* L. polysaccharides (Ac-PLCPs) on their immunoregulatory activities in bone-marrow-derived dendritic cells (DCs) model. Influences of natural *Plantago asiatica* L. crude polysaccharide (PLCP) and Ac-PLCPs on inducing phenotypic and functional maturation on DCs were determined. The results showed that Ac-PLCPs with degree of substitution (DS) from 0.06 to 0.1 could not only stimulate the expression of surface molecules major histocompatibility complex class II (MHC II), cluster of differentiation 86 (CD86), and CD80 on DCs ($P < 0.01$) but also increase the secretion of cytokine IL-12p70 ($P < 0.01$). The endocytosis activity of DCs was attenuated by Ac-PLCPs treatment ($P < 0.01$), while the mRNA expressions of chemokine receptors CCR7 and CXCR4 in DCs were significantly increased ($P < 0.01$). Besides, DCs treated with the Ac-PLCPs showed extremely strong T cell proliferation stimulating activity ($P < 0.01$). These data showed that Ac-PLCPs had higher maturation-stimulating activity on DCs than PLCP, which indicated that acetylation modification improved the immunoregulatory effect of PLCP.

1. Introduction

Bioactive polysaccharides have attracted extensive attention worldwide, and they are now considered as one functional component of traditional medicine. A great deal of researches indicate that bioactivity of polysaccharide is closely correlated with various structural parameters, such as molecular weight, monosaccharide composition, branches, conformation, and substituent group. Recently, a growing body of research implicates that appropriate structure modification can improve bioactivity of polysaccharide [1]. So far, lots of chemical modifications have been applied to modulate physicochemical or biological properties of polysaccharides, such as carboxymethylation [2, 3], sulfation [4], acetylation [5], phosphorylation [6], and oxidation [7] modification.

Acetyl content in some natural polysaccharides has been demonstrated to play an important role in their bioactivities. For example, the lymphocyte proliferation stimulating activity of polysaccharide from *Dendrobium nobile* mainly depended on the acetyl groups [8, 9]. Acetylated *Ulva pertusa* [10], *Laminaria japonica* polysaccharide [11], and *Enteromorpha linza* polysaccharide [12] have higher antioxidant activity. Acetylated *Grifola frondosa* polysaccharide peptides can inhibit glioma C6 cell line *in vitro*, and the combination treatment of cyclophosphamide and the acetylated polysaccharide peptides can significantly decrease total sialic acid levels in tumor-bear mice [13]. Acetyl fucoidan from *Cladosiphon okamuranus* could activate murine macrophage cell line, RAW 264.7 cells, through membrane receptor, toll-like receptor 4 (TLR4), cluster of differentiation 14 (CD14), and scavenger receptor class A [14].

Psyllium is a common name used for the plant genus *Plantago*. In western countries, dietary fiber from psyllium has been used extensively as both pharmacological supplements and food ingredients. Ripe seeds of *Plantago asiatica* L. are a traditional medicine used as antipyretic, diuretic, and expectorant treatment in China. Our previous study showed that polysaccharide from seeds of *Plantago asiatica* L. was a high branched heteroxylan consisting of β-1,4-linked Xyl*p* backbone with side chains attached to O-2 or O-3 position. And its side chains consist of β-T-linked Xyl*p*, α-T-linked Ara*f*, α-T-linked GlcA*p*, β-Xyl*p*-(1→3)-α-Ara*f*, and α-Ara*f*-(1→3)-β-Xyl*p* [15]. In particular, the polysaccharide was a potential immune adjuvant since it could promote maturation of mouse bone-marrow-derived dendritic cells *in vitro* [16].

In the present study, *Plantago asiatica* L. crude polysaccharide (PLCP) were acetylated using acetic anhydride as acetylation reagent, and bioactivities of the acetylated polysaccharides (Ac-PLCP) were investigated on dendritic cells model.

2. Materials and Methods

2.1. Preparation of Acetylated Plantago asiatica L. Polysaccharides. PLCP was prepared as previously described [17]. Briefly, water-soluble polysaccharide was extracted from the seeds of *Plantago asiatica* L. by boiling water for 2 h and precipitated with 80% (v/v) ethanol. Then the polysaccharide was deproteinised with sevage reagent, a mixture of chloroform and n-butyl alcohol at a ratio of 4 : 1 (v/v), dialysed against double distilled water for 48 h (Mw cut-off, 8000–14,000 kDa), and obtained by freeze-drying.

Acetylation modification of PLCP was carried out using acetic anhydride as acetylation reagent [18]. One hundred milligram PLCP was firstly dissolved in 10 mL distilled water (pH 9.0). The solution was maintained at 30°C with continuous stirring. Acetic anhydride (400–2000 μL) was added to the solution dropwise within 2 h (pH 8.5–9.0). After the modification reaction, the mixture was neutralized with 1 M HCl and dialysed against double distilled water for 48 h (Mw cut-off: 8000–14,000 kDa). The acetylated PLCP (AC-PLCP) was obtained by freeze-drying.

The degrees of substitution (DS) of the acetyl group in the Ac-PLCPs were determined as described [19]. Briefly, 40 mg Ac-PLCP was dissolved in 20 mL 0.01 M NaOH. The solution was maintained at 50°C for 2 h, then excess NaOH was back-titrated with 0.01 M HCl using phenolphthalein indicator. Four Ac-PLPCs with gradient degrees of substitution, that is, DS of 0.027 (Ac-PLCP I), 0.059 (Ac-PLCP II), 0.082 (Ac-PLCP III), and 0.11 (Ac-PLCP IV), were selected to investigate the immune regulatory activity in dendritic cell model.

In addition, the bacterial endotoxins in PLCP and Ac-PLCPs were determined using tachypleus amebocyte lysate test kit (Chinese Horseshoe Crab Reagent Manufactory, Co., Fujian Province, China). The quantity of endotoxin was estimated to be ≤0.015 endotoxin unit (EU) per mg in all polysaccharide samples.

2.2. Animals. 4- to 6-week-old BALB/c (H-2K^d and I-A^d) and C57BL/6 (H-2K^d and I-A^d) mice were purchased from Hunan SJA Laboratory Animal, Co. (Hunan Province, China). All animals were kept according to the Care and Use of Laboratory Animals Guidelines published by the United States National Institute of Health (NIH Publication 85-23, 1996). All experimental procedures were approved by the Animal Ethnics Committee, Nanchang University.

2.3. Generation of Bone-Marrow-Derived Dendritic Cells (DCs). Bone marrow of BALB/c mice was sacrificed to obtain dendritic cells. According to our previous report [16], bone-marrow cells were flushed out with RPMI 1640 medium, then recombinant murine granulocyte macrophage-colony stimulating factor (rmGM-CSF, 10 ng/mL) combined with recombinant murine interleukin-4 (rmIL-4, 10 ng/mL, both from R&D system) was used to generate immature dendritic cells.

2.4. Determination of Phenotypic and Functional Maturation of DCs. Immature DCs were stimulated with the PLCP or Ac-PLCPs (100 μg/mL) for 48 h, and then the phenotypic and functional maturation of DCs were determined according to the methods reported previously [20].

Briefly, the cells were collected and incubated with FITC-conjugated monoclonal antibody [anti-major histocompatibility complex (MHC) class II, anti-CD80 (B7-1), and anti-CD86 (B7-2), eBioscience] for 1 h, and then the mean fluorescence intensity was determined by using FACSCalibur flow cytometer (BD Biosciences, USA). The endocytosis activity of DCs was also determined by flow cytometry using FITC-dextran (40,000 Da, Sigma). The quantity of the cytokine IL-12p70 in the culture supernatant was measured using Mouse IL-12p70 Enzyme Immunoassay kit (Wuhan Boster Biological Technology, China).

The migration activity of DCs was investigated by determining the mRNA expression of chemokine mRNA of CCR7 and CXCR4. Total RNA was extracted using Trizol Reagent (Invitrogen, Carlsbad, CA, USA). Then cDNA was synthesized using RevertAid ™ First Strand cDNA Synthesis Kit (Thermo Scientific, Maryland, USA). PCR products were analyzed by GoldView-agarose gel electrophoresis on 2% (w/v) agarose gel and visualized under UV light. The signals of target genes were measured by scanning densitometry and normalized to β-actin using Quantity One software. The forward and reverse PCR primers used for CCR7 were 5′-GCCTTCCTGTGTGATTTCTACAG-3′ and 5-TCACCTTCTCTCCTTTCTGTCAC-3′; for CXCR4 were 5′-TGTTGCCATGGAACCGATCA-3′ and 5′-GGATCC-AGACGCCCACATAG-3′; for β-actin were 5′-TGGCAC-CACACCTTCTACAATG-3′ and 5′-CCTGCTTGCTGA-TCCACATCTG-3′, respectively. Condition of reverse transcription system was 91°C for 5 min, 4°C for 5 min, 30°C for 10 min, 42°C for 30 min, 99°C for 5 min, and 4°C for 5 min. Condition of PCR system was initial denaturation at 94°C × 3 min, followed by 25 cycles of denaturation at 94°C × 30 s, annealing at 60°C × 30 s, extension at 72°C × 2 min, and final extension at 72°C for 10 min.

Besides, splenic T lymphocytes were purified from C57BL/6 (H-2K^d and I-A^d) mice using Pan T Cell Isolation kit II (Miltenyi Biotec, Inc., Auburn, USA) and then cocultured with DCs for another 48 h. The proliferation of T cells was determined using MTT Cell Proliferation and Cytotoxicity Assay kit (Beyotime, Shanghai, China).

2.5. Statistical Analysis. Results data were analyzed using SPSS statistical software (version 17.0) and expressed as mean ± standard deviation (SD) of the indicated number of experiments. The statistical significance was estimated using a Student's *t*-test. $P < 0.05$ and $P < 0.01$ were considered as statistically significant and highly significant, respectively.

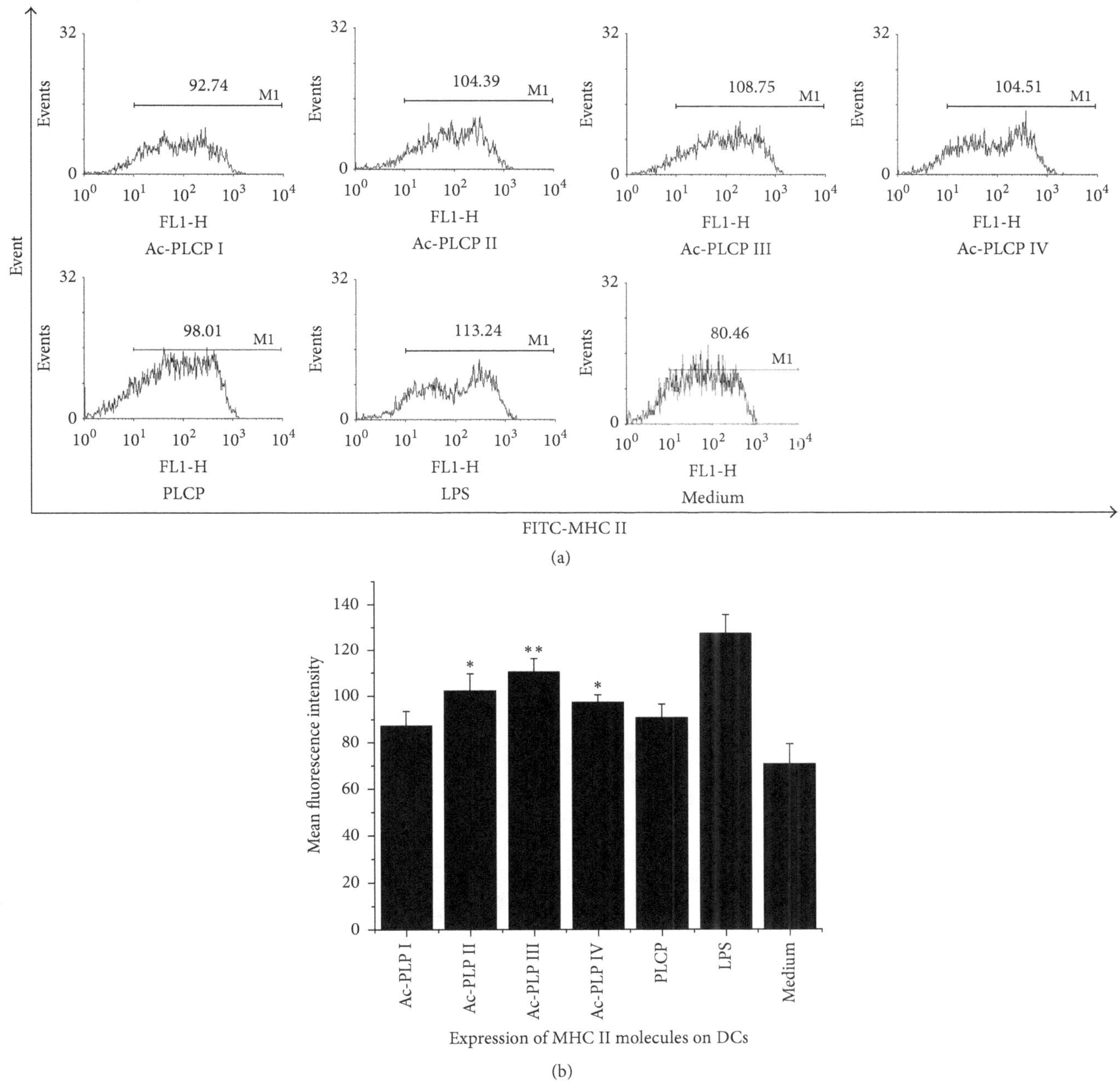

FIGURE 1: Ac-PLCP increased the expression of MHC II molecules on DCs. Immature DCs were treated with 100 μg/mL PLCP or Ac-PLCPs for 48 h. Then, the expression levels of MHC II molecules were determined by flow cytometry. The data were shown as mean fluorescence intensity (MFI). Compared with medium group, DCs treated with PLCP expressed higher level of MHC II molecules. While compared with the PLCP group, significant increases in MFI were observed in Ac-PLCP II, III, and IV groups. The results shown were from one representative experiment of three independent experiments performed. $^{*}P < 0.05$ compared to the PLCP group; $^{**}P < 0.01$ compared to the PLCP group.

3. Results

3.1. Ac-PLCP Enhanced Expression of MHC II, CD80, and CD86 Molecules in DCs. Immune responses were initiated through the specific recognition of antigens by lymphocytes, and this progress required stable adhesion of the T cells to the antigen-presenting cells (APCs), efficient antigen presentation, and transduction of signals from the cell surface to the nucleus of T cell. In particular, T cells could only recognize and respond to peptide antigens bound to and displayed by MHC molecules. Flow cytometry was applied to determine the expression of MHC II on DCs, and the data were shown as mean fluorescence intensity (MFI) in Figure 1. After incubation with PLCP, the expression of MHC II on DCs was increased from 80.46 to 98.01. Meanwhile, the expressions of MHC II in Ac-PLCP II, III, and IV groups were 104.39 ($P < 0.05$), 108.75 ($P < 0.01$), and 104.51 ($P < 0.05$), respectively, which were significantly higher than that of PLCP

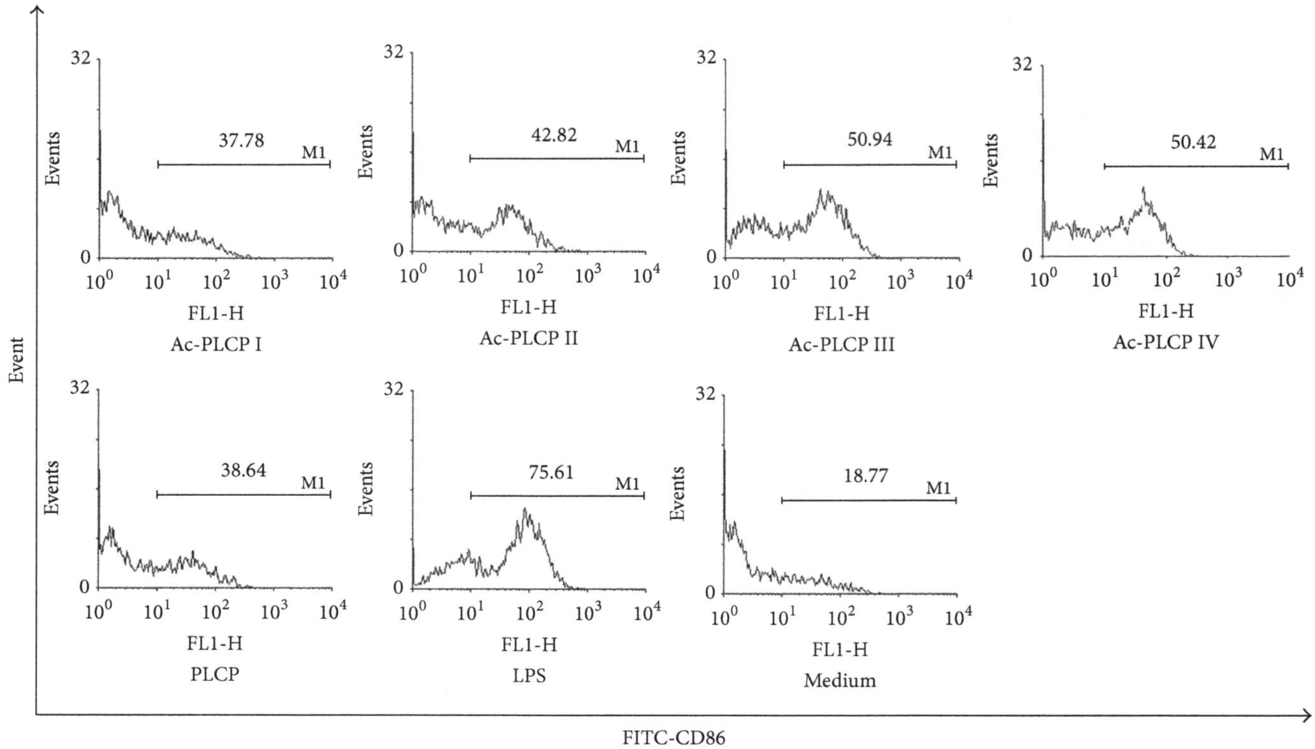

(a)

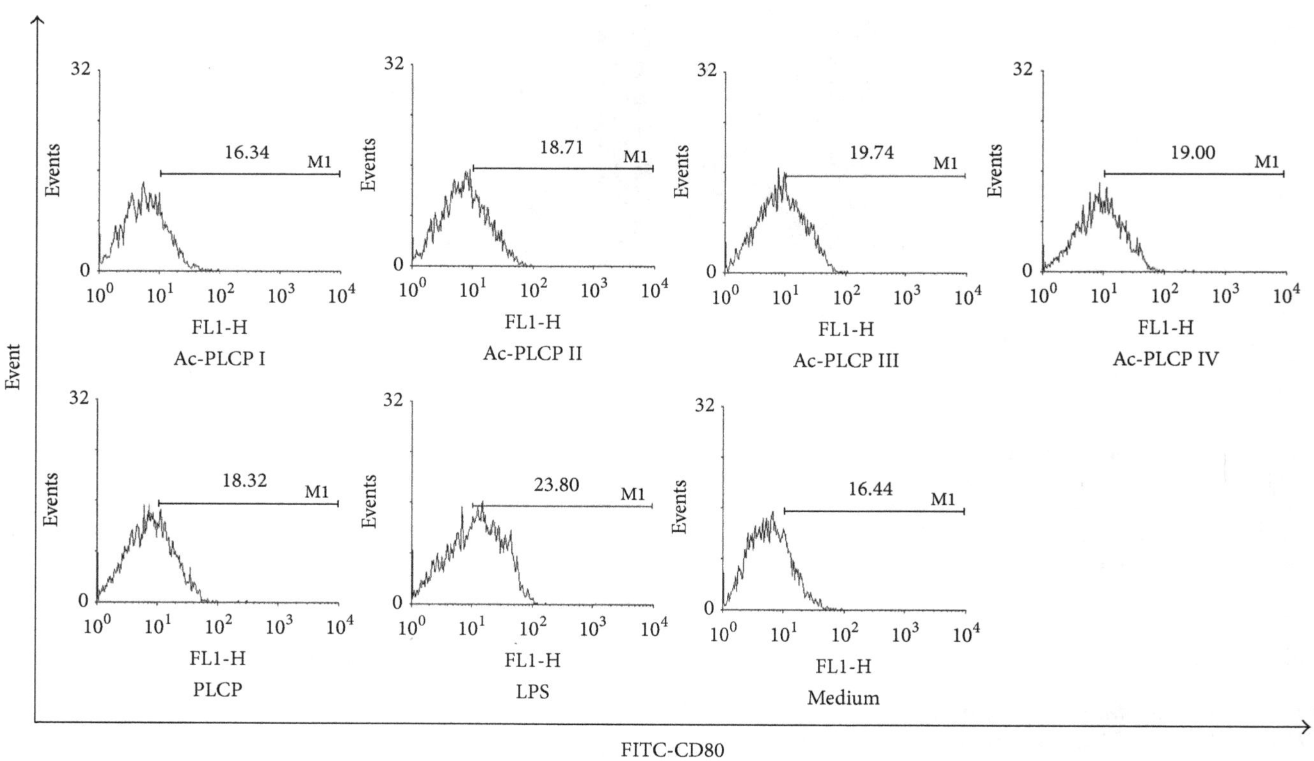

(b)

FIGURE 2: Continued.

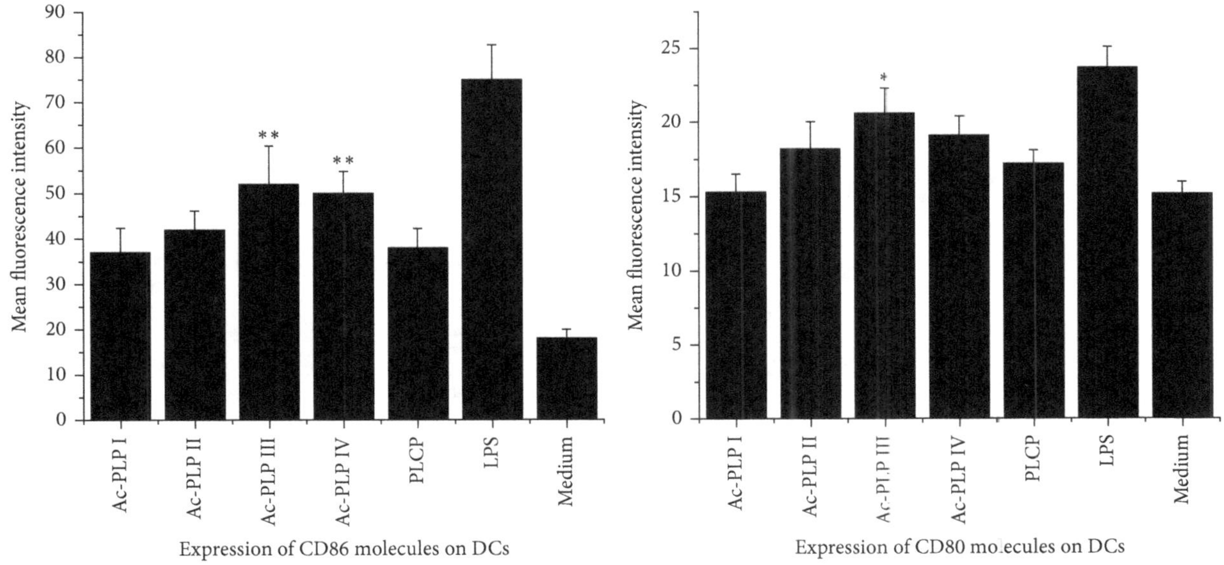

(c)

FIGURE 2: Ac-PLCP increased the expression of costimulating molecules on DC. DCs treated with Ac-PLCPs III and IV expressed a higher level of CD86 (a) and CD80 (b) molecules than PLCP group. The results shown were from one representative experiment of three independent experiments performed. $^{*}P < 0.05$ compared to the PLCP group; $^{**}P < 0.01$ compared to the PLCP group.

group. Costimulating molecules are necessary to ensure an effective immune response. The expressions of CD86 and CD80 costimulating molecules were also measured using flow cytometry. As shown in Figure 2(a), the expression levels of CD86 in DCs in Ac-PLCP I, II, III, and IV groups were 37.78, 42.82, 50.94, and 50.42, respectively, which were much higher than that in the control group (18.77) ($P < 0.01$). In particular, the expression levels of CD86 in Ac-PLCPs III and IV were significantly higher than PLCP group ($P < 0.01$). As shown in Figure 2(b), the expression of CD80 molecules was notably stimulated by Ac-PLCP III (19.74) compared with PLCP group (16.44) ($P < 0.05$).

3.2. Ac-PLCP Stimulated Secretion of IL-12p70 in DCs. Beyond interaction with T cells through surface molecules, DCs also regulate the immune response by secreting cytokines. As shown in Figure 3, PLCP increased the secretion of IL-12p70 in DCs from 20.38 pg/mL (control group) to 49.97 pg/mL (PLCP group), while the cytokine levels of IL-12p70 in the acetylated polysaccharides-treated groups were 33.11 pg/mL (Ac-PLCP I), 60.51 pg/mL (Ac-PLCP II), 96.43 pg/mL (Ac-PLCP III), and 81.06 pg/mL (Ac-PLCP IV), respectively. Compared with PLCP group, Ac-PLCP III and Ac-PLCP IV groups were significantly enhanced ($P < 0.01$).

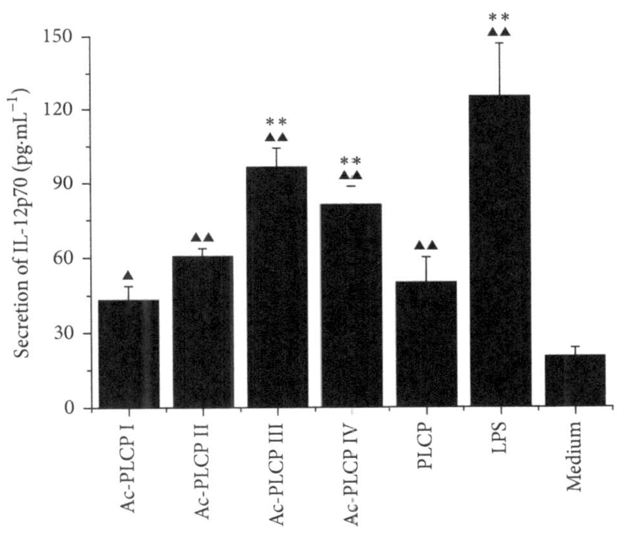

FIGURE 3: Ac-PLCP enhanced the secretion of IL-12p70 cytokine on DCs. Quantity of IL-12p70 in the culture supernatant was determined by ELISA method. Compared with PLCP group, the secretion of DCs in the Ac-PLCPs III and IV was significantly higher than that in PLCP group. $^{**}P < 0.01$ compared to the PLCP group; $^{▲}P < 0.05$ compared to the medium group; $^{▲▲}P < 0.01$ compared to the medium group.

3.3. Ac-PLCP Attenuated the Endocytosis of DCs. Although DCs are found in almost all tissues in the body, most of the cells exist in immature states. The immature DCs are not professional in antigen peptide presenting but antigen uptaking. However, once the immature DCs receive mature signal, they would develop maturation and transform from antigen-uptake cells into antigen-presenting cells. As shown in Figure 4, DCs in the control group (medium, 37°C) showed a dramatically high uptake of FITC-dextran. Compared with the control group, DCs in the PLCP and Ac-PLCP groups exhibited an attenuated endocytosis. Particularly, the MFI determined in the Ac-PLCP III (55.17) and IV groups (56.28) was significantly lower than PLCP group (67.60) ($P < 0.01$).

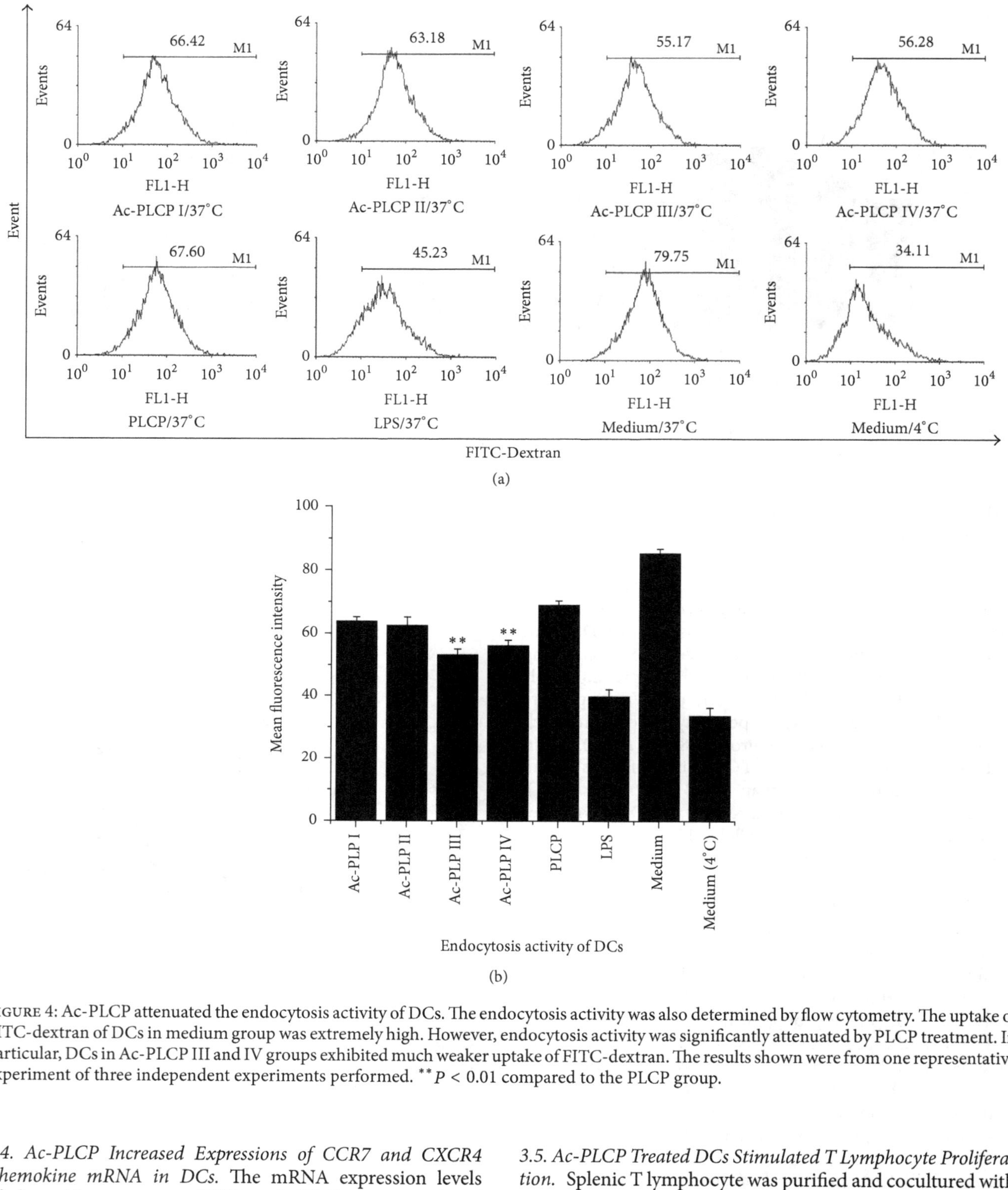

FIGURE 4: Ac-PLCP attenuated the endocytosis activity of DCs. The endocytosis activity was also determined by flow cytometry. The uptake of FITC-dextran of DCs in medium group was extremely high. However, endocytosis activity was significantly attenuated by PLCP treatment. In particular, DCs in Ac-PLCP III and IV groups exhibited much weaker uptake of FITC-dextran. The results shown were from one representative experiment of three independent experiments performed. $^{**}P < 0.01$ compared to the PLCP group.

3.4. Ac-PLCP Increased Expressions of CCR7 and CXCR4 Chemokine mRNA in DCs. The mRNA expression levels of chemokine receptors CCR7 and CXCR4 were analyzed using reverse transcription PCR to investigate the migration capability of DCs. As shown in Figure 5, notable increase was observed in the expressions of both CCR7 and CXCR4 mRNA in Ac-PLCP III and Ac-PLCP IV groups. In particular, the chemokine mRNA expression level in Ac-PLCP III or IV group was significantly higher than PLCP group ($P < 0.01$).

3.5. Ac-PLCP Treated DCs Stimulated T Lymphocyte Proliferation. Splenic T lymphocyte was purified and cocultured with DCs at a ratio of 10 : 1 for 48 h, and then the proliferation of T cells was determined by MTT assay. The proliferation index was calculated as absorbance ratio of polysaccharide group to control group. As shown in Figure 6, compared with control group, the proliferation index was significantly increased in all the polysaccharide groups ($P < 0.01$). The proliferation indexes in Ac-PLCP I and Ac-PLCP II groups were 124.21%

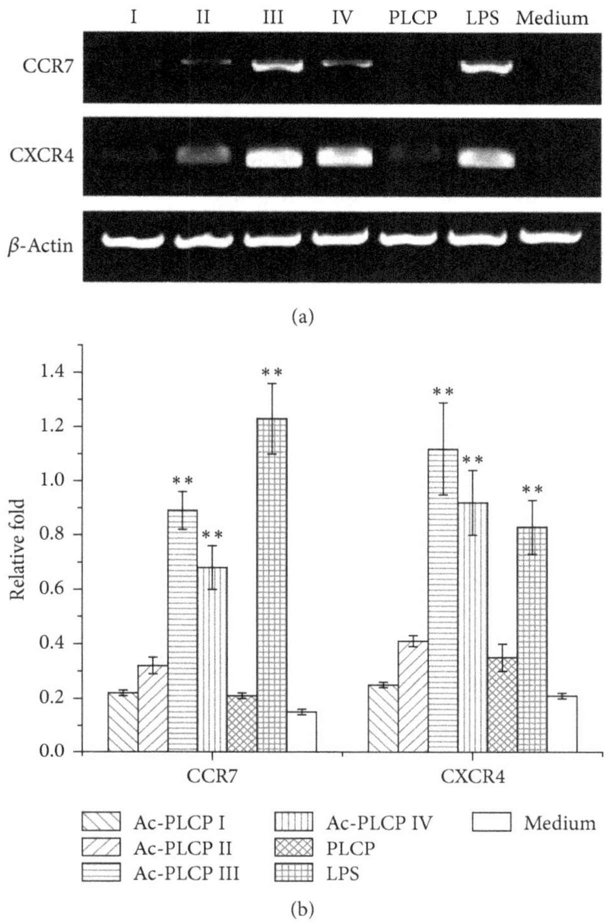

FIGURE 5: mRNA synthesis levels of CCR7 and CXCR4 were determined through reverse transcriptase PCR. The PCR product was analyzed by 1.5% agarose gel electrophoresis. The mRNA expression levels of both CCR7 and CXCR4 were significantly increased in Ac-PLCP III and IV groups. Besides, the expression of β-actin in each group was determined as internal control, and there was no significant difference among all the groups. $^{**}P < 0.01$ compared with PLCP group.

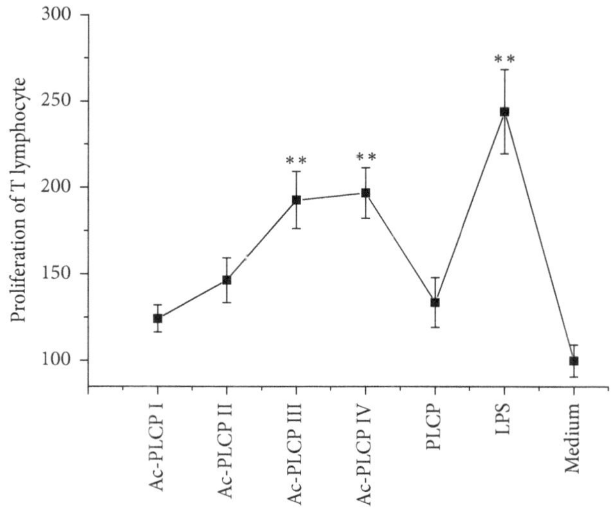

FIGURE 6: DCs treated with Ac-PLCPs showed stronger activity on stimulating T lymphocyte proliferation. T lymphocytes were obtained from C57BL/6 (H-2K^d and I-A^d) mice and cocultured with DCs in 96-well plate at the ratio of 10 : 1 for 48 h. The absorption value of the medium group was set at control values in the calculation of % proliferation. All the polysaccharide-treated DCs showed extremely strong T cell proliferation stimulating activity. Compared with PLCP group, DCs in Ac-PLCPs III and IV exhibited significantly enhanced T cell proliferation stimulating activity. $^{**}P < 0.01$ compared to the PLCP group.

and 146.52%. And the proliferation indexes in Ac-PLCP III and IV groups were 192.75% and 196.81%, respectively, which were dramatically higher than that in PLCP group (133.62%) ($P < 0.01$).

4. Discussion

DCs are professional antigen-presenting cells and essential mediators of innate and adaptive immune response, as well as tolerance [21]. Immature DCs are strategically located in dermis or mucosal system to capture antigen and then transport these antigens to regional lymph nodes, where they develop into mature and activate lymphocyte to initial immune response. However, immature DCs can weakly activate naïve T lymphocyte for lacking of antigen-presenting molecules. MHC molecules T cell receptor interaction and CD80(B7-1)/CD86(B7-2)-CD28 surface molecules interaction between APCs and T cells have been regarded as a key point to induce T cell activation [22]. Though immature dendritic cells synthesize large amounts of MHC II molecules, such MHC II molecules form $\alpha\beta$-dimers that exhibit intracellular distributions. The $\alpha\beta$-dimers are targeted to late endosomes and lysosomes where they reside unproductively with internalized antigens; however, after exposure to inflammatory mediators or microbial products, the newly formed immunogenic MHC II complexes are transported from lysosomes to plasma membrane [23]. In particular, the half-life of MHC II molecules increases from about 10 h to over 100 h during this process [24]. Ac-PLCPs III and IV were more effective in inducing DCs maturation than PLCP, since the expressions of MHC II molecules on DCs of the Ac-PLCP III and IV groups were significantly higher than that of PLCP group. Besides the primary signal delivered to T cells through MHC II molecules, the B7-CD28 interaction is one of the dominant costimulatory signals. Particularly, for naïve T cells that express only limited number of costimulatory receptors, CD28 is an indispensable receptor required for T cell priming [25]. The B7-CD28 costimulation signals could direct the development and function of T cells. Lenschow et al. found that effector and regulatory T cell responses were impaired in mice lacking CD28 or in mice lacking CD80 and CD86. [26]. Gimmi et al. reported that antigen presentation in absence of B7 costimulation results in human T cell clonal anergy [27]. We found that Ac-PLCP III and IV incubation significantly promoted the expression of CD86 molecules on DCs as well, which strongly suggests that the acetylated polysaccharides could induce phenotypic maturation on DCs. Meanwhile, an obvious decline of endocytosis activity on DCs was also observed, which indicates that the immature dendritic cells transform from antigen-capturing cells into the mature antigen-presenting cells.

Besides the ligand-receptor interaction, DCs could regulate the polarization of T lymphocytes by secreting various cytokines. IL-12 is a proinflammatory molecule produced primarily by monocytes/macrophages and DCs. And IL-12 principally activates natural killer cells and induces the differentiation of native $CD4^+$ T cells to become interferon-γ-producing T helper 1 (Th1) effectors in cell-mediated immune responses [28]. IL-12 could also synergize with B7/CD28 interaction in inducing efficient proliferation and cytokine production of human T cells [29]. It was reported that $CD1c^+$ myeloid DC but not plasmacytoid DC (pDC) in humans can induce high level of cytotoxic molecules in naïve T cells by producing high amount of IL-12 [30]. The secretion of IL-12p70 of DCs was promoted by Ac-PLPC III and IV treatments, and the levels of IL-12p70 secretion in these two groups were significantly higher than PLCP group. Consistently, we also found that DCs incubated with Ac-PLPC III or IV could significantly stimulate proliferation of naïve T cell *in vitro*. Furthermore, DCs in Ac-PLCP III and IV groups showed an extremely strong effect on provoking sizeable mixed lymphocyte reaction when cocultured with splenic T cells, which indicates that the acetylated polysaccharides stimulated functional maturation on DCs. All these findings support that DCs treated with Ac-PLCPs III and IV may possess much more powerful immune regulatory activity than the natural polysaccharide.

Optimal encounter with naïve T cells for the presentation of antigens requires DCs to migrate to secondary lymphoid organs, which is governed by chemokine. The key chemokines directing DCs migration are chemokine (C-C motif) ligand 19 (CCL19), CCL21, and chemokine (C-X-C motif) ligand 12 (CXCL12) [31, 32]. As receptor for CCL19 and CCL21, CC chemokine receptor CCR7 has been found to mediate the migration of DCs from skin into lymphatic vessels [33]. However, the migration of skin DC was initiated by CXCL12-CXCR4 engagement [34]. For plasmacytoid DCs, the migration into splenic white pulp was regulated by CCR7 signal coordinate with CXCR4 signal [35]. And the chemokine signal also enhanced DCs maturation and survival [36]. The migration capability of DCs was enhanced by Ac-PLCP III and IV treatments evidenced by the increased mRNA expression levels of CCR7 and CXCR4.

Dendritic cells are equipped with a battery of pattern-recognition receptors (PRRs) that can detect molecular patterns of invading microorganisms or endogenous signals and alter the immune response. The most widely studied family of PRRs on DCs is Toll-like receptors. In particular, TLR-4 expressed on the surface of DCs was demonstrated to be one to recognize polysaccharides [37, 38]. DCs in intestinal mucosa tissue were found to be able to open the tight junctions between epithelial cells without breaking the integrity of the epithelial barrier [39], which provides a probable route for the polysaccharide to interact with DCs. We have also found that TLR-4 mediates the maturation of DCs induced by PLCP [16]. The acetylation modification may enhance the immunoregulatory activities of PLCP by improving the solubility of PLCP and decreasing the viscosity of the polysaccharide.

In conclusion, Ac-PLCPs exhibited higher maturation-stimulating activities on DCs compared with the natural polysaccharide. In particular, the Ac-PLCPs of DS ranging from 0.06 to 0.1 showed the best immunoregulatory activities. The current study also provided a new insight into the structure-activity relationship of PLCP and the utilization of the seeds of *Plantago asiatica* L.

Abbreviations

CD: Cluster of differentiation
Ac: Acetylated
DCs: Dendritic cells
DS: Degree of substitution
MHC: Major histocompatibility complex
PLCP: *Plantago asiatica* L. crude polysaccharide
IL-12: Interleukin-12.

Conflicts of Interest

The authors declare that there are no conflicts of interest.

Acknowledgments

This study is supported by the National Key Technology R&D Program of China (2012BAD33B06), the National Natural Science Foundation of China (21062012, 31260364), and the Program for New Century Excellent Talents in University (NCET-12-0749), which are gratefully acknowledged.

References

[1] I. Cumpstey, "Chemical Modification of Polysaccharides," *ISRN Organic Chemistry*, vol. 2013, Article ID 417672, 27 pages, 2013.

[2] D. A. Silva, R. C. M. De Paula, J. P. A. Feitosa, A. C. F. De Brito, J. S. Maciel, and H. C. B. Paula, "Carboxymethylation of cashew tree exudate polysaccharide," *Carbohydrate Polymers*, vol. 58, no. 2, pp. 163–171, 2004.

[3] J. Xu, W. Liu, W. Yao, X. Pang, D. Yin, and X. Gao, "Carboxymethylation of a polysaccharide extracted from *Ganoderma lucidum* enhances its antioxidant activities *in vitro*," *Carbohydrate Polymers*, vol. 78, no. 2, pp. 227–234, 2009.

[4] J. Wang, L. Zhang, Y. Yu, and P. C. K. Cheung, "Enhancement of antitumor activities in sulfated and carboxymethylated polysaccharides of Ganoderma lucidum," *Journal of Agricultural and Food Chemistry*, vol. 57, no. 22, pp. 10565–10572, 2009.

[5] L. Ma, H. Chen, Y. Zhang, N. Zhang, and L. Fu, "Chemical modification and antioxidant activities of polysaccharide from mushroom *Inonotus obliquus*," *Carbohydrate Polymers*, vol. 89, no. 2, pp. 371–378, 2012.

[6] X.-X. Liu, Z.-J. Wan, L. Shi, and X.-X. Lu, "Preparation and antiherpetic activities of chemically modified polysaccharides from Polygonatum cyrtonema Hua," *Carbohydrate Polymers*, vol. 83, no. 2, pp. 737–742, 2011.

[7] I. Y. Bae, H. Y. Kim, S. Lee, and H. G. Lee, "Effect of the degree of oxidation on the physicochemical and biological properties of Grifola frondosa polysaccharides," *Carbohydrate Polymers*, vol. 83, no. 3, pp. 1298–1302, 2011.

[8] J.-H. Wang, J.-P. Luo, and X.-Q. Zha, "Structural features of a pectic polysaccharide from the stems of Dendrobium nobile Lindl," *Carbohydrate Polymers*, vol. 81, no. 1, pp. 1–7, 2010.

[9] X. Xing, S. W. Cui, S. Nie, G. O. Phillips, H. D. Goff, and Q. Wang, "Study on Dendrobium officinale O-acetyl-glucomannan (Dendronan®): part I. Extraction, purification, and partial structural characterization," *Bioactive Carbohydrates and Dietary Fibre*, vol. 4, no. 1, pp. 74–83, 2014.

[10] H. Qi, Q. Zhang, T. Zhao, R. Hu, K. Zhang, and Z. Li, "*In vitro* antioxidant activity of acetylated and benzoylated derivatives of polysaccharide extracted from *Ulva pertusa* (Chlorophyta)," *Bioorganic & Medicinal Chemistry Letters*, vol. 16, no. 9, pp. 2441–2445, 2006.

[11] J. Wang, L. Liu, Q. Zhang, Z. Zhang, H. Qi, and P. Li, "Synthesized oversulphated, acetylated and benzoylated derivatives of fucoidan extracted from Laminaria japonica and their potential antioxidant activity in vitro," *Food Chemistry*, vol. 114, no. 4, pp. 1285–1290, 2009.

[12] Z. Zhang, X. Wang, S. Yu, L. Yin, M. Zhao, and Z. Han, "Synthesized oversulfated and acetylated derivatives of polysaccharide extracted from Enteromorpha linza and their potential antioxidant activity," *International Journal of Biological Macromolecules*, vol. 49, no. 5, pp. 1012–1015, 2011.

[13] J. Y.-Y. Chan, E. Chan, S.-W. Chan et al., "Enhancement of in vitro and in vivo anticancer activities of polysaccharide peptide from Grifola frondosa by chemical modifications," *Pharmaceutical Biology*, vol. 49, no. 11, pp. 1114–1120, 2011.

[14] T. Teruya, H. Tatemoto, T. Konishi, and M. Tako, "Structural characteristics and in vitro macrophage activation of acetyl fucoidan from Cladosiphon okamuranus," *Glycoconjugate Journal*, vol. 26, no. 8, pp. 1019–1028, 2009.

[15] J. Yin, H. Lin, J. Li et al., "Structural characterization of a highly branched polysaccharide from the seeds of Plantago asiatica L.," *Carbohydrate Polymers*, vol. 87, no. 4, pp. 2416–2424, 2012.

[16] D. Huang, S. Nie, L. Jiang, and M. Xie, "A novel polysaccharide from the seeds of Plantago asiatica L. Induces dendritic cells maturation through toll-like receptor 4," *International Immunopharmacology*, vol. 18, no. 2, pp. 236–243, 2014.

[17] J.-Y. Yin, S.-P. Nie, C. Zhou, Y. Wan, and M.-Y. Xie, "Chemical characteristics and antioxidant activities of polysaccharide purified from the seeds of Plantago asiatica L," *Journal of the Science of Food and Agriculture*, vol. 90, no. 2, pp. 210–217, 2010.

[18] Y. Chen, H. Zhang, Y. Wang, S. Nie, C. Li, and M. Xie, "Acetylation and carboxymethylation of the polysaccharide from *Ganoderma atrum* and their antioxidant and immunomodulating activities," *Food Chemistry*, vol. 156, pp. 279–288, 2014.

[19] A. B. Das, G. Singh, S. Singh, and C. S. Riar, "Effect of acetylation and dual modification on physico-chemical, rheological and morphological characteristics of sweet potato (Ipomoea batatas) starch," *Carbohydrate Polymers*, vol. 80, no. 3, pp. 725–732, 2010.

[20] L.-M. Jiang, S.-P. Nie, H.-L. Zhou, D.-F. Huang, and M.-Y. Xie, "Carboxymethylation enhances the maturation-inducing activity in dendritic cells of polysaccharide from the seeds of Plantago asiatica L.," *International Immunopharmacology*, vol. 22, no. 2, pp. 324–331, 2014.

[21] J. Banchereau and R. M. Steinman, "Dendritic cells and the control of immunity," *Nature*, vol. 392, no. 6673, pp. 245–252, 1998.

[22] M. Girndt, H. Köhler, E. Schiedhelm-Weick, K.-H. Meyer Zum Büschenfelde, and B. Fleischer, "T cell activation defect in hemodialysis patients: Evidence for a role of the B7/CD28 pathway," *Kidney International*, vol. 44, no. 2, pp. 359–365, 1993.

[23] A. Chow, D. Toomre, W. Garrett, and I. Mellman, "Dendritic cell maturation triggers retrograde MHC class II transport from lysosomes to the plasma membrane," *Nature*, vol. 418, no. 6901, pp. 988–994, 2002.

[24] M. Cella, A. Engering, V. Pinet, J. Pieters, and A. Lanzavecchia, "Inflammatory stimuli induce accumulation of MHC class II complexes on dendritic cells," *Nature*, vol. 388, no. 6644, pp. 782–787, 1997.

[25] L. Chen and D. B. Flies, "Molecular mechanisms of T cell co-stimulation and co-inhibition," *Nature Reviews Immunology*, vol. 13, no. 4, pp. 227–242, 2013.

[26] D. J. Lenschow, T. L. Walunas, and J. A. Bluestone, "CD28/B7 system of T cell costimulation," *Annual Review of Immunology*, vol. 14, pp. 233–258, 1996.

[27] C. D. Gimmi, G. J. Freeman, J. G. Gribben, G. Gray, and L. M. Nadler, "Human T-cell clonal anergy is induced by antigen presentation in the absence of B7 costimulation," *Proceedings of the National Acadamy of Sciences of the United States of America*, vol. 90, no. 14, pp. 6586–6590, 1993.

[28] X. Ma, W. Yan, H. Zheng et al., "Regulation of IL-10 and IL-12 production and function in macrophages and dendritic cells," *F1000Research*, vol. 4, 2015.

[29] M. Kubin, M. Kamoun, and G. Trinchieri, "Interleukin 12 synergizes with B7/CD28 interaction in inducing efficient proliferation and cytokine production of human T cells," *The Journal of Experimental Medicine*, vol. 180, no. 1, pp. 211–222, 1994.

[30] G. Nizzoli, J. Krietsch, A. Weick et al., "Human CD1c$^+$ dendritic cells secrete high levels of IL-12 and potently prime cytotoxic T-cell responses," *Blood*, vol. 122, no. 6, pp. 932–942, 2013.

[31] B. G. Ricart, B. John, D. Lee, C. A. Hunter, and D. A. Hammer, "Dendritic cells distinguish individual chemokine signals through CCR7 and CXCR4," *The Journal of Immunology*, vol. 186, no. 1, pp. 53–61, 2011.

[32] A. M. Platt and G. J. Randolph, "Dendritic Cell Migration Through the Lymphatic Vasculature to Lymph Nodes," *Advances in Immunology*, vol. 120, pp. 51–68, 2013.

[33] L. Ohl, M. Mohaupt, N. Czeloth et al., "CCR7 governs skin dendritic cell migration under inflammatory and steady-state conditions," *Immunity*, vol. 21, no. 2, pp. 279–288, 2004.

[34] K. Kabashima, N. Shiraishi, K. Sugita et al., "CXCL12-CXCR4 engagement is required for migration of cutaneous dendritic cells," *The American Journal of Pathology*, vol. 171, no. 4, pp. 1249–1257, 2007.

[35] E. Umemoto, K. Otani, T. Ikeno et al., "Constitutive plasmacytoid dendritic cell migration to the splenic white pulp is cooperatively regulated by CCR7- and CXCR4-mediated signaling," *The Journal of Immunology*, vol. 189, no. 1, pp. 191–199, 2012.

[36] K. Kabashima, K. Sugita, N. Shiraishi, H. Tamamura, N. Fujii, and Y. Tokura, "CXCR4 engagement promotes dendritic cell survival and maturation," *Biochemical and Biophysical Research Communications*, vol. 361, no. 4, pp. 1012–1016, 2007.

[37] H. S. Kim, B. R. Shin, H. K. Lee et al., "Dendritic cell activation by polysaccharide isolated from Angelica dahurica," *Food and Chemical Toxicology*, vol. 55, pp. 241–247, 2013.

[38] M. J. Park, H. S. Ryu, J. S. Kim et al., "Platycodon grandiflorum polysaccharide induces dendritic cell maturation via TLR4 signaling," *Food and Chemical Toxicology*, vol. 72, pp. 212–220, 2014.

[39] M. Rescigno, M. Urbano, B. Valzasina et al., "Dendritic cells express tight junction proteins and penetrate gut epithelial monolayers to sample bacteria," *Nature Immunology*, vol. 2, no. 4, pp. 361–367, 2001.

3

A Novel and Chemoselective Process of *N*-Alkylation of Aromatic Nitrogen Compounds Using Quaternary Ammonium Salts as Starting Material

Carlos A. González-González,[1] Juan Javier Mejía Vega,[1] Ricardo García Monroy,[1] Davir González-Calderón,[1] David Corona-Becerril,[1] Aydeé Fuentes-Benítes,[1] Joaquín Tamariz Mascarúa,[2] and Carlos González-Romero[1]

[1]*Departamento de Química Orgánica, Facultad de Química, Universidad Autónoma del Estado de México, Paseo Colón Esquina Paseo Tollocan s/n, 50120 Toluca, MEX, Mexico*
[2]*Departamento de Química Orgánica, Escuela Nacional de Ciencias Biológicas, Instituto Politécnico Nacional, Prol. Carpio y Plan de Ayala s/n, 11340 Ciudad de México, Mexico*

Correspondence should be addressed to Carlos A. González-González; arnorian@gmail.com and Carlos González-Romero; cgonzalezr@uaemex.mx

Academic Editor: Jonathan White

The process of *N*-alkylation of several pyrroles, indoles, and derivative heterocycles is herein described, using quaternary ammonium salts as the source of an alkylating agent. These reactions were carried out on several heterocyclic rings with triethylbenzylammonium chloride or tetradecyltrimethylammonium bromide and an NaOH solution at 50%, leading to a chemoselective *N*-alkylated product and an average yield of 73%. This is an alternative process to the traditional benzylation and methylation of *N*-heterocycles with direct handling of alkyl halides.

1. Introduction

Quaternary ammonium salts are one of the most widely used reagents in the synthesis of organic compounds when seeking to develop certain vital products, including drugs, pesticides, monomers, and cleaning agents [1, 2]. In the area of materials development, they have been employed in recent years as surfactants, template agents, and alkaline anion exchange membranes (AAEM) [3, 4].

In organic chemistry, these salts have various applications in situations where inorganic and organic anions react with organic substrates that are separated into distinct phases known as phase-transfer catalysis. These anions need to be introduced in the organic phase as a lipophilic ion pair provided by a catalyst. Among the most commonly used catalysts are quaternary ammonium salts. Although there are a variety of articles about phase-transfer catalysts, little has been reported about them as an alkylating agent of nitrogen heterocyclic rings [5].

During one of our current research programs involving experiments with nitrogen heterocycles, utilizing reactions under conditions of phase-transfer catalysis between triethylbenzylammonium chloride (TEBAC) and a pyrrole ring, it was observed that the catalyst participated as an *N*-alkylating agent providing a benzylated product. By tracking this observation, we discovered that if the original alkylating agent was omitted from the reaction, employing instead more than a stoichiometric amount of TEBAC, a clean transfer of a unique benzyl group took place. We did not detect the transfer of an ethyl group from the TEBAC salt nor any long-chain alkylation in the case of tetradecyltrimethylammonium bromide. Therefore, we conducted a series of experiments with two quaternary ammonium salts (TEBAC and tetradecyltrimethylammonium bromide) and a series of nitrogenous

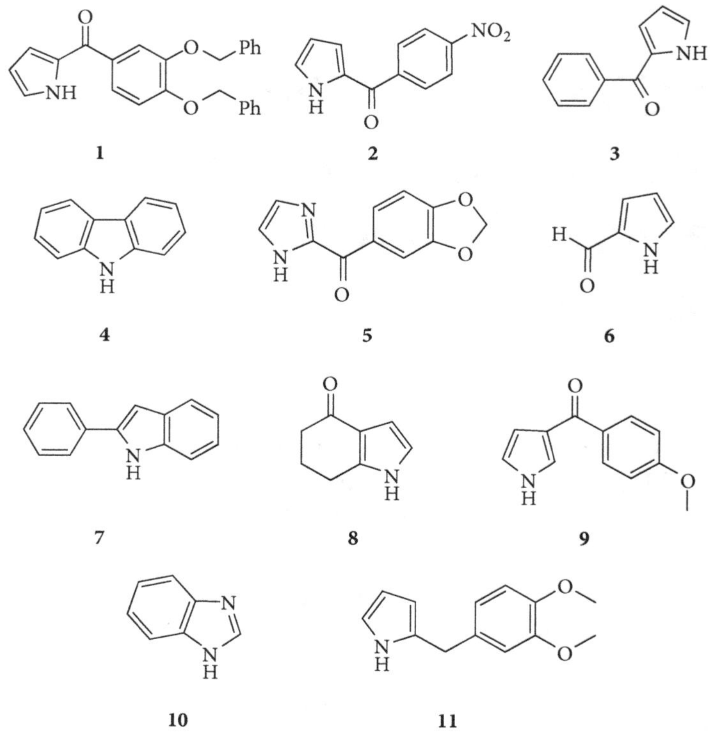

FIGURE 1

TABLE 1: Chemoselective alkylation of nitrogen heterocyclic rings with TEBAC.

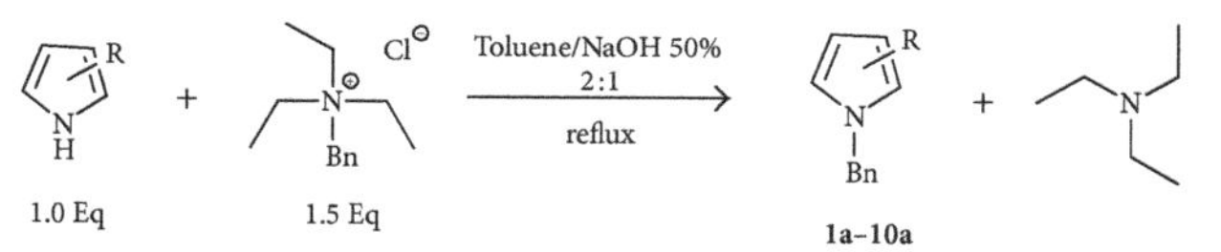

Starting material	Product	Time (h)	Yield* (%) R = Bn
1	1a	18	84
2	2a	6	90
3	3a	2	98
4	4a	17	88
5	5a	4	61
6	6a	3	86
7	7a	4	75
8	8a	8	52
9	9a	8	50
10	10a	2	64

*Yields refer to chromatographically pure isolated compounds.

aromatic heterocycles (Figure 1), obtaining the *N*-alkylation of these compounds as the main product (Table 1).

The transfer of an alkyl group from a quaternary ammonium salt has been briefly mentioned in the case of 2-chlorophenothiazine [6]. A study of the stability of three phase-transfer catalysts to afford five different nucleophiles suggests the feasibility of quaternary ammonium salts acting as alkylating agents for certain carbon and sulfur nucleophiles [7] and demonstrates the reaction of thiolate nucleophiles with a catalyst in preference to the substrate. However, there are no previous studies with aromatic nitrogen compounds which in comparison with the ones mentioned above are less acid.

2. Results and Discussion

In several TEBAC-promoted benzylation reactions, we observed the formation of diethylbenzylamine as a substantial side-product. Although its presence was not troubling insofar as the purification of the principal product was concerned, it did raise several mechanistic questions. The fact

TABLE 2: Chemoselective alkylation of nitrogen heterocyclic rings with tetradecyltrimethylammonium bromide.

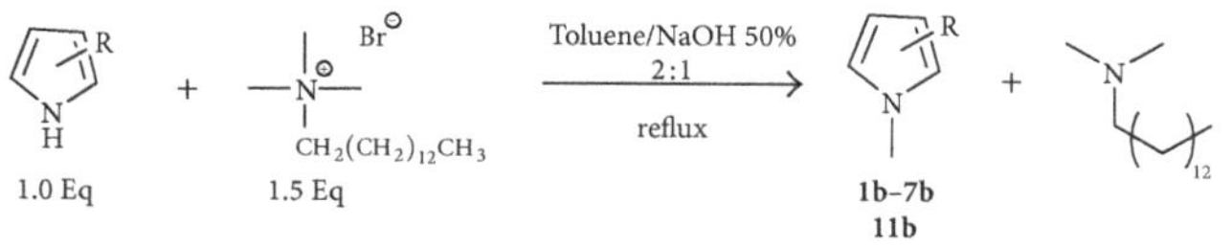

Starting material	Product	Time (h)	Yield* (%) R = CH_3
1	**1b**	3	73
2	**2b**	4	78
3	**3b**	5	76
4	**4b**	5	97
5	**5b**	18	74
6	**6b**	5	46
7	**7b**	3	73
11	**11b**	6	43

*Yields refer to chromatographically pure isolated compounds.

that no ethylated heterocycle was ever found argued against the attack of a heterocycle on the ethyl group.

Upon examining the behavior and stability of the catalyst at reaction temperature and in the absence of the heterocycle, it was found that diethylbenzylamine formed in high yield (76%) after only 4 hours, suggesting competitive catalyst/substrate decomposition under specific reaction conditions. The latter include the use of an aqueous NaOH solution at 50% in toluene, which means that the organic solvent does not need to be in an anhydrous condition, as would be necessary for a conventional alkylation reaction [8]. Two more reactions led to mechanistic information. First, when benzyltriethylammonium tetrafluoroborate was employed instead of the chloride salt, the yield of benzylated heterocycle in the case of 2-phenylindole (**7**) rose from 75% to 89% in less reaction time (3 h). In the same period, less than 2% formation of diethylbenzylamine was observed when the heterocycle was omitted from the reaction. This observation rules out the possibility that the quaternary ammonium chloride merely acted as an in situ source of benzyl chloride by chloride displacement of trimethylamine. Second, a predictably poor alkylating agent, isopropyl iodide, gave only *N*-isopropyl-2-phenylindole in low yield when a normal quantity (25% mole) of TEBAC was used in conjunction with 1 eq. of alkylating agent. When the quantity of TEBAC was increased from 1 to 2 eq. of alkylating agent, the *N*-isopropyl and *N*-benzyl products were formed in a 67 : 33 ratio. These observations led us to conclude that an S_N^2 attack on the quaternary ammonium salts is more likely than the reverse Menschutkin forming benzyl chloride [9, 10].

According to these results, a plausible mechanism is proposed in Scheme 1, in which the first step of phase-transfer catalysis proceeds to a nucleophilic attack by the pyrrolyl anion on the quaternary salt. Conditions for S_N^2 occurred in the case of TEBAC, attacking exclusively the methylene of the benzyl group. Chempath et al. described the degradation of trimethylbenzylammonium hydroxides, showing how a base extracts a proton from the methylene. They explained that the intrinsic stability of the benzyl cations is very close to that of $N(CH_3)_4^+$ [3]. Thus, *N*-benzylation was herein carried out on the pyrrole ring. Similarly, conditions for S_N^2 probably occurred with bromide tetradecyltrimethylammonium. Since it would be very difficult to carry out the attack on the long chain due to steric effects, the pyrrolyl anion would proceed to attack the less hindered electrophilic carbon, leading to *N*-methylation in the methyl group [11, 12].

Tables 1 and 2 indicate that the quaternary ammonium salts are the main substrate in the reaction that results in *N*-alkylation and the formation of a tertiary amine. Furthermore, in cases where TEBAC and tetradecyltrimethylammonium bromide acted as alkylating agents, *N*-substituted compounds were obtained by benzyl and methyl groups, respectively, as can be appreciated. Also, there was reduced performance with substrates that have low reactivity (those having electron-groups at position 3 on the ring), because they do not carry out the alkylation well. The present technique proved to be chemoselective, reacting specifically with only one substituent of the quaternary salt. One possible mechanism for these reactions of *N*-alkylation that take place through one of the two quaternary ammonium salts is shown in Scheme 1.

When comparing the conventional technique of homogeneous alkylation to that utilizing a quaternary ammonium salt, similar results are obtained. However, the benefit of the latter is the avoidance of an anhydrous solution that requires meticulous care. From the economic point of view, the use of anhydrous solutions in industry is generally seen as disadvantageous because of requiring greater investment. Additionally, the method herein employed does not require direct handling of an alkyl halide or dimethyl sulfate, therefore eliminating the respective health risks. We demonstrate that quaternary ammonium salts allow for feasible and simple alkylation reactions in organic synthesis [13, 14].

3. Conclusions

This method offers a convenient alternative to the traditional benzylation and methylation of *N*-heterocycles with alkyl halides. It is particularly useful in cases such as pyrroles which have a substantially more reactive anion compared to phenothiazine. Commercially available TEBAC and tetradecyltrimethylammonium bromide are considerably more convenient to handle than alkyl halides, the latter of which are lachrymators and carcinogenic reagents.

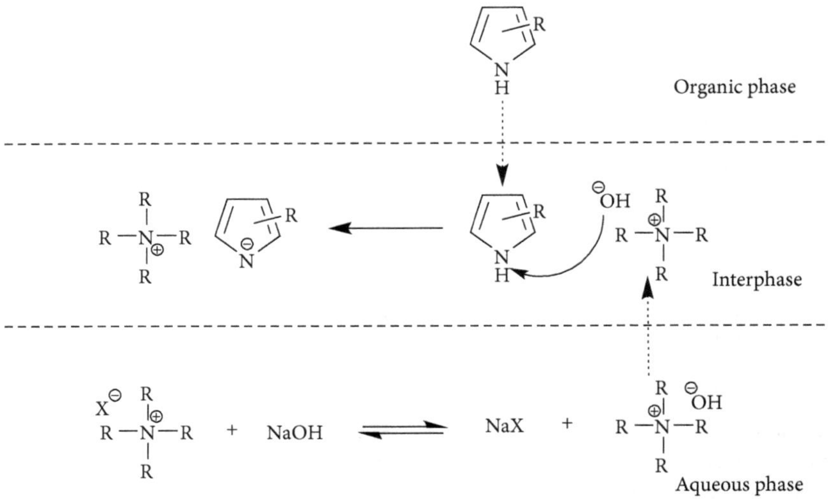

(a) General mechanism of phase-transfer catalysis

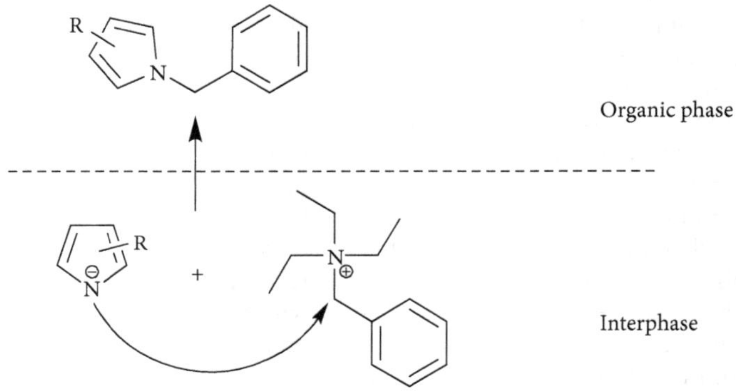

(b) Possible mechanism of *N*-benzylation

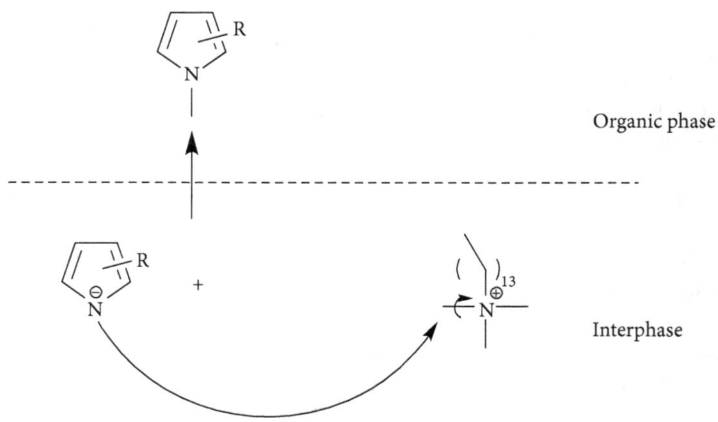

(c) Possible mechanism of *N*-methylation

Scheme 1

4. Material and Methods

General Details. The starting materials were purchased from Aldrich Chemical Co. and were used without further purification. Flash column chromatography was performed with silica gel 60 (230–400 mesh). Thin layer chromatography was carried out on silica plates of 0.20 mm thickness and visualized by UV light at 254 nm or by revelation with iodine. Melting points were determined on a Fischer-Johns Scientific melting point apparatus and are uncorrected. ^{1}H and ^{13}C NMR spectra were recorded using a Bruker Avance 300 MHz and a Varian 500 MHz. The chemical shifts (δ) are given in ppm relative to TMS as internal standard (0.00). For analytical purposes, the mass spectra were recorded on a Shimadzu GCMS-QP2010 Plus.

*4.1. General Procedure for the Synthesis of **1a–10a**.* A 100 mL round-bottom flask was equipped with a magnetic stir bar and a reflux condenser. To xylene (10.0 mL), TEBAC (1.1 mmol) and a heterocyclic compound (1.0 mmol) were added, followed by a solution of NaOH 50% (5.0 mL). The mixture was stirred at reflux temperature for 2–18 h. After completion of the reaction, the mixture was air-jet cooled to 25°C and TLC evidenced the disappearance of the starting material. The reaction mix was treated with AcOEt (4 × 20 mL), and the organic phase separated and removed under reduced pressure. The residue was purified to analytical purity by column chromatography.

*4.2. General Procedure for the Synthesis of **1b–7b** and **11b**.* A 100 mL round-bottom flask was equipped with a magnetic stir bar and a reflux condenser. To xylene (10.0 mL), tetradecyltrimethylammonium bromide (1.1 mmol) and a heterocyclic compound (1.0 mmol) were added, followed by a solution of NaOH 50% (5.0 mL). The mixture was stirred at reflux temperature for 2–18 h. After completion of the reaction, the mixture was air-jet cooled to 25°C and TLC indicated the disappearance of the starting material. The reaction mix was treated with AcOEt (4 × 20 mL), and the organic phase separated and removed under reduced pressure. The residue was purified to analytical purity by column chromatography.

(3a) *(1-Benzyl-1H-pyrrol-2-yl)(phenyl)methanone [15].* Using the standard procedure, to xylene (10.0 mL), 0.5003 g of TEBAC (1.1 Eq) and 0.2996 g of methanone **(3a)** (1.0 Eq) were added, followed by a solution of NaOH 50% (5.0 mL). The mixture was stirred at reflux temperature for 2 h. After cooling to r.t., TLC revealed the disappearance of the starting material. The reaction mix was treated with AcOEt (4 × 20 mL), and the organic phase separated and removed under reduced pressure. The compound was isolated as an off-yellow solid: ^{1}H NMR: (300 MHz, $CDCl_3$) δ = 7.81–7.68 (m, 2H), 7.53–7.34 (m, 3H), 7.34–7.10 (m, 5H), 7.00 (t, J = 2.1 Hz, 1H), 6.76 (dd, J = 4.1, 1.7 Hz, 1H), 6.19 (dd, J = 4.1, 2.6 Hz, 1H), 5.65 (s, 2H). ^{13}C NMR: (126 MHz, $CDCl_3$) δ = 186.20, 171.26, 139.83, 138.28, 131.46, 130.04, 129.24, 128.62, 128.42, 127.34, 127.16, 123.70, 108.73, 69.67, 60.44, 52.37. MS [EI$^+$] m/z (%): 306 (M$^+$), 289 (28), 259 (14), 229 (4), 156 (22), 91 (100), 65 (17).

Conflicts of Interest

The authors declare that there are no conflicts of interest regarding the publication of this paper.

Acknowledgments

The authors gratefully acknowledge the financial support from the Secretaría de Investigación y Estudios Avanzados-Universidad Autónoma del Estado de México (Project/Grant no. 3512/2013CHT) and CONACyT, Mexico (Postgraduate Scholarship no. 36422). The authors would like to thank Dr. Robert Greenhouse for his valuable comments and suggestions and M. S. Nieves Zavala-Segovia and M. S. Lizbeth-Triana Cruz (CCIQS UAEM-UNAM) for technical support.

References

[1] J. Tan and N. Yasuda, "Contemporary asymmetric phase transfer catalysis: large-scale industrial applications," *Organic Process Research and Development*, vol. 19, no. 11, pp. 1731–1746, 2015.

[2] L. Mathias, *Crown Ethers and Phase Transfer Catalysis in Polymer Science*, Springer, Berlin, Germany, 2012.

[3] S. Chempath, J. M. Boncella, L. R. Pratt, N. Henson, and B. S. Pivovar, "Density functional theory study of degradation of tetraalkylammonium hydroxides," *Journal of Physical Chemistry C*, vol. 114, no. 27, pp. 11977–11983, 2010.

[4] T. L. Y. Huynh, K. Poiroux, R. A. O'Brien, K. N. West, J. H. Davis, and C. W. West, "Fusion and thermal degradation behavior of symmetric sulfur-containing quaternary ammonium bromides," *Journal of Physical Chemistry B*, vol. 120, no. 7, pp. 1330–1335, 2016.

[5] S. Snipes and A. W. Herriott, "Use of tetraphenylborate as a phase transfer agent for acid catalysis," *Journal of the American Chemical Society*, vol. 101, no. 21, p. 6441, 1979.

[6] J. Masse, "Phase-transfer catalysis in the N-alkylation of 2-chlorophenothiazine," *Synthesis*, no. 5, pp. 341-342, 1977.

[7] H. J. M. Dou, R. Gallo, P. Hussanaly, and J. Metzger, "Behavior and stability of catalysts in bi- and triphase transfer catalysis," *The Journal of Organic Chemistry*, vol. 42, no. 26, pp. 4275-4276, 1977.

[8] J. E. Gordon and R. E. Kutina, "Theory of phase-transfer catalysis," *Journal of the American Chemical Society*, vol. 99, no. 12, pp. 3903–3909, 1977.

[9] C. González and R. Greenhouse, "Efficient synthesis of methylenebis pyrrole derivatives," *Heterocycles*, vol. 23, no. 5, pp. 1127–1130, 1985.

[10] C. Gonzalez, R. Greenhouse, R. Tallabs, and J. M. Muchowski, "Protecting groups for the pyrrole nitrogen atom. The 2-chloroethyl, 2-phenylsulfonylethyl, and related moieties," *Canadian Journal of Chemistry*, vol. 61, no. 8, pp. 1697–1702, 1983.

[11] Starks C. M., "Phase-transfer catalysis: an overview," *ACS Symposium Series*, vol. 659, pp. 10–28, 1997.

[12] M. Mąkosza and M. Fedoryński, "Phase transfer catalysis in dichlorocarbene chemistry: basic principles and specific features," *Russian Chemical Bulletin*, vol. 60, no. 11, pp. 2141–2146, 2011.

[13] Q. Zhao, J. Sun, and B. Liu, "Anion exchange cycle of catalyst in liquid–liquid phase-transfer catalysis reaction: novel autocatalysis," *Chemical Engineering Journal*, vol. 262, pp. 756–765, 2015.

[14] S. Zhao, J. Kang, Y. Du et al., "An efficient ultrasound-assisted synthesis of n-alkyl derivatives of carbazole, indole, and phenothiazine," *Journal of Heterocyclic Chemistry*, vol. 51, no. 3, pp. 683–689, 2014.

[15] J. E. Taylor, M. D. Jones, J. M. J. Williams, and S. D. Bull, "Friedel - Crafts acylation of pyrroles and indoles using 1,5-diazabicyclo[4.3.0]non-5-ene (DBN) as a nucleophilic catalyst," *Organic Letters*, vol. 12, no. 24, pp. 5740–5743, 2010.

4

Protective Effect of Sika Deer (*Cervus nippon*) Velvet Antler Extract against Cisplatin-Induced Kidney and Liver Injury in a Prostate Cancer PC-3 Cell Xenograft Model

Yujiao Tang,[1,2] Meiqi Fan,[2] Young-Jin Choi,[2] Eun-Ju Choi,[3] Sang-Ho Moon,[2] Trishna Debnath,[4] Yonghai Yu,[1] Il Nam Lee,[5] and Eun-Kyung Kim[2]

[1]*School of Bio-science and Food Engineering, Changchun University of Science and Technology, Changchun 130-600, China*
[2]*Division of Food Bioscience, College of Biomedical and Health Sciences, Konkuk University, Chungju 27478, Republic of Korea*
[3]*Department of Physical Education, College of Education, Daegu Catholic University, Gyeongsan 38430, Republic of Korea*
[4]*Department of Food Science and Biotechnology, Dongguk University, Goyang 10326, Republic of Korea*
[5]*R&D Department, Korea Mcnulty Co., Ltd., Cheonan 31009, Republic of Korea*

Correspondence should be addressed to Eun-Kyung Kim; eunkyungkim@kku.ac.kr

Guest Editor: Mariusz Korczyński

We previously discovered the antioxidant and antiprostate cancer effects of antler extract (AE), but whether it inhibits cisplatin- (Cis-) induced toxicity has not been investigated. In this study, the effect of AE on Cis-induced side effects in the kidney and liver using 3-(4,5-dimethyl-2-thiazolyl)-2,5-diphenyl-2H-tetrazolium bromide-based cytotoxicity and cell cycle assays in prostate cancer PC-3 cells in vitro is investigated. Furthermore, we used a xenograft mouse model of the same cells to examine the in vivo effects and mechanisms of action. Cis and Cis + AE treatment attenuated prostate cancer cell growth by inducing apoptosis in vitro. Cis + AE stimulated cleaved caspases 3, 7, and 9 and polyadenosine diphosphate ribose polymerase expression. Cis + AE treatment for 1 week significantly increased the superoxide dismutase and catalase antioxidant activity while thiobarbituric acid reactive substances decreased. The histopathological damage and tumor necrosis factor-α, interleukin- (IL-) 1β and IL-6, cyclooxygenase-2, and inducible nitric oxide synthase expression in the kidney and liver tissue decreased. Therefore, AE likely possesses antiprostate cancer activity and inhibits Cis toxicity.

1. Introduction

Prostate cancer (PC) is one of the most common malignancies and represents the second most common cause of cancer-associated mortalities among men in the US [1]. PC is divided into two types: hormone-dependent and hormone-independent. Hormonal therapy remains the most effective therapy for patients with advanced PC by inhibiting proliferation and inducing apoptosis of tumor cells (hormone-dependent) [2]. However, after short-term remissions (18–24 months), the growth of surviving tumor cells recurs with castrate-resistant prostate cancer (CRPC) with inevitable progression and death within 2 to 3 years in most men (hormone-independent) [3, 4].

Cisplatin (Cis) is a widely used anticancer drug and one of the most potent antitumor drugs used against a wide spectrum of malignancies including PC, testicular, bladder, head and neck, ovarian, breast, and lung cancer [5–10]. Indeed, it is used to treat 50% of all cancers [11]. Cis, a neutral inorganic and square planar complex, acts by binding with DNA to form adducts leading to unique cellular responses that mainly culminate in apoptosis induction [12]. Despite its use as a chemotherapeutic agent, Cis exerts serious side effects in several organs including the kidneys and liver [13, 14] mainly due to its high accumulation in these organs [14, 15]. There is evidence implicating oxidative stress in the pathogenesis of Cis-induced kidney and liver injury, mediated by increased reactive oxygen species [16–19]. Therefore, there is an urgent need to discover a novel, less toxic substance without potent antitumor activity.

Velvet antler, the unossified antler of *Cervus elaphus*, is well known as an animal-based folk medicine widely used in Asia as an alternative oriental medicine to treat various diseases including osteoarthritis, myocardial infarction, hypertension, breast cancer, and PC [20–24]. Our previous research showed that antler extract (AE) exhibited antioxidant and antiprostate cancer activity [24, 25]. However, the inhibitory effect of AE on Cis-induced toxicity has not been investigated, and therefore, the present study investigated this phenomenon for the first time.

2. Materials and Methods

2.1. Materials. Cis, protease inhibitor cocktail, radioimmunoprecipitation assay (RIPA) buffer, 2-propanol, 3-(4,5-dimethyl-2-thiazolyl)-2,5-diphenyl-2H-tetrazolium bromide (MTT), dimethyl sulfoxide (DMSO), chloroform, and propidium iodide were purchased from Sigma-Aldrich Chemicals (St. Louis, MO, USA). Cycle script reverse transcription (RT) premix (dT20), RT/polymerase chain reaction (PCR) premix, and the 100 bp DNA Ladder were obtained from BIONEER (Daejeon, Korea). Trizol reagent, diethyl pyrocarbonate- (DEPC-) treated water, bicinchoninic acid (BCA) protein assay reagents A and B, Ponceau S, and the lane marker sample buffers were purchased from Thermo Scientific (Waltham, MA, USA). RNase A and Tween-20 were supplied by Novagen (Darmstadt, Germany). The PC-3 human prostate cancer cell line was obtained from the Korean Cell Line Bank (Seoul, Korea, KCLB number: 21435). The Roswell Park Memorial Institute (RPMI) 1640 medium, fetal bovine serum (FBS), penicillin/streptomycin (P/S), 0.5% trypsin-ethylenediaminetetraacetic acid (EDTA), and phosphate-buffered saline (PBS) for the cells culture were from Invitrogen (Carlsbad, CA, USA).

2.2. Preparation of Extracts. The antler sample was extracted according to a previously reported method [25]. In brief, the antlers were harvested at approximately growing day 50 and then were divided into three segments: top, middle, and base. In this study, we used the top segments, which were lyophilized and homogenized with a grinder, and then, 100 g was extracted with 1000 mL boiling distilled water (DW) for 1 h. The AE was subsequently filtered (0.25 μm pore size) and then lyophilized using a freeze dryer for 5 days.

2.3. Cell Culture. The PC-3 cells were cultured in the RPMI 1640 medium supplemented with 10% FBS and 1% P/S exposed to a 5% CO_2 atmosphere at 37°C. For the migration assay, the cells were seeded at a density of 5.0×10^5 cells/well in a six-well culture plate, incubated for 24 h, treated with 125–1,000 μg/mL of AE for 24 h, and then harvested for analysis.

2.3.1. Cytotoxic Assessment Using MTT Assay. The cell cytotoxicity was determined using the MTT assay. The assay is based on the principle that the yellow tetrazolium salt is metabolized by viable cells to purple formazan crystals in a reaction catalyzed by mitochondrial succinyl dehydrogenase.

In brief, PC-3 cells were seeded at 1.9×10^4 cells/well in 96-well microtiter plates in the complete medium (RPMI with 10% FBS and 1% P/S) and incubated for 24 h exposed to a 5% CO_2 atmosphere at 37°C. Then, 200 μL samples of the solution in the medium were transferred to the wells, followed by a 24 h incubation, and then, the MTT solution (final concentration, 0.5 mg/mL) was added to each well. After a 4 h incubation, the medium was aspirated, the purple crystals were dissolved in DMSO, and the absorbance of the resulting solution in each well was measured at 540 nm using a microplate reader.

2.3.2. Cell Cycle. For the cell cycle analysis, the harvested cells were fixed with ethanol (with 0.5% Tween-20) for 8–24 h, washing with 1× PBS, incubated with 50 μg/mL propidium iodide and 1 μg/mL RNase A at 37°C for 30 min, and then analyzed using flow cytometry using the fluorescence-activated cell sorting (FACS) system (BD, Franklin Lakes, NJ, USA). The cells of the sub-G1 population were considered apoptotic, and the percentage of each cell cycle phase was determined.

2.4. Animal Experiments. The nude male BALB/c mice (15–17 g and 8-week-old) used in the present study were provided by Samtako Bio Co, (Osan, Korea). They were maintained in an air-conditioned room (20–25°C) under a 12 h light/dark cycle with free access to food and water. The mice were acclimatized to the new environment for 1 week before the commencement of the experiment (Figure 1). The mice were inoculated subcutaneously with 6.5×10^5 PC-3 cells suspended in 100 μL PBS thrice weekly. When the xenograft tumors reached a volume of approximately 100 mm^3, the mice were randomly assigned to four groups (n = 6/group) and were treated as follows. Group 1 (sham control) was administered PBS, Group 2 received 20 mg/kg body weight Cis, and Groups 3 and 4 each received 20 mg/kg body weight Cis, followed by 200 and 400 mg/kg body weight AE (AE low dose (AEL) and high dose (AEH)). The doses of Cis and the extract concentrations were selected based on previous studies [26–30]. At the end of the treatment, the gavage tube was left in place for several seconds to avoid regurgitation and ensure the total calculated dose was administered. The tumor sizes were monitored every 2 days using a digital caliper, and the tumor volumes were calculated using the formula: $L \times S^2 \times 0.5$, where L and S represent the longest and shortest tumor diameters, respectively [31]. The mice were also weighed. The extract was administered once daily at a fixed time for the entire 2-week experiment. At the end of the experiment, all the mice were fasted overnight, and the xenograft tumors were rapidly excised and weighed. One part of each excised tumor was immediately placed in liquid nitrogen for western

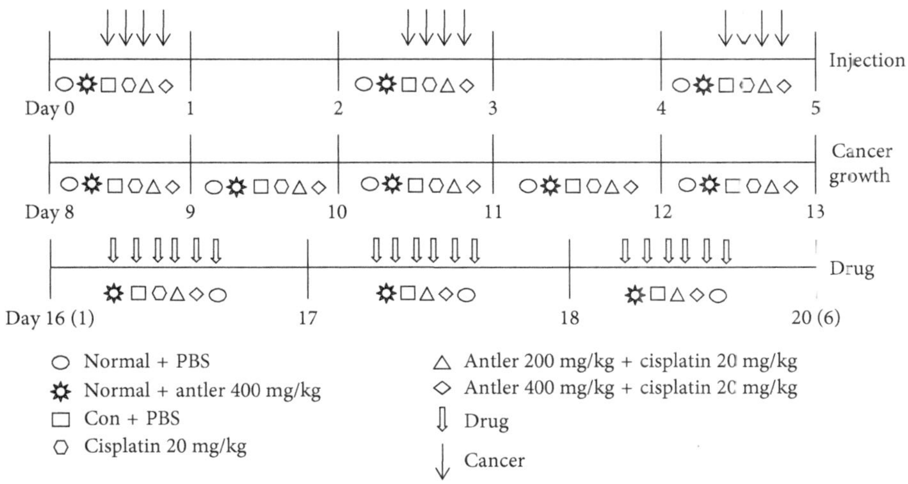

FIGURE 1: Scheme of the xenograft tumor model. Negative control (NC, phosphate-buffered saline (PBS), 200 μL), antler extract (AE 400 mg/kg), control (Con, PBS 200 μL + tumor cells), cisplatin (Cis 20 mg/kg + tumor cells), Cis + AE low dose (AEL, 200 mg/kg + Cis 20 mg/kg + tumor cells), and Cis + AE high dose (AEH, 400 mg/kg + Cis 20 mg/kg + tumor cells). From day 0, mice were injected thrice weekly, and CON, Cis, AEL, and AEH groups were left for 1 week. When the xenograft tumor volumes were approximately 100 mm^3, mice were randomly assigned to four groups and treated accordingly.

blot analysis, while the other was fixed in 10% neutral-buffered formalin for immunohistochemical analysis. Blood samples were collected from the heart. All animal care procedures and experiments were approved by the Institutional Animal Care and Use Committee of Konkuk University (KU15114).

2.5. RNA Isolation and mRNA Expression Analysis. RNA was isolated from the cells using Trizol according to the manufacturer's protocol. The first-strand complementary DNA (cDNA) was synthesized using Superscript II reverse transcriptase (Invitrogen). The PCR was performed as previously described, except for the following primer set: tumor necrosis factor- (TNF-) α (sense, 5′-ACC AGG AGA GAA AGT CAA CCT C-3′, and antisense, 5′-GGA CTC CGC AAA GTC TAA GT-3′), interleukin- (IL-) 1β (sense, 5′-TCT GTG ACT CAT GGG ATG AT-3′, and antisense, 5′-TAT TTT TGT CGT TGC TTG GTT-3′) and IL-6 (sense, 5′-GAG ACT TCC ATC CAG TTG C-3′, and antisense, 5′-CTC TTT TCT CAT TTC CAC GA-3′), cyclooxygenase-(COX-) 2 (sense, 5′-CCC CTC TCT ACG CAT TCT AT-3′, and antisense, 5′-AGG TCG TTT GTT GGG ATT AT-3′), inducible nitric oxide synthase (iNOS) (sense, 5′- ATC ATG AAC CCC AAG AGT TT-3′, and antisense, 5′-AGA GTG AGC TGG TAG GTT CC-3′), and glyceraldehyde 3-phosphate dehydrogenase (GAPDH, sense, 5′-GGT TGT CTC CTG CGA CTT CA-3′, and antisense, 5′- TAG GGC CTC TCT TGC TCA GT-3′), which was used as the internal control.

2.5. Western Blotting. The cell extracts were prepared using the detergent lysis procedure as described elsewhere [24]. Protein samples (40 μg) were electrophoresed using Novex 4%–12% Bis-Tris gels (Life Technologies, Carlsbad, CA, USA) and then transferred to nitrocellulose membranes for 7 min using the iBlot dry blotting system (Life Technologies). The membranes were blocked overnight at 4°C with clear milk (Thermo Scientific, IL, USA) and then subsequently incubated with primary antibodies (1 : 2000 in 1× Tris-buffered saline plus Tween (TBST) for 1 h. Antibodies against Cox-2, iNOS, and actin were purchased from Santa Cruz Biotechnology. Polyadenosine diphosphate (ADP) ribose polymerase (cleaved PARP; caspases 3, 7, and 9; cleaved caspases 3, 7, and 9; and B-cell lymphoma-2 (Bcl-2)) and Bcl-2-associated X protein (Bax) antibodies were from Cell Signaling Technology (Danvers, MA, USA). The goat anti-rabbit and goat anti-mouse horseradish peroxidase-conjugated secondary antibodies (Santa Cruz Biotechnology, Dallas, TX, USA) were used (1 : 2000, in 1× TBST). Equal protein loading was ascertained using Ponceau S staining of the blotted membranes and western blotting with β-actin bodies. Immunodetection was performed using an enhanced chemiluminescence detection kit (Amersham Pharmacia, Piscataway, NJ, USA). Goat anti-rabbit and goat anti-mouse horseradish peroxidase-conjugated secondary antibodies (Santa Cruz Biotechnology, Dallas, TX, USA) were used (1 : 2000, in 1× TBST).

2.5. Serum Biochemical Parameters. Serum was obtained by centrifuging the blood samples at 3,500 g for 15 min at 4°C. The superoxide dismutase (SOD), catalase (CAT), and thiobarbituric acid reactive substances (TBARS) assays were performed using a kit (Cayman Chemical Co., MI, USA).

2.6. Histological Analysis. The kidneys and livers of the mice were fixed with 10% paraformaldehyde and embedded in paraffin blocks, which were then cut into 5 μm thick sections, deparaffinized, and stained with hematoxylin and eosin (H&E).

2.7. Statistical Analysis. The data are presented as the mean ± standard error (SE) of triplicate experiments. The statistical analyses were performed using the statistical analysis software (SAS) program (SAS Institute, USA). The treatment effects were analyzed using a one-way analysis of variance (ANOVA), followed by Dunnett's multiple range tests. The statistical significance was defined at $p < 0.05$.

3. Results

3.1. Effect of Cis and AE on Growth and Morphology of PC-3 Cells In Vitro. The effects of Cis and AE on the proliferation of PC-3 cells were investigated. The cell line was cultured in vitro with the same concentration of Ci (200 μM) and different concentrations of AE (125–1,000 μg/mL) for 24 h, and cell viability was measured using the MTT assay. The results demonstrated that the proliferation of PC-3 cells decreased more in the Cis and Cis + AE groups than it did in the control group. Moreover, Cis + AE (500 and 1,000 μg/mL) significantly decreased to Cis (Figure 2(a)). Therefore, Cis + AE 500 and 1,000 μg/mL were selected for the apoptosis study.

The PC-3 cells line used was treated for 24 h. Light microscopy of the Cis-, Cis + AEL-, and Cis + AEH-treated PC-3 cells showed a significant morphological change. As shown in Figure 2(b), the control group cells did not show any apoptotic characteristics. However, Cis and AE treatment dramatically and dose-dependently induced cell morphology changes including shrinkage and detachment. Consistent with the cell viability results, PC-3 cells cultured with Cis + AEL and Cis + AEH showed more marked morphological changes than cells cultured with Cis alone, suggesting that AE may induce more apoptosis.

3.2. Apoptotic Effects of Cis and AE in PC-3 Cells. Flow cytometric analysis was used to determine the cell cycle distribution of the PC-3 cells in the absence and presence of Cis and AE for 24 h. As indicated in Figure 3, Cis, Cis + AEL, and Cis + AEH showed no significant changes in the percentage of cells in the G1, S, and G2 phases compared with the basal level (control). Specifically, there was a 51.4%, 62.6%, and 69.8% increase in the G1, S, and G2 apoptosis phases, respectively, compared with that of the control. The results indicated that Cis + AEL and Cis + AEH induced apoptosis more potently than Cis did.

3.3. Effect of AE on Apoptotic Protein Expression in PC-3 Cells. To determine whether the apoptotic effects of Cis and AE were mediated by the caspase-dependent apoptosis pathway, the Cis-, Cis + AEL-, and Cis + AEH-treated cells were analyzed using western blotting of intracellular full length and cleaved PARP, caspases 3, 7, and 9, and Bax and Bcl-2 (Figure 4). The results (Figures 4(b)–4(i)) showed that treatment with AEL and AEH decreased the inactive PARP and caspases 7 and 9, but the active cleaved PARP and caspases 3, 7, and 9 significantly increased gradually over the concentrations tested. This observation indicates that AE induced self-cleavage of caspases 3, 7, and 9 and PARP to activate the downstream signaling pathway. Bcl-2 family proteins affect cellular apoptosis by regulating cytochrome C release, which then mediates caspase activation. Therefore, the effects of Cis, Cis + AEL, and Cis + AEH on the protein expression of antiapoptotic Bcl-2 and proapoptotic Bax were evaluated. In Figures 4(j) and 4(k), treatment with AEL and AEH significantly decreased the protein expression of Bcl-2 and significantly increased that of Bax compared to the Cis group. Taken together, these data suggest that AE played a role in the induction of apoptosis.

3.4. Effect of AE on Cis-Induced Changes in Food Intake and Body Weight. To analyze the effects of AE and Cis on food intake and body weight, the animals were treated with the vehicle (NC), AEH, Cis + AEL, and Cis + AEH. Administration of Cis reduced the food intake and body weight, but AE treatment significantly increased the food intake and body weight (Figure 5). In addition, the AEH and control groups exhibited no significant effects on food intake and body weight. Therefore, high-dose AE did not show toxic effects in nude mice and was used for the subsequent experiments.

3.5. Effect of AE on Cis-Induced Alterations of Antioxidants. The effects of Cis and AE on the antioxidant enzymes, SOD and CAT, in mice are shown in Figures 6(a)–6(c). As illustrated in Figures 6(a) and 6(b), exposure to Cis decreased the SOD and CAT activities, but the activities significantly increased in the Cis + AE groups. In contrast, the TBARS contents significantly increased in the Cis-treated group but significantly decreased in the Cis + AE groups (Figure 6(c)). In mice administered with AE only, the SOD and CAT activities were highest, while the TBARS content was the lowest. Therefore, AE alleviated the changes induced by oxidative stress.

3.6. Effect of Cis and AE on Expression of Inflammation-Related Genes in Kidney and Liver. We also investigated the expression of inflammation-related genes in the kidney and liver. Initially, the mRNA expression levels of TNF-α, IL-1β, IL-6, COX-2, and iNOS in the kidney tissues were estimated. As shown in Figure 7, kidney tissues from Cis + AE-treated groups showed significantly decreased TNF-α, IL-1β, COX-2, and iNOS expression compared with that of the Cis-treated group. However, no significant change was observed in the expression of IL-6. The results of the liver tissue analysis (Figure 8) were similar to those of the kidney. Specifically, treatment with Cis + AE significantly decreased the kidney tissue expression of TNF-α, IL-1β, IL-6, COX-2, and iNOS compared with that of the Cis group.

3.7. Effect of Cis and AE on Histology of Kidney and Liver Tissues. As shown in Figure 9(a), the normal group showed normal histology of the mouse kidneys (the glomeruli,

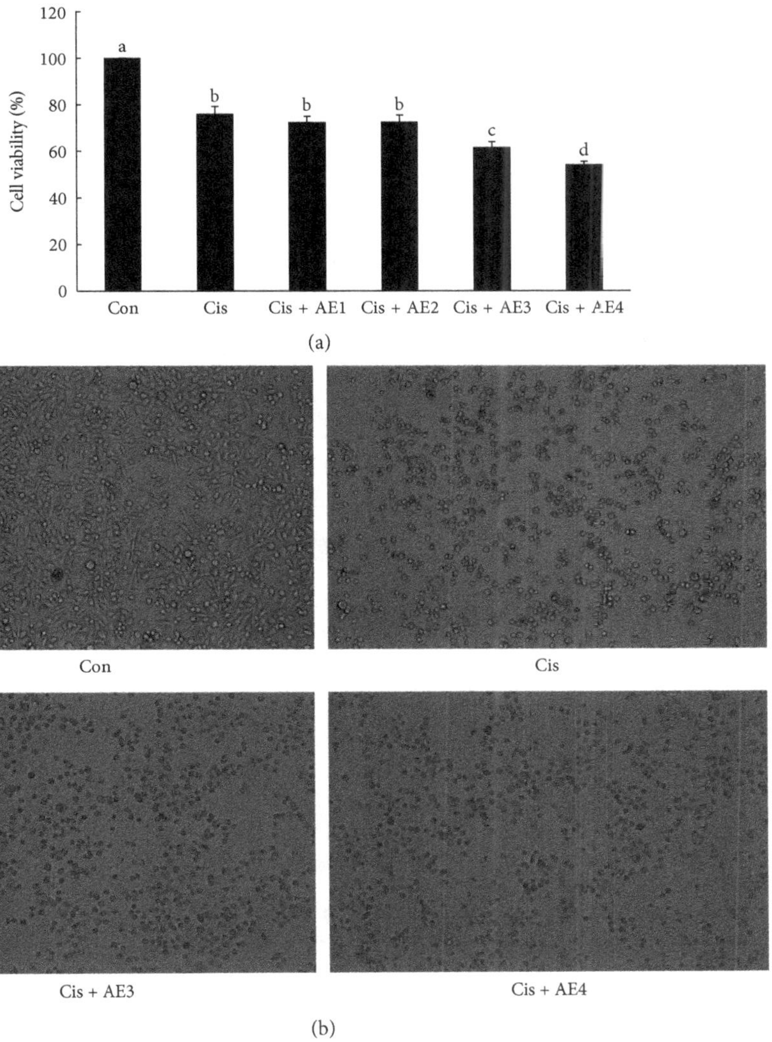

Figure 2: Antiproliferative effect of cisplatin (Cis) and Cis plus antler extract (AE, 125–1,000 μg/mL) in PC-3 prostate cancer cells. Cis (200 μM), Cis + AE1 (Cis 200 μM + AE 125 μg/mL), Cis + AE2 (Cis 200 μM + AE 250 μg/mL), Cis + AE3 (Cis 200 μM + AE 500 μg/mL), and Cis + AE4 (Cis 200 μM + AE 1,000 μg/mL). $^{a-d}$Means with different superscript in the same row are different ($p < 0.05$).

tubules, interstitium, and blood vessels). The Cis group exhibited proteinaceous casts in the tubular lumen. In addition, the Cis group showed inflammation of the blood vessels, increased thickness of Bowman's capsules, and decreased glomerulus size. Cis + AEH treatment improved the kidney histopathology by reducing the inflammation of the blood vessels and the glomerulus size, which was similar to that of the normal group.

The normal microscopic architecture of the liver is composed of hexagonal lobules and acini. Hexagonal lobules are centered on the central vein (CV) and have a portal triad containing branches of the portal vein (PV), hepatic artery (HA), and bile duct (BD). The liver sections from the Cis-treated group exhibited severe congestion of the CV and moderate disorganization of the hepatic cords (Figure 9(b)). In the Cis + AEH group, the liver sections showed mild congestion of the CV and mild disorganization of hepatic cords.

4. Discussion

AE has several pharmacological properties and has been used in the clinical treatment of numerous diseases [21–24, 32, 33]. Cis has been shown to cause hepatotoxicity, nephrotoxicity, and other side effects, which limit its use [34]. The increased incidence and mortality of PC and the current unsatisfactory treatment options for CRPC has increased the focus on the combination of Cis and natural-based medicines because of the advantage of decreased toxicity.

The present study reveals for the first time that AE treatment inhibited the apoptosis of PC-3 cells. Furthermore, AE treatment protected against Cis-induced kidney and liver injury in a PC-3 cell xenograft model. First, we investigated the effect of Cis and AE on the proliferation of PC-3 cells. AE treatment significantly reduced the viability of PC-3 cells and dramatically deteriorated the cell morphology, inducing cell shrinkage and detachment. Apoptosis is

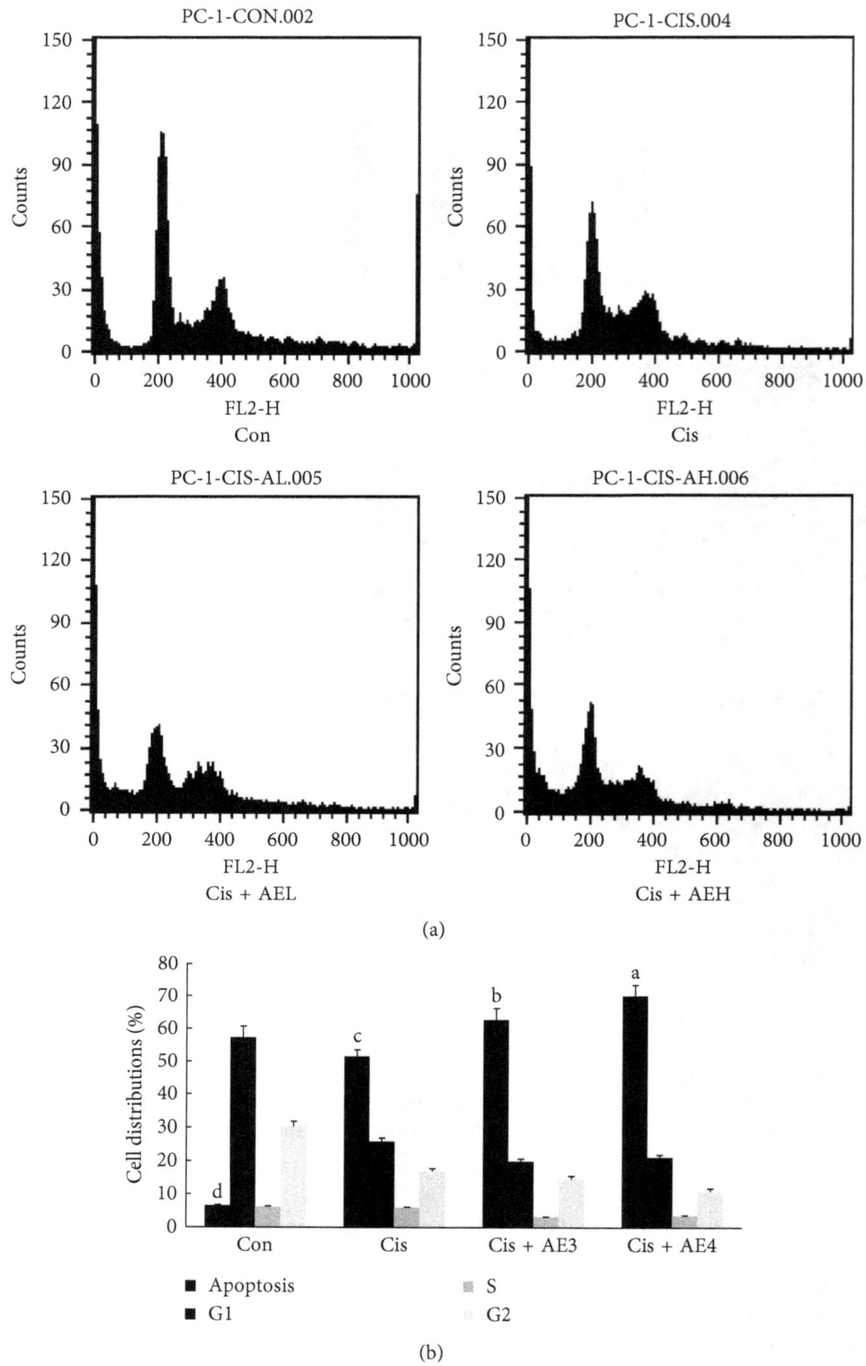

FIGURE 3: Cell cycle analysis of PC-3 prostate cancer cells. Cell cycle distribution was analyzed on a minimum of 1 × 104 cells. (a) PC-3 cells treated with cisplatin (Cis) and antler extract (AE). (b) Data from the cell cycle analysis were quantified and presented as a graph to illustrate percentage of cells in each cell cycle phase. Cis (Cis 200 μM), Cis + AEL (Cis 200 μM plus AE low dose, 500 μg/mL), and Cis + AEH (Cis 200 μM + AE high dose, 1,000 μg/mL).

characterized by a series of typical morphological events such as cell shrinkage and detachment, and our present results are similar to those of Sarastea and Pulkkic [35]. Thus, the effect of AE on cell viability and morphology may suggest that AE is likely to induce apoptosis of these cells.

In cancer, the normal mechanisms of cell cycle regulation are dysfunctional, with an over proliferation or decreased removal of cells or both [36]. In fact, the suppression of apoptosis during carcinogenesis is thought to play a central role in the development and progression of some cancers [37, 38]. Various molecular mechanisms mediate the suppression of apoptosis by tumor cells. The effect of Cis and AE on the cell cycle progression was evaluated in PC-3 cells, and our results are similar to those of

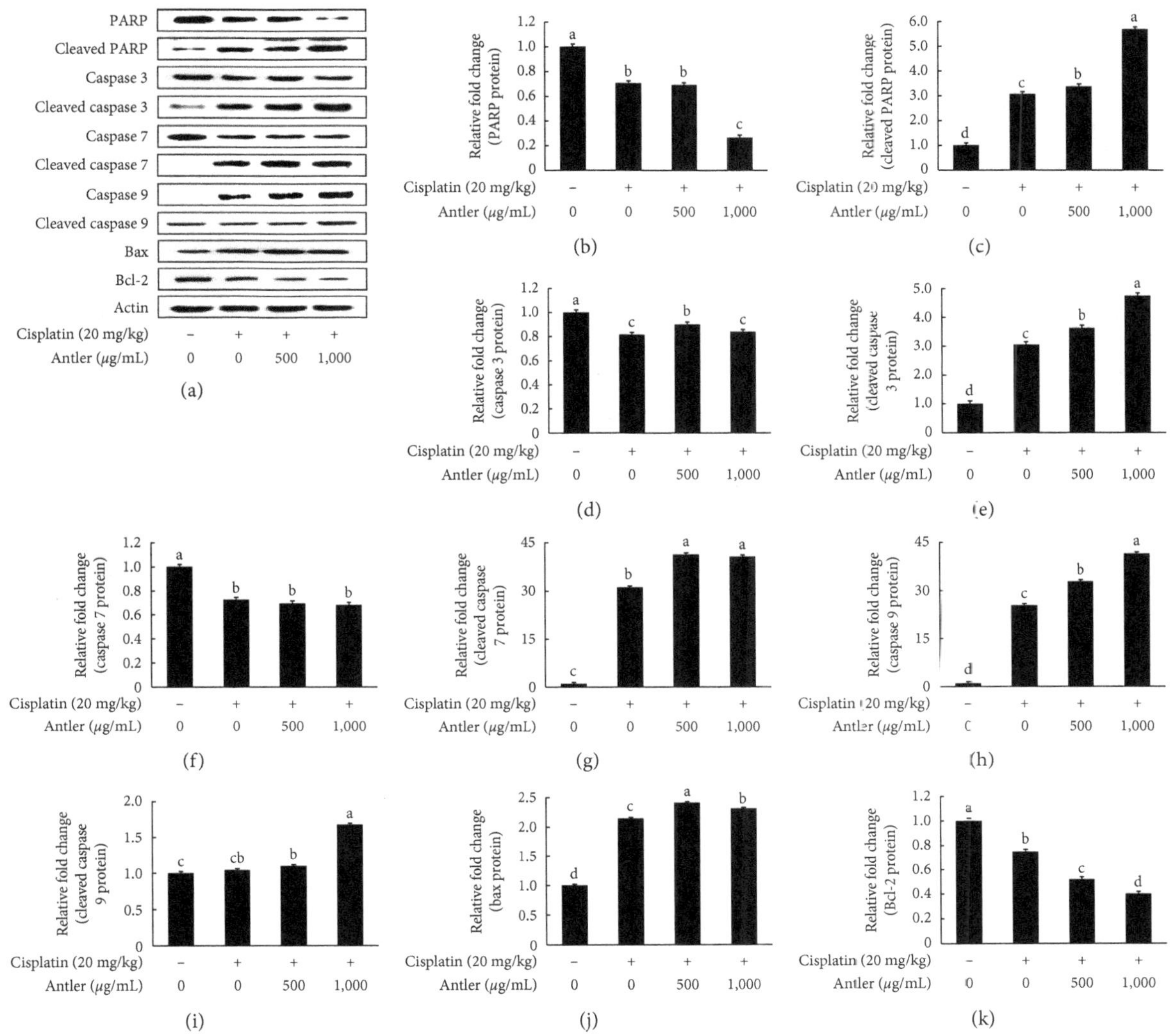

FIGURE 4: Apoptotic effect of cisplatin (Cis) and antler extract (AE) on PC-3 prostate cancer cells. β-Actin was used as a loading control. Results were similar in three independent experiments. $^{a-d}$Means with different superscript in the same row are different ($p < 0.05$).

Gumulec et al. [34] showing that Cis increased apoptosis of the PC cell model. To improve the current understanding of the apoptotic effect of Cis and AE, the expression levels of apoptosis-related genes were estimated. The released cytochrome c interacts with apoptotic protease-activating factor 1 and forms an apoptosome that activates caspase 9, leading to the activation of downstream caspases 3 and 7 and the apoptotic death response [39–42]. The extrinsic pathway is triggered in response to proapoptotic ligands, which bind and activate specific proapoptotic death receptors [36, 37]. In our study, treatment with AEL and AEH indicated that AE induces self-cleavage of PARP and caspases 3, 7, and 9 to activate downstream signaling pathway. Cis-induced cell death involves multiple pathways [43]. Khan et al. [44] reported that Cis alters the mitochondrial membrane potential, activates Bax, reduces Bcl-2, and shifts the Bax/Bcl-2 ratio in a proapoptotic direction in PC-3 cells. The AE + Cis groups exhibited significantly decreased and increased protein expression of Bcl-2 and Bax, respectively, compared to that of the Cis group.

Despite being widely used, Cis as chemotherapy is limited by its toxicity [45, 46]. In the present investigation, Cis-treated nude mice showed a significant decrease in body weight and percentage survival. Cis-induced weight loss might be due to gastrointestinal toxicity and reduction of food ingestion [47]. Pretreatment with AE markedly increased the food intake and body weight, indicating the amelioration of Cis toxicity in the present study, which is similar to the findings of Ahmed et al. [48].

Studies have provided evidence demonstrating that Cis-induced kidney and liver injury is mainly due to oxidative stress [14, 17, 18, 43, 49]. The result of the present study showed that Cis notably decreased the levels and activities of both SOD and CAT and, thereby, impaired the antioxidant defense mechanisms in the kidneys. Moreover, the lipid peroxidation marker, TBARS, significantly increased

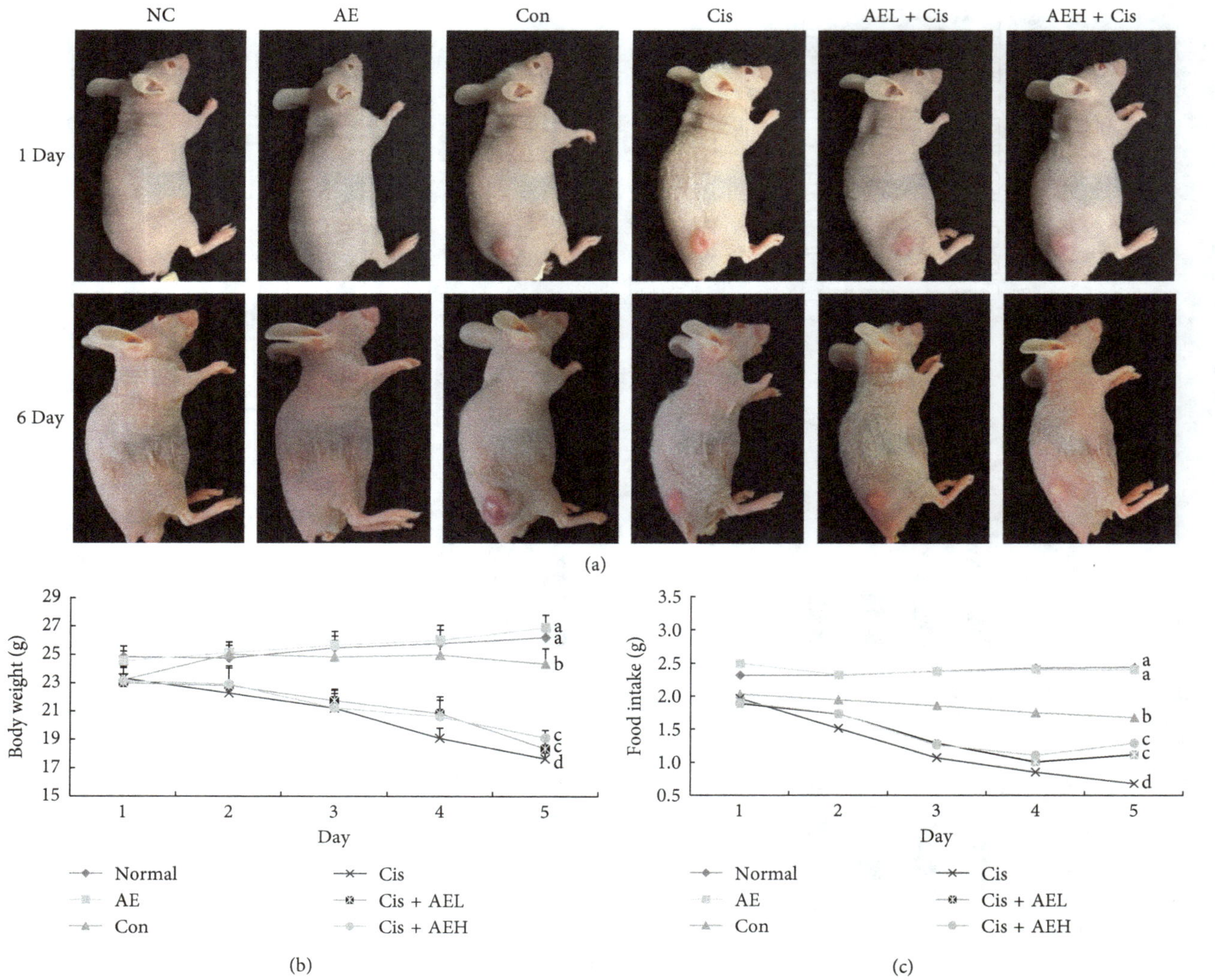

FIGURE 5: Effect of antler extract (AE) on cisplatin- (Cis-) induced changes in mouse food intake and body weight. (a) Mouse image, (b) food intake, and (c) body weight. Normal control (NC, phosphate-buffered saline (PBS) 200 μL), AE (AE 400 mg/kg), Con (PBS 200 μL + tumor cells), Cis (Cis 20 mg/kg + tumor cells), Cis + AEL (AE low dose 200 mg/kg + Cis 20 mg/kg + tumor cells), and Cis + AEH (AE high dose 400 mg/kg + Cis 20 mg/kg + tumor cells). $^{a-d}p < 0.05$ compared with the Cis group.

in the Cis group. Pretreatment with AE improved the Cis-induced nephrotoxicity by significantly enhancing SOD and CAT and significantly decreasing TBARS. Our results appeared to be consistent with many previous findings that indicated the nephrotoxic and hepatotoxic effect of Cis, its association with increased free radical formation, and the subsequent induction of oxidative and nitrosative stress [17, 50–52].

Inflammation plays an important role in the initiation and progression of Cis-induced kidney and liver damage [53–56]. Cis induces the release of a series of proinflammatory cytokines (TNF-α, IL-1β, COX2, and iNOS) and causes the infiltration of leukocytes and macrophages into damaged renal tissues [57]. In addition, apoptosis as well as proinflammatory genes, inducible COX (COX-2) and iNOS, may have critical roles in the mechanism of Cis-induced acute kidney and liver damage [16, 49, 58]. The present study shows that Cis-induced liver injury is accompanied by an inflammatory reaction, evidenced by increased formation of the proinflammatory cytokines TNF-α, IL-1β, and IL-6, as well as increased COX-2 and iNOS expression. Cis + AE significantly decreased the expression levels of the inflammation-related genes. These results are consistent with the findings of many recently published studies [48, 51, 52, 59–61].

Furthermore, inflammatory cells in the kidneys, as evidenced by light microscopic examination (H&E staining) of kidney tissues, were similar to those observed by Sahu et al. [16]. In addition, our data also suggest that Cis caused kidney pathologies including proteinaceous casts in the tubular lumen, detachment of tubular cells, from the basement membrane, and tubular necrosis.

The Cis + AE groups showed less damage than the Cis group did. An et al. [62] and Atasayar et al. [63] found that

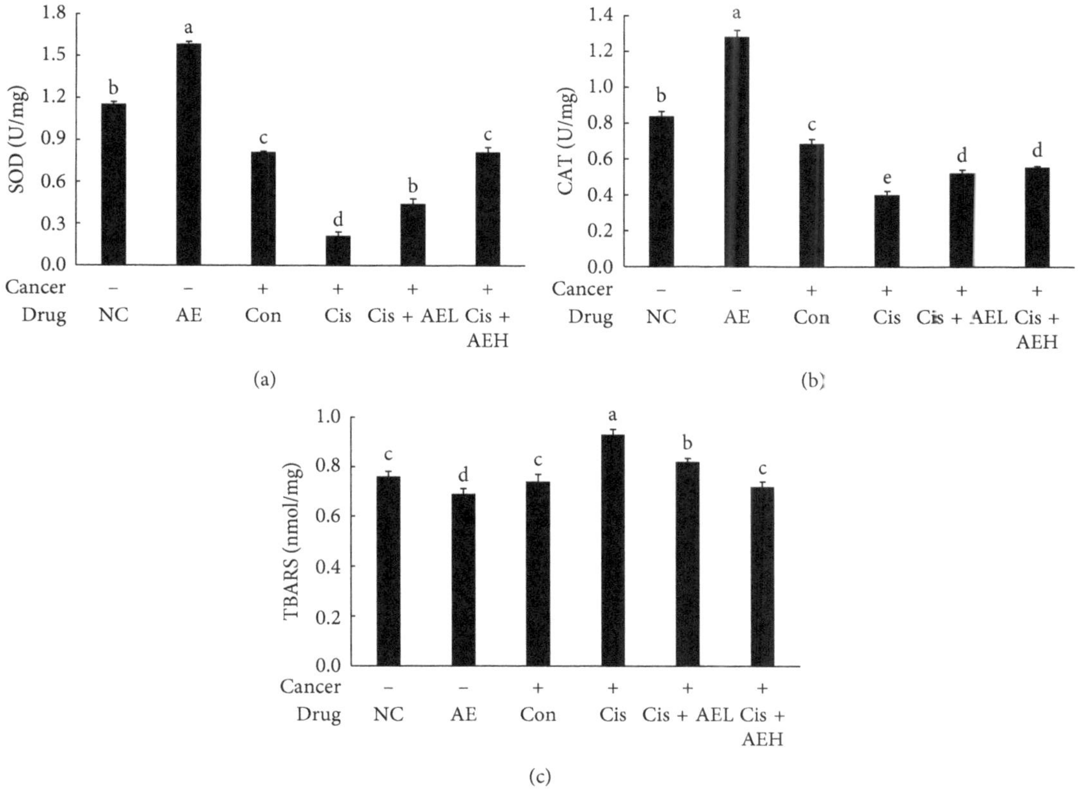

Figure 6: Effect of cisplatin (Cis) and antler extract (AE) treatment on superoxide dismutase (SOD), catalase (CAT), and thiobarbituric acid reactive substances (TBARS) activity. Normal control (NC, phosphate-buffered saline (PBS), 200 μL), AE (AE 400 mg/kg), Con (PBS 200 μL + tumor cells), Cis (Cis 20 mg/kg + tumor cells), Cis + AEL (AE low dose 200 mg/kg + Cis 20 mg/kg + tumor cells), and Cis + AEH (AE high-dose 400 mg/kg + Cis 20 mg/kg + tumor cells). $^{a\text{–}d}p < 0.05$ compared with the Cis group.

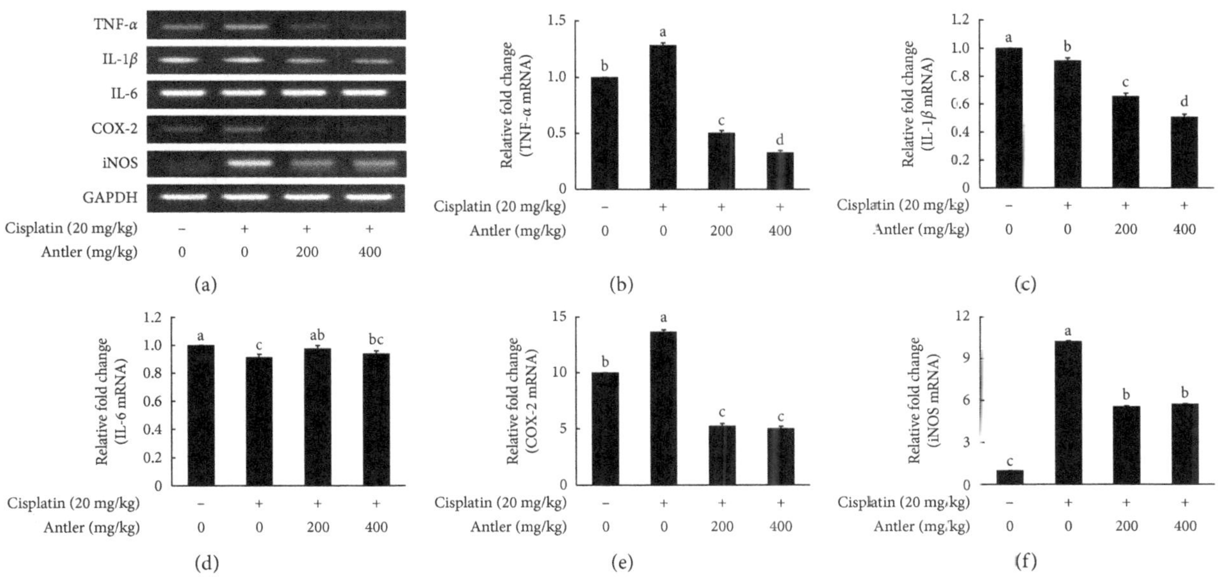

Figure 7: Continued.

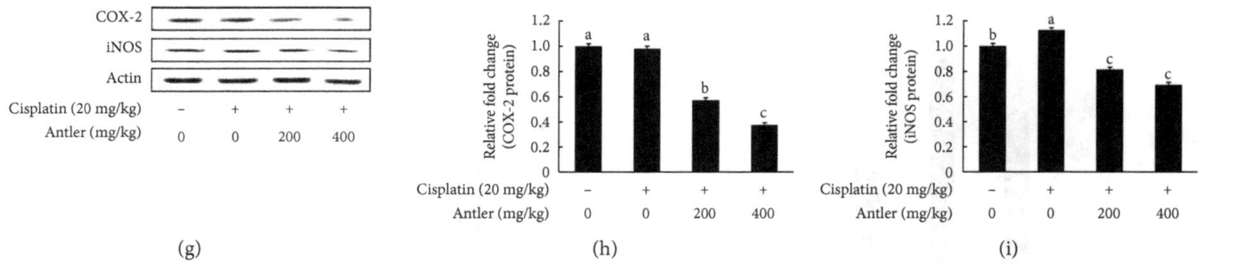

FIGURE 7: Expression of inflammation-related gene expression in kidney tissues using reverse transcription-polymerase chain reaction (RT-PCR) and western blotting. Glyceraldehyde 3-phosphate dehydrogenase (GAPDH) was used as a control. Results were similar in three independent experiments. Normal control (NC, phosphate-buffered saline (PBS), 200 μL), AE (AE 400 mg/kg), Con (PBS 200 μL + tumor cells), Cis (Cis 20 mg/kg + tumor cells), Cis + AEL (AE low dose 200 mg/kg + Cis 20 mg/kg + tumor cells), and Cis + AEH (AE high dose 400 mg/kg + Cis 20 mg/kg + tumor cells). $^{a-d}$Means with different superscript in the same row are different ($p < 0.05$).

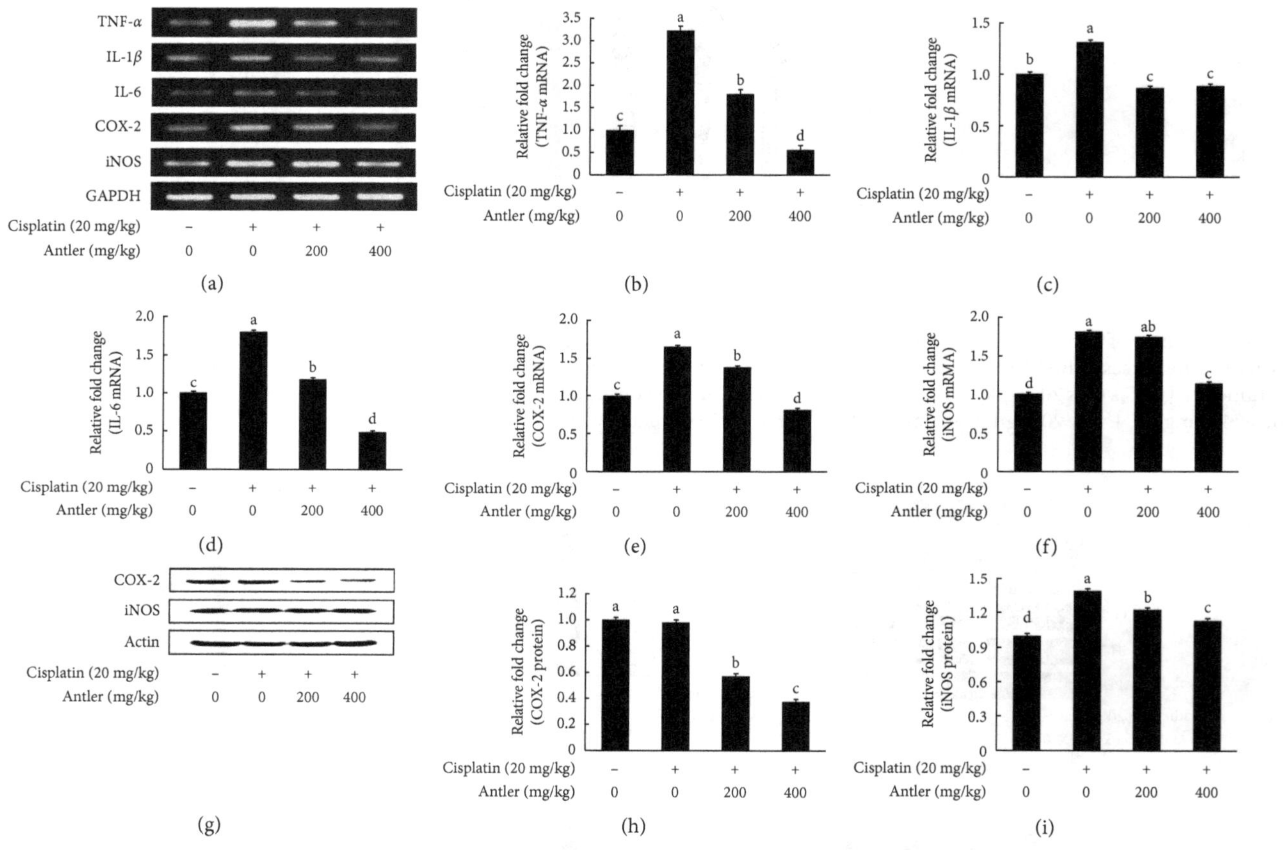

FIGURE 8: Expression of inflammation-related gene expression in the liver tissue using reverse transcription-polymerase chain reaction (RT-PCR) and western blotting. Glyceraldehyde 3-phosphate dehydrogenase (GAPDH) was used as a control. Results were similar in three independent experiments. Normal control (NC, phosphate-buffered saline (PBS) 200 μL), AE (AE 400 mg/kg), Con (PBS 200 μL + tumor cells), Cis (Cis 20 mg/kg + tumor cells), Cis + AEL (AE low dose 200 mg/kg + Cis 20 mg/kg + tumor cells), and Cis + AEH (AE high dose 400 mg/kg + Cis 20 mg/kg + tumor cells). $^{a-d}$Means with different superscript in the same row are different ($p < 0.05$).

the histology of renal sections exhibited remarkable vacuolation, necrosis, desquamation of epithelial cells, and proteinaceous casts in renal tubules after intraperitoneal treatment with single doses of Cis. Cis-induced hepatotoxicity is mainly characterized by multiple histopathological lesions, swelling of the hepatocytes with compressed blood sinusoid, fatty infiltration, prominence of Kupffer cells near the mild to moderate congestion of the HA, sinusoids, and PV, with dilatation of the PV and moderate disorganization of hepatic cords. A marked recovery was observed in the markers of liver function after combination treatment with Cis + AE. Our results are consistent with those results of Dkhil et al. [61] and Palipoch and Punsawad [19]. Therefore, we suggest

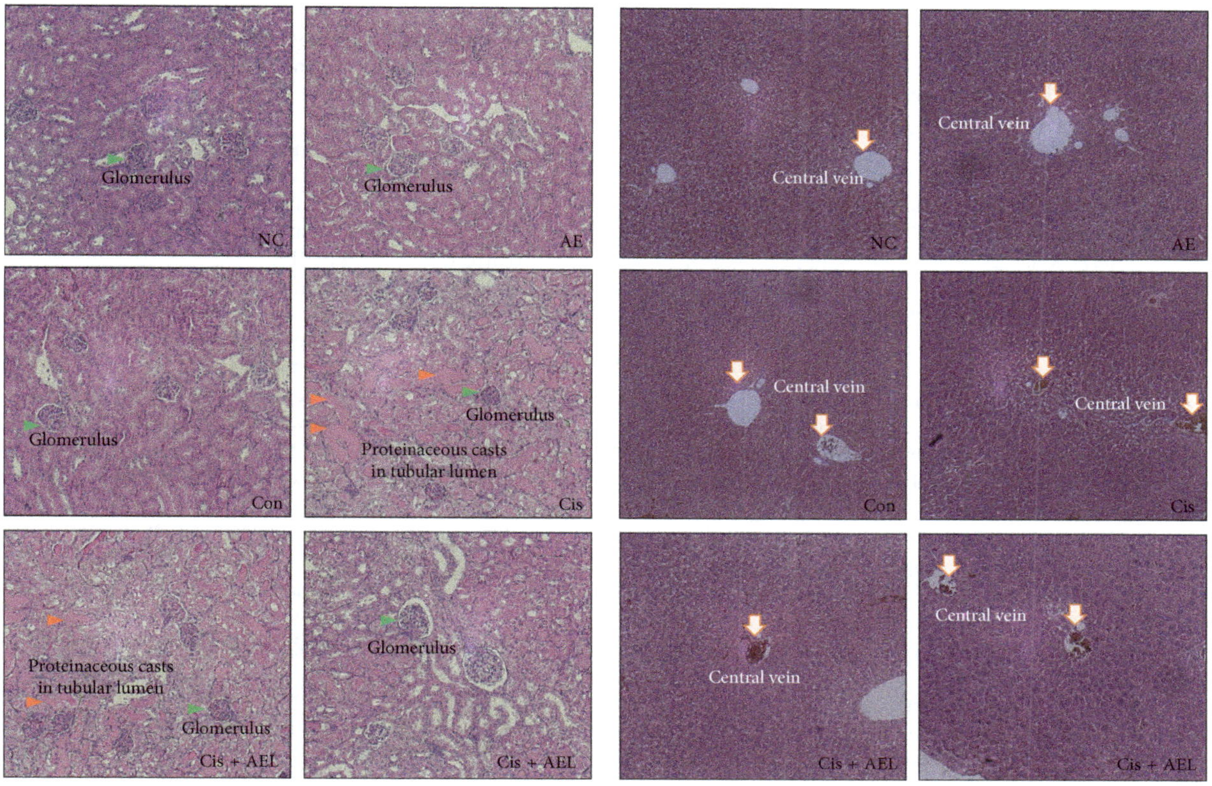

FIGURE 9: Histological alterations in the kidney and liver tissue of the cisplatin- (Cis-) treated group. Hematoxylin and eosin (H&E staining, 100× magnification). (a) Kidney and (b) liver. CV = central vein; normal control (NC, phosphate-buffered saline (PBS), 200 μL), AE (AE 400 mg/kg), Con (PBS 200 μL + tumor cells), Cis (Cis 20 mg/kg + tumor cells), Cis + AEL (AE low dose 200 mg/kg + Cis 20 mg/kg + tumor cells), and Cis + AEH (AE high dose 400 mg/kg + Cis 20 mg/kg + tumor cells).

that the use of AE in combination with Cis in antiprostate cancer therapy may contribute to reducing the Cis-induced toxicities.

5. Conclusions

In conclusion, the protective effects of AE against Cis-induced kidney and liver injury were likely mediated by the downregulation of apoptosis, oxidative stress, and inflammation and improvement of histological changes in vivo and in vitro. From these results, we suggest that the use of AE in combination with Cis in antiprostate cancer therapy may contribute to reducing the Cis-induced toxicities.

Disclosure

This research was awarded the Best Poster Award at the 2017 International Symposium and Annual Meeting of Korean Society of Food Science and Technology.

Conflicts of Interest

The authors declare that there are no conflicts of interests regarding the publication of this paper.

Acknowledgments

This research was supported by Basic Science Research Program through the National Research Foundation of Korea (NRF) funded by the Ministry of Education, Science and Technology (NRF-2017R1D1A1B03036247), the Korea Institute of Planning and Evaluation for Technology in Food, Agriculture, Forestry and Fisheries (IPET) through the Agri-Bio Industry Technology Development Program (116083-3) funded by the Ministry of Agriculture, Food and Rural Affairs (MAFRA) and the Functional Districts of the Science Belt support program, Ministry of Science and ICT, Republic of Korea (2017K000017).

References

[1] A. Snyder, J. E. Tepper, and S. Slovin, "Perspectives on immunotherapy in prostate cancer and solid tumors: where is the future?," *Seminars in Oncology*, vol. 40, no. 3, pp. 347–360, 2013.

[2] F. Lamoureux, C. Thomas, M. J. Yin et al., "A novel HSP90 inhibitor delays castrate-resistant prostate cancer without altering serum PSA levels and inhibits osteoclastogenesis," *Clinical Cancer Research*, vol. 17, no. 8, pp. 2301-2313, 2011.

[3] M. E. Gleave, N. Bruchovsky, M. J. Moore, and P. Venner, "Prostate cancer: 9. Treatment of advanced disease," *Canadian Medical Association Journal*, vol. 160, no. 2, pp. 225–232, 1999.

[4] J. P. Moreau, P. Delavault, and J. Blumberg, "Luteinizing hormone- releasing hormone agonists in the treatment of prostate cancer: a review of their discovery, development, and place in therapy," *Clinical Therapeutics*, vol. 28, no. 10, pp. 1485–1508, 2006.

[5] S. Huang, Z. Yang, Y. Ma, and S. Wang, "miR-101 enhances cisplatin- induced DNA damage through decreasing nicotinamide adenine dinucleotide phosphate levels by directly repressing Tp53-induced glycolysis and apoptosis regulator expression in prostate cancer cells," *DNA and Cell Biology*, vol. 36, no. 4, pp. 303–310, 2017.

[6] M. Chovanec, Z. Cierna, V. Miskovska et al., "Prognostic role of programmed-death ligand 1 (PD-L1) expressing tumor infiltrating lymphocytes in testicular germ cell tumors," *Onceotarget*, vol. 8, no. 13, pp. 21794–21805, 2017.

[7] G. L. Yang, L. H. Zhang, Q. Liu et al., "A novel treatment strategy for newly diagnosed high-grade T1 bladder cancer: gemcitabine and cisplatin adjuvant chemotherapy-A single-institution experience," *Urologic Oncology: Seminars and Original Investigations*, vol. 35, no. 2, pp. 38.e9–e15, 2017.

[8] T. Sugiyama, A. Okamoto, T. Enomoto et al., "Randomized phase III trial of irinotecan plus cisplatin compared with paclitaxel plus carboplatin as first-line chemotherapy for ovarian clear cell carcinoma: JGOG3017/GCIG trial," *Journal of Clinical Oncology*, vol. 34, no. 24, pp. 2881–2887, 2016.

[9] G. Cancello, V. Bagnardi, C. Sangalli et al., "Phase II study with epirubicin, cisplatin, and infusional fluorouracil followed by weekly paclitaxel with metronomic cyclophosphamide as a preoperative treatment of triple-negative breast cancer," *Clinical Breast Cancer*, vol. 15, no. 4, pp. 259–265, 2015.

[10] M. Joerger, J. von Pawel, S. Kraff et al., "Open-label, randomized study of individualized, pharmacokinetically (PK)-guided dosing of paclitaxel combined with carboplatin or cisplatin in patients with advanced non- small-cell lung cancer (NSCLC)," *Annals of Oncology*, vol. 27, no. 10, pp. 1895–1902, 2016.

[11] M. Galanski, M. A. Jakupec, and B. K. Keppler, "Update of the preclinical situation of anticancer platinum complexes: novel design strategies and innovative analytical approaches," *Current Medical Chemistry*, vol. 12, no. 18, pp. 2075–2094, 2005.

[12] L. Galluzzi, L. Senovilla, I. Vitale et al., "Molecular mechanisms of cisplatin resistance," *Oncogene*, vol. 31, no. 15, pp. 1869–1883, 2012.

[13] I. Arany and R. L. Safirstein, "Cisplatin nephrotoxicity," *Seminars in Nephrology*, vol. 23, no. 5, pp. 460–464, 2002.

[14] Y. Liao, X. Lu, C. Lu, G. Li, Y. Jin, and H. Tang, "Selection of agents for prevention of cisplatin-induced hepatotoxicity," *Pharmacology Research*, vol. 57, no. 2, pp. 125–131, 2008.

[15] P. D. Sanchez-Gonzalez, F. J. Lopez-Hernandez, J. M. Lopez-Novoa, and A. I. Morales, "An integrative view of the pathophysiological events leading to cisplatin nephrotoxicity," *Critical Reviews in Toxicology*, vol. 41, no. 10, pp. 803–821, 2011.

[16] B. D. Sahu, J. M. Kumar, and R. Sistla, "Baicalein, a bioflavonoid, prevents cisplatin-induced acute kidney injury by up-regulating antioxidant defenses and down-regulating the MAPKs and NF-κB pathways," *PLoS One*, vol. 10, no. 7, Article ID e0134139, 2015.

[17] B. Dl Sahu, M. Kuncha, G. J. Sindhura, and R. Sistla, "Hesperidin attenuates cisplatin-induced acute renal injury by decreasing oxidative stress, inflammation and DNA damage," *Phytomedicine*, vol. 20, no. 5, pp. 453–460, 2013.

[18] S. Işeri, F. Ercan, N. Gedik, M. Yüksel, and İ. Alican, "Simvastatin attenuates cisplatin-induced kidney and liver damage in rats," *Toxicology*, vol. 230, no. 2-3, pp. 256–264, 2007.

[19] S. Palipoch and C. Punsawad, "Biochemical and histological study of rat liver and kidney injury induced by cisplatin," *Journal of Toxicologic Pathology*, vol. 26, no. 3, pp. 293–299, 2013.

[20] Y. Li, Y. Zhao, R. Tang, and X. Qu, "Preventive and therapeutic effects of antler collagen on osteoporosis in ovariectomized rats," *African Journal of Biotechnology*, vol. 9, no. 38, pp. 6437–6441, 2010.

[21] M. J. Shao, S. R. Wang, M. J. Zhao et al., "The Effects of Velvet Antler of Deer on Cardiac Functions of Rats with Heart Failure following Myocardial Infarction," *Evidence Based Complementary and Alternative Medicine*, vol. 2012, Article ID 825056, 5 pages, 2012.

[22] R. Karawita, P. J. Park, N. Siriwardhana et al., "Angiotensin I-converting enzyme (ACE) inhibitory activity of elk (*Cervus elaphus*) velvet Antler," *Journal of Food Science and Nutrition*, vol. 10, no. 3, pp. 239–243, 2005.

[23] J. H. Kim, Y. I. Yang, J. H. Ahn, J. G. Lee, K. T. Lee, and J. H. Choi, "Deer (*Cervus elaphus*) antler extract suppresses adhesion and migration of endometriotic cells and regulates MMP-2 and MMP-9 expression," *Journal of Ethnopharmacology*, vol. 140, no. 2, pp. 391–397, 2012.

[24] Y. Tang, B. T. Jeon, Y. Wang et al., "First evidence that Sika deer (Cervus nippon) velvet antler extract suppresses migration of human prostate cancer cells," *Korean Journal for Food Science of Animal Resources*, vol. 35, no. 4, pp. 507–514, 2015.

[25] Y. Tang, B. T. Jeon, Y. Wang et al., "First evaluation of the biologically active substances and antioxidant potential of regrowth velvet antler by means of multiple biochemical assays," *Journal of Chemistry*, vol. 2015, Article ID 975292, 7 pages, 2015.

[26] M. R. Zirak, R. Rahimian, M. Ghazi-Khansari et al., "Tropisetron attenuates cisplatin-induced nephrotoxicity in mice," *European Journal of Pharmacology*, vol. 738, pp. 222–229, 2014.

[27] Y. C. Hung, G. S. Huang, L. W. Lin, M. Y. Hong, and P. S. Se, "Thea sinensis melanin prevents cisplatin-induced nephrotoxicity in mice," *Food and Chemical Toxicology*, vol. 45, no. 7, pp. 1123–1130, 2007.

[28] M. Ueki, M. Ueno, J. Morishita, and N. Maekawa, "Curcumin ameliorates cisplatin-induced nephrotoxicity by inhibiting renal inflammation in mice," *Journal of Bioscience and Bioengineering*, vol. 115, no. 5, pp. 547–551, 2013.

[29] Q. Wu, M. Kohli, H. R. Bergen et al., "Preclinical evaluation of the supercritical extract of azadirachta indica (neem) leaves in vitro and in vivo on inhibition of prostate cancer tumor

growth," *Molecular Cancer Therapeutics*, vol. 13, no. 5, pp. 1067–1077, 2014.
[30] F. Yang, L. Song, H. Wang, Z. Xu, and N. Xing, "Combination of quercetin and 2-methoxyestradiol enhances inhibition of human prostate cancer LNCaP and PC-3 cells xenograft tumor growth," *PLoS One*, vol. 10, no. 5, Article ID e0128277, 2015.
[31] A. Ehteda, P. Galettis, K. Pillai, and D. L. Morris, "Combination of albendazole and 2-methoxyestradiol significantly improves the survival of HCT-116 tumor-bearing nude mice," *BMC Cancer*, vol. 13, no. 1, p. 86, 2013.
[32] A. Gilbey and J. D. Perezgonzalez, "Health benefits of deer and elk velvet antlersupplements: a systematic review of randomised controlled studies," *New Zealand Medical Journal*, vol. 125, no. 1367, pp. 80–86, 2012.
[33] E. K. Kim, W. B. Lee, S. H. Moon et al., "Free radical scavenging activity by ESR spectroscopy and neuroprotective effect on H_2O_2-induced damage in PC-12 cells of enzymatic from Korean elk velvet antler," *Journal of Food Biochemistry*, vol. 33, no. 6, pp. 895–912, 2009.
[34] J. Gumulec, J. Balvan, M. Sztalmachova et al., "Cisplatin-resistant prostate cancer model: differences in antioxidant system, apoptosis and cell cycle," *International Journal of Oncology*, vol. 44, no. 3, pp. 923–933, 2014.
[35] A. Sarastea and K. Pulkkic, "Morphologic and biochemical hallmarks of apoptosis," *Cardiovascular Research*, vol. 45, no. 3, pp. 528–537, 2000.
[36] K. L. King and J. A. Cidlowski, "Cell cycle regulation and apoptosis," *Annual Review Physiology*, vol. 60, no. 1, pp. 601–617, 1998.
[37] J. F. Kerr, C. M. Winterford, and B. V. Harmon, "Apoptosis. Its significance in cancer and cancer therapy," *Cancer*, vol. 73, no. 8, pp. 2013–2026, 1994.
[38] D. Hanahan and R. A. Weinberg, "The hallmarks of cancer," *Cell*, vol. 100, no. 1, pp. 57–70, 2000.
[39] S. Gurumurthy, K. M. Vasudevan, and V. M. Rangnekar, "Regulation of apoptosis in prostate cancer," *Cancer Metastasis Review*, vol. 20, no. 3-4, pp. 225–243, 2001.
[40] S. W. Fesik, "Promoting apoptosis as a strategy for cancer drug discovery," *Nature Review Cancer*, vol. 5, no. 11, pp. 876–885, 2005.
[41] N. V. Guseva, A. F. Taghiyev, O. W. Rokhlin, and M. B. Cohen, "Death receptor-induced cell death in prostate cancer," *Journal of Cellular Biochemistry*, vol. 91, no. 1, pp. 70–99, 2004.
[42] A. Ashkenazi, "Targeting death and decoy receptors of the tumour-necrosis factor superfamily," *Nature Review Cancer*, vol. 2, no. 6, pp. 420–430, 2002.
[43] K. P. Kang, S. K. Park, D. H. Kim et al., "Luteolin ameliorates cisplatin-induced acute kidney injury in mice by regulation of p53-dependent renal tubular apoptosis," *Nephrology, Dialysis and Transplantation*, vol. 26, no. 3, pp. 814–822, 2011.
[44] M. A. H. Khan, J. Liu, G. Kumar, S. X. Skapek, J. R. Falck, and J. D. Imig, "Novel orally active epoxyeicosatrienoic acid (EET) analogs attenuate cisplatin nephrotoxicity," *FASEB Journal*, vol. 27, no. 8, pp. 2946–2956, 2013.
[45] N. A. Dos Santos, M. A. Carvalho Rodrigues, N. M. Martins, and A. C. dos Santos, "Cisplatin-induced nephrotoxicity and targets of nephroprotection: an update," *Archives of Toxicology*, vol. 86, no. 8, pp. 1233–1250, 2012.
[46] Y. Kidera, H. Kawakami, T., Sakiyama, K. Okamoto, and K. Tanaka, "Risk factors for cisplatin-induced nephrotoxicity and potential of magnesium supplementation for renal protection," *PLoS One*, vol. 9, no. 7, Article ID e101902, 2014.
[47] O. Mora Lde, L. M. Antunes, H. D. Francescato, and L. Bianchi Mde, "The effects of oral glutamine on cisplatin-induced nephrotoxicity in rats," *Pharmacological Research*, vol. 47, no. 6, pp. 517–522, 2003.
[48] L. A. Ahmed, N. I. Shehata, N. F. Abdelkader, and M. M. Khattab, "Tempol, a superoxide dismutase mimetic agent, ameliorates cisplatin-induced nephrotoxicity through alleviation of mitochondrial dysfunction in mice," *PLoS One*, vol. 9, no. 10, Article ID e108889, 2014.
[49] S. H. Kim, K. O. Hong, W. Y. Chung, J. K. Hwang, and K. K. Park, "Abrogation of cisplatin-induced hepatotoxicity in mice by xanthorrhizol is related to its effect on the regulation of gene transcription," *Toxicology and Applied Pharmacology*, vol. 196, no. 3, pp. 814–822, 2004.
[50] R. Bentli, H. Parlakpinar, A. Polat, E. Samdanci, M. E. Sarihan, and M. Sagir, "Molsidomine prevents cisplatin induced hepatotoxicity," *Archives of Medical Research*, vol. 44, no. 7, pp. 521–528, 2013.
[51] H. A. Omar, W. R. Mohamed, H. H. Arab, and S. A. Arafa, "Tangeretin alleviates cisplatin-induced acute hepatic injury in rats: targeting MAPKs and apoptosis," *PLoS One*, vol. 11, no. 3, Article ID e0151649, 2016.
[52] M. U. Rehman, N. Ali, and S. Rashid, "Alleviation of hepatic injury by chrysin in cisplatin administered rats: probable role of oxidative and inflammatory marker," *Pharmacological Report*, vol. 66, no. 6, pp. 1050–1059, 2014.
[53] N. A. El-Shitany and B. Eid, "Proanthocyanidin protects against cisplatin-induced oxidative liver damage through inhibition of inflammation and NF-κβ/TLR-4 pathway," *Environmental Toxicology*, vol. 32, no. 7, pp. 1952–1963, 2017.
[54] N. Pabla and Z. Dong, "Cisplatin nephrotoxicity: mechanisms and renoprotective strategies," *Kidney International*, vol. 73, no. 9, pp. 994–1007, 2008.
[55] G., Ramesh and W. B. Reeves, "TNF-alpha mediates chemokine and cytokine expression and renal injury in cisplatin nephrotoxicity," *Journal of Clinical Investigation*, vol. 110, no. 6, pp. 835–842, 2002.
[56] M. J. Sung, D. H. Kim, Y. J. Jung, K. P. Kang, A. S. Lee, and S. Lee, "Genistein protects the kidney from cisplatin-induced Injury," *Kidney International*, vol. 74, no. 12, pp. 1538–1547, 2008.
[57] C. E. Guerrero-Beltrán, P. Mukhopadhyay, B. Horváth, M. Rajesh, E. Tapia, and I. García-Torres, "Sulforaphane, a natural constituent of broccoli, prevents cell death and inflammation in nephropathy," *Journal of Nutritional Biochemistry*, vol. 23, no. 5, pp. 494–500, 2012.
[58] F. M. Kandemir, F. Benzer, M. Ozkaraca, S. Ceribasi, N. Cikcikoglu Yildirim, and N. Ozdemir, "Protective antioxidant effects of grape seed extract in a cisplatin-induced hepatotoxicity model in rabbits," *Revue De Medecine Veterinaire*, vol. 163, no. 11, pp. 539–545, 2012.
[59] B. I. Ognjanović, N. Z. Djordjević, M. M. Matić et al., "Lipid peroxidative damage on cisplatin exposure and alterations in antioxidant defense system in rat kidneys: a possible protective effect of selenium," *International Journal of Molecular Sciences*, vol. 13, no. 2, pp. 1790–1803, 2012.
[60] A. L. Al-Malki and A. A. Sayed, "Thymoquinone attenuates cisplatin induced hepatotoxicity via nuclear factor kappa-b," *BMC Complementary and Alternative Medicine*, vol. 14, no. 1, p. 282, 2014.
[61] M. A. Dkhil, S. Al-Quraishy, A. M. Aref, M. S. Othman, K. M. El-Deib, and A. E. Abdel Moneim, "The potential role of Azadirachta indica treatment on cisplatin-induced hepatotoxicity and oxidative stress in female rats," *Oxidative*

Medicine and Cell Longevity, vol. 2013, Article ID 741817, 9 pages, 2013.

[62] Y. An, H. Xin, W. Yan, and X. Zhou, "Amelioration of cisplatin-induced nephrotoxicity by pravastatin in mice," *Experimental and Txicologic Pthology*, vol. 63, no. 3, pp. 215–219, 2011.

[63] S. Atasayar, H. Gürer-Orhan, H. Orhan, B. Gürer, G. Girgin, and H. Özgüneş, "Preventive effect of amino guanidine compared to vitamin E and C on cisplatin-induced nephrotoxicity in rats," *Experimental and Toxicologic Pathology*, vol. 61, no. 1, pp. 23–32, 2009.

5

A Novel Green Synthesis of Thalidomide and Analogs

Ellis Benjamin[1] and Yousef M. Hijji[2]

[1]*Department of Chemistry, Stockton University, 101 Vera King Farris Drive, Galloway, NJ 08205-9441, USA*
[2]*Department of Chemistry and Earth Sciences, Qatar University, P.O. Box 2713, Doha, Qatar*

Correspondence should be addressed to Yousef M. Hijji; yousef.hijji@qu.edu.qa

Academic Editor: José L. Arias Mediano

Thalidomide and its derivatives are currently under investigation for their antiangiogenic, immunomodulative, and anticancer properties. Current methods used to synthesize these compounds involve multiple steps and extensive workup procedures. Described herein is an efficient microwave irradiation green synthesis method that allows preparation of thalidomide and its analogs in a one-pot multicomponent synthesis system. The multicomponent synthesis system developed involves an array of cyclic anhydrides, glutamic acid, and ammonium chloride in the presence of catalytic amounts of 4-N,N-dimethylaminopyridine (DMAP) to produce thalidomide and structurally related compounds within minutes in good isolated yields.

1. Introduction

Thalidomide has a long and tragic past that stems from its teratogenic effects. However, recent studies prove that it is an effective immunomodulatory, antiangiogenic, and anticancer pharmaceutical [1, 2]. This reemergence of thalidomide has led to a number of FDA approved derivatives including lenalidomide and pomalidomide (multiple myeloma) [3–5]. Currently thalidomide and its derivatives are used in the treatment of numerous diseases including Crohn [6–8], Leprosy [9–11], Graft-Versus-Host (GVHD) [12–14], and multiple forms of cancer [15–17]. They are also known to modulate the expression patterns of several proteins including TNFα [18–22], IL1β [23–27], and COX2 [28–31]. In fact, discovery of lenalidomide and pomalidomide had led scientists to explore the modification of thalidomide (Figure 1) to improve its efficacy and decrease its teratogenic and neurotoxic effects. Although there are several synthetic methods for making thalidomide many of the reported syntheses are often multistep processes which use exotic or expensive reagents, maintain long reaction times, and produce very low yields [32–43].

Our novel microwave assisted synthesis of thalidomide and its derivatives improves over current conventional and microwave assisted syntheses through the high yields of thalidomide and its systematic derivatization within minutes. Microwave application in organic synthesis has seen exponential growth over the last twenty years due to its easy experimental conditions, rapid turnaround, easy workups, and high yields. The speed at which microwave reactions are done matches well with combinational processes to synthesize derivatives for improved biological activity when compared to low-yielding conventional synthesis especially that of thalidomide and its analogs (24%). The power of microwave synthesis is utilized for speeding the reaction and providing an efficient convenient way of obtaining these thalidomide variations or related compounds [44–46].

Thalidomide and a series of its analogs (**2–5**) are shown in Figure 2 and synthesized in a one-pot reaction (Scheme 1), using mild conditions and inexpensive reagents, under green (microwave) conditions with relatively high yields. The strategy is based on utilizing the corresponding anhydrides, either commercially available anhydrides or constructed based on our earlier work of microwave assisted Diels Alder reaction of maleic anhydride with the required 1,3-diene. The synthesis of thalidomide (**1**) and derivatives (**2–5**) was accomplished in a one-pot reaction of glutamic acid with the corresponding anhydride and ammonium chloride with the use of DMAP as a base catalyst mixture at 150°C in 10 minutes. In this reaction we were able to form both substituted and unsubstituted

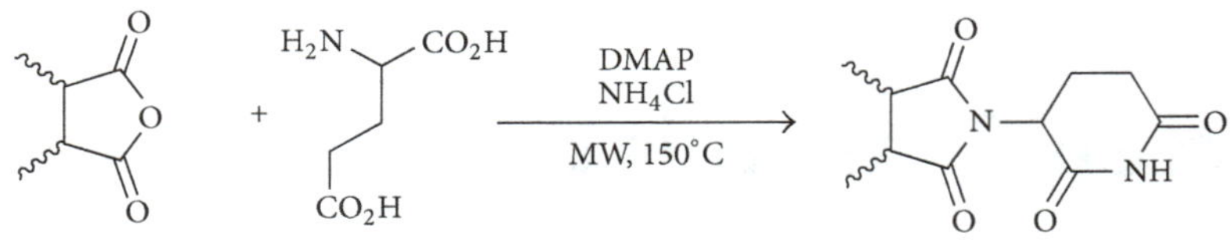

SCHEME 1: The synthesis of thalidomide and its analogs. Microwave conditions: 10 minutes at 150°C.

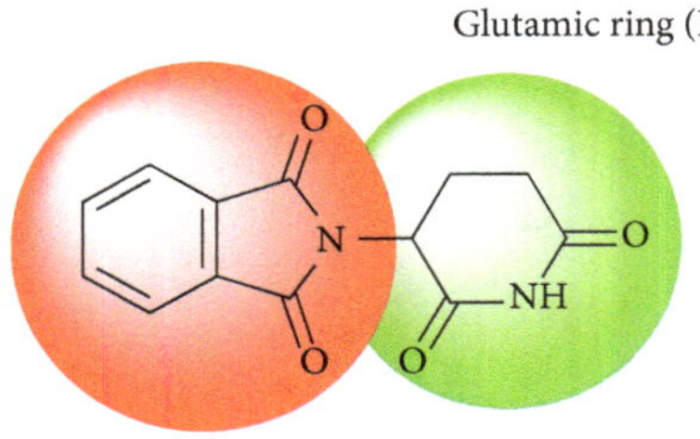

FIGURE 1: The structure of thalidomide.

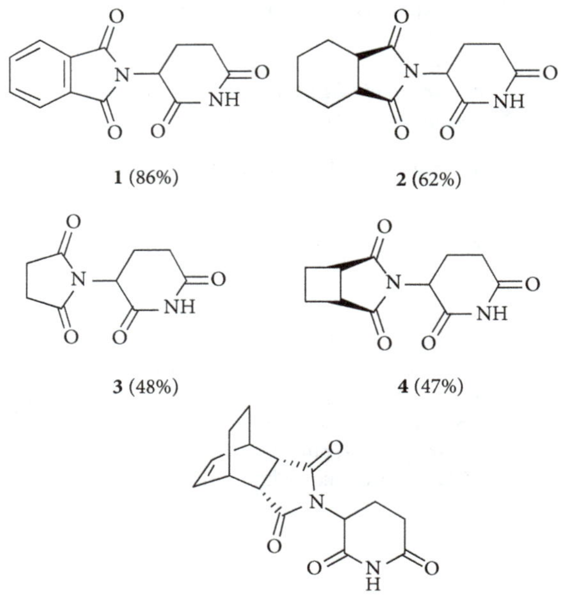

FIGURE 2: Thalidomide and thalidomide analogs synthesized.

cyclic imides through two condensation reactions of the cyclic anhydride and the glutamic acid moieties, respectively. The formation of unsubstituted imide is based on the generation of ammonia through the removal of proton from the ammonium chloride. The thalidomide derivatives produced a glutarimide ring with a diversity of cyclic anhydrides bound.

A primary benefit of our synthetic method over previous thalidomide syntheses methods includes the ability to modify the phthalic ring (Figure 1, moiety (A)). Although extensive work has focused on the modification of glutamic ring (Figure 1, moiety (B)) of thalidomide such as introduction of dithiocarbamate in order to enhance its biological activities, the carbon backbone is modified so that it can allow for variation in structure and functionality. This method focuses on the modification of the phthalic ring moiety of thalidomide (Figure 1), allowing for the addition of multiple carbon skeletons including bicyclic systems. Our synthesis allows for the exploration of the effects of ring size, ring strain, increased polarity, and structural complexity in addition to stereochemistry. These compounds may shed light on structure-activity relationship in thalidomide derivatives thereby improving their efficacy. Another benefit of this method is the ability to easily introduce labeled nitrogen through the use of ammonium-15N chloride in the reaction mixture.

Although this process occurs in one pot, it involves a multistep mechanism in which the amine of glutamic acid has to form an amide acid with the cyclic anhydride. If the cyclic amide-glutamic acid is not formed then reaction seems to proceed towards byproduct formation with the cyclic imide as the major product. A study to determine which catalyst could improve ammonia generation and product formation of thalidomide in the microwave was done using thiourea, ammonium acetate, and ammonium chloride/DMAP with phthalic anhydride using similar molar ratios. Previous research in our laboratory has found several ways to produce ammonia from ammonium acetate and hydroxylamine (HCl)/DMAP with ammonium acetate being the best [47, 48]. This study used GC-MS for determining the ratio of unsubstituted cyclic imides (phthalimide) to thalidomide. The data shows that ammonium acetate is highly efficient at forming phthalimide (90% yield) but not thalidomide (yield $< 7\%$). On the other hand, the reaction resulted in higher yields of thalidomide when DMAP/NH_4Cl (43%) and thiourea (40%) were used, Table 1.

The main byproduct formed when ammonium acetate is used is a cyclic imide and this may be explained by the rapid dissociation of ammonium acetate to give ammonia that initiates quick imide formation. Differences in the reactivity between ammonium acetate and ammonium chloride finds that ammonium chloride is highly stable and does not break down under microwave irradiation. The addition of DMAP as a base catalyst to ammonium chloride causes it to dissociate and generate ammonia at higher temperature, allowing glutamic acid to first react with the anhydrides to

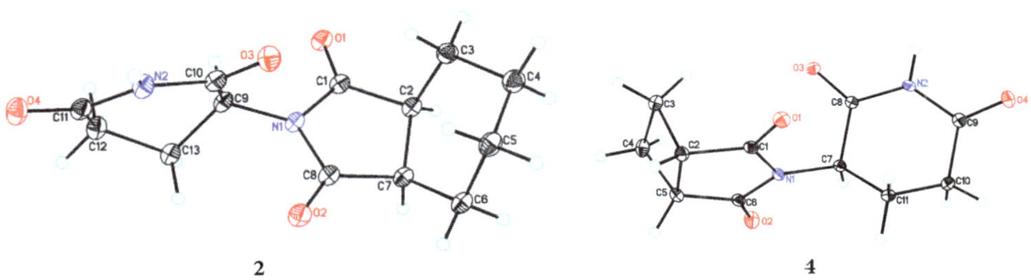

FIGURE 3: ORTEP drawings of compounds (3) and (4).

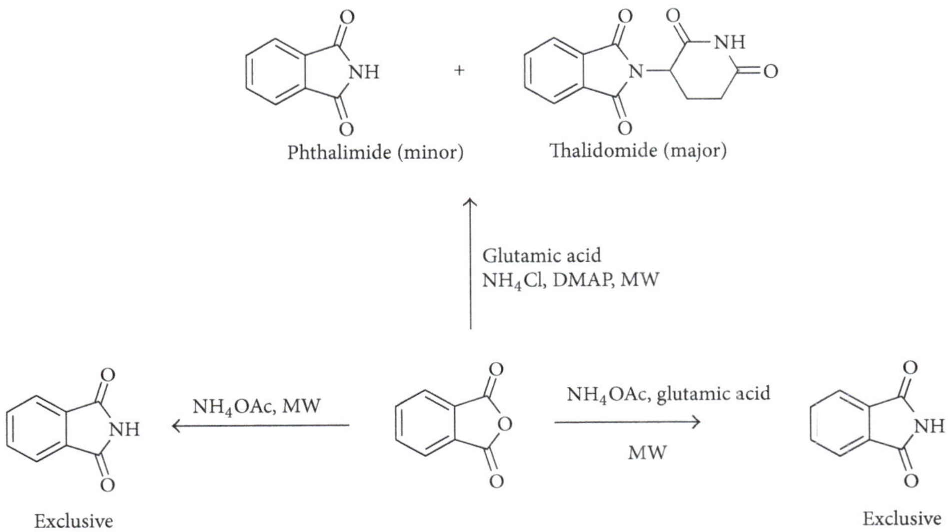

SCHEME 2: Synthesis of phthalimide and thalidomide.

TABLE 1: Comparison of thalidomide synthesis using DMAP/NH_4Cl, NH_4OAc, and thiourea in the monomode microwave at 150°C for 10 minutes.

Reagent	Phthalimide	Thalidomide
Thiourea	59%	40%
DMAP NH_4Cl	57%	43%
Ammonium acetate	90%	7%

form the imide (moiety (A)) followed by the generation of ammonia at higher temperature and the formation of the 2,6-dioxopiperidine, NH_4Cl ring later (moiety (B)). Long term tests (over 30 minutes) of NH_4Cl under microwave irradiation found that it does not readily break down and needs the presence of the DMAP as a base catalyst to initiate ammonia production allowing time for cyclic amide-glutamic acid formation (Scheme 2).

Compounds (**1–5**) were generated within minutes, even though the yields are moderate due the competition with ammonia generated in one-pot synthesis system. The yields are acceptable given the ease and simplicity of the reaction system. Compounds (**1–5**) were isolated, purified, and identified by spectroscopic analysis. (Detailed ^{1}H and ^{13}C NMR spectra and mass spectrometry data are presented in the Supplementary Material available online at https://doi.org/10.1155/2017/6436185.) Further structural confirmation was provided by single crystal X-ray structural identifications [49]. Compounds **3** and **4** ORTEP are presented in Figure 3. Compound **3** presents interesting structural features with 4- and 5-membered fused rings with the piperidinedione ring at almost right angle with the pyrrolidine ring containing dihedral angles of 67.6° and 73.9° for the cyclobutane and 2,6-dioxopiperidine rings, respectively. Compound **5** has an intriguing structure that contains a bicyclic system with 4 chiral centers, a double bond that is a handle for further functionalization and a possibility of two more chiral centers.

In conclusion, we have developed a novel green one-pot synthetic technique for the generation of thalidomide and its analogs. Although this article shows only a small subset of compounds, there are a limitless number of derivatives that can be produced/derived using this technique. This technique also has a wide variety of applications as it relates to the production of thalidomide analogs that may be used for the treatment of multiple diseases.

Appendix

Experimental

A CEM Discover monomode microwave was used for microwave enhanced organic synthesis. Isolation procedures use TLC as a basis for the column chromatography of the materials. Gas Chromatography Mass Spectrometry (GC-MS) will be used to quantify and identify the amounts of product and possible byproducts found in the purified materials.

Standardized methods for the identification of products consisted of GC-MS, ^{1}H NMR, ^{13}C NMR, DEPT-C NMR, melting point, and Infrared Spectroscopy (IR). Every new compound synthesized was completely analyzed by all corresponding techniques. Gas Chromatograph Mass Spectrometry was performed using either a Shimadzu GC-17A and GC-MS-QP5050A LabSolutions system or a Varian CP 3800 and Saturn 2200 system.

All IR spectra were performed on a Perkin-Elmer Spectrum RX I IR system. All melting points determinations were performed on a Laboratory Devices Mel-Temp II instrument. ^{1}H, ^{13}C, and DEPT-C NMR were performed on Bruker 400 MHz. All solvents (HPLC grade) were purchased from Fisher Scientific Corporation. All reagents were purchased from Aldrich Chemical Company and were used without purification.

Thalidomide Synthesis

2-(2,6-Dioxopiperidin-3-yl)isoindole-1,3-dione. (**1**): phthalic anhydride (0.10 g, 0.68 mmol), glutamic acid (0.10 g, 0.68 mmol), DMAP (0.02 g, 0.16 mmol), and NH_4Cl (0.04 g, 0.75 mmol) were mixed thoroughly in a CEM-sealed vial with a magnetic stirrer. The sample was heated for 10 min at 150°C in a CEM Discover microwave powered at 150 W. It was then cooled rapidly to 50°C. When at room temperature it was dissolved in 15 mL of (1 : 1) ethyl acetate : acetone. The organic layer was washed with 2x (10 mL) distilled water and then dried over sodium sulfate (anhydrous). The organic layer was concentrated under vacuum. The residue was treated with hexanes (30 mL) affording a white solid (0.14 g, 80%). mp 268–270°C. ^{1}H NMR (400 MHz, DMSO-d_6) δ 11.14 (s, 1 H, NH), 7.94 (m, 4 H, Ar), 5.17 (dd, 1 H, 12.5, 5.5 Hz), 2.92 (m, 1 H), 2.57 (m, 2), 2.09 (m, 1 H); ^{13}C NMR (100 MHz, DMSO-d_6) 172.7, 169.8, 167.1, 134.9, 131.2,123.4, 49.0, 30.9, 22.0; MS m/z 258 (M+); 230, 213, 202, 173, 148, 111, 76, 50.

2-(2,6-Dioxopiperidin-3-yl)-hexahydro-isoindole-1,3-dione. (**2**): a white solid (70% yield). ^{1}H NMR (400 MHz, DMSO-d_6) δ 11.0 (s, 1 H, NH), 4.9 (dd, 1 H, 12.5, 5.5 Hz), 3.0 (m, 1 H), 2.8 (m, 2 H), 2.5 (m, 1 H), 1.9 (m, 1 H), 1.7 (m, 3 H), 1.6 (m, 1 H), 1.4 (m, 4 H); ^{13}C NMR (100 MHz, DMSO-d_6) 178.8 (C=O), 178.7 (C=O), 172.7 (C=O), 169.4 (C=O), 48.7 (CH), 39.1 (CH), 38.8 (CH), 30.7 (CH_2), 23.1 (CH_2), 22.9 (CH_2), 21.1 (CH_2), 21.05 (CH_2), 21.00 (CH_2); MS m/z 264 (M+); 236, 210, 179, 154, 112, 82, 67, 54, 41.

3-(2,5-Dioxopyrrolidin-1-yl)-piperidine-2,6-dione. (**3**): white solid yield (0.10 g, 48%). ^{1}H NMR (400 MHz, DMSO-d_6) δ 11.0 (s, 1 H, NH), 4.9 (dd, 1 H, 12.5, 5.5 Hz), 3.3 (s, 1 H), 2.8 (m, 3 H), 2.5 (m, 1 H), 2.4 (m, 1 H), 1.9 (m, 1 H); 13C NMR (100 MHz, DMSO-d_6) 176.9, 172.7, 169.3, 49.0, 30.7, 28.0, 22.1; MS m/z 210 (M+) 182, 167, 125, 112, 83, 68, 56 41.

3-(2,6-Dioxopiperidin-3-yl)-3-aza-bicyclo[3.2.0]heptane-2,4-dione. (**4**): a white solid (54% yield). 1H NMR (400 MHz, DMSO-d_6) δ 11.06 (s, 1 H, NH), 4.95 (dd, 1 H, 12.5, 5.5 Hz), 2.84 (m, 2 H), 2.52 (m, 4 H), 2.02 (m, 2 H), 1.92 (m, 2 H); ^{13}C NMR (100 MHz, DMSO-d_6) 179.0 (C=O), 172.7 (C=O), 169.4 (C=O), 48.7 (CH), 49.1 (CH), 37.9 (CH), 37.7 (CH_2), 30.7 (CH_2), 22.1 (CH_2), 22.0 (CH_2), 21.0 (CH_2); MS m/z 236 (M+) 208, 151, 106, 112, 96, 83, 55, 41.

(2,6-Dioxopiperidin-3-yl)-3a,4,7,7a-tetrahydro-1H-4,7-ethanoisoindole-1,3(2H)-dione. (**5**): a white solid (62% yield). 1H NMR (400 MHz, DMSO-d6) δ 11.0 (s, 1 H, NH), 6.1 (m, 2 H), 4.8 (dd, 1 H, 12.5, 5.5 Hz), 3.0 (m, 4 H), 2.8 (m, 1 H), 2.5 (m, 1 H), 2.3 (m, 1 H), 1.7 (m, 1 H), 1.6 (d, 2 H, 7.5 Hz), 1.2 (d, 2 H, 7.5 Hz); 13C (100 MHz, DMSO-d_6) δ 177.75 (C=O), 177.71 (C=O), 172.57 (C=O), 168.73 (C=O), 132.09 (CH), 131.82 (CH), 48.75 (CH), 43.37 (CH), 31.39 (CH), 31.22 (CH), 30.48 (CH), 29.79 (CH), 23.17 (CH_2), 22.98 (CH_2), 22.87 (CH_2), 21.29 (CH_2); MS m/z 288 (M+) 260, 209, 178, 149, 136, 112, 99, 80, 78, 54, 41.

Disclosure

The statements made herein are solely the responsibility of the authors.

Competing Interests

The authors declare that they have no competing interests.

Acknowledgments

Yousef M. Hijji would like to thank QNRF for support: NPRP award from Qatar National Research Fund (a member of The Qatar Foundation) [NPRP-.7-495-1-094]. Ellis Benjamin would like to thank the School of Natural Sciences and Mathematics at Stockton University.

References

[1] R. Talaat, W. El-Sayed, H. S. Agwa, A. M. Gamal-Eldeen, S. Moawia, and M. A. H. Zahran, "Anti-inflammatory effect of thalidomide dithiocarbamate and dithioate analogs," *Chemico-Biological Interactions*, vol. 238, pp. 74–81, 2015.

[2] M. Y. Kim, R. Sposto, A. Swaika et al., "Pharmacoeconomic implications of lenalidomide maintenance therapy in multiple myeloma," *Oncology*, vol. 87, no. 4, pp. 224–231, 2014.

[3] Y. Liu, X. Huang, X. He et al., "A novel effect of thalidomide and its analogs: suppression of cereblon ubiquitination enhances ubiquitin ligase function," *The FASEB Journal*, vol. 29, no. 12, pp. 4829–4839, 2015.

[4] L. Roziaková, M. Mistrík, and A. Bátorová, "Pomalidomide in the treatment of relapsed and refractory multiple myeloma," *Klinicka Onkologie*, vol. 27, no. 5, pp. 318–325, 2014.

[5] G. Fouquet, C. Bories, S. Guidez et al., "Pomalidomide for multiple myeloma," *Expert Review of Hematology*, vol. 7, no. 6, pp. 719–731, 2014.

[6] C. Yang, P. Singh, H. Singh, M.-L. Le, and W. El-Matary, "Systematic review: thalidomide and thalidomide analogues for treatment of inflammatory bowel disease," *Alimentary Pharmacology & Therapeutics*, vol. 41, no. 11, pp. 1079–1093, 2015.

[7] S. Harbour and N. Brown, "Thalidomide improves clinical remission in children with Crohn's disease," *Archives of Disease in Childhood*, vol. 100, no. 2, article 111, 2015.

[8] S. C. Ng, "Thalidomide and refractory crohn's disease: what is in the future?" *Journal of Clinical Gastroenterology*, vol. 48, no. 6, pp. 476–477, 2014.

[9] S. Kamath, S. A. Vaccaro, T. H. Rea, and M. T. Ochoa, "Recognizing and managing the immunologic reactions in leprosy," *Journal of the American Academy of Dermatology*, vol. 71, no. 4, pp. 795–803, 2014.

[10] S. Zhou, F. Wang, T.-C. Hsieh, J. M. Wu, and E. Wu, "Thalidomide—a notorious sedative to a wonder anticancer drug," *Current Medicinal Chemistry*, vol. 20, no. 33, pp. 4102–4108, 2013.

[11] C. Mahony, L. Erskine, J. Niven, N. H. Greig, W. D. Figg, and N. Vargesson, "Pomalidomide is nonteratogenic in chicken and zebrafish embryos and nonneurotoxic in vitro," *Proceedings of the National Academy of Sciences of the United States of America*, vol. 110, no. 31, pp. 12703–12708, 2013.

[12] M. Alsina, P. S. Becker, X. Zhong et al., "Lenalidomide maintenance for high-risk multiple myeloma after allogeneic hematopoietic cell transplantation," *Biology of Blood and Marrow Transplantation*, vol. 20, no. 8, pp. 1183–1189, 2014.

[13] L. St John, S. M. Gordon, R. Childs et al., "Topical thalidomide gel in oral chronic GVHD and role of in situ cytokine expression in monitoring biological activity," *Bone Marrow Transplantation*, vol. 48, no. 4, pp. 610–611, 2013.

[14] P. J. Martin, Y. Inamoto, P. A. Carpenter, S. J. Lee, and M. E. D. Flowers, "Treatment of chronic graft-versus-host disease: past, present and future," *Korean Journal of Hematology*, vol. 46, no. 3, pp. 153–163, 2011.

[15] L. Song, X. Zhou, and X. Li, "Phase II trial of granulocyte-macrophage colony-stimulating factor plus thalidomide in older patients with castration-resistant prostate cancer," *Molecular and Clinical Oncology*, vol. 3, no. 4, pp. 865–868, 2015.

[16] Z. Qiao, J. Yuan, J. Shen et al., "Effect of thalidomide in combination with gemcitabine on human pancreatic carcinoma SW-1990 cell lines in vitro and in vivo," *Oncology Letters*, vol. 9, no. 5, pp. 2353–2360, 2015.

[17] P. M. Da Costa, M. P. Da Costa, A. A. Carvalho et al., "Improvement of in vivo anticancer and antiangiogenic potential of thalidomide derivatives," *Chemico-Biological Interactions*, vol. 239, article no. 7412, pp. 174–183, 2015.

[18] L. C. D. Coêlho, M. V. De Oliveira Cardoso, D. R. M. Moreira et al., "Novel phthalimide derivatives with TNF-α and IL-1β expression inhibitory and apoptotic inducing properties," *MedChemComm*, vol. 5, no. 6, pp. 758–765, 2014.

[19] M. V. De Oliveira Cardoso, D. R. Magalhães Moreira, G. B. Oliveira Filho et al., "Design, synthesis and structure—activity relationship of phthalimides endowed with dual antiproliferative and immunomodulatory activities," *European Journal of Medicinal Chemistry*, vol. 96, pp. 491–503, 2015.

[20] A. Diamanti, T. Capriati, B. Papadatou et al., "The clinical implications of thalidomide in inflammatory bowel diseases," *Expert Review of Clinical Immunology*, vol. 11, no. 6, pp. 699–708, 2015.

[21] Y. Huang, Y. Zang, L. Zhou, W. Gui, X. Liu, and Y. Zhong, "The role of TNF-α/NF-κB pathway on the up-regulation of voltage-gated sodium channel Nav1.7 in DRG neurons of rats with diabetic neuropathy," *Neurochemistry International*, vol. 75, pp. 112–119, 2014.

[22] M. S. Ali, R. M. Starke, P. M. Jabbour et al., "TNF-α induces phenotypic modulation in cerebral vascular smooth muscle cells: implications for cerebral aneurysm pathology," *Journal of Cerebral Blood Flow and Metabolism*, vol. 33, no. 10, pp. 1564–1573, 2013.

[23] A. Vangsted, T. W. Klausen, and U. Vogel, "Genetic variations in multiple myeloma II: association with effect of treatment," *European Journal of Haematology*, vol. 88, no. 2, pp. 93–117, 2012.

[24] L. Mazzoccoli, S. H. Cadoso, G. W. Amarante et al., "Novel thalidomide analogues from diamines inhibit pro-inflammatory cytokine production and CD80 expression while enhancing IL-10," *Biomedicine & Pharmacotherapy*, vol. 66, no. 5, pp. 323–329, 2012.

[25] M. Çizmecioğlu, V. Pabuççuoğlu, P. Ballar, A. Pabuççuoğlu, and Z. Soyer, "Synthesis and screening of cyclooxygenase inhibitory activity of some 1,3-dioxoisoindoline derivatives," *Arzneimittel-Forschung/Drug Research*, vol. 61, no. 3, pp. 186–190, 2011.

[26] G. D. Ferguson, K. Jensen-Pergakes, C. Wilkey et al., "Immunomodulatory drug CC-4047 is a cell-type and stimulus-selective transcriptional inhibitor of cyclooxygenase 2," *Journal of Clinical Immunology*, vol. 27, no. 2, pp. 210–220, 2007.

[27] J. L. Masferrer, "Methods of using a combination of cyclooxygenase-2 selective inhibitors and thalidomide for the treatment of neoplasia," US20030013739A1, 2003.

[28] E. D. Q. Crusoe, F. Higashi, M. P. N. C. Padilha et al., "Outcomes of autologous transplantation for multiple myeloma according to different induction regimens," *Revista Brasileira de Hematologia e Hemoterapia*, vol. 36, no. 1, pp. 19–24, 2014.

[29] M. Bolzoni, P. Storti, S. Bonomini et al., "Immunomodulatory drugs lenalidomide and pomalidomide inhibit multiple myeloma-induced osteoclast formation and the RANKL/OPG ratio in the myeloma microenvironment targeting the expression of adhesion molecules," *Experimental Hematology*, vol. 41, no. 4, pp. 387.e1–397.e1, 2013.

[30] M. Lazarini, F. Traina, S. M. Winnischofer, F. F. Costa, M. L. S. Queiroz, and S. T. O. Saad, "Effects of thalidomide on long-term bone marrow cultures from patients with myelodysplastic syndromes: induction of IL-10 expression in the stromal layers," *Leukemia Research*, vol. 35, no. 8, pp. 1102–1107, 2011.

[31] E. Shannon, R. Noveck, F. Sandoval, and B. Kamath, "Thalidomide suppressed IL-1β while enhancing TNF-α and IL-10, when cells in whole blood were stimulated with lipopolysaccharide," *Immunopharmacology and Immunotoxicology*, vol. 30, no. 3, pp. 447–457, 2008.

[32] J. Chen, K. Natte, A. Spannenberg, H. Neumann, M. Beller, and X.-F. Wu, "Efficient palladium-catalyzed double carbonylation of *o*-dibromobenzenes: synthesis of thalidomide," *Organic and Biomolecular Chemistry*, vol. 12, no. 30, pp. 5578–5581, 2014.

[33] Z. Xiao, K. Schaefer, S. Firestine, and P.-K. Li, "Solid-phase synthesis of thalidomide and its analogues," *Journal of Combinatorial Chemistry*, vol. 4, no. 2, pp. 149–153, 2002.

[34] B. Satyanarayana, P. Murali Krishna, and D. Ramachndran, "A novel and efficient synthesis of thalidomide," *International Journal of ChemTech Research*, vol. 3, no. 1, pp. 234–237, 2011.

[35] N. Shibata, T. Yamamoto, and T. Toru, "Synthesis of 2-(2, 6-dioxo-3-piperidinyl)-1h-isoindole-1, 3(2h)-dione (thalidomide)," in *Topics in Heterocyclic Chemistry*, vol. 8, pp. 73–97, Springer, Berlin, Germany, 2007.

[36] X.-H. Yuan, H.-Q. Guo, X.-P. Qiu, C.-Q. Kang, X.-D. Liu, and L.-X. Gao, "One-step synthesis of thalidomide and its important derivatives," *Chemical Journal of Chinese Universities*, vol. 26, no. 8, pp. 1477–1479, 2005.

[37] M.-Y. Chang, C.-H. Chang, S.-T. Chen, and N.-C. Chang, "A synthesis of thalidomide," *Journal of the Chinese Chemical Society*, vol. 49, no. 3, pp. 383–385, 2002.

[38] G. W. Muller, W. E. Konnecke, A. M. Smith, and V. D. Khetani, "A concise two-step synthesis of thalidomide," *Organic Process Research & Development*, vol. 3, no. 2, pp. 139–140, 1999.

[39] F. A. Luzzio, D. Y. Duveau, and W. D. Figg, "A chiral pool approach toward the synthesis of thalidomide metabolites," *Heterocycles*, vol. 70, pp. 321–334, 2006.

[40] H. B. Casazza Fanchini and A. Rodriguez Welton, "Novel process for the synthesis of thalidomide," MX2006008774A, 2008.

[41] M. Alpegiani, A. Mazzoni, D. Vergani, and W. Cabri, "Process for the one-pot synthesis of thalidomide directly from glutamine using phthaloylating agents and condensing agents," EP1602654A1, 2005.

[42] R. Varala and S. R. Adapa, "A practical and efficient synthesis of thalidomide via Na/liquid NH_3 methodology," *Organic Process Research and Development*, vol. 9, no. 6, pp. 853–856, 2005.

[43] N. Flaih, C. Pham-Huy, and H. Galons, "An expeditious synthesis of cyclic imides. Tetrahedron Lett. Tetrahedron," *Asymmetry*, vol. 6, pp. 3697–3698, 1999.

[44] J. A. Seijas, M. P. Vazquez-Tato, C. Gonzalez-Bande, M. M. Martinez, and B. Pacios-Lopez, "Microwave promoted synthesis of a rehabilitated drug: thalidomide," *Synthesis*, vol. 2001, no. 7, pp. 999–1000, 2001.

[45] S. You, Y. Li, D. Chen et al., "Application of microwave-assisted method in synthesis of thalidomide," *Huaxue Yanjiu Yu Yingyong*, vol. 23, pp. 652–656, 2011.

[46] J. Guo, Y. Zheng, D. Guo et al., "Application of the microwave-assisted method in synthesis of thalidomide amino acid derivatives," *Huaxue Yanjiu Yu Yingyong*, vol. 21, pp. 582–586, 2009.

[47] Y. M. Hijji and E. Benjamin, "Efficient microwave assisted syntheses of unsubstituted cyclic imides," *Heterocycles*, vol. 68, no. 11, pp. 2259–2267, 2006.

[48] E. Benjamin and Y. Hijji, "The synthesis of unsubstituted cyclic imides using hydroxylamine under microwave irradiation," *Molecules*, vol. 13, no. 1, pp. 157–169, 2008.

[49] Y. M. Hijji, E. Benjamin, E. Benjamin,R. J. Butcher, and J. P. Jasinski, "3-(2,6-dioxopiperidin-3-yl)-3-aza-bicyclo-[3.2.0]hep- tane-2,4-dione," *Acta Crystallographica*, vol. 65, pp. 394–395, 2009.

6

Dietary Inclusion of a Mixed Powder of Medicinal Plant Leaves Enhances the Feed Efficiency and Immune Function in Broiler Chickens

Roisbel Aroche,[1,2,3] Yordan Martínez,[2] Zheng Ruan,[1] Guiping Guan,[1,2] Samuel Waititu,[4] Charles M. Nyachoti,[5] Dairon Más,[6] and Shile Lan[2,3]

[1]*State Key Laboratory of Food Science and Technology, Nanchang University, Nanchang, Jiangxi 330047, China*
[2]*College of Bioscience and Biotechnology, Hunan Agricultural University, Changsha, Hunan 410128, China*
[3]*Faculty of Veterinary Medicine, University of Granma, Bayamo, 85100 Granma, Cuba*
[4]*Swine and Poultry Nutritionist, Fortified Nutrition Limited, Winnipeg, MB, Canada R2J OK4*
[5]*Animal Science Department, University of Manitoba, Winnipeg, MB, Canada R3T 2N2*
[6]*Laboratory of Animal Nutrition, Faculty of Natural Sciences, Autonomous University of Queretaro, 76230 Santiago de Querétaro, QRO, Mexico*

Correspondence should be addressed to Zheng Ruan; ruanzheng@ncu.edu.cn and Guiping Guan; guanguiping@hunau.edu.cn

Academic Editor: Sevgi Kolaylı

The trial was designed to determine the effect of the mixed powder of leaves of *Anacardium occidentale* (60%), *Psidium guajava* (20%), and *Morinda citrifolia* (20%) on growth performance, nutrient digestibility, and immunoglobulin concentration in broiler chickens. A total of 80 one-day-old Ross 308 broiler chickens were randomly assigned to two dietary treatments from days 1 to 21 of age, with 8 replicates by treatments and 5 birds per replicate. Treatments consisted of a control diet (T0) and the dietary inclusion of 0.5% of mixed powder of medicinal plant leaves (T1). The inclusion of herbs decreases the feed intake (FI) and feed conversion ratio (FCR; $P < 0.05$) in week 2 and in the period studied (1–21 days) compared with T0 but did not modify ($P < 0.05$) the body weight (BW) and average gain (AG). Also, the mixed powder has no influence ($P > 0.05$) on nutrient digestibility and IgA concentration; however, the IgG concentration increased by the effect of this experimental treatment. The results allow recommending the dietary inclusion of mixed powder of medicinal plant leaves as an alternative for obtaining acceptable performance in broilers.

1. Introduction

Medicinal plants have been used by humans from remote times to cure or to alleviate their illnesses or ailments. Also, they were used in the animals with therapeutic ends, for their antidiarrheal, antiseptic, antimicrobial, and anti-inflammatory properties [1]. Nowadays, the application of medicinal plants in the human health and animal production is increasing, due to the great existent concern at world level for the crossed possible resistance to the antibiotics for many microorganisms as a response to indiscriminate subtherapeutic use in animals [2].

Several studies have demonstrated that phytobiotics in the diets of farm animals improved bodily development, gut integrity, nutrient absorption, antioxidant activity, and immunity, with decreasing the diarrhea syndrome [3–5]; in consequence, these natural products have been considered as an effective alternative to feed antibiotics, mainly to diminish or decrease the residual effects in meat, egg, and milk. However, many researchers question these results, especially because of the variations found in biological indicators in *in vivo* studies; it is important to note that the positive effects will depend on the animal species, the productive category, environmental conditions, and characteristics of the plant material used [4]; this means that the best results are found in young birds under stressful environmental conditions in intensive production and with the use of phytobiotics rich in nontoxic secondary metabolites, with medicinal properties that attenuate these negative factors.

Good examples of medicinal plants are *Anacardium occidentale*, *Psidium guajava*, and *Morinda citrifolia*. In this sense, the leaves of *A. occidentale* have astringent, antidiarrheal, antioxidant, and antimicrobial properties, due to the presence of polyphenols (mainly tannins) and coumarins [6]; its use in poultry and swine diets improves the growth performance and the production and quality of eggs and decreases the incidence of diarrhea, respectively [7]. Also, *P. guajava*, a plant of great application in the treatment of the diarrheic syndrome in pigs (mainly the leaves), has hypoglycemic, antioxidant, and anti-inflammatory properties due to the presence of flavonoids (mainly gammapirons) and triterpenoids [8], and also its application as a medicinal drug in the diets of mice showed an antidiabetic effect [9]. Moreover, *M. citrifolia* is rich in alkaloids in its leaves and fruits, which can be recommended mainly for more than 20 illnesses to human and animals, due to its antioxidant, antimicrobial immune-stimulant properties [10]; specifically, in birds, Sunder et al. [11] have demonstrated that daily use of this plant in the diets of laying hens increases the egg weight and shell thickness.

In these three plants, separately, their effects have been investigated in animals of zootechnical interest with good results, mainly as nutraceutical supplements [9, 11, 12]. However, little is known about the use of these plants when they are mixed and supplemented or are included as powder in diets for broiler chickens. For this study, we hypothesized that inclusion in the diet with mixed powder, with a greater incorporation of *A. occidentale* in the powder, due to its beneficial secondary metabolites and its positive evidence on the production and health of young birds [12], might stimulate the immune system and, as a consequence, increase the productive behavior in broiler. Thus, the objective of this experiment is to determine the effect of the mixed powder of leaves of plants with medicinal properties on growth performance, nutrient digestibility, and immunoglobulin concentration in broiler chickens.

2. Materials and Methods

All experimental procedures used in this trial have been approved by the University of Manitoba, Animal Care Protocol Management and Review Committee, and birds were handled in accordance with the Canadian Council on Animal Care guidelines.

2.1. Animal, Housing, and Treatment. A total of 80 one-day-old Ross 308 male broiler chickens were divided into 16 uniform groups of 5 birds balanced for body weight and were randomly assigned to two dietary treatments (8 groups per treatment) from days 1 to 21 of age (Table 1).

The chicks in electrically heated cages in a battery brooder in a light- and temperature-controlled facility were housed. Upon arrival, the temperature inside the room was adjusted to 32°C on the first day and was then reduced by 2.5°C each week. The mean value of temperature in the experimental room throughout the whole experimental period was approximately 27°C.

Table 1: Composition and nutrient content.

Ingredients	Days 1 to 21	
	T0	T1
Corn	55.94	55.66
Soybean meal, 44% CP	35.00	34.83
Canola oil	4.00	3.98
Calcium carbonate	1.50	1.48
Monocalcium phosphate	1.70	1.69
DL-methionine	0.19	0.19
L-lysine	0.01	0.01
L-threonine	0.06	0.06
Common salt	0.30	0.30
Vitamin-mineral premix[1]	1.00	1.00
Titanium dioxide	0.30	0.30
Mixed powder of medicinal plants[2]	—	0.50
Calculated composition, % as fed		
Crude protein, %	22.01	21.94
Metabolizable energy, kcal/kg	3086	3074
Ca, %	1.11	1.09
Available P, %	0.48	0.46
SID[2] lysine, %	1.10	1.08
SID methionine, %	0.53	0.50

[1]Provided per kilogram of diet: manganese, 55 mg; zinc, 50 mg; iron, 80 mg; copper, 5 mg; selenium, 0.1 mg; iodine, 0.36 mg; sodium, 1.6 g; retinyl acetate, 8,250 IU; cholecalcipherol, 1,000 IU; DL-alpha-tocopherol, 11 IU; cyanocobalamin, 0.012 mg; phylloquinone, 1.1 mg; niacin, 53 mg; choline, 1,020 mg; folacin, 0.75 mg; biotin, 0.25 mg; riboflavin, 5.5 mg. [2]Mixed powder of medicinal plant leaves consisted in 60% of *Anacardium occidentale* 20% of *Psidium guajava*, and 20% of *Morinda citrifolia*.

They received water and feed on *ad libitum* basis throughout the experimental period. The two diets were made on the basis of standardized ileal digestible (SID) amino acids and consisted of a basal diet (T0) and dietary inclusion of 0.50 % of mixed powder of medicinal plant leaves (T1) (60% of *Anacardium occidentale*, 20% of *Psidium guajava*, and 20% of *Morinda citrifolia*), formulated to NRC [13] to meet nutrient requirements.

2.2. Growth Performance. In all experimental weeks, the BW (g), FI (g/bird), and viability were determined; then, based on these data, the AG (g) and FRC (kg/kg) were calculated according to the method of Aguilar et al. [14].

2.3. Nutrient Digestibility. Excreta from each pen during the last two days of each week were collected and frozen at −20°C until the laboratory analysis. At 21 days, all chicks within a pen by CO_2 asphyxiation were euthanized to collect ileal digesta, and then, the same conservation method was used. Then, excreta samples from each pen were pooled and weighed before and after being dehydrated in an oven at 60°C for 36 hours to determine excreta moisture content. Also, ileal digesta samples within a pen were pooled for apparent ileal digestibility (AID) calculation.

Diet samples were finely ground to pass through a 1 mm screen in a Thomas-Wiley mill (Thomas Scientific, Swedesboro, NJ). Also, ileal digesta samples were freeze-dried and finely ground (CBG5 Smart Grind; Applica Consumer Products Inc., Shelton, CT). Both samples were analyzed for dry matter (DM), gross energy (GE), nitrogen, crude fat (CF), and titanium concentration.

DM and CF were analyzed according to AOAC [15]. GE was determined using the Parr adiabatic oxygen bomb calorimeter (Parr Instrument Co., Moline, IL). Nitrogen was analyzed using an N analyzer (model NS-2000, Leco Corporation, St. Joseph, MI). Samples for titanium analysis were ashed, digested according the method described by Lomer et al. [16], and read on a Varian inductively coupled plasma mass spectrometer (Varian Inc., Palo Alto, CA).

The nutrient digestibility was calculated using the following equation:

$$\% \text{ apparent nutrient digestibility} = \left\{1 - \left[\left(\frac{T_d}{T_f}\right) \times \left(\frac{N_f}{N_d}\right)\right]\right\} \times 100, \quad (1)$$

where T_d = the titanium dioxide (TiO_2) concentration in the diet, T_f = the TiO_2 concentration in the excreta or ileal digesta, N_f = the nutrient concentration in the excreta or ileal digesta, and N_d = the nutrient concentration in the diet.

The AME and AMEn contents of experimental diets were calculated using the following equations:

AME (kcal/kg) = $GE_{\text{kcal/kg of the diet}} - [GE_{\text{kcal/kg of excreta}} \times (TiO_{2\ \%\ \text{diet}}/TiO_{2\ \%\ \text{excreta}})]$; and AMEn (kcal/kg) = $GE_{\text{kcal/kg of the diet}} - [GE_{\text{kcal/kg of excreta}} \times (TiO_{2\ \%\ \text{diet}}/TiO_{2\ \%\ \text{excreta}})] - 8.22 \times \{N_{\%\ \text{diet}} - [N_{\%\ \text{excreta}} \times (TiO_{2\ \%\ \text{diet}}/TiO_{2\ \%\ \text{excreta}})]\}$, where 8.22 is the energy equivalent of uric acid N.

2.4. Immunoglobulin Concentration. At 21 days of sacrifice, 5 mL of blood samples was collected from the jugular vein of 5 birds per treatment. Blood samples were placed in 2 mL tubes, and then, sodium heparin was added at a ratio of 2 : 1 and then stored at −15°C. Sandwich IgG ELISA Quantification Set (Bethyl Laboratories, Montgomery, TX) protocols were followed to determine natural immunoglobulin concentrations. A dilution of 1 : 10,000 of serum samples was used for natural IgG determination. Serum IgA concentrations in diluted samples were calculated by a Sandwich ELISA kit (Bethyl Laboratories Inc., Montgomery, TX) using 96-well microtiter plates.

Data of the experiment were subjected to unpaired t-test to independent samples of SAS (SAS software release 9.3; SAS Inst., Inc., Cary, NC). P values < 0.05 were taken to indicate significance.

3. Results

3.1. Growth Performance. Data on BW, FI, AG, and FCR during 21 days of broiler supplemented with mixed powder of medicinal plants are presented in Table 2. Viability was excellent in both treatments (100%; data not shown), and BW and AG did not show changes ($P > 0.05$) by the effect of the treatment with herbs. However, FI decreased and FCR improved ($P < 0.05$) with this natural product in week 2 and in the period of 1–21 days, where the chicks used 25.90% less feed than the control group, keeping no difference in the BW ($P > 0.05$).

Table 2: Effect of the mixed powder of leaves of plants with medicinal properties on growth performance in broiler.

Age (days)	Treatments T0	Treatments T1	±SEM	*P* value
BW (g)				
1	42.43	42.75	0.163	0.168
7	121.05	123.76	3.128	0.564
14	346.96	348.13	12.70	0.951
21	725.14	704.73	22.71	0.550
FI (g/bird)				
1–7	110.70	109.53	1.688	0.547
7–14	406.00	300.58	7.879	<0.001
14–21	509.96	489.98	15.93	0.407
0–21	1027.29	900.10	22.85	0.002
AG (g)				
1–7	78.62	81.09	3.167	0.616
7–14	225.91	224.37	10.79	0.919
14–21	378.17	356.60	11.75	0.232
0–21	682.71	661.98	22.75	0.544
FCR (kg/kg)				
1–7	1.42	1.37	0.030	0.502
7–14	1.81	1.35	0.066	<0.001
14–21	1.35	1.38	0.006	0.032
1–21	1.51	1.37	0.021	<0.001

The experiment lasted for 21 days; $n = 40$. BW: body weight; FI: feed intake; AG: average gain; FCR: feed conversion ratio.

Table 3: Effect of the mixed powder of leaves of plants with medicinal properties on nutrient digestibility in broiler.

Items	Treatments T0	Treatments T1	±SEM	*P* value
AMEn	2669	2822	61.82	0.244
NATTD	63.28	61.37	2.541	0.720
NAID	81.73	78.93	1.137	0.231
DMATTD	61.83	63.05	2.248	0.797
DMAID	67.01	62.99	1.406	0.190

The experiment lasted for 21 days; $n = 40$. AMEn: nitrogen-corrected apparent metabolizable energy; NATTD: nitrogen apparent total tract digestibility; NAID: nitrogen apparent ileal digestibility; DMATTD: dry matter apparent total tract digestibility; DMAID: dry matter apparent ileal digestibility.

3.2. Nutrient Digestibility. Table 3 shows the effect of the mixed powder of leaves of plants with medicinal properties on AMEn, NATTD, NAID, DMATTD, and DMAID in broilers. The inclusion of the medicinal powder did not show significant changes ($P > 0.05$) for any of these indicators shown in Table 3.

3.3. Immunoglobulin Concentrations. The effect of the mixed powder of leaves of plants with medicinal properties on IgG and sIgA serum concentration is shown in Table 4. Herb powder increased ($P < 0.05$) the serum concentration of IgG

Table 4: Effect of the mixed powder of leaves of plants with medicinal properties on IgG and sIgA serum concentration.

Items	Treatments		SEM±	*P* value
	T0	T1		
IgG (μg/ml)	326.43	555.14	11.37	0.001
sIgA (μg/ml)	91.70	126.17	41.34	0.140

The experiment lasted for 21 days; $n = 5$.

in 228.71 μg/ml compared with T0. Although it is possible to observe a numerical stimulation of secretory IgA (sIgA) serum concentration due to T1 (91.70 versus 126.17 μg/ml), this did not indicate statistical differences ($P > 0.05$).

4. Discussion

One of the aims of this experiment was to evaluate whether the inclusion of the mixed powder of medicinal plant leaves in broiler diets would influence the growth performance because their separate use has shown the regulatory effect on body weight and feed intake. Our study showed that dietary inclusion of 0.5% of mixed powder of leaves with medicinal properties improved the feed efficiency in broiler, because the FI and FCR decreased in week 2 and the experimental period (1–21 days; Table 2), without affecting the viability (data not shown) and normal digestive function of birds, which suggest that this natural product has beneficial phytochemical compounds that can improve the biological development of broiler chickens.

In this sense, Martínez et al. [7] reported that the medicinal effect of plants is due to the secondary metabolites; also, their effects will depend on the concentration and relationship of these compounds and its inclusion or supplementation on animal diets. Therefore, the herbs used in small concentrations rich in secondary metabolites such as tannins, coumarins, triterpenoids, flavonoids, and alkaloids may have influence on animal response due to their astringent, anti-inflammatory, antioxidant, and antimicrobial properties [17].

The powder was prepared to intensify the content of polyphenols, especially tannins obtained from the leaves of *A. occidentale*, which has the highest proportion in the mixture, mainly because this polyphenol has beneficial activity at intestinal level [7]. In this sense, this secondary metabolite is known for its astringent property, because they can bind to saliva lubricating proteins by hydrogen bonds [17]; therefore, an increase of this metabolite in the diet could reduce the passage of the digesta in the GIT and decrease the feed intake by a higher state of satiety in this period (Table 2). In addition, tannins have demonstrated *in vitro* antibacterial effects against strains of *Escherichia coli* and *Staphylococcus aureus*, with pathogenic bacteria being more frequent in the gastrointestinal tract (GIT) of poultry, which could reduce the population of these bacteria and disorders of the intestine [18]. However, an excess of tannins can provoke metabolic disturbances leading to an antinutritional influence [19], such as inhibiting the absorption of iron and sulphur-containing amino acids causing anaemia and depression of growth, respectively.

Moreover, the incorporation of the leaves of *M. citrifolia* and *P. guajava* into the mixed powder rich in secondary metabolites with antioxidant and anti-inflammatory effects [10, 20] could influence the better use of the nutrients with lower feed intake, mainly during the first moments of the bird's life that they are often exposed to different stressful conditions and presence of pathogenic agents, which increase the production of free radicals and postprandial inflammatory process and decrease the growth performance [21]. Taking into account our results, we can affirm that the secondary metabolites in adequate concentration have a direct effect on poultry development [18].

On the contrary, several studies have reported that medicinal plants (with emphasis on the leaves) in diets enhance the growth performance in birds. In the same way, Salami et al. [22] reported that use of medicinal herbs in the diets of broiler improved the FCR values at the end of the trial. Also, Buchanan et al. [23] stated that broiler chickens fed diets having plant extract blends had minimum FCR and had increased weight gain and breast yield.

In spite of the phytochemical components on mixed powder, the dietary inclusion of 0.5% has no change in the total and apparent ileal digestibility of both nitrogen (N) and dry matter (DM) in broilers. Our study did not indicate an adverse effect on nutrient digestibility because Savón et al. [19] have referred that the secondary metabolites as antinutritional factors affect the digestibility and therefore growth performance. Results of Hernández et al. [24] and Hassan et al. [25] showed an increase in digestibility coefficient of DM, ether extract, CP, and organic material with the dietary supplementation of Labiatae plants (*Salvia officinalis*, *Rosmarinus officinalis*, and *Thymus vulgaris*) and artichoke extracts in broilers, respectively. In our study, one reason for the nonpositive effect of the inclusion of mixed powder of medicinal plants in some productive indicators up to 21 days would be the controlled environment of the experiment and the use of the crude powder instead of the leaf extract. However, more studies are needed to corroborate this hypothesis.

At present, the scientific community is discussing the main mode of action of natural products. Some researches have reported that it is due to its anti-inflammatory properties that decrease inflammation of the small intestine and increase the absorption of nutrients [26], and other studies have shown that the inclusion of medicinal herbs reduces the proliferation of pathogenic bacteria in the GIT, which increases intestinal health and favours greater digestibility of nutrients and animal response [3, 25]. Similarly, other researchers using the same medicinal plants as our experiment have found both responses *in vitro* and *in vivo*. Although it was observed that T1 increased numerically the AMEn, maybe a higher concentration of this mixed powder of medicinal plant leaves may be necessary to modify these indicators in broilers.

The level of serum antibodies is currently an important indicator to know the effect of a new natural product on immune response in experimental animals [27]. An increased immunoglobulin concentration has been associated with a benefit in the immune status, and IgG and IgA are the main immunoglobulins protecting against pathogenic microorganisms [28], mainly to intestinal level. Our results showed that a small amount of powder inclusion (0.5%) exerts a beneficial immune effect, through an increase in the IgG concentration and with a synthesis of appropriate immune cells, without diminishing the growth performance [29].

IgG is one of the main defence barriers during the bacterial attack in the gastrointestinal tract (GIT), and the early proliferation of this cell is essential to improve the feed efficiency in these animals [30]. Thus, they are less exposed to bacterial attack and hence, to intestinal disorders. In addition, IgG is the main active antibody against infections due to its high degree of specificity and its bactericidal action and also is most abundant in the serum; compared to IgA and IgM, IgG has a longer life and can shuttle between serum and endothelium surfaces [31, 32]. Although the IgA concentration was higher compared with T0, this did not show significant difference ($P > 0.05$) among treatments. However, an increase of this immunoglobulin in a controlled environment shows that this powder can favour immunity in young birds, perhaps with higher efficacy under stress conditions. According to Merino-Guzmán [28], sIgA inhibits the uncontrolled absorption of macromolecules or the binding of allergens to target cells of the mucosa, has inflammatory and bactericidal effects, and improves nonspecific immunological defence mechanisms.

In general, this result indicates that the dietary inclusion of the mixed powder may assist a fast response of the immune system. In this sense, Tajodini et al. [33] using artichoke (*Cynara scolymus*) powder in broiler diets found that this product significantly increased antibody serum levels, resulting in a higher activity of immune system.

5. Conclusions

The experimental treatment in the diet decreased the FI and FCR during week 2 and throughout the trial period (1–21 days) and increased the IgG concentration compared with the control treatment (T0), without affecting the nutrient digestibility and IgA concentration. Taking into account the results, we recommend the inclusion of 0.50% of this mixed powder of medicinal plant leaves (60% *Anacardium occidentale*, 20% *Psidium guajava*, and 20% *Morinda citrifolia*) in broiler diets.

Conflicts of Interest

The authors declare that there are no conflicts of interest regarding the publication of this article.

Authors' Contributions

Roisbel Aroche and Yordan Martínez contributed equally to this manuscript.

Acknowledgments

This study was funded by the China Scholarship Council (201708430008), the National Natural Science Foundation of China (31402092), the Scientific Research Fund of Hunan Provincial Education Department (17K043), the Hunan Provincial Science and Technology Department (2017NK2322 and 2016NK2103), and the Open Project Program of State Key Laboratory of Food Science and Technology, Nanchang University (SKLF-KF-201416).

References

[1] S. Jarić, M. Mačukanović-Jocić, L. Djurdjević et al., "An ethnobotanical survey of traditionally used plants on Suva planina mountain (south-eastern Serbia)," *Journal of Ethnopharmacology*, vol. 175, pp. 93–108, 2015.

[2] P. R. Chowdhury, J. McKinnon, E. Wyrsch, J. M. Hammond, I. G. Charles, and S. P. Djordjevic, "Genomic interplay in bacterial communities: implications for growth promoting practices in animal husbandry," *Frontiers in Microbiology*, vol. 5, no. 394, pp. 1–13, 2014.

[3] Z. Zdunczyk, R. Gruzauskas, J. Juskiewicz et al., "Growth performance, gastrointestinal tract responses, and meat characteristics of broiler chickens fed a diet containing the natural alkaloid sanguinarine from *Macleaya cordata*," *Journal of Applied Poultry Research*, vol. 19, no. 4, pp. 393–400, 2010.

[4] J. Gong, F. Yin, Y. Hou, and Y. Yin, "Chinese herbs as alternatives to antibiotics in feed for swine and poultry production: potential and challenges in application," *Canadian Journal of Animal Science*, vol. 94, no. 2, pp. 223–241, 2013.

[5] Z. Zeng, S. Zhang, H. Wang, and X. Piao, "Essential oil and aromatic plants as feed additives in non-ruminant nutrition: a review," *Journal of Animal Science and Biotechnology*, vol. 6, no. 1, p. 7, 2015.

[6] V. E. Okpashi, B. P. R. Bayim, and M. Obi-Abang, "Comparative effects of some medicinal plants: *Anacardium occidentale, Eucalyptus globulus, Psidium guajava*, and *Xylopia aethiopica* extracts in alloxan-induced diabetic male wistar albino rats," *Biochemistry Research International*, vol. 2014, Article ID 203051, 13 pages, 2014.

[7] Y. Martínez, O. Martínez, G. Liu et al., "Effect of dietary supplementation with *Anacardium occidentale* on growth performance and immune and visceral organ weights in replacement laying pullets," *Journal of Food, Agriculture and Environment*, vol. 11, no. 3-4, pp. 1352–1357, 2013.

[8] M. R. V. Fernandes, A. L. T. Días, R. R. Carvalho, C. R. F. Souza, and W. P. Oliveira, "Antioxidant and antimicrobial activities of *Psidium guajava* L. spray dried extracts," *Industrial Crops and Products*, vol. 60, pp. 39–44, 2014.

[9] W. K. Oh, C. H. Lee, M. S. Lee et al., "Antidiabetic effects of extracts from *Psidium guajava*," *Journal of Ethnopharmacology*, vol. 96, pp. 411–415, 2005.

[10] M. Ali, M. Kenganora, and S. N. Manjula, "Health benefits of *Morinda citrifolia* (Noni): a review," *Pharmacognosy Journal*, vol. 8, no. 4, pp. 321–334, 2016.

[11] J. Sunder, T. Sujatha, and A. Kundu, "Effect of *Morinda citrifolia* in growth, production and immunomodulatory properties in livestock and poultry: a review," *Journal of Experimental Biology and Agricultural Sciences*, vol. 2016, pp. 249–265, 2016.

[12] Y. Martínez, O. Martínez, E. Olmos, S. Siza, and C. Betancur, "Nutraceutical effect of *Anacardium occidentale* (AO) in diets of replacement laying pullets," *Revista MVZ Córdoba*, vol. 17, no. 3, pp. 3125–3132, 2012.

[13] NRC, *Nutrient Requirements of Poultry: 9th Revision Edited*, National Academic Press, Washington, DC, USA, 1994.

[14] Y. M. Aguilar, J. C. Becerra, R. R. Bertot, J. C. Peláez, G. Liu, and C. B. Hurtado, "Growth performance, carcass traits and lipid profile of broiler chicks fed with an exogenous emulsifier

and increasing levels of energy provided by palm oil," *Journal of Food, Agriculture & Environment*, vol. 11, no. 1, pp. 629–633, 2013.
[15] AOAC, *Official Methods of Analysis of AOAC: Edition 18*, Association of Official Analytical Chemists, Gaithersburg, MD, USA, 2000.
[16] M. C. E. Lomer, R. P. H. Thompson, J. Commisso, C. L. Keen, and J. J. Powell, "Determination of titanium dioxide in foods using inductively coupled plasma optical emission spectrometry," *Analyst*, vol. 125, no. 12, pp. 2339–2343, 2000.
[17] M. R. Perez-Gregorio, N. Mateus, and V. De Freitas, "Rapid screening and identification of new soluble tannin–salivary protein aggregates in saliva by mass spectrometry (MALDI-TOF-TOF and FIA-ESI-MS)," *Langmuir*, vol. 30, no. 28, pp. 8528–8537, 2014.
[18] S. Khalaji, M. Zaghari, K. H. Hatami, S. Hedari-Dastjerdi, L. Lotfi, and H. Nazarian, "Black cumin seeds, *Artemisia* leaves (*Artemisia sieberi*), and *Camellia* L. plant extract as phytogenic products in broiler diets and their effects on performance, blood constituents, immunity, and cecal microbial population," *Poultry Science*, vol. 90, no. 11, pp. 2500–2510, 2011.
[19] L. Savón, I. Scull, and M. Martinez, "Integral foliage meal for poultry feeding. Chemical composition, physical properties and phytochemical screening," *Cuban Journal of Agricultural Science*, vol. 41, pp. 359–369, 2007.
[20] H. D. S. M. Perera, J. K. R. R. Samarasekera, S. M. Handunnetti, and O. V. D. S. J. Weerasena, "*In vitro* anti-inflammatory and anti-oxidant activities of Sri Lankan medicinal plants," *Industrial Crops and Products*, vol. 94, pp. 610–620, 2016.
[21] K. W. Bai, Q. Huang, J. F. Zhang, J. T. He, L. L. Zhang, and T. Wang, "Supplemental effects of probiotic *Bacillus subtilis* fmbJ on growth performance, antioxidant capacity, and meat quality of broiler chickens," *Poultry Science*, vol. 96, no. 1, pp. 74–82, 2016.
[22] S. A. Salami, M. A. Majoka, S. Saha, A. Garber, and J. F. Gabarrou, "Efficacy of dietary antioxidants on broiler oxidative stress, performance and meat quality: science and market," *Avian Biology Research*, vol. 8, no. 2, pp. 65–78, 2015.
[23] N. P. Buchanan, J. M. Hott, S. E. Cutlip, A. L. Rack, A. Asamer, and J. S. Moritz, "The effects of a natural antibiotic alternative and a natural growth promoter feed additive on broiler performance and carcass quality," *Journal of Applied Poultry Research*, vol. 17, no. 2, pp. 202–210, 2008.
[24] F. Hernández, J. Madrid, V. García, J. Orengo, and M. D. Megías, "Influence of two plant extracts on broilers performance, digestibility, and digestive organ size," *Poultry Science*, vol. 83, no. 2, pp. 169–174, 2004.
[25] H. M. A. Hassan, A.W. Youssef, H. M. Ali, and M. A. Mohamed, "Adding phytogenic material and/or organic acids to broiler diets: effect on performance, nutrient digestibility and net profit," *Asian Journal of Poultry Science*, vol. 9, no. 2, pp. 97–105, 2015.
[26] Y. Xiong, X. Tang, Q. Meng, and H. Zhang, "Differential expression analysis of the broiler tracheal proteins responsible for the immune response and muscle contraction induced by high concentration of ammonia using iTRAQ-coupled 2D LC-MS/MS," *Science China Life Sciences*, vol. 59, no. 11, pp. 1166–1176, 2016.
[27] M. Iser, Y. Martínez, H. Ni et al., "Effects of *Agave fourcroydes* powder as a dietary supplement on growth performance, gut morphology, concentration of IgG and hematology parameters of broiler rabbits," *BioMed Research International*, vol. 2016, Article ID 3414319, 7 pages, 2016.
[28] R. Merino-Guzmán, J. D. Latorre, R. Delgado et al., "Comparison of total immunoglobulin A levels in different samples in Leghorn and broiler chickens," *Asian Pacific Journal of Tropical Biomedicine*, vol. 7, no. 2, pp. 116–120, 2017.
[29] M. L. Moraes, A. M. L. Ribeiro, E. Santin, and K. C. Klasing, "Effects of conjugated linoleic acid and lutein on the growth performance and immune response of broiler chickens," *Poultry Science*, vol. 95, no. 2, pp. 237–246, 2015.
[30] L. Antoni, S. Nuding, J. Wehkamp, and E. F. Stange, "Intestinal barrier in inflammatory bowel disease," *World Journal of Gastroenterology*, vol. 20, no. 5, p. 1165, 2014.
[31] C. Perez-Carbajal, D. Caldwell, M. Farnell et al., "Immune response of broiler chickens fed different levels of arginine and vitamin E to a coccidiosis vaccine and Eimeria challenge," *Poultry Science*, vol. 89, no. 9, pp. 1870–1877, 2010.
[32] R. Yang and M. C. Hung, "The role of T-cell immunoglobulin mucin-3 and its ligand galectin-9 in antitumor immunity and cancer immunotherapy," *Science China Life Sciences*, vol. 60, no. 10, pp. 1058–1064, 2017.
[33] M. Tajodini, F. Samadi, S. R. Hashemi, S. Hassani, and M. Shams-Shargh, "Effect of different levels of Artichoke (*Cynara scolymus* L.) powder on the performance and immune response of broiler chickens," *International Journal of AgriScience*, vol. 4, no. 1, pp. 66–73, 2014.

7

New Caffeic Acid Phenylethyl Ester Analogs Bearing Substituted Triazole: Synthesis and Structure-Activity Relationship Study towards 5-Lipoxygenase Inhibition

Pierre-Philipe Roy, Diene Faye, Sébastien Blanchard, Marc Cormier, Jérémie A. Doiron, Marc E. Surette, and Mohamed Touaibia

Department of Chemistry and Biochemistry, Université de Moncton, Moncton, NB, Canada E1A 3E9

Correspondence should be addressed to Mohamed Touaibia; mohamed.touaibia@umoncton.ca

Academic Editor: Naoki Toyooka

Leukotrienes are biosynthesized by the conversion of arachidonic acid by 5-Lipoxygenase and play a key role in many inflammatory disorders. Inspired by caffeic acid phenylethyl ester (CAPE) (**2**) and an analog carrying a triazole substituted by cinnamoyl and 5-LO inhibitors recently reported by our team, sixteen new CAPE analogs bearing substituted triazole were synthesized by copper catalyzed Huisgen 1,3-dipolar cycloaddition. Compound **10e**, an analog bearing *p*-CF3 phenethyl substituted triazole, was equivalent to CAPE (**2**) but clearly surpassed Zileuton (**2**), the only approved 5-LO inhibitor. Substitution of the phenethyl moiety by cyclohexylethyl, as with **12g**, clearly increased 5-LO inhibition which confirms the importance of hydrophobic interactions. Molecular docking revealed new hydrogen bonds and π-π interactions between the enzyme and some of the investigated compounds. Overall, this work highlights the relevance of exploring polyphenolic compounds as leukotrienes biosynthesis inhibitors.

1. Introduction

Leukotrienes (LTs) are a class of lipid mediators implicated in inflammatory responses [1]. In recent years, many studies have developed new pharmacological agent that influence the biosynthesis or action of LTs [2]. Despite their implication in inflammatory and allergic responses, LTs are also involved in many inflammatory diseases like asthma, arthritis, irritable bowel syndrome, and even cancer [3, 4].

5-Lipoxygenase (5-LO) is an enzyme that plays a crucial role for the formation of LTs as it catalyzes the first two steps of the conversion of arachidonic acid to the bioactive LTs [5]. Therefore, 5-LO is the preferred target for pharmaceutical control of LTs biosynthesis [2]. Moreover, the inhibition of 5-LO is a validated target for the treatment of asthma [6, 7]. 5-LO inhibitors have also shown promising results in blocking the proliferation of several tumors cell lines [8, 9].

Recently, caffeic acid phenethyl ester (CAPE, **1**) (Figure 1) was found to be a potent 5-LO inhibitor [10–12] by our research group. Many studies reported by our laboratory have also shown the potential of CAPE and its analogs as potential anti-inflammatory compounds.

Contrary to the belief that polyphenols inhibit 5-LO exclusively through their antioxidant properties, our previous results clearly show that the presence or absence of specific functional groups influences 5-LO inhibition independent of their impact on antioxidant activity [10, 11]. Structural modifications which resulted from changes in the ester structure showed significantly improved inhibition of 5-LO when compared to Zileuton (**2**) [13], the only clinically available 5-LO inhibitor. However, Zileuton (**2**) was shown to have hepatotoxic effects, low potency, and short half-life [14]; therefore research continues for the development of new 5-LO inhibitors.

In a previous work, an analog of caffeic acid bearing a triazole (compound **3**, Figure 1) emerged as the best inhibitor of a series of analogs with 1 to 6 caffeoyl moieties [15]. Further structure-activity relationship was performed later where

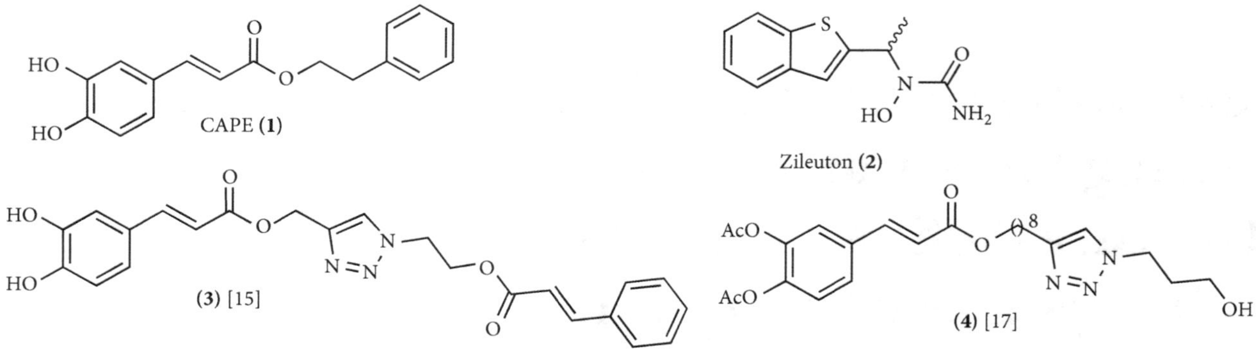

FIGURE 1: Structures of CAPE (**1**), Zileuton (**2**), compound (**3**) [15], and compound **4** [13].

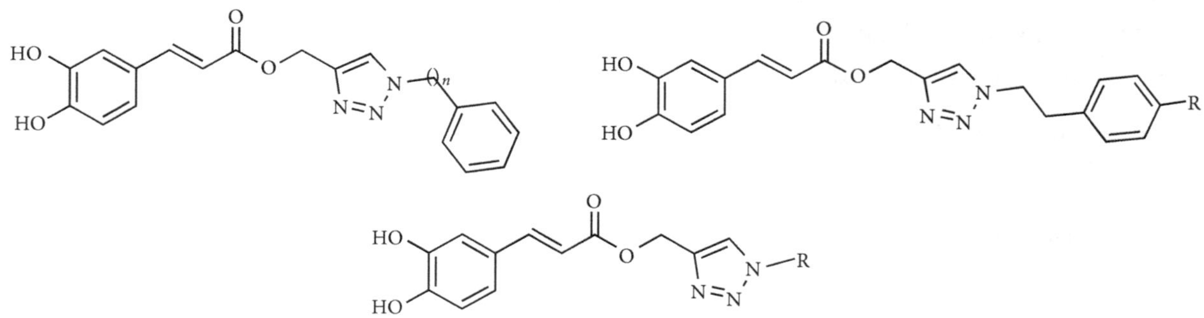

FIGURE 2: Caffeic acid phenylethyl ester analogs bearing substituted triazole designed in this study.

different triazoles containing cinnamic acid and caffeic acid analogs were investigated for 5-LO inhibition [16]. De Lucia also synthesized additional analogs [13] related to compound **3**. Interestingly, they found that the length of the linker was critical for the activity of the molecule; a longer linker had better inhibition capacity [13]. While the importance of the linker length is now relatively established, it is not clear if the triazole needs to be far away from the caffeic acid moiety and which type of substitution on the triazole moiety will impact the 5-LO inhibition. It is also noteworthy that the mechanism of inhibition of 5-LO by these CAPE analogs bearing substituted triazole analogs is not well established [16, 17]. Moreover, structure-activity relationship studies (SAR) are required to learn more about the potential mode of action of these molecules.

In the present study, we describe the design and synthesis of new caffeic acid phenylethyl ester analogs bearing substituted triazole derived from compound **3** (Figure 2). With these analogs, the effect of the substitution of the triazole moiety by a phenyl as well as the linker length between the latter and triazole was investigated. The substitution of the phenyl with electron-withdrawing and electron-donating groups was also explored.

Finally, the substitution of the triazole with various allyl groups of various sizes was also investigated. After anti-LTs activity screening in a HEK293 cell model, 5-LO inhibition potential of selected compounds was determined in human polymorphonuclear leukocytes (PMNL). Moreover, a molecular docking study was conducted to shine a new light on the mechanism of inhibition of these compounds.

2. Results and Discussion

2.1. Chemistry. To investigate the structure-activity relationship of compound **3**, closely related analogs were synthesized. We examined the molecular interaction that would have an effect on 5-LO inhibition by linking with the Huisgen 1,3-dipolar cycloaddition (commonly known as "click reaction") reaction different molecular groups on the caffeoyl moiety. All analogs were synthesized by an efficient sequential procedure that directly links the azides, which can be potentially explosive, obtained from the commercially available organic halides to the alkyne moiety of the caffeic propargylic ester.

As shown in Scheme 1, a first series of analogs was synthesized to examine the effect of the length of the linker between the triazole core and a nonsubstituted phenyl moiety.

Sequential synthesis of the appropriate organic azide synthesized from the commercially available organic halides (**7a–c**) followed by the copper catalyzed 1,3-dipolar cycloaddition of these azides with (**6**) provided analogs **8a–8c** with one to three methylenes between the triazole and the nonsubstituted phenyl moiety in moderate yield (Scheme 1).

To investigate the effect of substitution of the phenyl of series 1 with electron-withdrawing and electron-donating groups, new analogs were designed as shown in Scheme 2. As compounds **8a-b** were essentially equipotent with Zileuton (**2**) and compound (**3**) (see results of LT inhibition below), analogs of **8b** with substituted phenyl were designed and synthesized. It should also be mentioned that the accessibility of the various halogenated derivatives with two methylenes allowed us to make this series more easily

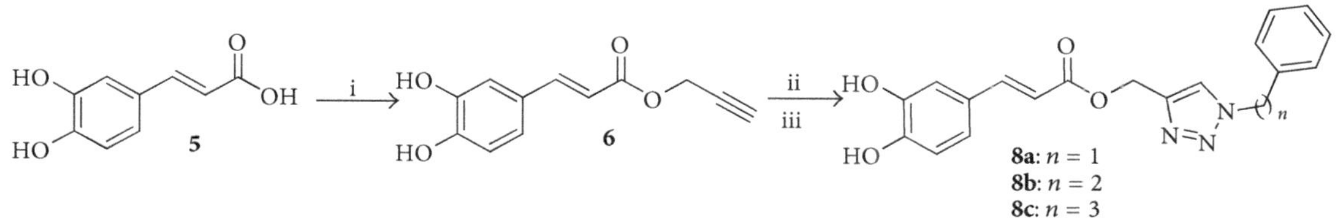

SCHEME 1: Synthesis of series 1: (i): propargyl bromide, Na_2CO_3, KI, HMPA, 0°C-rt, overnight, 43%; (ii): appropriate aryl bromide (**7a–c**), NaN_3, DMSO, rt, overnight; (iii): **6**, $CuSO_4 \cdot 5H_2O$, sodium ascorbate, H_2O, rt, overnight (**8a**: 42%; **8b**: 57%; **8c**: 53%).

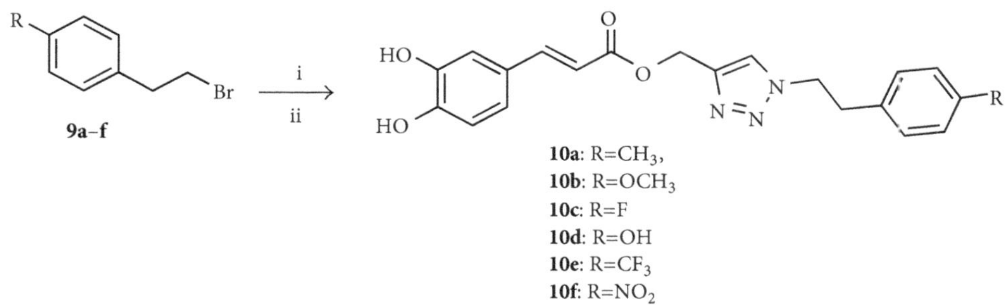

SCHEME 2: Synthesis of series 2: (i): NaN_3, DMSO, rt, overnight; (ii) **6**, $CuSO_4 \cdot 5H_2O$, sodium ascorbate, H_2O, rt, overnight (**10a**: 65%; **10b**: 65%; **10c**: 63%; **10d**: 47%; **10e**: 68%; **10f**: 53%).

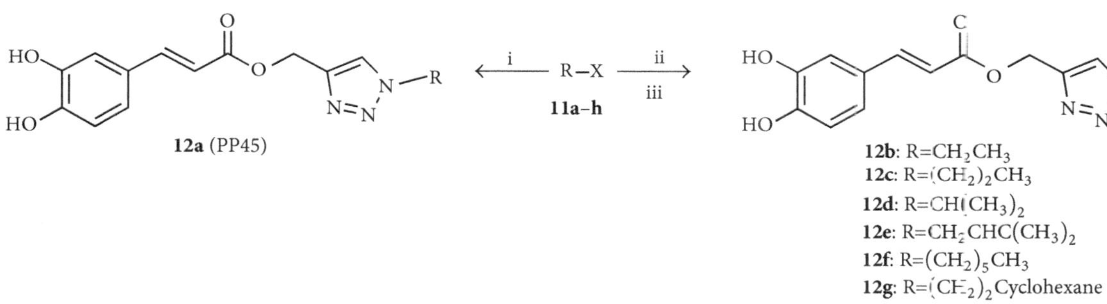

SCHEME 3: Synthesis of series 3: (i): **6**, NaN_3, MeI, CuI, acetone/H_2O (1/1), rt, ultrasound, overnight (**12a**: 24%). (ii): NaN_3, DMSO, rt, overnight; (iii): **6**, $CuSO_4 \cdot 5H_2O$, Sodium ascorbate, H_2O, rt, overnight (**12b**: 59%; **12c**: 55%; **12d**: 54%; **12e**: 63%; **12f**: 22%; **12g**: 51%).

following our sequential procedure (Scheme 2). Copper catalyzed 1,3-dipolar cycloaddition of substituted phenethyl azides obtained from the corresponding bromide analogs (**9a–f**) with compound **6** provided analogs **10a–f** with electron-withdrawing and electron-donating substituents on the phenyl moiety (Scheme 2).

To investigate the effect of the presence of hydrophobic carbon chains, a third series of analogs was designed. In this series, the triazole was substituted by various alkyls, from the simpler (methyl) to branched alkyls. As shown in Scheme 3 and following our sequential procedure, analogs (**12b–g**) containing various alkyl groups linked to the triazole were synthesized. Compound **12a** could not be obtained using the same method. This is probably due to the gaseous nature of methyl azide. Therefore, another method was used to obtain the methyl-substituted triazole analog (**12a**). With a one-pot procedure, we were able to obtain the desired methyl-substituted triazole by copper catalyzed 1,3-dipolar cycloaddition of the methyl azide, formed in situ by the substitution reaction of methyl iodine and sodium azide, with the caffeic propargylic ester **6** (Scheme 3).

2.2. Biological Evaluation

2.2.1. Inhibition of 5-LO Products Synthesis in Whole HEK293 Cells. Inhibition of leukotriene biosynthesis was assayed in intact HEK293 cells that are stably transfected with human 5-LO. This model serves as a highly reproducible model of 5-LO product biosynthesis in which compounds can be preliminarily screened for anti-LTs activity before moving on to more complex systems [11]. All compounds were assayed at 1 μM, including known inhibitors CAPE (**1**) and Zileuton (**2**) whose inhibitory potencies were used as reference points (Figure 3). As shown in Figure 3, analogs **8a–c** having one- to three-methylene linker between the triazole and the nonsubstituted phenyl moiety are equivalent to Zileuton but remain less active than CAPE (**1**). Analog **8c**, with three-methylene linker, is the most active of the three compounds with an inhibition at around 50%, which clearly surpasses Zileuton (**2**). The small difference in lipophilicity between these molecules can probably explain the small difference in activity.

In order to investigate the effect of substitution of the **8b** benzylic ring, analogs with electron-withdrawing

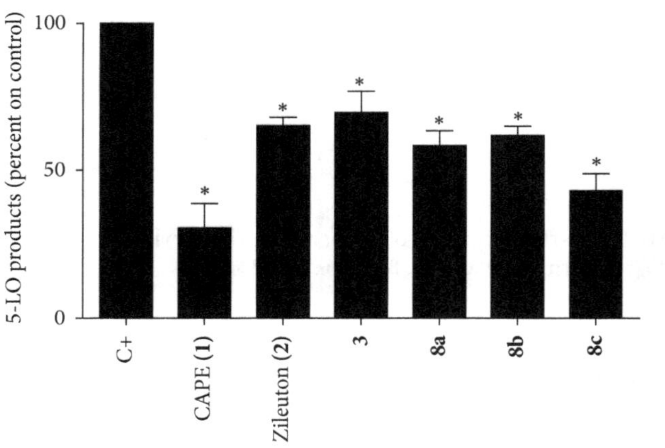

FIGURE 3: Inhibition of 5-LO products synthesis by analogs (**8a–c**) in whole HEK293 cells. Values are the means ± SEM of three independent experiments, each performed in quadruplicate; *different from control (diluent) as determined by one-way ANOVA with Dunnett's multiple comparison test ($p < 0.05$).

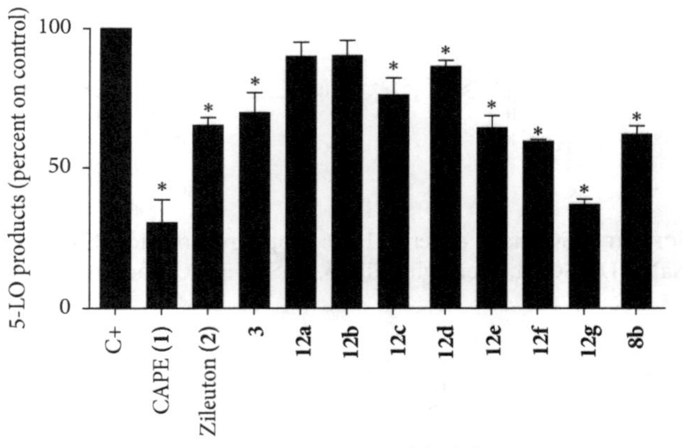

FIGURE 5: Inhibition of 5-LO products synthesis by **12a–c** and **8b** in whole HEK293 cells. Values are the means ± SEM of three independent experiments, each performed in quadruplicate; *different from control (diluent) as determined by one-way ANOVA with Dunnett's multiple comparison test ($p < 0.05$).

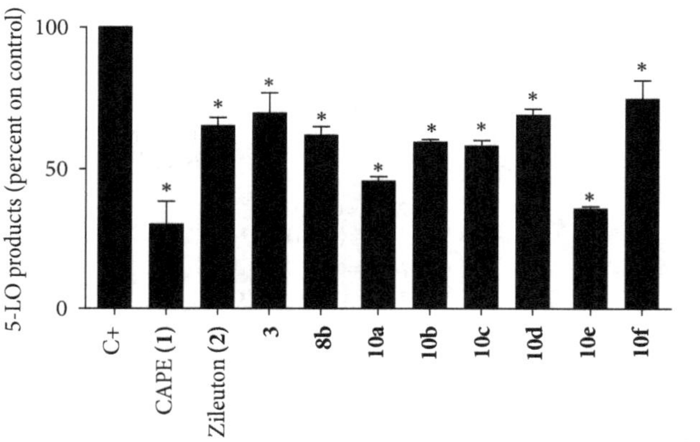

FIGURE 4: Inhibition of 5-LO products synthesis by **10a–f** in whole HEK293 cells. Values are the means ± SEM of three independent experiments, each performed in quadruplicate; *different from control (diluent) as determined by one-way ANOVA with Dunnett's multiple comparison test ($p < 0.05$).

and electron-donating substituents were tested as shown in Figure 4. All analogs were either equivalent to CAPE (**1**) and analog **3** or significantly more active than Zileuton (**2**). As shown in Figure 4, the substitution of the phenyl in the para position with methyl, a weak donating substituent, seems to have a positive impact on 5-LO inhibition which was increased by almost 20%. The effect of the hydrophobic substituent seems to be decisive for increasing the inhibition of 5-LO.

The introduction in the para position of the phenyl moiety of moderate (-OCH_3 and -F) or even strong (-OH) donating groups appears to have a moderate effect on activity since the inhibitory activity of **10b–d** is equivalent to that of **8b**. On the other hand, the substitution of phenyl by weak electron-withdrawing substituent seems to be very favorable for 5-LO inhibition; the induction of -CF_3 analog increases the inhibitory capacity of **8b** by almost 30%. Compound **10e** is equivalent to CAPE (**1**) but clearly surpasses the Zileuton (**2**) (Figure 4). The difference of lipophilicity is not as important in explaining the difference of 5-LO inhibition of **8b**, **10a,** and **10e**; an electronic effect associated with favorable interactions due to the presence of fluorine atoms can explain this difference.

It would have been supposed that the accentuation of the deactivating effect on phenyl by strong withdrawing substituents such as nitro group could increase the activity even more. However, as shown in Figure 4, the substitution of phenyl with a nitro considerably reduces the effect of **10f**, which becomes less active than **10e**. Analog **10f** was even less potent than the unsubstituted analog **8b** (Figure 4).

In order to investigate the presence of a hydrophobic group, hydrophobic chains of different sizes and degrees of substitution were tested in the third series. As shown in Figure 5, some molecules were significantly less active than CAPE (**1**), Zileuton (**2**), and analog **3**. Analogs **12e** and **12f** were equivalent to Zileuton (**2**) and analog **3**. Analog **12g** was the most active of the whole while being also much more active than the Zileuton and the analog **3**.

Substitution of triazole with methyl, as in **12a**, or two- to three-carbon alkyl chain (**12b** and **12c**), was not conducive to good inhibition of 5-LO. Substitution of triazole with a more substituted three-carbon alkyl chain such as an isopropyl (analog **12d**) or a substituted alkyl chain with a less flexible end, such as 3-methylbut-2-enyl (analog **12e**), does not seem to be favorable for 5-LO inhibition. Substitution of the triazole with a hexyl chain slightly increases the 5-LO inhibition. As shown in Figure 5, analog **12f**, bearing the hexyl chain, was more active than **12a** but equivalent to **8b**, the analog with triazole substituted with phenethyl. Substitution of the triazole with cyclohexylethyl was very favorable for the inhibition of 5-LO as shown in Figure 5.

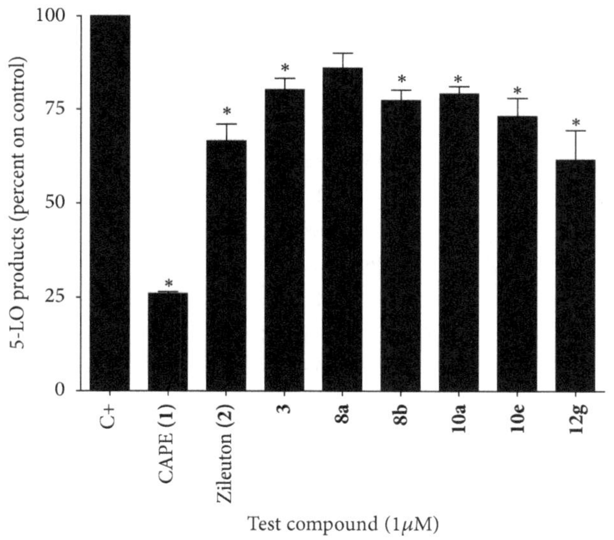

FIGURE 6: Inhibition of 5-LO products synthesis by analogs **8a-b**, **10a**, **10e**, and **12g** in stimulated human PMNL. Values are the means ± SEM of three independent experiments, each performed in quadruplicate; *different from control (diluent) as determined by one-way ANOVA with Dunnett's multiple comparison test ($p < 0.05$).

Favorable interactions of the cyclohexylethyl moiety can explain this good activity since the molecule was significantly more active than Zileuton (**2**). Analog **12g** was also more active than analogs **12f** and **8b** having substituted triazole with hexyl or phenethyl, respectively (Figure 5). The difference in lipophilicity can not explain this difference since the estimated Logp of the three molecules (**8b**, **12f**, and **12g**) was very close.

2.2.2. Inhibition of 5-LO Products in Simulated Human PMNL. To further probe inhibitory activity of lead compounds selected from preliminary HEK293 screenings (Figures 3, 4, and 5), 5-LO inhibition assays were undertaken in stimulated human PMNL in the presence of 1 μM of inhibitors. 5-LO is highly expressed in PMNL and these cells are important physiological producers of LTB_4 [11, 18].

As shown in Figure 6, analogs **8a** and **8b** having a triazole substituted by benzyl or ethynylphenyl were less active than CAPE (**1**) but were equivalent to Zileuton (**2**) and analog **3**. Bearing an ethenylcyclohexyl-substituted triazole, **12g** was the best inhibitor of the selected molecules. **12g** was more active than its analogs **3**, **10a**, and **10b** which have methyl- or trifluoromethyl-substituted triazole, respectively.

Analog **12g** was also equivalent to Zileuton (**2**) but remains less active than CAPE (**1**); the favorable effect of the presence of the cyclohexylethyl moiety seems to be confirmed in the PMNL test. The difference in inhibitory activity of our molecules in HEK293 cells and in stimulated human PMNL may reflect the availability or cell permeation of our inhibitors. The substitution of the terminal phenyl group in CAPE (**1**) with monosubstituted triazoles leads to a loss of activity in PMNL at 1μM. Moreover, while **10e** is more lipophilic than **12g**, the three fluorine atoms and the aromatic ring may form hydrogen bonds or π-π interactions that could diminish the availability and/or cell permeability. Furthermore, Van der Waals interaction of the cyclohexylethyl moiety of **12g** seems to be favorable for 5-LO inhibition as **12g** was the best inhibitor and was equipotent to Zileuton (**2**).

2.3. Molecular Docking. All designed and synthesized compounds were docked to 5-LO in this study. From the obtained docking results, the binding energy was determined for each ligand as well as hydrogen bonds, interactions, and coordination (Table 1). Molecular docking was performed on a modified 5-LO protein (PDB code: 3O8Y).

Analogs **8b-c**, **10a–d**, and **12g** were almost overlaid with slight shifts in position for the R group. The catechol group was positioned near the iron atom and its coordinating residues while the other extremity of the ligand was located near the end of the cavity, near His600. This group seems to have the best classed analogs (Table 1).

As shown in Figure 7, it is interesting to see the proximity/interaction of the cyclohexyl moiety of analog **12g** with several amino acids of the protein, in particular with Tyr181 Al603 and Asn425. This proximity/interaction can explain the favorable effect of the presence of this group observed in the inhibition tests of 5-LO with **12g** in HEK293 and PMNL cells (Figures 5 and 6). Comparison of the results of the modeling of analogs **12g** and **8b** clearly shows that hydrophobic interactions and the cyclohexyl flexibility appear to be crucial for 5-LO inhibition.

Analogs **8a**, **10e-f**, and **12e-f** differ, in some minor way, from the previous group. For **10e-f**, the catechol was perpendicular compared to the previous molecules. **12f** takes a more direct route due to its short length.

Analogs **12b–d** were overlaid; only the R group differed slightly. This group has the characteristic of being inverted compared to the previous molecules, having the catechol section near the end of the cavity while the R group is near the iron atom. It does keep the same general overlap even if inverted. This group seems to have the least stable ligand.

CAPE (**1**) analogs **3** and **4** differ in their positioning. In general, they take similar positions to the previous group. Analog **4**, due to its long length, must contort much more to fit inside the cavity and reaches the furthest it can in both extremities. It also presents the triazole ring (nitrogen atoms side) to the iron atom compared to the R group for previous ligands. CAPE's (**1**) shorter length and position make it so that it cannot reach the iron atom. Analog **3** presents its second ester to the iron atom instead of presenting a triazole.

Analog **12a** was shifted slightly deeper inside the cavity due to its shorter length, meaning that the R group was placed near the end of the cavity and the catechol group does not reach the section of the cavity containing the iron atom. Its position resembles more Zileuton (**2**) than the other analogs. Similarly, both Zileuton (**2**) enantiomers occupy the end of the cavity and do not reach the iron containing section of the cavity.

In most cases, the tested ligands seem to fill most of the cavity. **12a** seems to leave a slightly more open space than the

Table 1: Molecular modeling data*.

Molecule	Affinity (kcal/mol)	Hydrogen bonds	π-π interactions	Coordination
10e	−9.1	Asn554	Phe177, His367 × 2	Fe2 × 2
10c	−8.6	—	His367, Phe421	Fe2
CAPE (**1**)	−8.5	—	His372	—
8b	−8.4	—	His367, Phe421	Fe2
8c	−8.4	—	His367, Phe421	Fe2
4 [17]	−8.4	—	His372	Fe2 × 2
10f	−8.4	Leu607	Phe177, His367 × 2	Fe2 × 2
10a	−8.4	—	His367, Phe421	Fe2
12g	−8.4	—	His367, Phe421	Fe2
12a	−8.3	—	His372	—
10d	−8.2	Ile673	His367, Phe421	Fe2
8a	−8.1	Tyr181	His600	Fe2
10b	−8.1	—	His367, Phe421	Fe2
12b	−8.0	—	His367, Phe421	—
12d	−8.0	—	His367	—
12e	−8.0	Tyr181	Phe421	—
12f	−7.9	—	Phe177, His367	Fe2
12c	−7.9	—	His367, Phe421	—
3 [15]	−7.6	—	Phe177	Fe2
(*R*)-Zileuton (**2**)	−6.6	Leu420 × 2	Phe421	—
(*S*)-Zileuton (**2**)	−6.5	—	—	—

*Molecular docking was performed on a modified 5-LO protein (PDB code: 3O8Y). Figure 7 showing the double bond in "*Z*"-form is a result of the software used to generate the 2D image, LigPlot. The file given by the molecular docking software has the "*E*"-form conserved.

others while analog **4** seems to fill the cavity particularly well. Zileuton (**2**) fills the end of the cavity while leaving the area near the iron atom open due to its shorter length.

In agreement with what was previously said, Zileuton (**2**) and the first series have similar affinity, with the first series being favored compared to Zileuton (**2**). For the second series, the tendency seems to also agree except for **10c** and **10f** being slightly more stable than what would be expected when compared to the inhibitory potential. For the third series, the results seem to differ more than the previous series. For analog **12g** and to some extent **12f**, the affinity seems to be less stable than what would be expected.

When looking at the interactions made by the most stable ligands or that have the strongest measured inhibition, it seems that a π-π interaction with either His367 or His372 is helpful. Both are iron coordinating residues [19]. It also seems that not doing a metal coordination with the iron atom is detrimental to both its affinity and measured activity.

3. Conclusion

Despite its undesirable effects, particularly its hepatotoxicity, Zileuton is still the only approved 5-LO inhibitor for clinical use to date. In this study, sixteen new 5-LO inhibitors derived from CAPE (**1**) and compound **3**, recently reported by our team, were synthesized, characterized, and tested for their potential to inhibit 5-LO in both HEK293 and stimulated human PMNL cells. Analog **12g**, bearing an ethenylcyclohexyl-substituted triazole, was the best inhibitor of all the series. Analog **12g** was also equipotent to Zileuton (**2**) but remains less active than CAPE (**1**). The favorable effect of the presence of the cyclohexylethyl moiety seems to be confirmed in the PMNL test. Hydrophobic interactions appear to be crucial for good inhibition of 5-LO as shown by compound **12g**. Overall, this work highlights novel compounds with interesting properties against 5-LO and further exemplifies the likely importance of exploring the impact of polyphenolic compounds as leukotriene biosynthesis inhibitors.

4. Experimental Section

4.1. Methods and Materials. All chemicals used were purchased from Aldrich (CA). Purification of compounds was carried out by flash chromatography (CombiFlash®, Separation System SG 100C, ISCO) or by silica gel circular chromatography (chromatotron®, model 7924, Harrison Research). TLC was run on silica gel coated aluminum sheets (SiliaPlate TLC, Silicycle®) with detection by UV light (254 nm, UVS-11, Mineralight® shortwave UV lamp). Melting points were obtained using a MELTEMP (model 1001D) melting point apparatus. FTIR spectra were recorded on a Cary 630 FTIR spectrometer equipped with a single reflection diamond ATR. NMR spectra were recorded on a Bruker® Avance III 400 MHz spectrometer. High-resolution

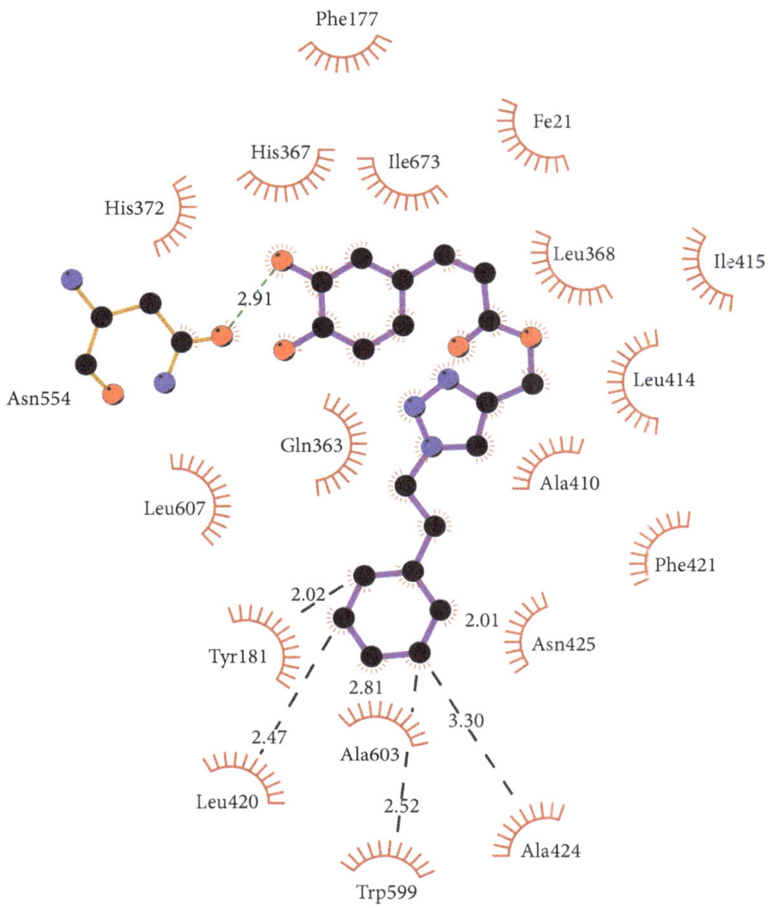

FIGURE 7: Molecular docking results of **12g** displayed in LigPlot.

mass measurements were performed on a Bruker Daltonics' micrOTOF instrument in positive or negative electrospray.

2-Propynyl (E)-3-(3,4-dihydroxyphenyl)acrylate **6**. To a solution of caffeic acid (1 g, 5.61 mmol) in 5 ml of hexamethylphosphoramide (HMPA) under argon atmosphere 1.2 equivalent of Na_2CO_3 (698.5 mg, 6.59 mmol) is added and the solution is stirred at 0°C for an hour. Then a solution of propargyl bromide in 1 ml of HMPA is added to the solution in a span of 30 min. A catalytic amount of KI is added to the solution and the mixture was stirred overnight. The reaction mixture is then poured in ice. The solution is extracted 3 times by ethyl acetate; then the aqueous phase is neutralized by a solution of HCl 1 M and extracted one last time with EtOAc. The combined organic phase is washed three times with brine, dried with $MgSO_4$, and concentrated under vacuum to afford the crude product. Compound **6** was obtained as a white solid is obtained after flash chromatography (EtOAc/Hexane 20 : 80). Yield: 528 mg (43%); R_f = 0.57 (EtOAc/hexane 50 : 50); m.p. = 133–135°C; IR (ν (cm^{-1})): 3503–3479 (C_{ar}O-H), 3270 (C_{sp}-H), 2100 (C≡C), 1683 (C=O), 1629–1440 (C=C), 1290, 1162 (C-O); ^1H NMR (400 MHz, DMSO-d_6, 25°C), δ (ppm): 9.66 (s, 1H, C_{ar}OH), 9.18 (s, 1H, C_{ar}OH), 7.55–7.51 (d, 1H, J = 15.9 Hz, =CHC_{ar}), 7.08–7.02 (m, 2H, H_{ar}), 6.79–6.77 (m, 1H, H_{ar}), 6.33–6.29 (d, 1H, J = 15.9 Hz, =CHCO), 4.80–4.79 (d, 2H, J = 2.44 Hz, OCH_2C), 3.55–3.54 (t, 1H, J = 1.80 Hz, CCH); ^{13}C NMR (100 MHz, DMSO-d_6, 25°C), δ (ppm): 166.23, 149.14, 146.61, 146.05, 125.82, 122.15, 116.20, 115.39, 113.40, 79.21, 77.97, 51.93; HRMS *m/s* calc. for $C_{12}H_{10}O_4$+ (H^+): 219.0652; found: 219.0651.

*4.1.1. General Procedure for the Preparation Compounds **8a–8c**, **10a–10f**, and **12b–12h**.* To a solution of the appropriate halide (**7a–c**, **9a–f or 11b–h**) in 2 ml of DMSO NaN_3 is added and the solution was stirred overnight. Then, **6**, $CuSO_4 \cdot 5H_2O$ (0.2 eq), sodium ascorbate (0.10 eq), and 2 ml of distilled water are added to the solution and the solution was stirred overnight. The solution is diluted in water and the mixture is extracted 3 times with EtOAc. The combined organic phase is washed twice with a solution of NH_4Cl 5 M and then twice with brine. The combined organic phase is dried with $MgSO_4$, to afford the crude product. The purification was performed as stated for each compound.

(1-Benzyl-1H-1,2,3-triazol-4-yl)methyl (E)-3-(3,4-dihydroxyphenyl)acrylate **8a**. Benzyl bromide (**7a**) (54.43 μl, 0.46 mmol), NaN_3 (33.61 mg, 0.52 mmol), **6** (118.59 mg, 0.54 mmol), $CuSO_4 \cdot 5H_2O$ (53.59 mg, 0.21 mmol), and

sodium ascorbate (10.38 mg, 0.052 mmol). A beige solid is obtained after flash chromatography (EtOAc/hexane 50 : 50). Yield: 79.4 mg (42%); R_f = 0.14 (EtOAc/hexane 50 : 50); m.p. = 152-153°C; IR (ν (cm^{-1})): 3300–3000 (C_{ar}O-H), 1687 (C=O), 1630–1443 (C=C), 1280, 1179 (C-O); ^{1}H NMR (4000 MHz, DMSO-d_6, 25°C), δ (ppm): 9.63 (s, 1H, C_{ar}OH), 9.15 (s, 1H, C_{ar}OH), 8.24 (s, 1H, =CHN), 7.51–7.48 (d, 1H, J = 15.8 Hz, =CHC_{ar}), 7.41–7.32 (m, 5H, H_{ar}), 7.05–7.00 (m, 2H, H_{ar}), 6.77–6.75 (m, 1H, H_{ar}), 6.30–6.26 (d, 1H, J = 15.9 Hz, =CHCO), 5.61 (s, 2H, -OCH_2-), 5.22 (s, 2H, -NCH_2C_{ar}); ^{13}C NMR (100 MHz, DMSO-d_6, 25°C), δ (ppm): 166.73, 149.00, 146.18, 146.02, 142.85, 136.42, 129.25, 128.65, 128.50, 125.87, 125.32, 121.99, 116.17, 115.36, 113.88, 57.46, 53.29; HRMS m/s calc. for $C_{19}H_{17}N_3O_4$+ (H^+): 352.1292; found: 352.1289.

(1-Phenethyl-1H-1,2,3-triazol-4-yl)methyl (E)-3-(3,4-dihydroxyphenyl)acrylate ***8b***. (2-Bromoethyl)benzene (**7b**) (65 μl, 0.48 mmol), NaN_3 (37.97 mg, 0.58 mmol), **6** (107.76 mg, 0.49 mmol), $CuSO_4{\cdot}5H_2O$ (45 mg, 0.18 mmol), and sodium ascorbate (16.20 mg, 0.082 mmol). A beige solid is obtained after flash chromatography (EtOAc/hexane 50 : 50). Yield: 102.4 mg (57%); R_f = 0.12 (EtOAc/hexane 50 : 50); m.p. = 108–110°C; IR (ν (cm^{-1})): 3300–3000 (C_{ar}O-H), 1693 (C=O), 1598–1450 (C=C), 1280, 1153 (C-O); ^{1}H NMR (400 MHz, DMSO-d_6, 25°C), δ (ppm): 9.36 (s, 1H, C_{ar}OH), 9.19 (s, 1H, C_{ar}OH), 8.10 (s, 1H, =CHN), 7.51–7.47 (d, 1H, J = 15.9 Hz, =CHC_{ar}), 7.29–7.19 (m, 5H, H_{ar}), 7.06–7.00 (m, 2H, H_{ar}), 6.78–6.76 (m, 1H, H_{ar}), 6.29–6.25 (d, 1H, J = 15.9 Hz, =CHCO), 5.20 (s, 2H, -OCH_2C), 4.64–4.60 (t, 2H, J = 7.2 Hz, -NCH_2C), 3.18–3.15 (t, 2H, J = 7.2 Hz, -CH_2C_{ar}); ^{13}C NMR (100 MHz, DMSO-d_6, 25°C), δ (ppm): 166.68, 149.01, 146.13, 146.04, 142.38, 138.04, 129.14, 128.87, 127.04, 125.88, 125.20, 121.97, 116.19, 115.34, 113.93, 57.45, 50.90, 36.13; HRMS m/s calc. for $C_{19}H_{17}N_3O_4$+ (H^+): 366.1448; found: 366.1454.

[1-(3-Phenylpropyl)-1H-1,2,3-triazol-4-yl]methyl (E)-3-(3,4 dihydroxyphenyl) acrylate ***8c***. 1-Bromo-3-phenylpropane (**7c**) (72 μl, 0.47 mmol), NaN_3 (40.66 mg, 0.63 mmol), **6** (100.18 mg, 0.46 mmol), $CuSO_4{\cdot}5H_2O$ (60.94 mg, 0.24 mmol), and sodium ascorbate (10.66 mg, 0.054 mmol). A beige solid is obtained after flash chromatography (EtOAc/hexane 50 : 50). Yield: 92.9 mg (53%); R_f = 0.12 (EtOAc/hexane 50 : 50); m.p. = 122–125°C; IR (ν (cm^{-1})): 3486, 3309 (C_{ar}O-H), 1685 (C=O), 1603–1450 (C=C), 1277, 1169 (C-O); ^{1}H NMR (400 MHz, DMSO-d_6, 25°C), δ (ppm): 9.63 (s, 1H, C_{ar}OH), 9.17 (s, 1H, C_{ar}OH), 8.22 (s, 1H, =CHN), 7.53–7.49 (d, 1H, J = 15.9 Hz, =CHC_{ar}), 7.31–7.18 (m, 5H, H_{ar}), 7.06–7.00 (m, 2H, H_{ar}), 6.77–6.75 (m, 1H, H_{ar}), 6.31–6.27 (d, 1H J = 15.9 Hz, =CHCO), 5.24 (s, 2H, OCH_2C=), 4.40–4.36 (t, 2H, J = 7.1 Hz, NCH_2-), 2.59–2.55 (t, 2H J = 7.4 Hz, -CH_2C_{ar}), 2.18–2.12 (quint, 2H, J = 7.4 Hz, -CH_2-); ^{13}C NMR (100 MHz, DMSO-d_6, 25°C), δ (ppm): 166.74, 149.01, 146.15, 146.04, 142.58, 141.18, 128.86, 128.78, 126.47, 125.88, 125.12, 122.00, 116.18, 115.32, 113.92, 57.51, 49.40, 32.32, 31.75; HRMS m/s calc. for $C_{19}H_{17}N_3O_4$+ (H^+): 380.1605; found: 380.1622.

{1-[2-(p-Tolyl)ethyl]-1H-1,2,3-triazol-4-yl}methyl (E)-3-(3,4-dihydroxyphenyl) acrylate ***10a***. *p*-(2-Bromoethyl)toluene (**9a**) (157.18 μl, 1.03 mmol), NaN_3 (75.92 mg, 1.16 mmol), **6** (154.02 mg, 0.71 mmol), $CuSO_4{\cdot}5H_2O$ (38.42 mg, 0.15 mmol), and sodium ascorbate (15.49 mg, 0.08 mmol). A beige solid is obtained after flash chromatography (EtOAc/hexane 50 : 50). Yield: 173.8 mg (65%); R_f = 0.13 (EtOAc/hexane 50 : 50); m.p. = 123–126°C; IR (ν (cm^{-1})): 3477 (C_{ar}O-H), 1702 (C=O), 1601–1447 (C=C), 1281, 1160 (C-O); ^{1}H NMR (400 MHz, DMSO-d_6, 25°C), δ (ppm): 9.64 (s, 1H, C_{ar}OH), 9.18 (s, 1H, C_{ar}OH), 8.08 (s, 1H, =CHN), 7.51–7.47 (d, 1H, J = 15.9 Hz, =CHC_{ar}), 7.06–7.00 (m, 6H, H_{ar}), 6.78–6.76 (m, 1H, H_{ar}), 6.29–6.25 (d, 1H, J = 15.9 Hz, =CHCO), 5.20 (s, 2H, -OCH_2C), 4.60–4.56 (t, 2H, J = 7.2 Hz, -NCH_2C), 3.13–3.09 (t, 2H, J = 7.2 Hz, -CH_2C_{ar}), 2.23 (s, 3H, $C_{ar}CH_3$); ^{13}C NMR (100 MHz, DMSO-d_6, 25°C), δ (ppm): 166.68, 149.01, 146.12, 146.04, 142.37, 136.02, 134.89, 129.43, 129.00, 125.88, 125.20, 122.00, 116.19, 115.29, 113.92, 57.44, 51.01, 35.71, 21.07; HRMS m/s calc. for $C_{21}H_{22}N_3O_4$+ (H^+): 380.1605; found: 380.1623.

{1-[2-(p-Methoxyphenyl)ethyl]-1H-1,2,3-triazol-4-yl}methyl (E)-3-(3,4-dihydroxyphenyl) acrylate ***10b***. *p*-(2-Bromoethyl)methoxybenzene (**9b**) (108 μl, 0.69 mmol), NaN_3 (49.77 mg, 0.77 mmol), **6** (150.38 mg, 0.69 mmol), $CuSO_4{\cdot}5H_2O$ (36.85 mg, 0.15 mmol), and sodium ascorbate (16.17 mg, 0.082 mmol). A white solid is obtained after flash chromatography (EtOAc/hexane 60 : 40). Yield: 176.2 mg (65%); R_f = 0.09 (EtOAc/hexane 50 : 50); m.p. = 153-154°C; IR (ν (cm^{-1})): 3471 (C_{ar}O-H), 1700 (C=O), 1620–1440 (C=C), 1280, 1250, 1160 (C-O); ^{1}H NMR (400 MHz, DMSO-d_6, 25°C), δ (ppm): 9.62 (s, 1H, C_{ar}OH), 9.15 (s, 1H, C_{ar}OH), 8.09 (s, 1H, =CHN), 7.52–7.48 (d, 1H, J = 15.9 Hz, =CHC_{ar}), 7.11–7.00 (m, 4H, H_{ar}), 6.84–6.76 (m, 3H, H_{ar}), 6.30–6.26 (d, 1H, J = 15.9 Hz, =CHCO), 5.20 (s, 2H, OCH_2C=), 4.59–4.55 (t, 2H, J = 7.2 Hz, NCH_2C), 3.69 (s, 3H, OCH_3), 3.11–3.07 (t, 2H, J = 7.2 Hz, -CH_2C_{ar}); ^{13}C NMR (100 MHz, DMSO-d_6, 25°C), δ (ppm): 166.68, 158.43, 149.01, 146.13, 146.04, 142.36, 130.16, 129.83, 125.88, 125.18, 121.98, 116.18, 115.34, 114.27, 113.92, 57.46, 55.41, 51.18, 35.29; HRMS m/s calc. for $C_{19}H_{17}N_3O_4$+ (H^+): 396.1554; found: 396.1530.

{1-[2-(p-Fluorophenyl)ethyl]-1H-1,2,3-triazol-4-yl}methyl (E)-3-(3,4-dihydroxyphenyl) acrylate ***10c***. *p*-(2-Bromoethyl)fluorobenzene (**9c**) (144.41 μl, 1.03 mmol), NaN_3 (70.05 mg, 1.08 mmol), **6** (151.17 mg, 0.69 mmol), $CuSO_4{\cdot}5H_2O$ (45.65 mg, 0.18 mmol), and sodium ascorbate (16.07 mg, 0.08 mmol). A pale yellow solid is obtained after flash chromatography (EtOAc/hexane 50 : 50). Yield: 168.0 mg (63%); R_f = 0.11 (EtOAc/hexane 50 : 50); m.p. = 48–53°C; IR (ν (cm^{-1})): 3300–3000 (C_{ar}O-H), 1702 (C=O), 1600–1440 (C=C), 1220, 1151 (C-O), 1210 (C-F); ^{1}H NMR (400 MHz, DMSO-d_6, 25°C), δ (ppm): 9.26 (m, 3H, C_{ar}OH), 8.09 (s, 1H, =CHN), 7.52–7.48 (d, 1H, J = 15.9 Hz, =CHC_{ar}), 7.06–6.97 (m, 4H, H_{ar}), 6.78–6.76 (m, 1H, H_{ar}), 6.67–6.65 (m, 2H, H_{ar}), 6.30–6.26 (d, 1H, J = 15.9 Hz, =CHCO), 5.20 (s, 2H, -OCH_2C), 4.55–4.52 (t, 2H, J = 7.3 Hz, NCH_2C), 3.05–3.02 (t, 2H, J = 7.3 Hz, -CH_2C_{ar}); ^{13}C NMR (100 MHz, DMSO-d_6, 25°C), δ (ppm): 166.71, 156.44, 149.00, 146.13, 146.03, 142.33, 130.07, 128.01, 125.88, 125.14, 122.01, 116.19115.66, 115.30, 113.92, 57.47, 51.30, 35.41; HRMS m/s calc. for $C_{20}H_{19}N_3O_4$+ (H^+): 384.1354; found: 384.1375.

{1-[2-(p-Tolyl)ethyl]-1H-1,2,3-triazol-4-yl}methyl (E)-3-(3,4-dihydroxyphenyl) acrylate ***10d***. *p*-(2-Bromoethyl)phenol (**9d**) (212.59 mg, 1.06 mmol), NaN_3 (78.20 mg, 1.20 mmol), **6** (156.05 mg, 0.72 mmol), $CuSO_4{\cdot}5H_2O$ (42.50 mg, 0.17 mmol), and sodium ascorbate (20.41 mg, 0.11 mmol). A white solid is obtained after flash chromatography (EtOAc/hexane 70 : 30). Yield = 129.4 mg (47%); R_f = 0.19 (EtOAc/hexane 70 : 30); m.p. = 70–75°C; IR (ν (cm^{-1})): 3400–3000 (C_{ar}O-H), 2922 (C_{sp3}-H), 1687 (C=O), 1598–1440 (C=C), 1270, 1230, 1156 (C-O); ^{1}H NMR (400 MHz, DMSO-d$_6$, 25°C), δ (ppm): 9.26 (m, 3H, C_{ar}OH), 8.09 (s, 1H, =CHN), 7.52–7.48 (d, 1H, J = 15.9 Hz, =CHC_{ar}), 7.06–6.97 (m, 4H, H_{ar}), 6.78–6.76 (m, 1H, H_{ar}), 6.67–6.65 (m, 2H, H_{ar}), 6.30–6.26 (d, 1H, J = 15.9 Hz, =CHCO), 5.20 (s, 2H, -OCH_2C), 4.55–4.52 (t, 2H, J = 7.3 Hz, NCH_2C), 3.05–3.02 (t, 2H, J = 7.3 Hz, -CH_2C_{ar}); ^{13}C NMR (100 MHz, DMSO-d$_6$, 25°C), δ (ppm): 166.71, 156.44, 149.00, 146.13, 146.03, 142.33, 130.07, 128.01, 125.88, 125.14, 122.01, 116.19, 115.66, 115.30, 113.92, 57.47, 51.30, 35.41; HRMS *m/s* calc. for $C_{20}H_{20}N_3O_5$+ (H^+): 382.1397; found: 382.1392.

(1-{2-[p-(Trifluoromethyl)phenyl]ethyl}-1H-1,2,3-triazol-4-yl)methyl (E)-3-(3,4-dihydroxyphenyl)acrylate ***10e***. *p*-(2-Bromoethyl)(trifluoromethyl)benzene (**9e**) (120 μl, 0.71 mmol), NaN_3 (49.71 mg, 0.77 mmol), **6** (153.03 mg, 0.70 mmol), $CuSO_4{\cdot}5H_2O$ (40.57 mg, 0.16 mmol), and sodium ascorbate (15.18 mg, 0.077 mmol). A white solid is obtained after flash chromatography (EtOAc/hexane 55 : 45). Yield: 205.6 mg (68%); R_f = 0.10 (EtOAc/hexane 50 : 50); m.p. = 150-151°C; IR (ν (cm^{-1})): 3400–3000 (C_{ar}O-H), 1695 (C=O), 1600–1440 (C=C), 1270, 1156 (C-O), 1220 (C-F); ^{1}H NMR (400 MHz, DMSO-d$_6$, 25°C), δ (ppm): 9.67 (s, 1H, C_{ar}OH), 9.19 (s, 1H, C_{ar}OH), 8.15 (s, 1H, =CHN), 7.64–7.63 (m, 2H, H_{ar}), 7.52–7.48 (d, 1H, J = 15.9 Hz, =CHC_{ar}), 7.43–7.41 (m, 2H, H_{ar}), 7.06–7.00 (m, 2H, H_{ar}), 6.78–6.76 (m, 1H, H_{ar}), 6.29–6.26 (d, 1H, J = 15.9 Hz, =CHCO), 5.20 (s, 2H, -OCH_2C), 4.69–4.65 (t, 2H, J = 7.2 Hz, NCH_2C), 3.29–3.26 (t, 2H, J = 7.2 Hz, -CCH_2C_{ar}); ^{13}C NMR (100 MHz, DMSO-d$_6$, 25°C), δ (ppm): 166.68, 149.01, 146.16, 146.03, 143.05, 142.49, 130.04, 125.86, 125.67, 125.64, 125.28, 121.97, 116.16, 115.32, 113.86, 57.43, 55.37, 50.40, 35.77; HRMS *m/s* calc. for $C_{19}H_{17}N_3O_4$+ (H^+): 434.1322; found: 434.1314.

{1-[2-(p-Nitrophenyl)ethyl]-1H-1,2,3-triazol-4-yl}methyl (E)-3-(3,4-dihydroxy-phenyl) acrylate ***10f***. *p*-(2-Bromoethyl)nitrobenzene (**9f**) (161.25 mg, 0.70 mmol), NaN_3 (46.81 mg, 0.72 mmol), **6** (149.83 mg, 0.69 mmol), $CuSO_4{\cdot}5H_2O$ (40.68 mg, 0.16 mmol), and sodium ascorbate (23.51 mg, 0.12 mmol). A yellow solid is obtained after flash chromatography (EtOAc/hexane 70 : 30). Yield = 149.7 mg (53%); R_f = 0.19 (EtOAc/hexane 7 : 3); m.p. = 159–161°C; IR (ν (cm^{-1})): 3445 (C_{ar}O-H), 2920 (C_{sp3}-H), 1719 (C=O), 1603–1440 (C=C), 1520, 1340 (NO_2), 1255, 1151 (C-O); ^{1}H NMR (400 MHz, DMSO-d$_6$, 25°C), δ (ppm): 9.64 (s, 1H, C_{ar}OH), 9.17 (s, 1H, C_{ar}OH), 8.14 (s, 1H, =CHN), 8.12–8.11 (m, 2H, H_{ar}), 7.51–7.47 (d, 1H, J = 15.9 Hz, =CHC_{ar}), 7.49 (m, 2H, H_{ar}), 7.05–6.99 (m, 2H, H_{ar}), 6.77–6.75 (m, 1H, H_{ar}), 6.28–6.24 (d, 1H, J = 15.9 Hz, =CHCO), 5.19 (s, 2H, -OCH_2C), 4.71–4.68 (t, 2H, J = 7.2 Hz, NCH_2C), 3.33–3.31 (t, 2H, J = 7.2 Hz, -CCH_2C_{ar}); ^{13}C NMR (100 MHz, DMSO-d$_6$, 25°C), δ (ppm): 166.67, 149.02, 146.84, 146.39, 146.16, 146.04, 142.52, 130.56, 125.87, 125.30, 123.90, 121.97, 116.18, 115.35, 113.86, 57.43, 50.17, 35.73; HRMS *m/s* calc. for $C_{19}H_{17}N_3O_4$+ (H^+): 411.1299; found: 411.1276.

(1-Methyl-1H-1,2,3-triazol-4-yl)methyl (E)-3-(3,4-dihydroxy-phenyl)acrylate ***12a***. A solution of iodomethane (**11a**) (64.19 μl, 1.03 mmol), NaN_3 (72.03 mg, 1.11 mmol), **6** (154.45 mg, 0.71 mmol), and copper iodine (27.87 mg, 0.15 mmol) in 5 ml of a mixture of water and acetone (1 : 1) was placed in an ultrasound bath and was left to react under ultrasound at room temperature overnight. The solution was then extracted 3 times with EtOAc. The combined organic phases are washed twice with a solution of NH_4Cl 5 M and then twice with brine. The combined organic phase is dried with $MgSO_4$ and concentrated under vacuum to afford the crude product. Compound **12a** was obtained as brown solid after flash chromatography (EtOAc/Hexane 70 : 30). Yield = 45.9 mg (24%); R_f = 0.14 (EtOAc/hexane 70 : 30); m.p. = 49–55°C; IR (ν (cm^{-1})): 3400–3000 (C_{ar}O-H), 2922 (C_{sp3}-H), 1696 (C=O), 1598–1440 (C=C), 1251, 1147 (C-O); ^{1}H NMR (400 MHz, CD_3OD, 25°C), δ (ppm): 8.00 (s, 1H, =CHN), 7.61–7.57 (d, 1H, J = 15.9 Hz, =CHC_{ar}), 7.04–6.94 (m, 2H, H_{ar}), 6.80–6.78 (m, 1H, H_{ar}), 6.29–6.25 (d, 1H, J = 15.9 Hz, =CHCO), 5.29 (s, 2H, -OCH_2C), 4.13 (s, 3H, NCH_3); ^{13}C NMR (100 MHz, CD_3OD, 25°C), δ (ppm): 167.24, 148.33, 146.09, 145.43, 142.98, 126.20, 125.32, 121.65, 115.09, 113.73, 113.12, 56.66, 35.62; HRMS *m/s* calc. for $C_{13}H_{13}N_3O_4$+ (H^+): 276.0979; found: 276.0997.

(1-Ethyl-1H-1,2,3-triazol-4-yl)methyl (E)-3-(3,4-dihydroxy-phenyl)acrylate ***12b***. Bromoethane (**11b**) (80 μl, 1.08 mmol), NaN_3 (61.01 mg, 0.88 mmol), **6** (148.72 mg, 0.68 mmol), $CuSO_4{\cdot}5H_2O$ (41.62 mg, 0.17 mmol), and sodium ascorbate (20.94 mg, 0.11 mmol). A beige solid is obtained after flash chromatography (EtOAc/hexane 70 : 30). Yield: 127.5 mg (59%); R_f = 0.19 (EtOAc/hexane 70 : 30); m.p. = 188–191°C; IR (ν (cm^{-1})): 3400–3000 (C_{ar}O-H), 2924 (C_{sp3}-H), 1687 (C=O), 1596–1440 (C=C), 1257, 1147 (C-O); ^{1}H NMR (400 MHz, DMSO-d$_6$, 25°C), δ (ppm): 9.62 (s, 1H, C_{ar}OH), 9.19 (s, 1H, C_{ar}OH), 8.19 (s, 1H, =CHN), 7.52–7.48 (d, 1H, J = 15.9 Hz, =CHC_{ar}), 7.05–7.00 (m, 2H, H_{ar}), 6.77–6.75 (m, 1H, H_{ar}), 6.30–6.26 (d, 1H, J = 15.9 Hz, =CHCO), 5.22 (s, 2H, -OCH_2C), 4.42–4.36 (q, 2H, J = 7.3 Hz, -CH_2C), 1.45–1.41 (t, 3H, J = 7.3 Hz, -CH_3); ^{13}C NMR (100 MHz, DMSO-d$_6$, 25°C), δ (ppm): 166.74, 149.00, 146.15, 146.03, 142.51, 125.87, 124.63, 122.00, 116.18, 115.30, 113.91, 57.50, 45.06, 15.82; HRMS *m/s* calc. for $C_{19}H_{17}N_3O_4$+ (H^+): 320.0877; found: 320.0874.

(1-Propyl-1H-1,2,3-triazol-4-yl)methyl (E)-3-(3,4-dihydroxy-phenyl)acrylate ***12c***. 1-Bromopropane (**11c**) (88.43 μl, 1.03 mmol), NaN_3 (67.37 mg, 1.04 mmol), **6** (155.09 mg, 0.71 mmol), $CuSO_4{\cdot}5H_2O$ (44.51 mg, 0.18 mmol), and sodium ascorbate (23.50 mg, 0.12 mmol). A darker beige solid is obtained after flash chromatography (EtOAc/hexane 50 : 50). Yield: 118.7 mg (55%); R_f = 0.11 (EtOAc/hexane

50 : 50); m.p. = 45–50°C; IR (ν (cm^{-1})): 3334 (C_{ar}O-H), 2926 (C_{sp3}-H), 1687 (C=O), 1605–1440 (C=C), 1274, 1150 (C-O); ^{1}H NMR (400 MHz, DMSO-d_6, 25°C), δ (ppm): 9.62 (s, 1H, C_{ar}OH), 9.14 (s, 1H, C_{ar}OH), 8.18 (s, 1H, =CHN), 7.53–7.49 (d, 1H, J = 15.9 Hz, =CHC_{ar}), 7.05–7.00 (m, 2H, H_{ar}), 6.77–6.75 (m, 1H, H_{ar}), 6.30–6.26 (d, 1H, J = 15.9 Hz, =CHCO), 5.23 (s, 2H, -OCH_2C), 4.34–4.31 (t, 2H, J = 7.0 Hz, NCH_2C), 1.88–1.79 (sext, 2H, J = 7.2 Hz, -CH_2C), 0.87–0.83 (t, 3H, J = 7.4 Hz, -CH_3); ^{13}C NMR (100 MHz, DMSO-d_6, 25°C), δ (ppm): 166.73, 149.00, 146.15, 146.03, 142.46, 125.88, 125.08, 121.97, 116.18, 115.35, 113.92, 57.51, 51.40, 23.61, 11.27; HRMS *m/s* calc. for $C_{15}H_{17}N_3O_4$+ (H^+): 304.1292; found: 304.1291.

(1-Isopropyl-1H-1,2,3-triazol-4-yl)methyl (E)-3-(3,4-dihydroxyphenyl)acrylate ***12d***. 2-Bromopropane (**11d**) (96.81 μl, 1.03 mmol), NaN_3 (72.32 mg, 1.11 mmol), **6** (155.78 mg, 0.71 mmol), $CuSO_4{\cdot}5H_2O$ (57.31 mg, 0.23 mmol), and sodium ascorbate (19.14 mg, 0.10 mmol). A white solid is obtained after flash chromatography (EtOAc/hexane 50 : 50). Yield: 117.3 mg (54%); R_f = 0.11 (EtOAc/hexane 50 : 50); m.p. = 50–55°C; IR (ν (cm^{-1})): 3376 (C_{ar}O-H), 2930 (C_{sp3}-H), 1693 (C=O), 1633–1440 (C=C), 1285, 1162 (C-O); ^{1}H NMR (400 MHz, DMSO-d_6, 25°C), δ (ppm): 9.61 (s, 1H, C_{ar}OH), 9.14 (s, 1H, C_{ar}OH), 8.24 (s, 1H, =CHN), 7.53–7.49 (d, 1H, J = 15.9 Hz, =CHC_{ar}), 7.06–7.00 (m, 2H, H_{ar}), 6.77–6.75 (m, 1H, H_{ar}), 6.31–6.27 (d, 1H, J = 15.9 Hz, =CHCO), 5.22 (s, 2H, -OCH_2C), 4.86–4.79 (sept, 1H, J = 6.6 Hz, NCH-), 1.50–1.49 (d, 6H, J = 6.7 Hz, -$(CH_3)_2$); ^{13}C NMR (100 MHz, DMSO-d_6, 25°C), δ (ppm): 166.73, 149.00, 146.16, 146.03, 142.34, 125.89, 123.02, 121.97, 116.18, 115.37, 113.92, 57.55, 52.83, 23.08; HRMS *m/s* calc. for $C_{15}H_{17}N_3O_4$+ (H^+): 304.1292; found: 304.1334.

[1-(3-Methyl-2-butenyl)-1H-1,2,3-triazol-4-yl]methyl (E)-3-(3, 4-dihydroxy phenyl) acrylate ***12e***. 1-Bromo-3-methyl-2-butene (**11e**) (95.30 μl, 0.82 mmol), NaN_3 (79.84 mg, 1.23 mmol), **6** (153.88 mg, 0.71 mmol), $CuSO_4{\cdot}5H_2O$ (46.63 mg, 0.19 mmol), and sodium ascorbate (18.40 mg, 0.09 mmol). A dark beige solid is obtained after flash chromatography (EtOAc/hexane 40 : 60). Yield: 147.1 mg (63%); R_f = 0.18 (EtOAc/hexane 50 : 50); m.p. = 123–126°C; IR (ν (cm^{-1})): 3460 (C_{ar}O-H), 2969 (C_{sp3}-H), 1702 (C=O), 1603–1440 (C=C), 1264, 1153 (C-O); ^{1}H NMR (400 MHz, DMSO-d_6, 25°C), δ (ppm): 9.62 (s, 1H, C_{ar}OH), 9.14 (s, 1H, C_{ar}OH), 8.10 (s, 1H, =CHN), 7.52–7.48 (d, 1H, J = 15.9 Hz, =CHC_{ar}), 7.05–7.00 (m, 2H, H_{ar}), 6.77–6.75 (m, 1H, H_{ar}), 6.30–6.26 (d, 1H, J = 15.9 Hz, =CHCO), 5.43–5.40 (m, 1H, -$C\underline{H}C(CH_3)_2$), 5.22 (s, 2H, -OCH_2C), 4.99–4.97 (d, 2H, J = 7.3 Hz, NCH_2-), 1.77–1.74 (d, 6H, J = 13.5 Hz, -$C(CH_3)_2$); ^{13}C NMR (100 MHz, DMSO-d_6, 25°C), δ (ppm): 166.73, 149.01, 146.15, 146.03, 142.57, 138.63, 125.88, 124.69, 121.98, 118.82, 116.18, 115.35, 113.91, 57.48, 55.28, 47.71, 25.76, 18.30; HRMS *m/s* calc. for $C_{17}H_{20}N_3O_4$+ (H^+): 330.1448; found: 330.1419.

(1-Hexyl-1H-1,2,3-triazol-4-yl)methyl (E)-3-(3,4-dihydroxyphenyl)acrylate ***12f***. 1-Chlorohexane (**11f**) (100 μl, 0.73 mmol), NaN_3 (52.12 mg, 0.80 mmol), **6** (155 mg, 0.71 mmol), $CuSO_4{\cdot}5H_2O$ (37.26 mg, 0.15 mmol), and sodium ascorbate (15.37 mg, 0.078 mmol). A brown solid is obtained after flash chromatography (EtOAc/hexane 40 : 60). Yield: 53.6 mg (22%); R_f = 0.20 (EtOAc/hexane 50 : 50); m.p. = 51–55°C; IR (ν (cm^{-1})): 3500–3000 (C_{ar}O-H), 2924, 2855 (C_{sp3}-H), 1702 (C=O), 1598–1440 (C=C), 1257, 1149 (C-O); ^{1}H NMR (400 MHz, DMSO-d_6, 25°C), δ (ppm): 9.63 (s, 1H, C_{ar}OH), 9.17 (s, 1H, C_{ar}OH), 8.18 (s, 1H, =CHN), 7.52–7.48 (d, 1H, J = 15.9 Hz, =CHC_{ar}), 7.05–6.99 (m, 2H, H_{ar}), 6.78–6.72 (m, 1H, H_{ar}), 6.31–6.26 (d, 1H, J = 15.9 Hz, =CHCO), 5.22 (s, 2H, -OCH_2C), 4.37–4.33 (t, 2H, J = 7.1 Hz, NCH_2C), 1.84–177 (quint, 2H, J = 7.2 Hz, CCH_2C), 1.25–1.21 (m, 6H, H_{sp3}), 0.86–0.83 (t, 3H, J = 6.7 Hz, -CCH_3); ^{13}C NMR (100 MHz, DMSO-d_6, 25°C), δ (ppm): 166.73, 149.00, 146.14, 146.03, 142.46, 125.87, 125.07, 121.99, 116.18, 115.30, 113.92, 57.50, 49.82, 31.00, 30.08, 25.93, 22.36, 14.28; HRMS *m/s* calc. for $C_{19}H_{17}N_3O_4$+ (H^+): 346.1761; found: 346.1764.

[1-(2-Cyclohexylethyl)-1H-1,2,3-triazol-4-yl]methyl (E)-3-(3,4-dihydroxy phenyl) acrylate ***12g***. 1-(2-Bromoethyl)cyclohexane (**11g**) (108 μl, 0.69 mmol), NaN_3 (55.15 mg, 0.85 mmol), **6** (152.30 mg, 0.70 mmol), $CuSO_4{\cdot}5H_2O$ (59.72 mg, 0.24 mmol), and sodium ascorbate (20.80 mg, 0.10 mmol). A pale yellow solid is obtained after flash chromatography (EtOAc/hexane 40 : 60). Yield = 131.8 mg (51%); R_f = 0.19 (EtOAc/hexane 50 : 50); m.p. = 66–72°C; IR (ν (cm^{-1})): 3356 (C_{ar}O-H), 2924, 2853 (C_{sp3}-H), 1687 (C=O), 1600–1443 (C=C), 1292, 1160 (C-O); ^{1}H NMR (400 MHz, CD_3OD, 25°C), δ (ppm): 8.04 (s, 1H, =CHN), 7.61–7.57 (d, 1H, J = 15.9 Hz, =CHC_{ar}), 7.05–6.78 (m, 3H, H_{ar}), 6.30–6.23 (d, 1H, J = 15.9 Hz, =CHCO), 5.30 (s, 2H, OCH_2C=), 4.47–4.43 (t, 2H, J = 7.4 Hz, NCH_2C), 2.03–1.01 (m, 13H, H_{sp3}); ^{13}C NMR (100 MHz, CD_3OD, 25°C), δ (ppm): 167.26, 148.33, 146.10, 145.43, 142.81, 126.21, 124.28, 121.66, 115.09, 113.75, 113.15, 56.73, 47.93, 37.31, 34.77, 32.54, 26.08, 25.81; HRMS *m/s* calc. for $C_{19}H_{17}N_3O_4$+ (H^+): 372.1918; found: 372.1922.

4.2. 5-LO Product Biosynthesis Inhibition Assays

4.2.1. 5-LO Products Biosynthesis Inhibition in HEK293 Cells. HEK293 cells stably cotransfected with a pcDNA3.1 vector expressing human 5-LO and a pBUDCE4.1 vector expressing human 5-LO activating protein (FLAP) were utilized [15, 20] to screen compounds for 5-LO inhibition.

For cell stimulation of 5-LO products, transfected HEK293 cells were collected following trypsinization and washed and the cell pellet was resuspended in Hank's balanced salt solution (HBSS) (Lonza, Walkerville, MD) containing 1.6 mM $CaCl_2$ at a concentration of 5×10^5 cells mL^{-1}. Cells were preincubated with each test compound at 1 μM for 5 min at 37°C. Cells were then stimulated for 15 minutes at 37°C with the addition of 10 μM calcium ionophore A23187 (Sigma-Aldrich, Oakville, ON, Canada) and 10 μM arachidonic acid (Cayman Chemical, Ann Arbour, MI). Stimulations were stopped by the addition of 0.5 volume of cold MeOH : CH_3CN (1 : 1) containing 50 ng of PGB_2 as internal standard and samples were stored at −20°C until processing on octadecyl (C18) columns and analysis by RP-HPLC as described previously [15, 20]. Data are expressed as means ± SEM of 3 independent experiments, each performed in duplicate (Figures 3–5).

4.2.2. 5-LO Products Biosynthesis Inhibition in PMNL Cells PMNL. Human PMNL prepared from peripheral blood as described [11] were suspended in HBSS containing 1.6 mM $CaCl_2$ (10^7 cells/ml) and preincubated with test compounds for 5 min at 37°C in the presence of 0.4 U/ml of adenosine deaminase (Sigma-Aldrich, Oakville, On, Canada). Cells were then stimulated for 15 min at 37°C with 1 μM thapsigargin (Sigma-Aldrich) and 10 μM AA (Cayman Chemical, Ann Arbour, MI) as previously described [11]. Reactions were stopped, processed, and analyzed by RP-HPLC as indicated above. Data are expressed as means ± SEM of 3 independent experiments, each performed in duplicate (Figure 6).

4.3. Molecular Docking. Molecular docking was performed with AutoDock Vina 1.1.2 on 5-LO (PDB code: 3O8Y) following the procedure [17] outlined in "Design, Synthesis and Evaluation of Semi-Synthetic Triazole-Containing Caffeic Acid Analogs as 5-Lipoxygenase Inhibitors" by Daniela De Lucia et al. For optimization, the MMFF94 force field was used. Results were analyzed with Maestro 11.0 [21].

Conflicts of Interest

The authors declare that there are no conflicts of interest regarding the publication of this paper.

Acknowledgments

Mohamed Touaibia would like to acknowledge the contribution of the Natural Sciences and Engineering Research Council of Canada (NSERC), the New Brunswick Innovation Foundation (NBIF), the Canadian Foundation for Innovation (CFI), and Université de Moncton. Marc E. Surette is supported by a New Brunswick Innovation Research Chair awarded by the NBIF.

References

[1] W. R. Henderson Jr., "The role of leukotrienes in inflammation," *Annals of Internal Medicine*, vol. 121, no. 9, pp. 684–697, 1994.

[2] C. Pergola and O. Werz, "5-Lipoxygenase inhibitors: a review of recent developments and patents," *Expert Opinion on Therapeutic Patents*, vol. 20, no. 3, pp. 355–375, 2010.

[3] O. Kowal-Bielecka, K. Kowal, O. Distler, and S. Gay, "Mechanisms of disease: leukotrienes and lipoxins in scleroderma lung disease—insights and potential therapeutic implications," *Nature Clinical Practice Rheumatology*, vol. 3, no. 1, pp. 43–51, 2007.

[4] J. N. Sharma and L. A. Mohammed, "The role of leukotrienes in the pathophysiology of inflammatory disorders: is there a case for revisiting leukotrienes as therapeutic targets?" *Inflammopharmacology*, vol. 14, no. 1-2, pp. 10–16, 2006.

[5] B. Samuelsson, S.-E. Dahlen, J. A. Lindgren, C. A. Rouzer, and C. N. Serhan, "Leukotrienes and lipoxins: structures, biosynthesis, and biological effects," *Science*, vol. 237, no. 4819, pp. 1171–1176, 1987.

[6] E. Israel, J. Cohn, L. Dubé, and J. M. Drazen, "Effect of treatment with zileuton, a 5-lipoxygenase inhibitor, in patients with asthma: a randomized controlled trial," *Journal of the American Medical Association*, vol. 275, no. 12, pp. 931–936, 1996.

[7] O. Werz and D. Steinhilber, "Therapeutic options for 5-lipoxygenase inhibitors," *Pharmacology & Therapeutics*, vol. 112, no. 3, pp. 701–718, 2006.

[8] A. S. Fischer, J. Metzner, S. D. Steinbrink et al., "5-Lipoxygenase inhibitors induce potent anti-proliferative and cytotoxic effects in human tumour cells independently of suppression of 5-lipoxygenase activity," *British Journal of Pharmacology*, vol. 161, no. 4, pp. 936–949, 2010.

[9] J. Ghosh and C. E. Myers, "Inhibition of arachidonate 5-lipoxygenase triggers massive apoptosis in human prostate cancer cells," *Proceedings of the National Academy of Sciences of the United States of America*, vol. 95, no. 22, pp. 13182–13187, 1998.

[10] A.-P. Beauregard, J. Harquail, G. Lassalle-Claux et al., "CAPE analogs induce growth arrest and apoptosis in breast cancer cells," *Molecules*, vol. 20, no. 7, pp. 12576–12589, 2015.

[11] Jérémie A. Doiron, Luc M. Leblanc, Martin J. G. Hébert et al., "Structure activity relationship of caffeic acid phenethyl ester analogs as new 5-lipoxygenase inhibitors," *Chemical Biology & Drug Design*, vol. 89, no. 4, pp. 514–528, 2017.

[12] L. H. Boudreau, J. Maillet, L. M. LeBlanc et al., "Caffeic acid phenethyl ester and its amide analogue are potent inhibitors of leukotriene biosynthesis in human polymorphonuclear leukocytes," *PLoS ONE*, vol. 7, no. 2, Article ID e31833, 2012.

[13] D. De Lucia, O. M. Lucio, B. Musio et al., "Design, synthesis and evaluation of semi-synthetic triazole-containing caffeic acid analogues as 5-lipoxygenase inhibitors," *European Journal of Medicinal Chemistry*, vol. 101, pp. 573–583, 2015.

[14] E. M. Joshi, B. H. Heasley, M. D. Chordia, and T. L. Macdonald, "In vitro metabolism of 2-acetylbenzothiophene: relevance to zileuton hepatotoxicity," *Chemical Research in Toxicology*, vol. 17, no. 2, pp. 137–143, 2004.

[15] L. H. Boudreau, N. Picot, J. Doiron et al., "Caffeoyl and cinnamoyl clusters with anti-inflammatory and anti-cancer effects. Synthesis and structure-activity relationship," *New Journal of Chemistry*, vol. 33, no. 9, pp. 1932–1940, 2009.

[16] J. A. Doiron, B. Métayer, R. R. Richard et al., "Clicked cinnamic/caffeic esters and amides as radical scavengers and 5-lipoxygenase inhibitors," *International Journal of Medicinal Chemistry*, vol. 2014, 12 pages, 2014.

[17] C. E. Walsh, B. M. Waite, M. J. Thomas, and L. R. DeChatelet, "Release and metabolism of arachidonic acid in human neutrophils," *Journal of Biological Chemistry*, vol. 256, no. 14, pp. 7228–7234, 1981.

[18] E. Shang et al., "Benzo[d]isothiazole 1, 1-dioxide derivatives as dual functional inhibitors of 5-lipoxygenase and microsomal prostaglandin E2 synthase-1," *Bioorganic & Medicinal Chemistry Letters*, vol. 24, no. 12, pp. 2764–2767, 2014.

[19] N. C. Gilbert, S. G. Bartlett, M. T. Waight et al., "The structure of human 5-lipoxygenase," *Science*, vol. 331, no. 6014, pp. 217–219, 2011.

[20] J. Doiron, L. H. Boudreau, N. Picot, B. Villebonet, M. E. Surette, and M. Touaibia, "Synthesis and 5-lipoxygenase inhibitory activity of new cinnamoyl and caffeoylclusters," *Bioorganic and Medicinal Chemistry Letters*, vol. 19, no. 4, pp. 1118–1121, 2009.

[21] Maestro, *Schrödinger Release 2016-4*, Maestro, Schrödinger, LLC, New York, NY, USA, 2016.

Drug Effect of Thulium(III)-Arsenazo III Complex on Herring Sperm DNA

Hongbo Li, Shutao Lei, Yongqiong Tang, Liyuan Zou, Hongqin Hu, and Suqin Wang

Key Laboratory of Functional Small Organic Molecule, Ministry of Education, College of Chemistry and Chemical Engineering, Jiangxi Normal University, Nanchang 330022, China

Correspondence should be addressed to Hongbo Li; lihongbo112@126.com and Suqin Wang; wangsuqin163@163.com

Academic Editor: Marcelino Maneiro

To assess the potential cytostatic properties of the thulium(III)-arsenazo III complex as a probe of rare earth complex antitumor drugs, the interaction information of the thulium(III)-arsenazo III complex with DNA was obtained by using spectroscopy, viscosity measurements, and voltammetric methods. The thermodynamic functions demonstrated that the binding constants of the thulium(III)-arsenazo III complex with DNA were $K^{\theta}_{298.15K} = 4.84 \times 10^6\ L{\cdot}mol^{-1}$ and $K^{\theta}_{308.15K} = 4.48 \times 10^6\ L{\cdot}mol^{-1}$, and the binding process was enthalpy driven. The increase in relative viscosity of DNA with the addition of the thulium(III)-arsenazo III complex and the results from Scatchard and voltammetric methods showed that the interaction mode between the thulium(III)-arsenazo III complex and DNA was groove binding along with weak intercalative binding.

1. Introduction

As is known to all, DNA is an important genetic substance in organisms which controls the synthesis of protein and expression of the gene. [1] Due to advances in medical technology, DNA has become the target molecule for many antitumor and anticancer drugs in recent years [2]. Interaction of DNA with small molecules is of current interest and significance in many research fields, such as biochemistry, medicinal chemistry, and cancer therapy [3]. A large number of researches have shown that the pharmacological activity of most antitumor drugs is related to the binding mode of DNA. Generally, the interaction modes of small molecules with DNA have been broadly categorized into the following three types including intercalative binding, groove binding, and electrostatic binding [4]. In the majority of situations, the intercalative binding is stronger than the other two binding modes because the intercalative molecule is sandwiched between the aromatic heterocyclic base pairs of DNA [5]. However, depending on particular properties of the ligands and other conditions of binding, groove binders may bind to DNA more tightly than typical intercalators. Thus, due to the interaction mechanism of small molecules with DNA being revealed, it will help to improve the efficacy of antitumor basing on molecular biological theory and provide a guidance for the development of the novel, highly efficient, and low toxic antitumor drugs [6, 7].

Since the rare earth ion has anti-inflammatory, bactericidal, and analgesic properties, it can effectively combine with nucleic acids which are superior to transition metals and small molecules [6]. In addition, their toxicities are decreased and the pharmaceutical properties have been improved when they coordinate with a ligand [8]. So, the interaction of rare earth metal complexes with DNA has received much attention in recent years, and these results are promising for their application in the development of new antitumor drugs [1]. Arsenazo III (ASA) is a bis-azo metal-chromogenic reagent which has been in use since 1960 for spectrophotometric determination of metal ions. And now, it continues to attract oncologists' attention [9, 10], and the chemical structure of ASA is shown in Figure 1. As the researchers all know, ASA has been widely used in the analysis of rare earth elements [11]. A large number of studies have demonstrated the complexation behavior of ASA with rare earth elements [12, 13], but the interaction of

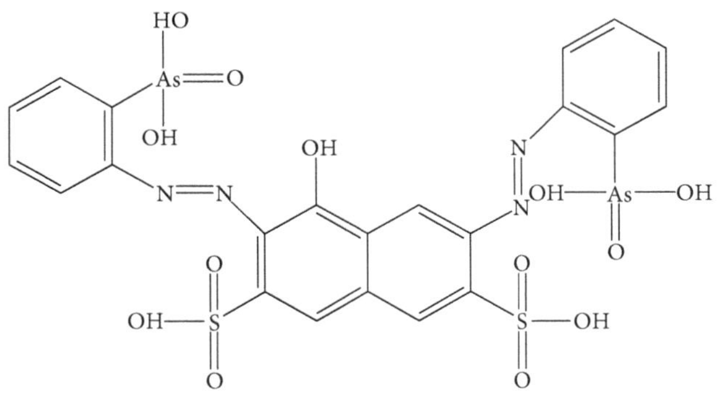

FIGURE 1: The chemical structure of arsenazo III.

ASA-rare earth metal complexes with DNA has seldom been reported until now. In addition, formation of earth metal complexes can not only improve the pharmaceutical properties of rare earth metals but also produce better synergistic effects of rare earth elements and small organic molecules. Hence, the researchers have done more meticulous research about the interaction of the thulium(III)-arsenazo III (Tm(III)-ASA) complex with DNA.

A number of techniques have been applied for investigation of the interaction of small molecules with DNA, including UV spectrophotometry [14–16], fluorescence spectroscopy [17–20], circular dichroism spectroscopy [21, 22], mass spectrometry [23], electrochemical methods [24], voltammetry [25], X-ray diffraction [26], and dynamic viscosity measurements [27]. Among these techniques, UV spectrophotometry and fluorescence spectroscopy are regarded as the more effective methods because they are simple, rapid, and sensitive [28]. Furthermore, UV-vis and fluorescence measurements could obtain information related to the binding mechanisms between small molecules and DNA, such as molar ratio, binding modes, binding constants, quenching rate constants, and thermodynamic parameters [29].

In this article, the researchers have characterized the composition of Tm(III)-ASA complex and the interaction parameter of Tm(III)-ASA complex with DNA by UV-vis spectrophotometry, fluorescence spectroscopy, viscosity measurements, and voltammetry. The binding mode, binding ratio, thermodynamic parameters, and binding constants were also obtained. Additionally, we selected acridine orange (AO) and 4S green plus nucleic acid stain (4S Green) as the probe, which can intercalate into the base pairs of double-stranded DNA, and the fluorescence intensity is enhanced, and they offer lower toxicity, higher stability, and convenience of using [27]. From this study, the researchers believe that it will provide useful information on the interaction mechanism of rare earth metal complexes with DNA and guide for the synthesis of new anticancer and antitumor drugs.

2. Experiments and Methods

2.1. Materials and Reagents. Herring sperm DNA (Sigma Chem. Co., USA) was used without further purification, and its stock solution was prepared by dissolving an appropriate amount of DNA in 50.0 mM tris-HCl buffer solution (pH 7.40) and storing at 4°C before being used. The concentration of DNA in stock solution was determined by UV absorption at 260 nm using a molar absorption coefficient, $\varepsilon_{260} = 6600\ L{\cdot}mol^{-1}{\cdot}cm^{-1}$ [30]. Purity of DNA was checked by monitoring the absorbance at 260 nm and 280 nm. When it was the ratio >1.8 of A_{260}/A_{280}, it indicates that DNA was free from protein [31]. All samples were dissolved in 50.0 mM tris-HCl buffer solution (pH 7.40). The stock solution of Tm^{3+} ($2.00 \times 10^{-3}\ mol{\cdot}L^{-1}$) was prepared by dissolving Tm_2O_3 (99.95%; Fourth Reagent Factory, Shanghai, China) in hydrochloric acid and heating it until nearly being dried, then diluting with ultrapure water. A stock solution ($5.00 \times 10^{-4}\ mol{\cdot}L^{-1}$) of arsenazo III (Jinsui Bio-Technology, Shanghai, China) was prepared by dissolving its crystals in ultrapure water. Acridine orange (AO) dye stock solution ($5.00 \times 10^{-4}\ mol{\cdot}L^{-1}$) was prepared by dissolving its crystals (Sigma Chem. Co., USA) in ultrapure water. 4S green plus nucleic acid stain was purchased from Sigma-Aldrich Corporation (America). All chemicals were of the analytical grade without further purification. All solutions were prepared with ultrapure water which was obtained through a Millipore Milli-Q water purification system (Billerica, Bedford, MA) and had an electrical resistivity of > 18.25 MΩ·cm.

2.2. Apparatus. The absorption spectra were measured on a Lambda 35 spectrophotometer (Perkin Elmer Company, America) using a 1.0 cm cuvette. The fluorescence spectra were measured using a LS-55 fluorophotometer (Perkin Elmer Company, America). The slits for excitation and emission were both set at 5.0 nm and the scan rate at $240\ nm{\cdot}min^{-1}$. The viscosity measurements were carried out using an NDJ-79 viscosity meter (Yinhua Flowmeter Co. Ltd., Hangzhou, China). The cyclic voltammetric (CV) experiments were carried out with a CHI820B electrochemical workstation (CH Instrument Company, Texas, America). A FE20 pH meter (Mettler-Toledo Instruments, Shanghai, China) was used to pH measurements. All experiments, unless otherwise specified, were carried out at room temperature (25 ± 1°C).

2.3. Procedures

2.3.1. UV-Vis Spectral Measurements. The UV-vis absorption spectra of ASA, Tm(III), and its mixture with DNA were measured in the wavelength range of 300 to 800 nm at room temperature. The solution in the cuvette was mixed thoroughly prior to each scan, and the volume affection was so small that it could be ignored. In addition, all experiments used the 50.0 mM tris-HCl buffer solution (pH 7.40) as the reference solution.

2.3.2. Fluorescence Spectral Measurements. The different concentrations of the Tm(III)-ASA complex, 4S Green, or AO were added directly into the quartz cuvette that contains the DNA solution. The solution in the cuvette was mixed uniformly before each being scanned, and the volume affection could be ignored for small change in volume. The slits for excitation and emission were both set at 5.0 nm and $\lambda_{ex} = 490$ nm, respectively, and the scan rate at 240 nm·min^{-1}.

2.3.3. Viscosity Measurement. Viscosity measurement was measured by using a viscometer at room temperature, and the viscometer was placed in the thermostat for keeping a constant temperature. Different amounts of the Tm(III)-ASA complex was added into the viscometer while keeping DNA concentration constant. The flow times of the samples were repeatedly measured with an accuracy of ±0.2°C, and each sample was measured three times at least. The data were presented as $\ln(\eta/\eta_0)$ versus $c_{Tm(III)\text{-}ASA}/c_{DNA}$, where η and η_0 are the viscosity of DNA in the presence and the absence of the Tm(III)-ASA complex.

2.3.4. Cyclic Voltammetric Measurement. The cyclic voltammetric (CV) experiment was carried out with an electrochemical workstation which used three-electrode system at room temperature. A glassy carbon (GC) disk electrode of 3.0 mm diameter was used as the working electrode, a platinum sheet as the counter electrode, and a saturated calomel electrode (SCE) as the reference electrode. Cyclic voltammograms of electrodes were obtained in 50.0 mM tris-HCl buffer solution (pH 7.40) with the potential range between −0.5 V and 0.5 V and a sweep rate of 50.0 mV·s^{-1}.

3. Results and Discussion

3.1. UV-Vis Spectroscopic Characteristics. Numerous research studies showed that the absorption spectrum changed when the small molecule interacted with DNA and formed a new complex [32]. Figure 2 shows that the absorption spectra of ASA, DNA, Tm(III), DNA-ASA, DNA-Tm(III), Tm(III)-ASA complex, and Tm(III)-ASA-DNA system at a certain concentration were obtained using a UV-vis spectrometer. By seeing the spectrogram, there were not absorption peaks of DNA (curve f), Tm(III) (curve g), and DNA-Tm(III) system (curve e) in the wavelength of the visible region. The maximum absorption of ASA (curve c) was located at approximately 550 nm, but the absorbance of DNA-ASA (curve d) decreased apparently, it indicated the formation of DNA-ASA supramolecular complexes. However, when Tm(III) was added in ASA solution, the maximum absorption of Tm(III)-ASA (curve b) was located in about 605 nm with an apparently red shift of 55 nm, and peak height increased remarkably, which indicated the formation of the Tm(III)-ASA complex and then, added the same amount of DNA to the system, the absorbance of DNA-Tm(III)-ASA (curve a) system increased remarkably at 603 nm, with a slight blue shift of 2 nm. The result showed that a binding interaction between the Tm(III)-ASA complex and DNA had occurred.

3.2. Spectroscopic Study of Interaction between Thulium(III) and ASA III. The absorption spectra of the ASA with the addition of Tm(III) are shown in Figure 3(a). When the concentration of Tm(III) was increased, the absorption peak intensity of ASA at approximately 550 nm exhibited that it gradually decreased with a slight red shift. In addition, two new absorption peaks were, respectively, formed at about 605 nm and 655 nm. An isosbestic point at 575 nm also provided an evidence for the new Tm(III)-ASA complex formation.

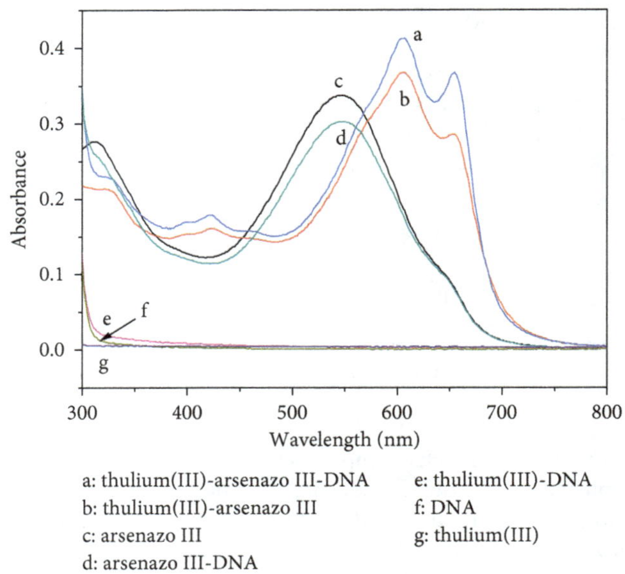

FIGURE 2: Absorption spectra of seven kinds of samples (pH 7.40). $c_{\text{Arsenazo III}} = 5 \times 10^{-5}$ mol·L^{-1}; $c_{\text{Thulium (III)}} = 1 \times 10^{-4}$ mol·L^{-1}; $c_{DNA} = 1 \times 10^{-5}$ mol·L^{-1}.

In order to obtain the stoichiometry of Tm(III)-ASA complex, the molar ratio method was executed based on the absorbance intensity of Tm(III)-ASA complex at 605 nm. The mole ratio [33] plots of ASA with Tm(III) are shown in Figure 3(b), and the binding ratio of the complex was $n_{Tm(III)}$: $n_{ASA} = 2{:}3$. According to the Lambert–Beer law, $A = \varepsilon bc$, where A is the absorbance intensity of Tm(III)-ASA and c is the concentration of Tm(III)-ASA. The apparent molar absorption coefficient of Tm(III)-ASA complex was determined, $\varepsilon = 7.94\times10^3$ L·mol^{-1}·cm^{-1}.

3.3. Spectroscopic Study of Thulium(III)-Arsenazo III Complex Interaction with DNA. The absorption spectra of Tm(III)-ASA complex in the presence of different concentrations of DNA are shown in Figure 4(a). From Figure 4(a), the absorption peak of the Tm(III)-ASA complex at about 550 nm. With the addition of DNA, the absorbance peak of Tm(III)-ASA complex decreased gradually, and a red shift was observed from 550 to 560 nm which leaded to hypochromic effect. In addition, two new absorption peaks, respectively, appeared at about 605 nm and 655 nm. It indicated that the Tm(III)-ASA complex could strongly interact with DNA.

The molar ratio method was used to obtained the stoichiometry of the Tm(III)-ASA complex with DNA at 655 nm. Figure 4(b) showed that the binding ratio of the complex was $n_{Tm(III)\text{-}ASA}$: $n_{DNA} = 12{:}1$. According to the

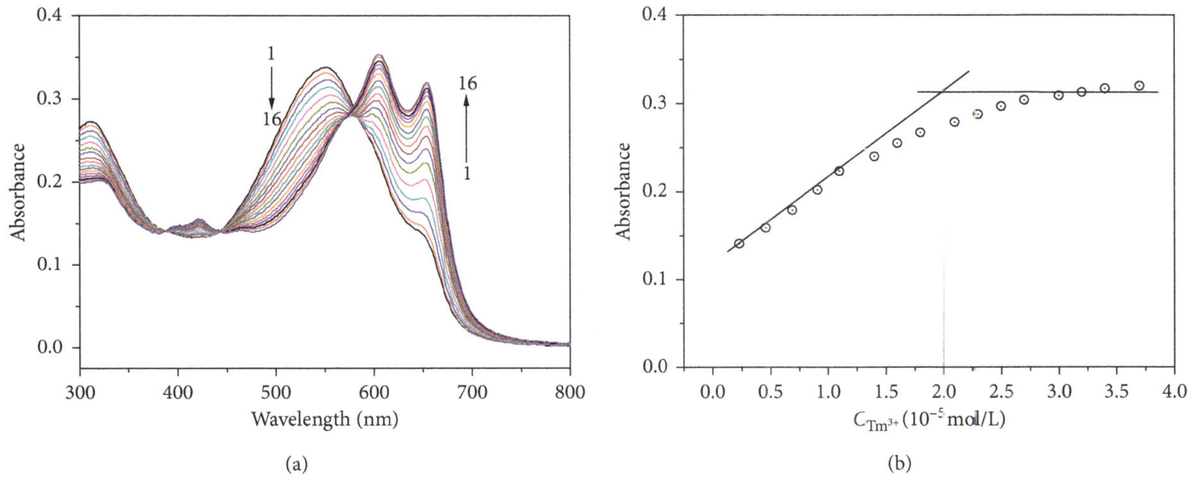

FIGURE 3: (a) Absorption spectra of ASA in different concentrations of thulium (III); (b) molar ratio of thulium(III)-arsenazo III complex at $\lambda = 605$ nm (pH 7.40). $c_{\text{Arsenazo III}} = 5\times10^{-5}$ mol·L^{-1}; $c_{\text{Thulium (III)}} = 1\times10^{-4}$ mol·L^{-1}; $c_{\text{DNA}} = 1\times10^{-5}$ mol·L^{-1} (10 μL per scan, 1–16: 0–150 μL).

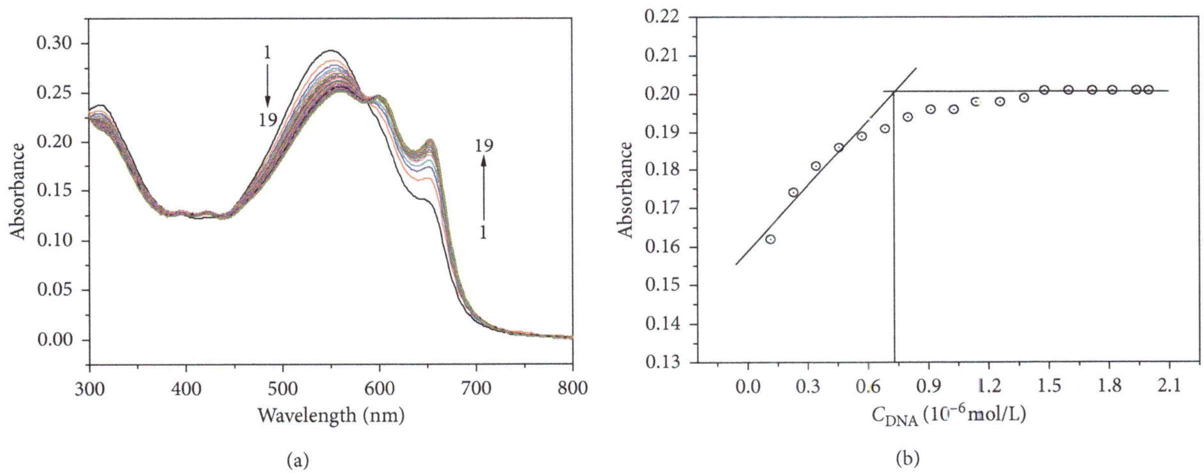

FIGURE 4: (a) Absorption spectra of thulium(III)-arsenazo III complex in different concentrations of DNA; (b) molar ratio of DNA-thulium (III)-arsenazo III system at $\lambda = 655$ nm (pH 7.40). $c_{\text{Arsenazo III}} = 5\times10^{-5}$ mol·L^{-1}; $c_{\text{Thulium (III)}} = 1\times10^{-4}$ mol·L^{-1}; $c_{\text{DNA}} = 1\times10^{-5}$ mol·L^{-1}. (10 μL per scan, 1–19: 0–180 μL).

Lambert–Beer law $A = \varepsilon bc$, and the apparent molar absorption coefficient, $\varepsilon = 4.71\times10^5$ L·mol^{-1}·cm^{-1} was calculated.

3.4. Thermodynamic Analysis. The value of the binding constant K of Tm(III)-ASA complex with DNA was expressed by the double reciprocal equation [34]:

$$\frac{1}{(A_0 - A)} = \frac{1}{A_0} + \frac{1}{(K \times A_0 \times c_{\text{DNA}})}, \tag{1}$$

where A_0 and A are the absorbance of Tm(III)-ASA complex in the absence and in the presence of DNA, respectively. K is the binding constant of Tm(III)-ASA complex with DNA and c_{DNA} is the concentration of DNA. The double-reciprocal plots of $1/(A_0 - A)$ versus $1/c_{\text{DNA}}$ were linear at 298.15K and 308.15K, respectively. The binding constants K were calculated from the ratio of the intercept on the vertical (Figure 5): $K^\theta_{298.15\text{K}} = 4.84\times10^6$ L·mol^{-1} and $K^\theta_{308.15\text{K}} = 4.48\times10^6$ L·mol^{-1}. The interaction forces between small molecules and DNA mainly include hydrophobic force, electrostatic interaction, Van der Waals force, and hydrogen bonds [35]. According to Ross' view [36], the main binding force can be judged by the standard enthalpy change ($\Delta_r H^\theta_m$) and the standard entropy change ($\Delta_r S^\theta_m$). If the temperature was a little changed, the $\Delta_r H^\theta_m$ could be regarded as a constant, and then, the value of $\Delta_r H^\theta_m$ is calculated from the van't Hoff equation [37]:

$$\ln \frac{K^\theta_2}{K^\theta_1} = \frac{-\Delta_r H^\theta_m (1/T_2 - 1/T_1)}{R}, \tag{2}$$

where K^θ_1 and K^θ_2 are the standard binding constants of Tm (III)-ASA complex with DNA at 298.15K and 308.15K, respectively and R is gas constant. So, the value of $\Delta_r H^\theta_m$ is obtained, $\Delta_r H^\theta_m = -5.903\times10^3$ J·mol^{-1}. According to the

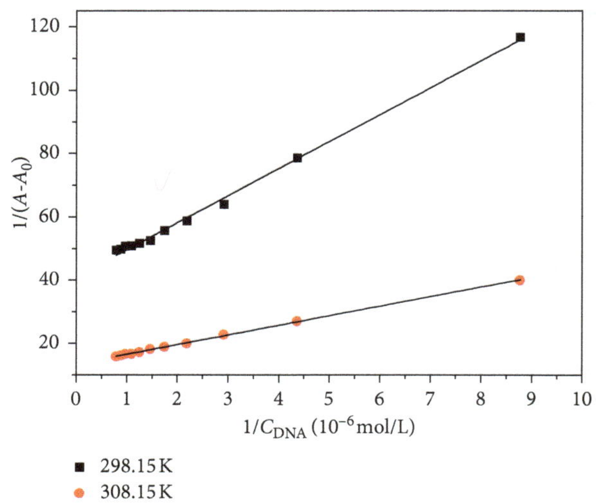

Figure 5: Double reciprocal plots of DNA-thulium(III)-arsenazo III system (pH 7.40; $\lambda = 655$ nm). $c_{\text{Arsenazo III}} = 5 \times 10^{-5}$ mol·L^{-1}; $c_{\text{Thulium (III)}} = 1 \times 10^{-4}$ mol·L^{-1}; $c_{\text{DNA}} = 1 \times 10^{-5}$ mol·L^{-1} (10 μL per scan, 1–19: 0–180 μL).

Equations (3) and (4), the standard free energy change $\Delta_r G_m^\theta$ and the standard entropy change $\Delta_r S_m^\theta$ are calculated, respectively.

$$\Delta_r G_m^\theta = -RT \ln K^\theta, \quad (3)$$

$$\Delta_r G_m^\theta = \Delta_r H_m^\theta - T\Delta_r S_m^\theta. \quad (4)$$

From the Equation (3), the standard molar reaction Gibbs free energy $\Delta_r G_m^\theta$ can be obtained, $\Delta_r G_m^\theta = -3.82 \times 10^4$ J·mol^{-1}. So, the value of $\Delta_r S_m^\theta$ is 108.33 J·mol^{-1}·K^{-1}. Due to the $\Delta_r G_m^\theta < 0$, it showed that the binding process of Tm(III)-ASA complex with DNA is always spontaneous. However, the enthalpy contribution was calculated as −5.903 kJ/mol for dG = −38.2 kJ/mol. It follows that the entropic contribution equals to 38–5.9 kJ/mol which therefore appears to be dominating over the enthalpic contribution.

3.5. Fluorescence Spectrophotometry for Thulium(III)-Arsenazo III Complex and 4S Green Probe Competitive Interaction with DNA. The direct use of the fluorescence emission of DNA has a major limitation in the study of its related properties because the fluorescence of DNA is very weak. 4S green plus nucleic acid stain (4S Green) is a kind of cationic dye, for its planar aromatic chromophore, which can intercalate into the base pairs of double-stranded DNA and the fluorescence intensity is greatly enhanced. So, it can act as an indicator of intercalation. On excitation at 490 nm, 4S Green has a strong fluorescence emission peaked at 525 nm. Figure 6(a) showed that the maximum emission intensity of 4S Green gradually increased with increasing concentration of DNA, and a slight blue shift took place. This result indicated that 4S Green could intercalate into the base pair of DNA and form a new complex.

The fluorescence intensity of the DNA-4S Green system gradually decreased with the increased concentration of the Tm(III)-ASA complex at 525 nm (Figure 6(b)). Thus, there was a competition mode between 4S Green and the Tm(III)-ASA complex with DNA, which led to the change of fluorescence of spectrum [38]. Figure 6(b) shows that the fluorescence of the DNA-4S Green system is efficiently quenched by adding the Tm(III)-ASA complex. These phenomena proved that the Tm(III)-ASA complex intercalated into the double helix of the DNA by exchanging with 4S Green. In addition, it also indicated that the binding mode of the Tm(III)-ASA complex with DNA was the same as 4S Green with DNA. [39].

3.6. Scatchard Method. The fluorescence Scatchard analysis is usually used to study the influence of AO to the DNA-Tm (III)-ASA system and confirm the binding modes [40]. Scatchard equation [41] expresses the binding mode of the DNA-Tm(III)-ASA complex in the presence of different concentrations of AO probe.

$$\frac{r_{\text{AO}}}{c_{\text{AO}}} = K(n - r_{\text{AO}}), \quad (5)$$

where r_{AO} refers to the molecular amount of bound AO to total DNA concentration; c_{AO} is the concentration of free AO; n is the number of binding sites of DNA; and K is the intrinsic binding constant of DNA-AO. There are different concentrations of the Tm(III)-ASA complex ($R_t = c_{\text{Tm (III)-ASA}}/c_{\text{DNA}}$; a, $R_t = 0.00$; b, $R_t = 0.20$; c, $R_t = 0.40$; and d, $R_t = 0.60$, respectively), and then the samples were titrated by AO solution, and the fluorescence intensity was measured. If the values of n were equal at different R_t, the Tm(III)-ASA complex would be regarded as an intercalation binding mode, and it would be regarded as a disintercalation binding mode if the values of K were equal. However, there were mix binding modes that contained disintercalation and intercalation binding if both the values of n and K were changed. [42, 43] In order to testify if there was electrostatic binding between the Tm(III)-ASA complex and DNA, the researcher used 0.05 mol/L NaCl in the Scatchard studies. The Scatchard plots in the absence of NaCl and the presence of NaCl are shown in Figures. 7(a) and 7(b), respectively.

According to the Scatchard equation (5) and from the Scatchard plots, the values of n and K were calculated. From Table 1, if the value of n is increased in the presence of NaCl, these results indicated that there was no electrostatic binding [44], but disintercalation binding mode-groove interaction existed in the DNA-Tm(III)ASA system because of the variation of the parameters n and K that are different from each other. In addition, a part of the free AO molecules would combine with DNA molecules, the binding between Tm(III)-ASA and DNA was weak, and the values of n increased. Thus, the interaction mode between the Tm(III)-ASA complex and DNA contained groove and weak intercalation bindings.

3.7. Viscosity Measurements. Viscosity measurement is one of the most effective methods to judge the binding modes between small molecules and DNA than spectral measurements [45, 46]. Generally speaking, classical intercalation

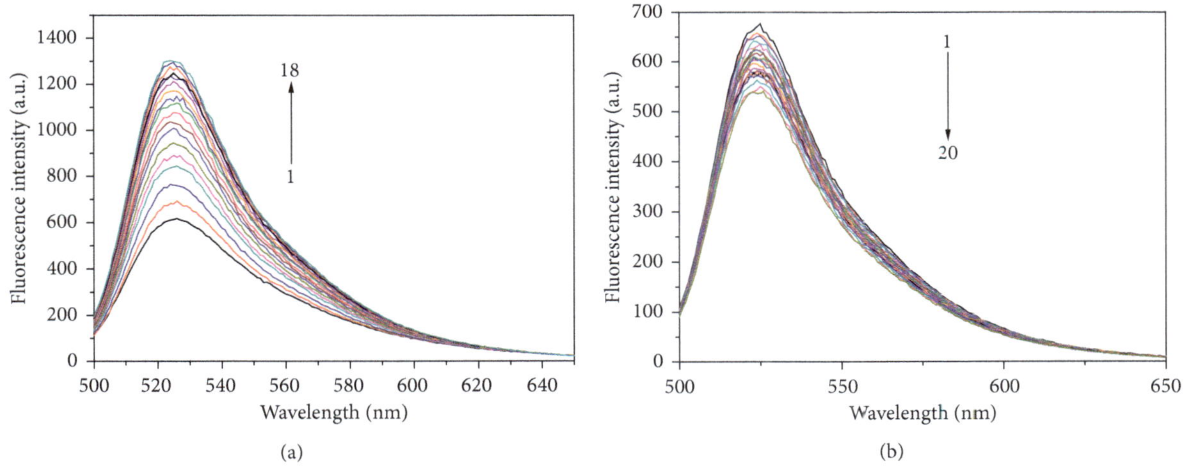

FIGURE 6: (a) Fluorescence spectra of 4S Green in different concentrations of DNA (pH 7.40; $\lambda_{ex} = 490$ nm, $\lambda_{em} = 525$ nm). $c_{4S\ Green} = 2.5\times10^{-6}$ mol·L^{-1}; $c_{DNA} = 1\times10^{-5}$ mol·L^{-1} (10 μL per scan, 1–18: 0–170 μL); (b) fluorescence spectra of DNA-4S Green system in different concentrations of thulium(III)-arsenazo III complex (pH 7.40; $\lambda_{ex} = 490$ nm, $\lambda_{em} = 525$ nm). $c_{4S\ Green} = 2.5\times10^{-6}$ mol·L^{-1}; $c_{DNA} = 1\times10^{-6}$ mol·L^{-1}; $c_{Thulium\ (III)\text{-}arsenazo\ III} = 5.00\times10^{-4}$ mol·L^{-1} (10 μL per scan, 1–18: 0–170 μL).

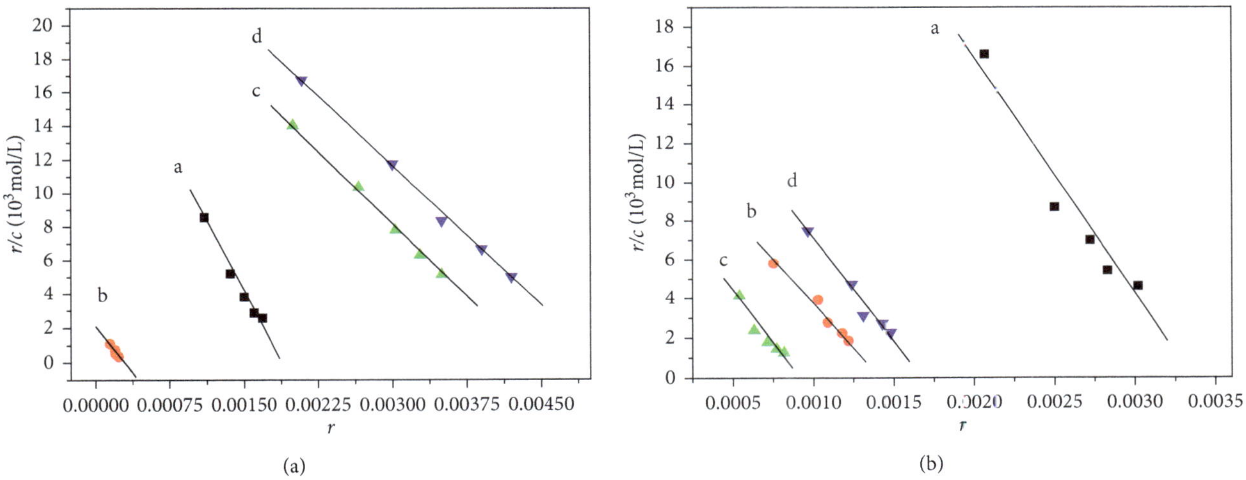

FIGURE 7: Scatchard plots of the interaction between thulium(III)-arsenazo III and DNA in different concentrations of AO. (a) Without NaCl; (b) with NaCl (pH = 7.40). $c_{Thulium\ (III)\text{-}arsenazo\ III} = 8.30\times10^{-6}$ mol/L; $c_{DNA} = 4.00\times10^{-6}$ mol/L; $c_{AO} = 4.00\times10^{-5}$ mol/L (10 μL per can); $Rt = c_{Thulium\ (III)\text{-}arsenazo\ III}/c_{DNA}$; a, Rt = 0.00; b, Rt = 0.20; c, Rt = 0.40; d, Rt = 0.60.

TABLE 1: Data from the Scatchard equation on the interaction between thulium(III)-arsenazo III complex and DNA.

Curve	$c_{Thulium(III)\text{-}arsenazo\ III}/c_{DNA}$	NaCl (mol/L)	Scatchard plot values	K (L/mol)	n
a	0.00	0.00	$r/c = 1.99\times10^4 - 1.05\times10^7$ r	1.05×10^7	0.00169
		0.05	$r/c = 4.23\times10^4 - 1.28\times10^7$ r	1.28×10^7	0.00207
b	0.20	0.00	$r/c = 2.35\times10^3 - 8.63\times10^6$ r	8.63×10^6	0.00027
		0.05	$r/c = 1.24\times10^4 - 8.58\times10^6$ r	8.58×10^6	0.00122
c	0.40	0.00	$r/c = 2.61\times10^4 - 6.01\times10^6$ r	6.01×10^6	0.00404
		0.05	$r/c = 9.14\times10^3 - 9.93\times10^6$ r	9.93×10^6	0.00082
d	0.60	0.00	$r/c = 2.85\times10^4 - 5.62\times10^6$ r	5.62×10^6	0.00420
		0.05	$r/c = 1.74\times10^4 - 1.04\times10^7$ r	1.04×110^7	0.00149

leads to an increase in the viscosity of DNA solution because the length of the double-helical structure of the DNA is increased for the insertion ligand. Groove and electrostatic bindings cause no obvious increase in DNA viscosity. Figure 8 shows that the value of relative viscosity was enhanced by increasing the concentrations of the Tm(III)-ASA complex.

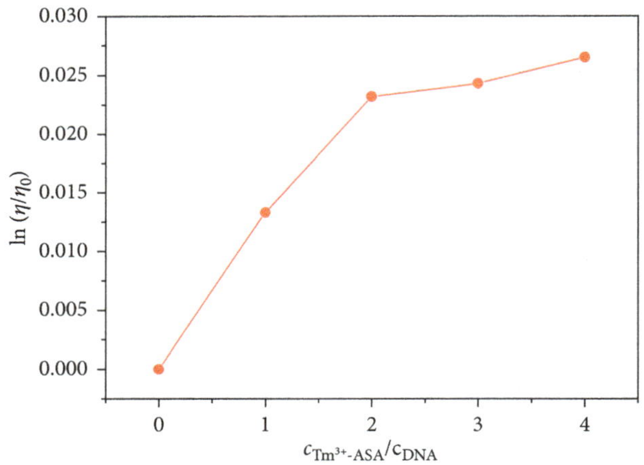

FIGURE 8: Effect viscosity of thulium(III)-arsenazo III on DNA (pH = 7.40). $c_{DNA} = 2.00 \times 10^{-5}$ mol/L; $c_{\text{Thulium (III)-arsenazo III}}/c_{DNA} = 0$, 1, 2, 3, 4.

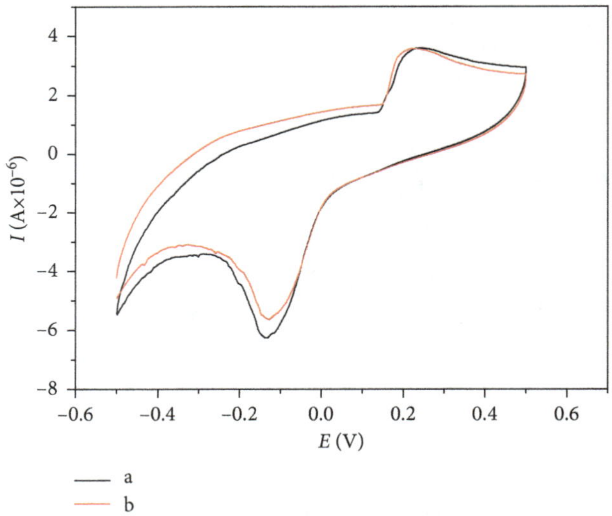

FIGURE 9: Cyclic voltammograms of thulium(III)-arsenazo III complex in absence of and presence of DNA (50.0 mM tris-HCl buffer, pH 7.40). $c_{DNA} = 1 \times 10^{-5}$ mol·L^{-1}; $c_{\text{Thulium (III)-arsenazo III}} = 5.00 \times 10^{-6}$ mol·L^{-1}; scan rate = 50 mV/s.

This result indicated that the interaction method of the Tm (III)-ASA complex with DNA should be intercalation binding.

3.8. *Cyclic Voltammetric Measurements.* The electrochemical study on the interaction of Tm(III)-ASA complex with DNA has seldom been presented although numerous studies on the interaction of some drugs with DNA have been reported [47]. In order to investigate the interaction of DNA with the Tm(III)-ASA complex, cyclic voltammograms of electrodes in 50.0 mM tris-HCl buffer solution (pH 7.40) which contains a certain concentration of Tm(III)-ASA complex in the presence of DNA and absence of DNA were used for a voltammetric test. As seen in Figure 9, it was noteworthy that a pair redox peaks appeared, and the peaks potential at about −0.136 V and 0.231 V where produced by the Tm(III)-ASA complex (curve a). After DNA addition, the oxidation peak potential of the corresponding complex shifted towards the more positive potential, and the peak current increased (curve b). This suggested that the Tm(III)-ASA complex interacts with DNA. Because numerous studies showed that if there was a shift towards the negative potential, interaction between DNA and complex is mainly electrostatic. On the contrary, when the shift of peak happens towards the positive potential, there is the interaction of small molecules and DNA [48, 49]. In this study, the peak potential appears at more positive values in the presence of DNA, which also indicated that an intercalation binding should be the interaction mode of the Tm(III)-ASA complex with DNA.

4. Conclusion

In summary, the interaction of Tm(III)-ASA complex with DNA was demonstrated in detail. Hypochromicity and red shift of the absorption spectra of Tm(III)-ASA complex were obtained in the presence of DNA which provided an evidence for the new Tm(III)-ASA-DNA complex formation. A competitive reaction of the Tm(III)-ASA complex and 4S Green with DNA was monitored by fluorescence spectroscopy, and it indicated that the Tm(III)-ASA complex could intercalate into the double helix of DNA, but the intercalation binding mode was weaker than 4S Green. The researchers believed that the interaction information between Tm(III)-ASA complex and DNA would provide a useful reference on the interaction mechanism of rare earth complexes with DNA, which would be beneficial to design highly efficient anticancer and antitumor drugs for most of the researchers.

Conflicts of Interest

The authors declare that they have no conflicts of interest.

Acknowledgments

This work was supported by NSFC (Grant nos. 21565015 and 21663014), the Foundation of Jiangxi Educational Committee (GJJ150363), the State Key Laboratory of Chemical Biosensing & Chemometrics (Z2015022), and the Key Laboratory of Functional Small organic molecule (KLFS-KF-201425).

References

[1] G. W. Zhang, J. B. Guo, N. Zhao, and J. R. Wang, "Study of interaction between kaempferol-Eu^{3+} complex and DNA with the use of the Neutral Red dye as a fluorescence probe," *Sensors and Actuators B: Chemical*, vol. 144, no. 1, pp. 239–246, 2010.

[2] A. Lauria, R. Bonsignore, A. Terenzi et al., "Nickel(II), copper (II) and zinc(II) metallo-intercalators: structural details of the DNA-binding by a combined experimental and computational investigation," *Dalton Transactions*, vol. 43, no. 16, pp. 6108–6119, 2014.

[3] Z. Asadi, N. Nasrollahi, H. Karbalaei-Heidari et al., "Investigation of the complex structure, comparative DNA-binding and DNA cleavage of two water-soluble mono-nuclear lanthanum(III) complexes and cytotoxic activity of chitosan-coated magnetic nanoparticles as drug delivery for the complexes," *Spectrochimica Acta Part A: Molecular and Biomolecular Spectroscopy*, vol. 178, pp. 125–135, 2017.

[4] N. Zhao, X. M. Wang, H. Z. A. Pan, Y. M. Hu, and L. S. Ding, "Spectroscopic studies on the interaction between tryptophan-erbium(III) complex and herring sperm DNA," *Spectrochimica Acta Part A: Molecular and Biomolecular Spectroscopy*, vol. 75, no. 5, pp. 1435–1442, 2010.

[5] H. J. Sun, A. L. Wang, H. B. Chu, and Y. L. Zhao, "Fluorescent studies on the interaction of DNA and ternary lanthanide complexes with cinnamic acid-phenanthroline and antibacterial activities testing," *Luminescence*, vol. 30, no. 2, pp. 131–136, 2015.

[6] G. Zhang, J. Guo, J. Pan, X. Chen, and J. Wang, "Spectroscopic studies on the interaction of morin–Eu(III) complex with calf thymus DNA," *Journal of Molecular Structure*, vol. 923, no. 1–3, pp. 114–119, 2009.

[7] G. J. Chen, X. Qiao, C. Y. Gao et al., "DNA binding, photo-induced DNA cleavage and cell cytotoxicity studies of a family of light rare earth complexes," *Journal of Inorganic Biochemistry*, vol. 109, pp. 90–96, 2012.

[8] J. Sangster, "C-Na (Carbon-Sodium) system," *Journal of Phase Equilibria and Diffusion*, vol. 28, no. 6, pp. 571–579, 2007.

[9] E. Singer and M. Matucha, "Erfahrungen mit der bestimmung von Uran in Erzen und Gesteinen mit Arsenazo III," *Fresenius' Zeitschrift für analytische Chemie*, vol. 191, no. 4, pp. 248–253, 1962.

[10] K. Matharu, S. K. Mittal, S. K. Ashok Kumar, and S. K. Sahoo, "Selectivity enhancement of Arsenazo(III) reagent towards heavier lanthanides using polyaminocarboxylic acids: a spectrophotometric study," *Spectrochimica Acta Part A: Molecular and Biomolecular Spectroscopy*, vol. 145, pp. 165–175, 2015.

[11] H. Rohwer, N. Collier, and E. Hosten, "Spectrophotometric study of arsenazo III and its interactions with lanthanides," *Analytica Chimica Acta*, vol. 314, no. 3, pp. 219–223, 1995.

[12] E. Hosten and H. Rohwer, "Interaction of anions with arsenazo III-lanthanide (III) complexes," *Analytica Chimica Acta*, vol. 345, no. 1–3, pp. 227–233, 1997.

[13] N. N. Basargin, V. M. Ivanov, V. V. Kuznetsov, and A. V. Mikhailova, "40 years since the discovery of the arsenazo III reagent," *Journal of Analytical Chemistry*, vol. 55, no. 3, pp. 204–210, 2000.

[14] N. K. Modukuru, K. J. Snow, B. S. Perrin, A. Bhambhani, M. Duff, and C. V. Kumar, "Tuning the DNA binding modes of an anthracene derivative with salt," *Journal of Photochemistry and Photobiology A: Chemistry*, vol. 177, no. 1, pp. 43–54, 2006.

[15] X. L. Xiong, N. Zhao, and X. M. Wang, "Interaction between tryptophan-Sm(III) complex and DNA with the use of a acridine orange dye fluorophor probe," *Luminescence*, vol. 31, no. 1, pp. 210–216, 2016.

[16] N. Ganji, A. Rambabu, N. Vamsikrishna, S. Daravath, and Shivaraj, "Copper(II) complexes with isoxazole Schiff bases: synthesis, spectroscopic investigation, DNA binding and nuclease activities, antioxidant and antimicrobial studies," *Journal of Molecular Structure*, vol. 1173, pp. 173–182, 2018.

[17] J. F. Li and C. Dong, "Study on the interaction of morphine chloride with deoxyribonucleic acid by fluorescence method," *Spectrochimica Acta Part A: Molecular and Biomolecular Spectroscopy*, vol. 71, no. 5, pp. 1938–1943, 2009.

[18] N. Ataci, E. Ozcelik, and N. Arsu, "Spectrophotometric study on binding of 2-thioxanthone acetic acid with ct-DNA," *Spectrochimica Acta Part A: Molecular and Biomolecular Spectroscopy*, vol. 204, pp. 281–286, 2018.

[19] E. Grueso, P. Perez-Tejeda, R. M. Giráldez-Pérez, R. Prado-Gotor, and F. Muriel-Delgado, "Ethanol effect on gold nanoparticle aggregation state and its implication in the interaction mechanism with DNA," *Journal of Colloid and Interface Science*, vol. 529, pp. 65–76, 2018.

[20] T. Thirunavukkarasu, H. A. Sparkes, and K. Natarajan, "Quinoline based Pd(II) complexes: synthesis, characterization and evaluation of DNA/protein binding, molecular docking and in vitro anticancer activity," *Inorganica Chimica Acta*, vol. 482, pp. 229–239, 2018.

[21] Mudasir, N. Yoshioka, and H. Inoue, "DNA binding of iron (II) mixed-ligand complexes containing 1,10-phenanthroline and 4,7-diphenyl-1,10-phenanthroline," *Journal of Inorganic Biochemistry*, vol. 77, no. 3-4, pp. 239–247, 1999.

[22] Y. Zhao, Z. Li, H. Li, S. Wang, and M. Niu, "Synthesis, crystal structure, DNA binding and in vitro cytotoxicity studies of Zn (II) complexes derived from amino-alcohol schiff-bases," *Inorganica Chimica Acta*, vol. 482, pp. 136–143, 2018.

[23] A. Kapur, J. L. Beck, and M. M. Sheil, "Observation of daunomycin and nogalamycin complexes with duplex DNA using electrospray ionisation mass spectrometry," *Rapid Communications in Mass Spectrometry*, vol. 13, no. 24, pp. 2489–2497, 1999.

[24] X. Jiang and X. Lin, "Voltammetry of the interaction of metronidazole with DNA and its analytical applications," *Bioelectrochemistry*, vol. 68, no. 2, pp. 206–212, 2006.

[25] S. Q. Liu, M. L. Cao, and S. L. Dong, "Electrochemical and Ultraviolet–visible spectroscopic studies on the interaction of deoxyribonucleic acid with vitamin B6," *Bioelectrochemistry*, vol. 74, no. 1, pp. 164–169, 2008.

[26] M. Carcelli, P. Mazza, C. Pelizzi, G. Pelizzi, and F. Zani, "Antimicrobial and genotoxic activity of 2,6-diacetylpyridine bis(acylhydrazones) and their complexes with some first transition series metal-ions - X-ray crystal-structure of a dinuclear copper(II) complex," *Journal of Inorganic Biochemistry*, vol. 57, no. 1, pp. 43–62, 1995.

[27] S. Bi, C. Qiao, D. Song et al., "Study of interactions of flavonoids with DNA using acridine orange as a fluorescence probe," *Sensors and Actuators B: Chemical*, vol. 119, no. 1, pp. 199–208, 2006.

[28] C. Y. Qiao, S. Y. Bi, Y. Sun, D. Q. Song, H. Q. Zhang, and W. H. Zhou, "Study of interactions of anthraquinones with DNA using ethidium bromide as a fluorescence probe," *Spectrochimica Acta Part A: Molecular and Biomolecular Spectroscopy*, vol. 70, no. 1, pp. 136–143, 2008.

[29] M. Xu, F. J. Chen, L. Huang, P. X. Xi, and Z. Z. Zeng, "Binding of rare earth metal complexes with an ofloxacin derivative to bovine serum albumin and its effect on the conformation of protein," *Journal of Luminescence*, vol. 131, no. 8, pp. 1557–1565, 2011.

[30] Y. Ni, S. Du, and S. Kokot, "Interaction between quercetin–copper(II) complex and DNA with the use of the Neutral Red dye fluorophor probe," *Analytica Chimica Acta*, vol. 584, no. 1, pp. 19–27, 2007.

[31] S. Mahadevan and M. Palaniandavar, "Spectroscopic and voltammetric studies on copper complexes of 2,9-dimethyl-1,10-phenanthrolines bound to calf thymus DNA," *Inorganic Chemistry*, vol. 37, no. 4, pp. 693–700, 1998.

[32] L. Z. Zhang and G. Q. Tang, "The binding properties of photosensitizer methylene blue to herring sperm DNA: a spectroscopic study," *Journal of Photochemistry and Photobiology B: Biology*, vol. 74, no. 2-3, pp. 119–125, 2004.

[33] P. X. Xi, Z. H. Xu, X. H. Liu et al., "Synthesis, characterization, antioxidant activity and DNA-binding studies of three rare Earth (III) complexes with 1-(4-aminoantipyrine)-3-tosylurea ligand," *Journal of Fluorescence*, vol. 19, no. 1, pp. 63–72, 2009.

[34] A. A. Ouameur, R. Marty, and H. A. Tajmir-Riahi, "Human serum albumin complexes with chlorophyll and chlorophyllin," *Biopolymers*, vol. 77, no. 3, pp. 129–136, 2005.

[35] H. Ojha, B. M. Murari, S. Anand, M. I. Hassan, F. Ahmad, and N. K. Chaudhury, "Interaction of DNA minor groove binder hoechst 33258 with bovine serum albumin," *Chemical & Pharmaceutical Bulletin*, vol. 57, no. 5, pp. 481–486, 2009.

[36] P. D. Ross and S. Subramanian, "Thermodynamics of protein association reactions: forces contributing to stability," *Biochemistry*, vol. 20, no. 11, pp. 3096–3102, 1981.

[37] G. W. Zhang, X. Hu, and J. H. Pan, "Spectroscopic studies of the interaction between pirimicarb and calf thymus DNA," *Spectrochimica Acta Part A: Molecular and Biomolecular Spectroscopy*, vol. 78, no. 2, pp. 687–694, 2011.

[38] N. Shahabadi, S. Kashanian, and M. Purfoulad, "DNA interaction studies of a platinum(II) complex, $PtCl_2(NN)$ (NN=4,7-dimethyl-1,10-phenanthroline), using different instrumental methods," *Spectrochimica Acta Part A: Molecular and Biomolecular Spectroscopy*, vol. 72, no. 4, pp. 757–761, 2009.

[39] B. Jing, M. Zhang, and T. Shen, "[Ruthenium(II)$(bpy)_2L]^{2+}$, where L are imidazo[f]-1,10-phenanthrolines: synthesis, photophysics and binding with DNA," *Spectrochimica Acta Part A: Molecular and Biomolecular Spectroscopy*, vol. 60, no. 11, pp. 2635–2641, 2004.

[40] Y. Gao, J. S. Li, G. X. Huang, L. J. Yan, and Z. Dong, "Spectroscopic studies on the interaction between anthragallol and DNA using of ethidium bromide as a fluorescence probe," *Spectrochimica Acta Part A: Molecular and Biomolecular Spectroscopy*, vol. 141, pp. 239–243, 2015.

[41] H. Mansuri-Torshizi, R. Mital, T. S. Srivastava, H. Parekh, and M. P. Chitnis, "Synthesis, characterization, and cytotoxic studies of α-diimine/1,2-diamine platinum(II) and palladium (II) complexes of selenite and tellurite and binding of some of these complexes to DNA," *Journal of Inorganic Biochemistry*, vol. 44, no. 4, pp. 239–247, 1991.

[42] Z. G. Zhang and X. D. Dong, "Interaction of DNA with a novel photoactive platinum diimine complex," *Biometals*, vol. 22, no. 2, pp. 283–288, 2009.

[43] Y. L. Zhou and Y. Z. Li, "Studies of the interaction between poly(diallyldimethyl ammonium chloride) and DNA by spectroscopic methods," *Colloids and Surfaces A: Physicochemical and Engineering Aspects*, vol. 233, no. 1–3, pp. 129–135, 2004.

[44] S. Tabassum, S. Parveen, and F. Arjmand, "New modulated metallic macrocycles: electrochemistry and their interaction with calf thymus DNA," *Acta Biomaterialia*, vol. 1, no. 6, pp. 677–689, 2005.

[45] S. Satyanarayana, J. C. Dabrowiak, and J. B. Chaires, "Tris (phenanthroline)ruthenium(II) enantiomer interactions with DNA - mode and specificity of binding," *Biochemistry*, vol. 32, no. 10, pp. 2573–2584, 1993.

[46] L. B. Liao, H. Y. Zhou, and X. M. Xiao, "Spectroscopic and viscosity study of doxorubicin interaction with DNA," *Journal of Molecular Structure*, vol. 749, no. 1–3, pp. 108–113, 2005.

[47] H. E. Satana, A. D. R. Pontinha, V. C. Diculescu, and A. M. Oliveira-Brett, "Nucleoside analogue electrochemical behaviour and in situ evaluation of DNA–clofarabine interaction," *Bioelectrochemistry*, vol. 87, pp. 3–8, 2012.

[48] A. Noorbakhsh and A. Salimi, "Development of DNA electrochemical biosensor based on immobilization of ssDNA on the surface of nickel oxide nanoparticles modified glassy carbon electrode," *Biosensors and Bioelectronics*, vol. 30, no. 1, pp. 188–196, 2011.

[49] I. Kocak, U. Yildiz, B. Coban, and A. Sengul, "DNA-binding studies of complex of $Pt(bpy)(pip)]^{2+}$ and $[Pt(bpy)(hpip)]^{2+}$ by electrochemical methods: development of an electrochemical DNA biosensor," *Journal of Solid State Electrochemistry*, vol. 19, no. 7, pp. 2189–2197, 2015.

Kinetic Study of the Bioadsorption of Methylene Blue on the Surface of the Biomass Obtained from the Algae *D. antarctica*

Jhonatan R. Guarín,[1] Juan Carlos Moreno-Pirajan,[1] and Liliana Giraldo[2]

[1]*Department of Chemistry, Sciences Faculty, Research Group in Porous Solids and Calorimetry, University of the Andes, Bogota 111711, Colombia*
[2]*Departamento de Química, Facultad de Ciencias, Universidad Nacional de Colombia, Bogota 111321, Colombia*

Correspondence should be addressed to Juan Carlos Moreno-Pirajan; jumoreno@uniandes.edu.co

Academic Editor: Carlos A. Martínez-Huitle

Currently, there is a great pollution of water by the dyes; due to this, several studies have been carried out to remove these compounds. However, the total elimination of these pollutants from the aquatic effluents has represented a great challenge for the scientific community, for which it is necessary to carry out investigations that allow the purification of water. In this work, we studied the bioadsorption of methylene blue on the surface of the biomass obtained from the algae *D. antarctica*. This material was characterized by SEM and FTIR. To the data obtained in the biosorption experiments, different models of biosorption and kinetics were applied, finding that the best fit to the obtained data is given by applying the pseudo-second-order models and the Toth model, respectively. It was also determined that the maximum adsorption capacity of MB on the surface of the biomass is 702.9 mg/g, which shows that this material has great properties as a bioadsorbent.

1. Introduction

Water pollution by various industries is the main problem the world is facing these times. Among these pollutants are the dyes [1, 2]; their world production per year is approximately 7×10^5 MT, and around 10 to 15% of the dyes used are dumped in waste water [3]. Synthetic dyes are more difficult to remove than natural dyes [4], and this is due to their complex aromatic structure, which gives them stability against air and light [1]. They can easily transport within the aqueous environment due to their high solubility in water [5]. They affect the aquatic autotrophs by restricting their photosynthetic efficiency because only limited sunlight is allowed to penetrate due to their colour; their toxic effect spreads along the food chain, and they are conserved for long periods deteriorating animal and human health [1].

Methylene blue (MB) is a cationic dye with broad applications such as dye for paper, hair, and cotton and filters for medical surgery [2] among others. This pollutant not only deteriorates water quality but also significantly affects the environment and human health [6]. It has been shown that, in humans, it causes vomiting, lightheadedness, cyanosis, jaundice, tissue necrosis [2], increased heart rate, vomiting, shock, and tetraplegia [7]. Therefore, there is a need to eliminate this dye present in wastewater or industrial effluents to ensure a safer environment [2]. This and the other dyes are difficult to degrade biologically, so the studies aimed at removing these contaminants present in aqueous solution are important and must be carried out [8, 9] to reduce environmental damage and human health.

Various technologies, including adsorption, flocculation, membrane filtration, electrolysis, biological treatments, oxidation [10], sedimentation, membrane separation, reverse osmosis [8], and photocatalytic degradation [11], have been adopted to remove dyes from sewage water. Among the mentioned treatments, adsorption has some advantages because the adsorbents are inexpensive, are available, are simple to use, and have high efficiency [8] and their design is simple [3]. However, the total removal of MB from the wastewater has not been achieved; therefore, it is necessary to continue with the studies to discover new bioadsorbents [6].

Many adsorbents have been studied to reduce the concentration of the dyes present in aqueous solutions, among which the activated carbon stands out for its high efficiency; however, its manufacture and regeneration involve a high cost. So, they have discovered and applied other natural adsorbents such as natural coal, chitin, chitosan, fly ash, wood sawdust, and rice husk [8].

Adsorption using biomass is considered a potential technique for the removal of contaminants present in water, and the advantage of using this adsorbent is that it is more economical than commercial adsorbents [12]. The scope of this research is to perform the kinetic study of the MB adsorption on the surface of the biomass obtained from the algae *D. antarctica* and to determine the capacity of adsorption of the contaminant (MB) by the bioadsorbent. To understand the mechanism of adsorption of this process, different kinetic and adsorption models were applied to the obtained data. Additionally, traditional instrumental techniques were used to characterize the bioadsorbent.

2. Materials and Methods

2.1. Preparation of the Biomass. Biomass brown alga was collected from the Chilean coast, for preparing samples of fresh alga washed with distilled water to remove impurities, and dried at 40°C using an oven (Memmert, Germany) for 72 hours; once the drying process was finished, the material was cut into small pieces, before being grounded, and the dried biomass was sieved to a particle size of 500–1000 μm. This material was used in all experiments [13].

2.2. Characterization. In order to evaluate the properties of the biomass surfaces obtained from the algae *D. antarctica*, which allow MB adsorption to be effective, this material was characterized by the techniques provided below.

2.2.1. Infrared Spectroscopy. The data of the IR spectra of the biomass before and after the adsorption of MB were recorded using an instrument SHIMADZU IRTracer-100. The pressed granules were prepared by grinding the biomass and mixing it with KBr in an agate mortar. The spectral data were recorded at wavenumber values between 4000 and 500 cm^{-1}.

2.2.2. Scanning Electron Microscopy. This technique was used to observe the morphology of the biomass surface before and after the MB adsorption process in a JEOL JSM-6490LV unit. For the SEM analysis, the sample obtained after the MB adsorption process was separated from the solution and dried at a temperature of 120°C for 5 h. Before carrying out this analysis, the surface of the biomass was coated with gold using the sputtering method to obtain a conductive surface.

2.3. Studies of the Biosorption of the Function of the pH. For the determination of the optimum pH of biosorption, methylene blue solutions of 50 ppm were prepared, whose pH was adjusted to values of 3, 5, 7, 9, and 11, after which the biomass was added in a ratio of 1 g/L, the samples were left in constant agitation for a period of 15 hours, and the MB concentration was determined at the beginning and end of the bioadsorption process by UV-Vis spectroscopy at a wavelength of 664 nm in a GENESYS 5 spectrophotometer.

2.4. Studies of Adsorption Kinetics. To prepare the kinetic studies, solutions of 25, 50, 100, 150, and 200 ppm of MB were prepared. According to the results obtained in Section 3.2, the optimum pH value of adsorption must be above 5; therefore, the pH of the solutions was adjusted to a value of 10 using 0.1 M·HCl and NaOH solutions. After this, the biomass was added in a ratio of 1 g/L, the samples were left in agitation, and aliquots were taken at 0, 5, 10, 15, 30, 60, 120, 240, 360, 480, and 560 minutes, to track the decrease in MB concentration present in the solution. Using (1), the values of q_t were determined for the different kinetic experiments carried out:

$$q_t = \frac{V(C_0 - C_t)}{m}, \tag{1}$$

where q_t is the amount of MB adsorbed per gram of the sample (mg/g) in time (t) of the adsorption process, C_0 is the initial concentration of the solution (mg/L), C_t is the concentration of the solution in time (t) of the adsorption process (mg/L), V is the initial volume of the solution, and m is the biomass added to the different solutions [14].

2.5. Studies of Equilibrium in the Solid-Liquid Interface. To understand the mechanism of adsorption of methylene blue on the surface of the biomass, solutions were prepared with initial concentrations of 25, 50, 100, 150, 200, 400, 600, 800, 100, and 1400 ppm of this contaminant; the pH of the solutions that were in the concentration range of 25 to 200 ppm was adjusted to a value of 10, and the pH value of the solutions that were in the concentration range of 400 to 1400 ppm was not adjusted since the pH value of these solutions was above 5. In Section 3.2, it was determined that, at basic pH values, the adsorption capacity of MB on the surface of the biomass is constant, so it was decided not to make adjustments to the pH value of these solutions.

The samples were left in agitation for 18 hours; using UV-Vis spectroscopy, the concentration of these solutions was determined at the beginning and at the end of the process. Using (2), the MB's adsorption capacity was determined by the biomass when the solution is in equilibrium at the end of this process:

$$q_e = \frac{V(C_0 - C_e)}{m}, \tag{2}$$

where q_e is the amount of MB adsorbed per gram of the sample (mg/g) at equilibrium and C_e is the concentration of MB in equilibrium (mg/L). The data obtained were adjusted to the models of Freundlich, Langmuir, Redlich–Peterson, Sips, and Toth, to understand the mechanism of adsorption of MB which takes place on the surface of biomass [14].

2.6. Thermodynamic Studies of the Kinetics of Adsorption. Speed laws are essential in any study of biosorption because they provide an exact expression during the duration of the reaction [15]. Then, the models used in the present study are described below.

2.6.1. Pseudo-First-Order and Pseudo-Second-Order Models. There are several models of kinetics, but in general, the most used and compared ones are the pseudo-first-order and pseudo-second-order kinetic models. The pseudo-first-order kinetic model was proposed for the first time at the end of the 19th century by Lagergren, and the speed constant when using this model is denoted as K_1 [16]. The pseudo-second-order kinetic model was introduced for the first time in the mid-80s, but its recognition increased in the year of 1999 when Ho and McKay [17] took data from experiments reported in the literature and determined that the best fit to the data coming from all the systems studied was given by applying the pseudo-second-order model. After this publication came to light, the speed constant K_2 has become more popular since in most studies, it has been found that the data obtained fit this model; for this reason, the original article has more than 6000 citations, a figure that has increased exponentially over the course of the time [18], and the equations of these two models are presented below.

A pseudo-first-order model is given as follows:

$$\log\left(1-\frac{q_t}{q_e}\right)=\frac{k_1}{2.303}t, \quad (3)$$

where q_t is the concentration of the adsorbed phase at a certain time of the adsorption process, q_e is the concentration of the adsorbed phase at the end of equilibrium with the solid present in suspension, t is the time that has elapsed since the process of adsorption began, and k_L is the Lagergren constant (s^{-1}). The pollutant adsorption rate is proportional to the time that has elapsed since the adsorption process began; for $T=0$, the value of $q_t=0$; once equilibrium has been reached, the value of $q_t \approx q_e$ [19].

A pseudo-second-order model is given as follows:

$$\frac{t}{q_t}=\frac{1}{k_2 q_e^2}+\frac{1}{q_e}t. \quad (4)$$

When t approaches zero, the bioadsorption velocity q_t/t becomes the initial bioadsorption velocity, and the data obtained in the kinetic study are adapted to the pseudo-second-order model by plotting the graph of t/q_t versus t. The linear relationship obtained will allow the calculation of q_e, k_2, and $k_2 q_e^2$ without previously knowing any parameter [20].

2.6.2. Modified Pseudo-First-Order Model. This kinetic model is useful when the experimentally obtained data do not conform to the pseudo-first-order and pseudo-second-order kinetic models and the model of intraparticle diffusion. In this new model, the pseudo-first-order equation is modified by the speed constant as observed in the following equation [21]:

$$\ln\left(1-\frac{q_t}{q_e}\right)+\frac{q_t}{q_e}=-k_m t. \quad (5)$$

Since $q_t < q_e$, this equation implies that the velocity constant k_1 is minimal when equilibrium has been reached [21].

2.6.3. Elovich Model. It has been found that Elovich's empirical adsorption model has broad applicability for numerous adsorption systems. This model is based on the assumption of energetic heterogeneity of the adsorption sites in the form of rectangular distribution [22].

The kinetic model of Elovich is reported as follows:

$$q_t=\frac{1}{\beta}\ln(\alpha\beta)+\frac{1}{\beta}\ln t, \quad (6)$$

where α is the initial adsorption rate of the Elovich equation ($mg{\cdot}g^{-1}\,min^{-1}$) and β is the adsorption constant of the model ($g{\cdot}mg^{-1}$) [15]; it is related to the adsorption energy [23].

The Elovich equation can be linearized by assuming $\alpha\beta t >> t$ and that $Q=0$ at $t=0$ and that $Q=Q_t$ for a time $t=t_t$ [24]. Under these conditions, a graph of q_t versus ln (t) shows a linear relationship with a slope of ($1/\beta$) and an intercept of ($1/\beta$) ln ($\alpha\beta$) [15].

This equation has been used by some authors to describe the kinetics of adsorption of contaminants in natural adsorbents, prepared or modified, which found that the best behavior of the data obtained is given when applying this model. But in spite of the utility of the Elovich equation in the estimation of the biosorption kinetics, the study of other kinetic models must be carried out, since these can present, in some cases, marginally superior correlation coefficients in the kinetics of adsorption of chemical substances [15].

This equation is also used to describe reaction mechanisms such as solute diffusion in the solution or interface phase, surface activation, and deactivation. It is adequate to elucidate the process with significant changes of activation energy. In addition, the Elovich equation can also reveal the irregularity of data that other dynamic equations have ignored [25].

2.6.4. Intraparticle Diffusion Model. According to chemical reaction fundamental theory, intraparticle diffusion becomes a limiting factor to the overall reaction rate when the catalyst particle size is too large and/or the intrinsic reaction rate is too fast compared to the diffusion rate of reacting molecules inside the catalyst pores. The degree of inhibition resulting from the overall velocity depends on the combination of particle size and shape, intrinsic reaction velocity, and diffusion rate [26].

According to literature [26], if the rate-limiting step is intraparticle diffusion, then the amount adsorbed at any time t should be directly proportional to the square root of the contact time t and pass through the origin.

Table 1: The names and equations of applied error functions [7].

Error function	Definition	Equation
RMSE	$\sqrt{\frac{1}{n-2}\sum_{I=1}^{N}\left(q_{e,exp}-q_{e,isotherm}\right)^2}$	(9)
X^2	$X^2=\sum_{1}^{N}\frac{\left(q_{e,exp}-q_{e,isotherm}\right)^2}{q_{e,isotherm}}$	(10)
ERRSQ	$\sum_{i=1}^{N}\left(q_{e,exp}-q_{e,isotherm}\right)_i^2$	(11)
HYBRD	$\sum_{i=1}^{N}\left[\frac{\left(q_{e,exp}-q_{e,isotherm}\right)^2}{q_{e,exp}}\right]_i$	(12)
MPSD	$\sum_{i=1}^{N}\left[\frac{\left(q_{e,exp}-q_{e,isotherm}\right)}{q_{e,exp}}\right]_i^2$	(13)

The intraparticle diffusion model is reported as follows:

$$\text{Log}\, q_t = \log k_{ip} + 0.5 \log t, \quad (7)$$

where k_{ip} is the intraparticle diffusion rate constant (mg/g/min$^{-0.5}$) [14].

The graph of q_t against 0.5 log t should produce a straight line with a positive intercept for the adsorption process controlled by intraparticle diffusion. The constant K is determined from the intersection with the Y-axis, and the high values of this parameter are related to a great adsorption speed [27].

2.6.5. Film Diffusion Model. The film diffusion model is reported as follows:

$$\ln\left(1-\frac{q_t}{q_e}\right) = -k_F t + A, \quad (8)$$

where k_F (1/min) and A are the constants of the film diffusion model [28].

In order to minimize the error distribution between experimental equilibrium data and values obtained by isotherm correlations, different error functions are generally applied. This goal of minimizing the error distribution is achieved by finding the minimum value of certain error functions or maximizing them depending on the definition of the selected error function. Therefore, choosing an error function seems indispensable in order to evaluate the best fit isotherm to the experimental equilibrium data [7]. In the present investigation, five different error functions (Table 1) were used to establish with certainty the isotherm of the kinetic model that presents the best adjustment of the adsorption of methylene blue on the surface of the algae *D. antarctica*.

2.7. Thermodynamic Studies of Bioadsorption

2.7.1. Model of Freundlich. The Freundlich model is described by an empirical equation used for systems with a high degree of heterogeneity. In this model, it is assumed that adsorption occurs at different sites and the formation of multilayers occurs with different adsorption energies. This leads to an exponential decrease in energy as the coverage of the surface originates [29].

The model of the Freundlich isotherm is reported as follows:

$$q_e = K_f C_e^{(1/n_f)}, \quad (14)$$

where q_e (mg * g^{-1}) is the capacity of bioadsorption of the metal by the algae *D. antarctica* in the balance, C_e (mg * L^{-1}) is the concentration of the metal in the equilibrium, and K_f (mg * g^{-1} (L * mg^{-1})$^{(1/n_f)}$) and n_f (dimensionless) are the constants of the Freundlich isotherm and are related to the capacity of biosorption and intensity, respectively; Freundlich's constant (n_f) depends on the adsorption capacity, and it is used to evaluate the favouribility of adsorption. The values of n_f between 2 and 10 indicate a high adsorption capacity, while values between 1 and 2 indicate a moderate adsorption capacity and values less than 1 indicate a small adsorption capacity [30].

2.7.2. Langmuir Model. The Langmuir model is based on the assumption that adsorption occurs on a surface with homogeneous sites, where each molecule of the adsorbate occupies an adsorption site, and maximum adsorption occurs with the formation of the complete monolayer, in which there is migration of the adsorbate molecules on the surface of the adsorbent [31].

The model of the Langmuir isotherm is reported as follows:

$$q_e = \frac{q_m K_L C_e}{1 + K_L C_e}, \quad (15)$$

where q_m (mg * g^{-1}) is the maximum adsorption capacity of the bioadsorbent and K_L (L * mg^{-1}) is the Langmuir equilibrium counterrelated to bioadsorption energy.

2.7.3. Sips Model. The two models of previous isotherms have been widely used to analyze the biosorption of various pollutants present in aqueous solutions. The Sips model simultaneously involves the Freundlich and Langmuir models, has three parameters taken from the theory of these models [32], and has a greater capacity to describe the biosorption equilibrium.

The model of the Sips isotherm is reported as follows:

$$q_e = \frac{q_m \left(K_s C_e\right)^{(1/n_s)}}{1 + \left(K_s C_e\right)^{(1/n_s)}}, \quad (16)$$

where K_s is the Sips count in equilibrium and n_s (dimensionless) is the heterogeneity factor; a value of n_s close to or equal to 1 is given in bioadsorbents with homogeneous active sites [33], while a value of n_s close to 0 is given in biosorbents with heterogeneous active sites [34].

2.7.4. Toth Model. In the Toth model, an asymmetric quasi-Gaussian energy distribution is assumed, where most of the sites have a lower adsorption energy than the energy peak of maximum adsorption [29]. Toth modifies the Langmuir equation in order to reduce the error between the experimental and predicted values of adsorption equilibrium. This model is best suited for heterogeneous systems in which multiple layer adsorption occurs. The exponent of the isotherm of Toth (n_{T_0}) is related to the heterogeneity of the surface; if it is equal to the unit, the Toth model is reduced to the Langmuir model. This correlation is applied for liquid-solid adsorption [35].

The model of the Toth isotherm is reported as follows:

$$q_e = Q_m C_e \left(b_{T_0} + C_e^{n_{T_0}}\right)^{\left(-1/n_{T_0}\right)}, \quad (17)$$

where b_{T_0} and n_{T_0} are constants of the model.

2.7.5. Redlich–Peterson Model. The Redlich–Peterson isotherm is an empirical model with no theoretical basis, which contains undefined empirical parameters (K_R and α_R). This model can be applied for homogeneous and heterogeneous surfaces and does not follow the ideal monolayer adsorption; as a consequence, this model provides the best approximation of the experimental data. However, the main disadvantage of this model is the lack of estimation of the maximum adsorption capacity. Another disadvantage when applying this model is that it does not adjust to the experimental data; in the trials, the dose of some variables has varied, since the constant α_R depends on the dose of the variables studied [22]. It should also be noted that this model includes the characteristics of the Langmuir and Freundlich isotherms in a single equation.

The Redlich–Peterson isotherm model is reported as follows:

$$q_e = \frac{K_{RP} C_e}{1 + \alpha_{RP} C_e^{g}}, \quad (18)$$

where K_{RP} is the constant of the Redlich–Peterson model (L/g), α_{RP} is another constant of the Redlich–Peterson model $(\mathrm{mg/L})^{-g}$, and g is the exponent of this model and must be between $0 < g \leq 1$.

When $g = 1$, the Redlich–Peterson equation becomes the Langmuir equation:

$$q_e = \frac{K_{RP} C_e}{1 + a_{RP} C_e}. \quad (19)$$

When $g = 0$, the Redlich–Peterson equation becomes Henry's law (20) [35]:

$$q_e = \frac{K_{RP} C_e}{1 + a_{RP}}. \quad (20)$$

3. Results

3.1. Characterization

3.1.1. Infrared Spectroscopy. The FTIR spectra before and after MB adsorption can be seen in Figure 1. In this analysis, several peaks assigned to the vibrations of the biomass functional groups were detected. Some of them are involved in MB adsorption, while others did not disappear, or their wavenumber shifted when this process was completed. The appearance of new peaks after the bioadsorption process was also observed due to the interaction between the functional groups of the bioadsorbent and the MB.

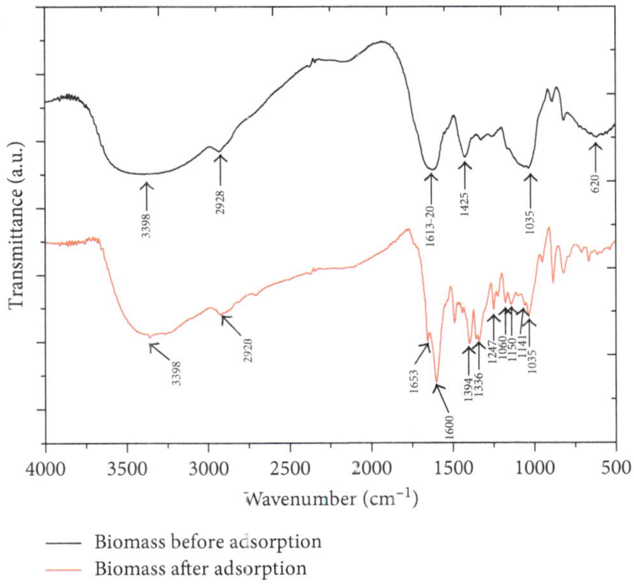

FIGURE 1: Infrared spectrum of the biomass of *D. antarctica* before and after the bioadsorption of MB.

The peaks at 3398 and 1613 cm^{-1} of the biomass before the adsorption process correspond to antisymmetric OH stretch vibrations [36, 37] and to the C=O bond present in the acids [36, 38], the peak located at 1425 cm^{-1} corresponds to primary alcohols [38], the peak at 2.936 cm^{-1} is due to the stretching vibrations of OH of carboxylic acids [39], the biomass adsorption band before and after adsorption of MB at a wavenumber of 1035 cm^{-1} is assigned to the CO present in the esters, and the peak at 1060 cm^{-1} can also be assigned to CO stretches [40].

The IR peaks of the MB that appeared after the adsorption process are described as follows: the peaks at 1600 and 1394 cm^{-1} correspond to vibrations of the aromatic ring and the CN bond, respectively, the peaks at 1336, 1141, and 1247 cm^{-1} are attributed to symmetric deformation of $-CH_3$ and the CN bond, respectively [37], the peaks at 1653 [39, 40], 1394, 1336, and 1141 cm^{-1} can be assigned to the vibrations of –C=C– of the stretch of the alkenes, to the C–C stretching in the aromatic ring, to the balancing of the CH alkanes, and to the wagging of the C–H ($-CH_2Cl$) of alkyl halides, respectively, the peak at 1247 cm^{-1} corresponds to the stretching of aliphatic amines [39], and the peak observed at 1150 cm^{-1} belongs to the vibration stretch of $-SO_3$ [41].

These observations clearly showed that the MB molecules were adsorbed on the surface of the biomass obtained from the algae *D. antarctica* and that the active functional groups of the biomass that participated in the binding of the MB molecules were –OH, CO, C=O, and –COO– representing, respectively, hydroxyl group, inorganic carbonates, and cyclic anhydride and carboxylic groups [38].

3.1.2. Scanning Electron Microscopy. This analysis was carried out in order to obtain a view of the morphology of the biosorbent surface. The obtained micrographs are presented in Figure 2.

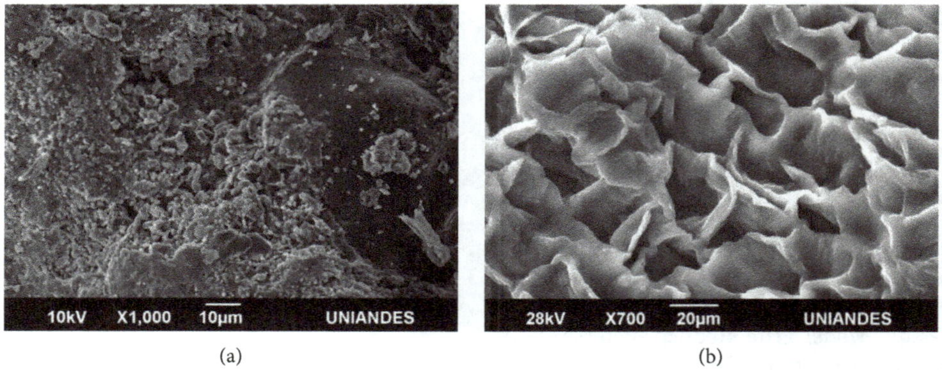

Figure 2: SEM micrographs of the biomass (a) before and (b) after the biosorption process of MB.

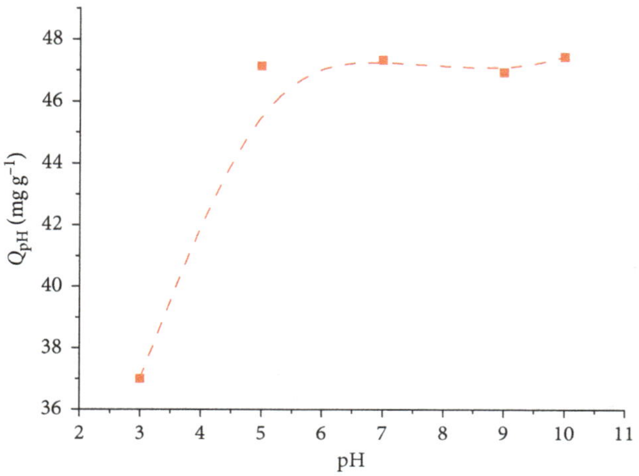

Figure 3: Effect of pH on bioadsorption of MB on the surface of the biomass of *D. antarctica*.

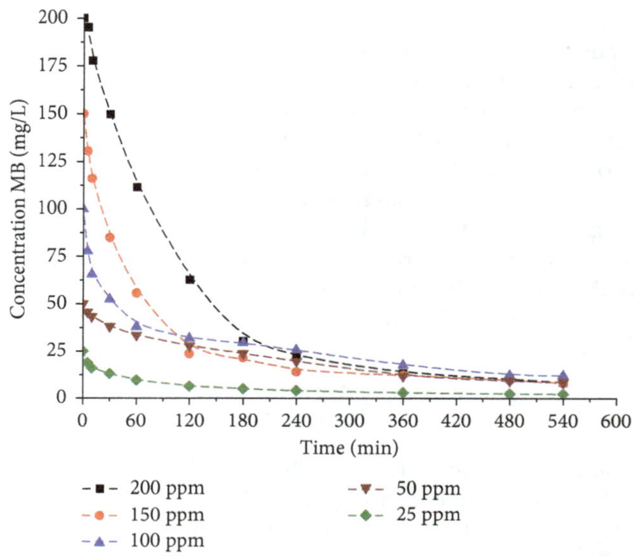

Figure 4: Effect of the concentration on the adsorption speed of MB on the surface of the biomass of *D. antarctica*.

Using this technique, it was determined that the pores present in the surface of the biomass are homogeneously distributed; however, there are areas in the surface that are not porous, so it is valid to say that a percentage of the porosity is distributed heterogeneously and it is worth mentioning. In addition, the MB adsorption process did not modify the surface of the biomass. In Figure 2(a), a part of the nonporous surface of the biomass is observed, while in Figure 2(b), the porous part of the biomass is observed, which retained its texture after the MB adsorption process. This is attributed mainly to the size of the methylene blue molecule, which is 0.84 nm [42], since each molecule of this contaminant interacts with a single active site present on the surface of the biomass; therefore, the pores of the biomass, having a diameter size greater than 50 μm, were not blocked during the biosorption process.

3.2. Determination of the Optimum pH for the Adsorption of Methylene Blue. The effect of pH on the adsorption of dyes is an important parameter. In Figure 3, the capacity of adsorption of MB by the biomass increased significantly when changing the pH value from 3 to 5, and this value was kept constant up to a basic pH corresponding to 11.

The chemical surface of an adsorbent has a great influence on the adsorption of an adsorbate, and this can be altered by modifying the initial pH value. Some authors have reported that, at low pH values, there is an electrostatic repulsion between the surface of the positively charged adsorbent and the cationic MB molecules, which inhibits the adsorption of this contaminant [4, 37, 43], because the ions of hydrogen compete with the MB molecule for the adsorption sites. As the pH increases, the surface of the biomass acquires a negative charge and a deprotonation of the surface is carried out which facilitates the adsorption of MB and therefore the capacity of adsorption of this pollutant by the biomass increases [37]. This is mainly because in the solution, the MB is positively charged; hence, its adsorption is more effective when the surface charge presents a negative value [44, 45]. Therefore, it was considered that a pH value of 10 was appropriate for carrying out the kinetic experiments.

3.3. Kinetics Models. In Figure 4, the larger the value of the initial concentration of MB, the more drastic the decrease in

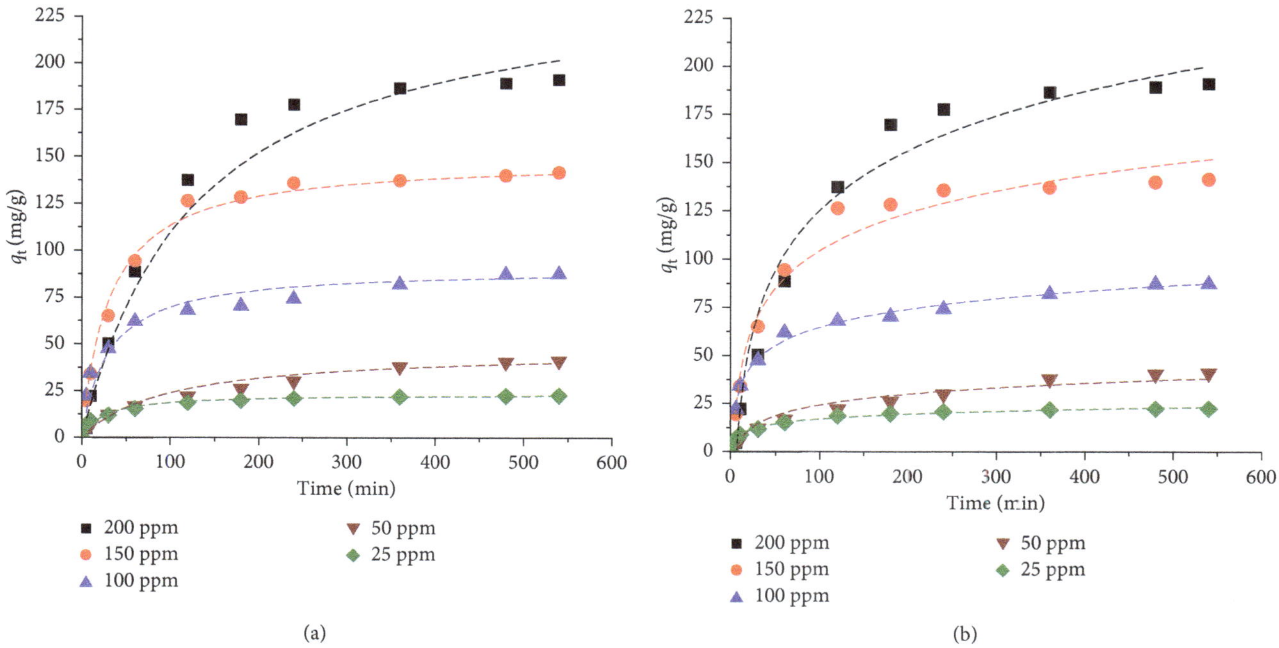

FIGURE 5: Result of applying the kinetic models used in the adsorption of MB on the surface of the biomass: (a) modified pseudo-first-order model; (b) Elovich model.

the value of its concentration, and as the time elapses, the change in concentration is less noticeable. In the first 180 minutes, the decrease in the MB concentration in the samples whose initial concentration was 200, 150, and 100 ppm is quite fast because the MB concentration remained above 20 ppm. However, after this time and smaller MB concentration values, the concentration changes are less noticeable. This is corroborated when observing that the decrease of the concentration of the contaminant in the solution whose initial concentration was 25 ppm is slow, in comparison with the other solutions.

Different models were applied to the results obtained in the kinetics of MB adsorption on the surface of the biomass studied. The pseudo-first-order model can describe the initial phase and the progress of the adsorption process; the pseudo-second-order equation describes that the chemisorption controls the speed of this process; According to the intraparticle diffusion model, several mechanisms are involved in the adsorption process, and this process can be described in three steps: external surface adsorption, intraparticle diffusion that is the speed-limiting step, and the final equilibrium that is very fast [46].

Next, the order in which the studied kinetic models were adjusted to the bioadsorption data obtained experimentally is mentioned.

Pseudo-second-order > Elovich > film diffusion = pseudo-first-order > intraparticle diffusion > modified pseudo-first-order.

Figure 5 shows the pseudo-second-order and Elovich isotherms obtained by adjusting the data kinetic experiments to these models; as can be seen, both models have a high correlation and therefore can describe the adsorption mechanism of methylene blue on the surface of *D. antarctica* algae.

In Table 2, it is observed that regardless of the value of the initial concentration of MB in the range studied, all bioadsorption data were adjusted to the pseudo-second-order model, this being the only model in which all the values of R^2 were above 0.95. P or which the initial adsorption rate depends exponentially on the concentration of methylene blue. The fact that this process is accurately described by a pseudo-second-order model further indicates that film diffusion, intraparticle diffusion, and surface adsorption coexisted during the adsorption of MB on the surface of the biomass, all of which factors affected the adsorption process and influence that the adsorption kinetics present the best fit when applying this model [47].

Given that the pseudo-second-order model presented the best fit of the data obtained, the value of q_e was determined, using this model; as can be seen in Table 2, the adsorption capacity values obtained experimentally are quite close to the values obtained theoretically.

The Elovich model presents the second best fit to the data obtained; this model occurs when the adsorption processes involve the chemisorption in the solid surface, and the speed of adsorption decreases with time due to the coverage of the surface layer [48]. The other models studied have a lower adjustment to the data studied, so it cannot explain with the same reliability the adsorption process that is being carried out.

To determine with greater precision the adjustment of the data obtained to the pseudo-second-order and Elovich kinetic models, five types of additional errors were determined. Table 3 reports the model that presents the best fit according to the type of error determined and the value of the initial concentration of methylene blue.

Regardless of the error studied, the best adjustment to the initial concentrations of MB of 200, 150, and 50 ppm is

Table 2: Adsorption kinetic parameters obtained by adsorption of MB onto *D. antarctica*.

Kinetic model	Parameter	Initial concentration of MB (mg/L)				
	r	25	50	100	150	200
Pseudo-first-order	K_1	0.00967	0.00783	0.00967	0.00921	0.00990
	R^2	0.986	0.952	0.893	0.937	0.992
Pseudo-second-order	K_2	0.00178	0.000287	0.000385	0.000234	0.0000730
	q_e, exp.	22.565	41.221	87.754	141.763	191.294
	q_e, cal.	22.402	40.003	86.024	140.955	196.349
	R^2	0.999	0.972	0.995	0.999	0.994
Modified pseudo-first-order	K_m	0.008	0.0059	0.0081	0.0074	0.0078
	R^2	0.985	0.875	0.839	0.971	0.991
Elovich	α	3.908	1.634	15.783	11.672	7.450
	β	0.272	0.122	0.0733	0.0356	0.0224
	R^2	0.991	0.943	0.992	0.967	0.970
Intraparticle diffusion	k_i	4.547	2.212	17.519	13.766	3.259
	R^2	0.963	0.996	0.949	0.914	0.897
Film diffusion	A	0.346	0.0591	0.212	0.416	0.0664
	k_f	0.0097	0.0078	0.0096	0.0092	0.01
	R^2	0.986	0.952	0.893	0.937	0.992

Table 3: Kinetic model that presents the best adjustment according to the initial concentration of MB and the type of error determined.

Error function	Initial concentration of MB (mg/L)				
	25	50	100	150	200
RMSE	Elovich	Pseudo-second-order	Elovich	Pseudo-second-order	Pseudo-second-order
X^2	Elovich	Pseudo-second-order	Elovich	Pseudo-second-order	Pseudo-second-order
ERRSQ	Elovich	Pseudo-second-order	Elovich	Pseudo-second-order	Pseudo-second-order
HYBRD	Elovich	Pseudo-second-order	Elovich	Pseudo-second-order	Pseudo-second-order
MPSD	Elovich	Pseudo-second-order	Elovich	Pseudo-second-order	Pseudo-second-order

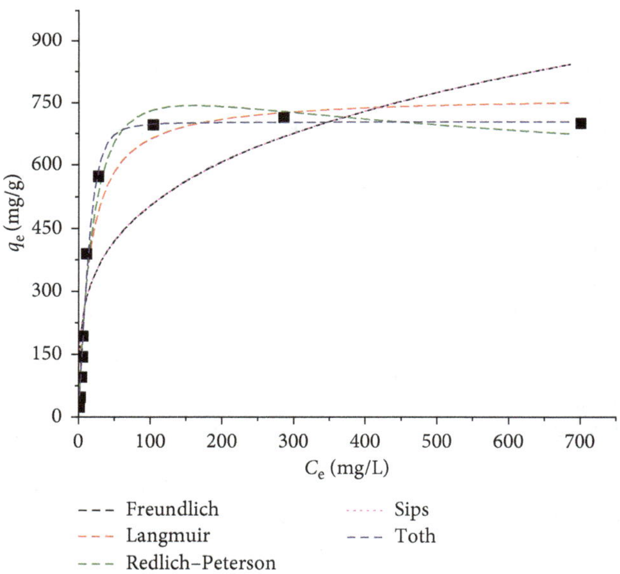

Figure 6: Bioadsorption models of methylene blue in *D. antarctica* biomass at 18 hours.

Table 4: Parameters of the models of the Freundlich, Redlich–Peterson, Sips, and Toth isotherms for the adsorption of MB on the surface of *D. antarctica*.

Models	Parameters	MB
Freundlich	K_f	148.309
	n_f	3.762
	R^2	0.873
Langmuir	Q_m	765.750
	K	0.064
	R^2	0.980
Redlich–Peterson	K_{RP}	38.133
	α_{RP}	0.022
	g	1.139
	R^2	0.989
Sips	Q_m	148.311
	K_s	−31.367
	n_s	0.266
	R^2	0.873
Toth	Q_m	702.911
	b_{T_0}	3735.703
	n_{T_0}	2.633
	R^2	0.995

given in the pseudo-second-order model, while at 100 and 25 ppm, the Elovich model presents the best correlation. It can be deduced that, at high initial concentrations, the obtained data are adjusted to the pseudo-second-order model given that the adsorption speed is related to the value of the initial concentration.

TABLE 5: Comparison of the adsorption capacity of MB on the surface of the biomass obtained from the algae *D. antarctica* and other compounds.

Compound	q_e (mg/g)	T (°C)	Concentration range (mg/L)	Time (min)	Reference
Mesoporous activated carbon	200.0	30	25–250	1440	[49]
GO/silicates	230	20	40–120	100	[50]
Nonionic surfactant modified montmorillonite	740.7	25	100	180	[51]
Fe_3O_4@SiO_2@CS-TETA-GO	529.1	30	1000	20	[11]
Cellulose nanocrystal	217.4	30	50–400	300	[52]
TDAW	285.7	25	40–1000	120	[53]
κ-carrageenan/graphene oxide gel beads	658.4	25	100–800	1440	[54]
Chitosan flakes	143.5	50	25–400	420	[4]
Algae *Gelidium*	171		40–800	160	[55]
Defatted algal biomass	7.8	25	1–5	60	[56]
Defatted *Laminaria japonica* biomass modified by sulfuric acid	549.45	35	20–250	60	[57]
Durvillaea antarctica	702.9	25	25–1400	540	This work

3.4. Bioadsorption Isotherms. Figure 6 shows the data of C_e and q_e obtained when determining the bioadsorption capacity of MB on the surface of the biomass obtained from the algae *D. antarctica*; in this figure, the graphs are obtained by adjusting the data of bioadsorption to the models of Freundlich, Langmuir, Redlich–Peterson, Sips, and Toth.

The best fit to the data obtained by the bioadsorption of MB on the surface of the biomass obtained from the algae *D. antarctica* is given by applying the Toth model, which indicates that the adsorption occurs on a heterogeneous surface that consists of multiple layers. This also agrees with the good fit presented by the kinetic data when applying the Elovich model since this model also describes the adsorption of the contaminant on a heterogeneous surface.

The values of the parameters obtained by adjusting the methylene blue adsorption data on the surface of *D. antarctica* biomass to the bioadsorption models studied are presented in Table 4.

The Langmuir model presented the second best fit to the data obtained in this biosorption process, and this equation is valid for systems where the biosorption occurs on a homogeneous surface.

The best fit to the obtained data is in a three-parameter model that describes that the adsorption occurs on a heterogeneous surface in multiple layers and in a model of two parameters that describe an adsorption process on a homogeneous surface. It is mainly due to the fact that the surface of the biomass has an irregular surface in which there are areas where the pores are homogeneously distributed and others in which the surface has no porosity; this agrees with the analysis of the SEM micrographs that were presented in Section 3.1.

The value of Q_e when applying the Toth model is 702.91 mg/g. This shows that the capacity of bioadsorption of MB on the surface of the biomass is high in comparison with other works reported in the literature. The exponent of the isotherm of this model (n_{T_0}) is related to the heterogeneity of the surfaces, and if it is equal to unity, the Toth model is reduced to the Langmuir model [35]. In these biosorption experiments, it was found that the value of this constant is 2.63; this value is different from one, which corroborates that this adsorption does not conform to the Langmuir model, and therefore, the majority of active sites are distributed heterogeneously. The Sips and Redlich–Peterson models are of three parameters and present characteristics of the models of Langmuir (homogeneous) and Freundlich (heterogeneous). The correlation coefficient when adjusting these statistical models to the bioadsorption data obtained was 0.873 and 0.989, respectively; these values are below the value of the correlation coefficient when applying the Toth and Langmuir models corresponding to 0.995 and 0.980, respectively. However, the value of R^2 when applying the Redlich–Peterson model is high, which suggests that the adsorption does not occur in an ideal monolayer, corroborating that this adsorption process occurred in multilayers as described by the Toth model.

Although the value of R^2 when applying the Sips model is small (0.873), the value of the constant n_s corresponding to 0.266 is closer to zero than to one, which corroborates that most active sites on the surface of the biomass which are involved in this adsorption process are distributed heterogeneously.

The value of R^2 when applying the Freundlich model was 0.873. This value is small; however, it should be noted that the value of the constant n_f when applying this model was 3.76, and according to the theory of this model, when the value of this constant is between 2 and 10, the adsorption capacity of the contaminant by the adsorbent is high [30]. This statement is corroborated when comparing the values of the methylene blue adsorption capacity presented in Table 5 for some adsorbents reported in the literature and the biomass obtained from the algae *D. antarctica*.

4. Conclusions

The bioadsorption capacity of the contaminant on the surface of the biomass increases with increasing pH, always when working at acid pH value intervals. Since the MB solution is at basic pH values, the MB adsorption capacity remains constant; in addition to this, it was determined that, in initial MB concentrations above 400 ppm, the initial pH value of these solutions is basic, which is why it is necessary to make modifications of the pH value which increase the efficiency of this process.

The pseudo-second-order model presented the best fit to the data obtained from the adsorption kinetics of MB on the surface of the biomass obtained from algae *D. antarctica*, which shows that the adsorption speed of this process is proportional to the exponential form of the initial concentration of MB present in the solution and to the time that has elapsed since the beginning of this process, as long as it is not greater than 540 min.

The Toth model presented the best fit to the data obtained from bioadsorption. By this model, it was determined that the adsorption capacity of MB on the surface of the biomass is 702.9 mg/g; this demonstrates the high efficiency that this material possesses to adsorb this contaminant.

Conflicts of Interest

The authors declare that there are no conflicts of interest.

Acknowledgments

The authors thank the Framework Agreement between the Universidad de los Andes and the Universidad Nacional de Colombia and the act of agreement established between the Chemistry Departments of the two universities. The authors also appreciate the grant for the funding of research programs for Associate Professors, Full Professors, and Emeritus Professors, announced by the Faculty of Sciences of the Universidad de los Andes (Colombia), 11-28-2017, 2018-2019, according to the project "Thermodynamic Characterization of the Adsorption of Contaminants on Porous Adsorbents." This work was financed by the Department of Chemistry and the Faculty of Sciences of the Andes University.

References

[1] F. Marrakchi, W. A. Khanday, M. Asif, and B. H. Hameed, "Cross-linked chitosan/sepiolite composite for the adsorption of methylene blue and reactive orange 16," *International Journal of Biological Macromolecules*, vol. 93, pp. 1231–1239, 2016.

[2] S. Thakur, S. Pandey, and O. A. Arotiba, "Development of a sodium alginate-based organic/inorganic superabsorbent composite hydrogel for adsorption of methylene blue," *Carbohydrate Polymers*, vol. 153, pp. 34–46, 2016.

[3] D. Pathania, S. Sharma, and P. Singh, "Removal of methylene blue by adsorption onto activated carbon developed from *Ficus carica* bast," *Arabian Journal of Chemistry*, vol. 10, pp. S1445–S1451, 2017.

[4] F. Marrakchi, M. J. Ahmed, W. A. Khanday, M. Asif, and B. H. Hameed, "Mesoporous-activated carbon prepared from chitosan flakes via single-step sodium hydroxide activation for the adsorption of methylene blue," *International Journal of Biological Macromolecules*, vol. 98, pp. 233–239, 2017.

[5] M. Heidarizad and S. S. Şengör, "Synthesis of graphene oxide/magnesium oxide nanocomposites with high-rate adsorption of methylene blue," *Journal of Molecular Liquids*, vol. 224, pp. 607–617, 2016.

[6] S. Aslam, J. Zeng, F. Subhan et al., "In situ one-step synthesis of Fe_3O_4@MIL-100(Fe) core-shells for adsorption of methylene blue from water," *Journal of Colloid and Interface Science*, vol. 505, pp. 186–195, 2017.

[7] S. M. Miraboutalebi, S. K. Nikouzad, M. Peydayesh, N. Allahgholi, L. Vafajoo, and G. McKay, "Methylene blue adsorption via maize silk powder: kinetic, equilibrium, thermodynamic studies and residual error analysis," *Process Safety and Environmental Protection*, vol. 106, pp. 191–202, 2017.

[8] H. Aysan, S. Edebali, C. Ozdemir, M. Celik Karakaya, and N. Karakaya, "Use of chabazite, a naturally abundant zeolite, for the investigation of the adsorption kinetics and mechanism of methylene blue dye," *Microporous and Mesoporous Materials*, vol. 235, pp. 78–86, 2016.

[9] Q. Huang, M. Liu, J. Chen et al., "Marrying the mussel inspired chemistry and Kabachnik–Fields reaction for preparation of SiO_2 polymer composites and enhancement removal of methylene blue," *Applied Surface Science*, vol. 422, pp. 17–27, 2017.

[10] X.-P. Luo, S.-Y. Fu, Y.-M. Du, J.-Z. Guo, and B. Li, "Adsorption of methylene blue and malachite green from aqueous solution by sulfonic acid group modified MIL-101," *Microporous and Mesoporous Materials*, vol. 237, pp. 268–274, 2017.

[11] F. Wang, L. Zhang, Y. Wang, X. Liu, S. Rohani, and J. Lu, "Fe_3O_4@SiO_2@CS-TETA functionalized graphene oxide for the adsorption of methylene blue (MB) and Cu(II)," *Applied Surface Science*, vol. 420, pp. 970–981, 2017.

[12] N. Tahir, H. N. Bhatti, M. Iqbal, and S. Noreen, "Biopolymers composites with peanut hull waste biomass and application for Crystal Violet adsorption," *International Journal of Biological Macromolecules*, vol. 94, pp. 210–220, 2017.

[13] H. Cid, C. Ortiz, J. Pizarro et al., "Characterization of copper (II) biosorption by brown algae *Durvillaea antarctica* dead biomass," *Adsorption*, vol. 21, no. 8, pp. 645–658, 2015.

[14] M. A. Hossain, H. H. Ngo, W. S. Guo, and T. Setiadi, "Adsorption and desorption of copper(II) ions onto garden grass," *Bioresource Technology*, vol. 121, pp. 386–395, 2012.

[15] H. I. Inyang, A. Onwawoma, and S. Bae, "The Elovich equation as a predictor of lead and cadmium sorption rates on contaminant barrier minerals," *Soil and Tillage Research*, vol. 155, pp. 124–132, 2016.

[16] S. Lagergren, "About the theory of so-called adsorption of soluble substances, Kungliga Svenska Vetenskapsakad," *Handlingar*, vol. 24, pp. 1–39, 1898.

[17] Y. S. Ho and G. McKay, "Pseudo-second order model for sorption processes," *Process Biochemistry*, vol. 34, no. 5, pp. 451–465, 1999.

[18] J.-P. Simonin, "On the comparison of pseudo-first order and pseudo-second order rate laws in the modeling of adsorption kinetics," *Chemical Engineering Journal*, vol. 300, pp. 254–263, 2016.

[19] A. E. Rodrigues and C. M. Silva, "What's wrong with Lagergreen pseudo first order model for adsorption kinetics?," *Chemical Engineering Journal*, vol. 306, pp. 1138–1142, 2016.

[20] A. E. Ofomaja, E. B. Naidoo, and S. J. Modise, "Dynamic studies and pseudo-second order modeling of copper(II) biosorption onto pine cone powder," *Desalination*, vol. 251, no. 1–3, pp. 112–122, 2010.

[21] X. Yang and B. Al-Duri, "Kinetic modeling of liquid-phase adsorption of reactive dyes on activated carbon," *Journal of*

Colloid and Interface Science, vol. 287, no. 1, pp. 25–34, 2005.

[22] M. Baghdadi, "UT (University of Tehran) isotherm as a novel and useful adsorption isotherm for investigation of adsorptive removal of pollutants," *Journal of Environmental Chemical Engineering*, vol. 5, no. 2, pp. 1906–1919, 2017.

[23] C. Rill, Z. I. Kolar, G. Kickelbick, H. T. Wolterbeek, and J. A. Peters, "Kinetics and thermodynamics of adsorption on hydroxyapatite of the [160Tb] terbium complexes of the bone-targeting ligands DOTP and BPPED," *Langmuir*, vol. 25, no. 4, pp. 2294–2301, 2009.

[24] H. Uslu and S. Majumder, "Adsorption studies of lactic acid by polymeric adsorbent amberlite XAD-7: equilibrium and kinetics," *Journal of Chemical & Engineering Data*, vol. 62, no. 4, pp. 1501–1506, 2017.

[25] H. Zhang, Y. Wang, P. Bai, X. Guo, and X. Ni, "Adsorptive separation of acetic acid from dilute aqueous solutions: adsorption kinetic, isotherms, and thermodynamic studies," *Journal of Chemical & Engineering Data*, vol. 61, no. 1, pp. 213–219, 2015.

[26] J. Chen, H. Yang, and Z. Ring, "Study of intra-particle diffusion effect on hydrodesulphurization of dibenzothiophenic compounds," *Catalysis Today*, vol. 109, no. 1–4, pp. 93–98, 2005.

[27] A. Okoye, P. Ejikeme, and O. Onukwuli, "Lead removal from wastewater using fluted pumpkin seed shell activated carbon: adsorption modeling and kinetics," *International Journal of Environmental Science & Technology*, vol. 7, no. 4, pp. 793–800, 2010.

[28] E. M. Kalhori, T. J. Al-Musawi, E. Ghahramani, H. Kazemian, and M. Zarrabi, "Enhancement of the adsorption capacity of the light-weight expanded clay aggregate surface for the metronidazole antibiotic by coating with MgO nanoparticles: studies on the kinetic, isotherm, and effects of environmental parameters," *Chemosphere*, vol. 175, pp. 8–20, 2017.

[29] N. Barka, K. Ouzaouit, M. Abdennouri, and M. E. Makhfouk, "Dried prickly pear cactus (*Opuntia ficus indica*) cladodes as a low-cost and eco-friendly biosorbent for dyes removal from aqueous solutions," *Journal of the Taiwan Institute of Chemical Engineers*, vol. 44, no. 1, pp. 52–60, 2013.

[30] M. Brdar, M. Šćiban, A. Takači, and T. Došenović, "Comparison of two and three parameters adsorption isotherm for Cr(VI) onto Kraft lignin," *Chemical Engineering Journal*, vol. 183, pp. 108–111, 2012.

[31] G. McKay, A. Mesdaghinia, S. Nasseri, M. Hadi, and M. Solaimany Aminabad, "Optimum isotherms of dyes sorption by activated carbon: fractional theoretical capacity & error analysis," *Chemical Engineering Journal*, vol. 251, pp. 236–247, 2014.

[32] F. Deniz and E. T. Ersanli, "Simultaneous bioremoval of two unsafe dyes from aqueous solution using a novel green composite biosorbent," *Microchemical Journal*, vol. 128, pp. 312–319, 2016.

[33] S. Chatterjee, M. W. Lee, and S. H. Woo, "Adsorption of congo red by chitosan hydrogel beads impregnated with carbon nanotubes," *Bioresource Technology*, vol. 101, no. 6, pp. 1800–1806, 2010.

[34] A. Bera, T. Kumar, K. Ojha, and A. Mandal, "Adsorption of surfactants on sand surface in enhanced oil recovery: isotherms, kinetics and thermodynamic studies," *Applied Surface Science*, vol. 284, pp. 87–99, 2013.

[35] S. Rangabhashiyam, N. Anu, M. S. Giri Nandagopal, and N. Selvaraju, "Relevance of isotherm models in biosorption of pollutants by agricultural byproducts," *Journal of Environmental Chemical Engineering*, vol. 2, no. 1, pp. 398–414, 2014.

[36] D.-q Li, J. Wang, Z.-g Guo, J. Li, and J. Shuai, "Pectin gels cross-linked by Ca^{2+}: an efficient material for methylene blue removal," *Journal of Molecular Liquids*, vol. 238, pp. 36–42, 2017.

[37] J. Pang, F. Fu, Z. Ding, J. Lu, N. Li, and B. Tang, "Adsorption behaviors of methylene blue from aqueous solution on mesoporous birnessite," *Journal of the Taiwan Institute of Chemical Engineers*, vol. 77, pp. 168–176, 2017.

[38] F. Marrakchi, M. Auta, W. A. Khanday, and B. H. Hameed, "High-surface-area and nitrogen-rich mesoporous carbon material from fishery waste for effective adsorption of methylene blue," *Powder Technology*, vol. 321, pp. 428–434, 2017.

[39] S. Anisuzzaman, C. G. Joseph, D. Krishnaiah, A. Bono, and L. Ooi, "Parametric and adsorption kinetic studies of methylene blue removal from simulated textile water using durian (*Durio zibethinus* Murray) skin," *Water Science and Technology*, vol. 72, no. 6, pp. 896–907, 2015.

[40] D. L. Postai, C. A. Demarchi, F. Zanatta, D. C. C. Melo, and C. A. Rodrigues, "Adsorption of rhodamine B and methylene blue dyes using waste of seeds of *Aleurites moluccana*, a low cost adsorbent," *Alexandria Engineering Journal*, vol. 55, no. 2, pp. 1713–1723, 2016.

[41] A. Naskar and R. Majumder, "Understanding the adsorption behaviour of acid yellow 99 on *Aspergillus niger* biomass," *Journal of Molecular Liquids*, vol. 242, pp. 892–899, 2017.

[42] L. Giraldo, V. García, and J. C. Moreno, "Caracterización superficial en fase gas y líquida de carbones activados," *Revista de Ingeniería*, vol. 27, pp. 07–16, 2008.

[43] R. Pandimurugan and S. Thambidurai, "Synthesis of seaweed-ZnO-PANI hybrid composite for adsorption of methylene blue dye," *Journal of Environmental Chemical Engineering*, vol. 4, no. 1, pp. 1332–1347, 2016.

[44] S. Zhang, Z. Wang, Y. Zhang, H. Pan, and L. Tao, "Adsorption of methylene blue on organosolv lignin from rice straw," *Procedia Environmental Sciences*, vol. 31, pp. 3–11, 2016.

[45] W. Gao, S. Zhao, H. Wu, W. Deligeer, and S. Asuha, "Direct acid activation of kaolinite and its effects on the adsorption of methylene blue," *Applied Clay Science*, vol. 126, pp. 98–106, 2016.

[46] L.-g. Yan, L.-l. Qin, H.-q. Yu, S. Li, R.-r. Shan, and B. Du, "Adsorption of acid dyes from aqueous solution by CTMAB modified bentonite: kinetic and isotherm modeling," *Journal of Molecular Liquids*, vol. 211, pp. 1074–1081, 2015.

[47] S. Fan, Y. Wang, Z. Wang, J. Tang, J. Tang, and X. Li, "Removal of methylene blue from aqueous solution by sewage sludge-derived biochar: adsorption kinetics, equilibrium, thermodynamics and mechanism," *Journal of Environmental Chemical Engineering*, vol. 5, no. 1, pp. 601–611, 2017.

[48] J. S. Piccin, G. L. Dotto, M. L. Vieira, and L. A. Pinto, "Kinetics and mechanism of the food dye FD&C Red 40 adsorption onto chitosan," *Journal of Chemical & Engineering Data*, vol. 56, no. 10, pp. 3759–3765, 2011.

[49] M. A. Islam, M. J. Ahmed, W. A. Khanday, M. Asif, and B. H. Hameed, "Mesoporous activated coconut shell-derived hydrochar prepared via hydrothermal carbonization-NaOH activation for methylene blue adsorption," *Journal of Environmental Management*, vol. 203, pp. 237–244, 2017.

[50] P. Chen, Z.-f. Cao, X. Wen et al., "*In situ* nano-silicate functionalized graphene oxide composites to improve MB removal," *Journal of the Taiwan Institute of Chemical Engineers*, vol. 81, pp. 87–94, 2017.

[51] G. Wang, S. Wang, Z. Sun, S. Zheng, and Y. Xi, "Structures of nonionic surfactant modified montmorillonites and their enhanced adsorption capacities towards a cationic organic dye," *Applied Clay Science*, vol. 148, pp. 1–10, 2017.

[52] X. Yang, H. Liu, F. Han, S. Jiang, L. Liu, and Z. Xia, "Fabrication of cellulose nanocrystal from *Carex meyeriana* Kunth and its application in the adsorption of methylene blue," *Carbohydrate Polymers*, vol. 175, pp. 464–472, 2017.

[53] Z. Anfar, R. El Haouti, S. Lhanafi, M. Benafqir, Y. Azougarh, and N. El Alem, "Treated digested residue during anaerobic co-digestion of Agri-food organic waste: methylene blue adsorption, mechanism and CCD-RSM design," *Journal of Environmental Chemical Engineering*, vol. 5, no. 6, pp. 5857–5867, 2017.

[54] M. Yang, X. Liu, Y. Qi, W. Sun, and Y. Men, "Preparation of κ-carrageenan/graphene oxide gel beads and their efficient adsorption for methylene blue," *Journal of Colloid and Interface Science*, vol. 506, pp. 669–677, 2017.

[55] V. J. Vilar, C. M. Botelho, and R. A. Boaventura, "Methylene blue adsorption by algal biomass based materials: biosorbents characterization and process behaviour," *Journal of hazardous materials*, vol. 147, no. 1-2, pp. 120–132, 2007.

[56] T. S. Chandra, S. Mudliar, S. Vidyashankar et al., "Defatted algal biomass as a non-conventional low-cost adsorbent: surface characterization and methylene blue adsorption characteristics," *Bioresource technology*, vol. 184, pp. 395–404, 2015.

[57] H. Shao, Y. Li, L. Zheng, T. Chen, and J. Liu, "Removal of methylene blue by chemically modified defatted brown algae *Laminaria japonica*," *Journal of the Taiwan Institute of Chemical Engineers*, vol. 80, pp. 525–532, 2017.

FTIR Spectrophotometry as a Green Tool for Quantitative Analysis of Drugs: Practical Application to Amoxicillin

Stefany Fanelli,[1] Alexander Zimmermann,[2] Eliane Gandolpho Totóli,[1] and Hérida Regina Nunes Salgado[1]

[1]*School of Pharmaceutical Sciences, Universidade Estadual Paulista, Araraquara, SP, Brazil*
[2]*Institute of Biochemistry and Biotechnology, Faculty of Natural Science I, Martin-Luther-University Halle-Wittenberg, Halle (Saale), Germany*

Correspondence should be addressed to Eliane Gandolpho Totóli; eliane.totoli@gmail.com

Academic Editor: Serkos A. Haroutounian

Amoxicillin is an antimicrobial agent that belongs to the penicillin compounds. Its bactericidal action causes a destruction of the cell wall of bacteria. It is widely used in clinical practice, and it belongs to the Brazilian National List of Essential Drugs (RENAME). In literature, there are some green analytical methods for the amoxicillin analysis; however, none of them is focused on its quantification in capsules. Therefore, the aim of this study was to validate an environmentally friendly analytical method for the analysis of this antimicrobial action in capsules, using spectrophotometry in the mid-infrared region. The analyses were performed in the spectral range of 1815–1736 cm^{-1}, and the samples were analyzed as potassium bromide pellets. The method was validated according to the ICH guidelines and Brazilian legislation. Linearity, selectivity, precision, accuracy, and robustness were evaluated and showed adequate results for method validation, in a concentration range of 0.5–1.5 mg/pellet. Thus, it is concluded that the validated spectrophotometric method is able to quantify amoxicillin in capsules. In addition, it is a fast, economical, and environmentally friendly method, since it does not use organic solvents, and it can be used for quality control of routine analysis of this drug.

1. Introduction

Amoxicillin is a broad-spectrum antimicrobial agent belonging to the class of penicillins and is widely used in clinical practice. Furthermore, it belongs to the Brazilian National List of Essential Drugs (RENAME). Amoxicillin is available in tablets, capsules, and powder for oral suspension. It is also available associated with potassium clavulanate, for the treatment of infections caused by β-lactamase-producing bacteria, and presented in the pharmaceutical dosage forms of tablets and oral suspension [1].

This penicillin is generally indicated for treating infections caused by microorganisms responsible for gonorrhea, pneumonia, bronchitis, otitis, or endocarditis. Furthermore, it is applied after infections of the urinary tract caused by *Haemophilus influenzae*. It can be used as an alternative treatment in cases of gastroenteritis and typhoid fever caused by *Shigella*. Amoxicillin is considered the drug of choice for the prophylaxis of bacterial endocarditis, and it can be combined with omeprazole and clarithromycin for combating *Helicobacter pylori* in the patients with peptic ulcer [2].

This drug has a rapid bactericidal action, acting on the cell wall of bacteria. It is very well-absorbed orally, and 60 to 70% are excreted renally. Generally, amoxicillin is administered three times daily as a 500 mg capsule, or even twice daily, and it is well tolerated by the organism. However, it may have some undesirable effects such as nausea, vomiting, and gastrointestinal irritation [3].

In literature, there are several published studies describing the development of the analytical methods for analysis of amoxicillin in different matrices. However, most of the present methods use highly potential toxic solvents for the operators and environment, for example, high-performance liquid chromatography (HPLC) [4–8], ultraperformance liquid chromatography (UPLC) [9], or iodometry [10].

The quality of drugs has always been a concern of the World Health Organization (WHO). Since without the guarantee these products meet the required standards of quality, safety, and effectiveness, any health service is evidently affected [11]. Because of this, attention on the quality control of drugs is important. Pharmaceutical companies must ensure that their marketed products have adequate safety and efficacy for the population.

Therefore, the validation of the efficient analytical methods for use in the quality control of the marketed drugs is essential. The validation of an analytical methodology aims at verifying whether a tested method is suitable for a particular purpose, in other words, whether the method is able to analyze qualitatively and/or quantitatively a drug or related substances in pharmaceutical dosage forms [12–14].

The trend is that industries seek ways to reduce the impacts of their activities on the environment. Thus, they can adopt the position of reduction, prevention, or elimination of process waste. In this regard, there are some actions that can be taken, such as the replacement of the analytical methods that employs a high amount of organic solvents with others that do not use them, the replacement of a synthesis process by a greener one, or the exchange of raw materials or inputs by other less toxic ones [15].

In this context, the infrared spectrophotometry stands out. This method allows a quantification of substances without using organic solvents. This technique does not require any extraction step of the sample and therefore can also be used for substances with solubility problems.

Considering the importance of amoxicillin for clinical practice and the relevance of seeking green methods for the analysis of drugs, this work describes the development and validation of an innovative and environmental friendly analytical method, by spectrophotometry in the mid-infrared region, for quantification of amoxicillin in capsules.

2. Experiment

2.1. Equipment. For spectrophotometric analysis, a spectrophotometer FTIR Shimadzu (Kyoto, Japan), IR Prestige-21 model, was used. This equipment was connected to a computer to use the "IR Solution" software for analysis of the spectra. The construction of calibration curves was performed using Microsoft Excel (2013).

The other equipment used is as follows: H51 analytical balance (Mettler Toledo™, Barueri, Brazil) and oven ECB 1.2 Digital (Odontobrás™, Ribeirão Preto, Brazil).

2.2. Chemicals and Reagents. The pharmaceutical company União Química (São Paulo, Brazil) kindly donated amoxicillin reference standard (AMX RS) (declared a purity of 98.9%) and amoxicillin (AMX) in the pharmaceutical dosage form of the capsule (Uni Amox™, 500 mg/capsule, União Química, São Paulo, Brazil). Adjuvants present in this pharmaceutical dosage form are sodium lauryl sulfate, croscarmellose sodium, and magnesium stearate. These substances were purchased from Sigma-Aldrich (São Paulo, Brazil). Potassium bromide (KBr) analytical grade was also used (Synth™, Diadema, Brazil).

2.3. Qualitative Analysis. To perform the method, the spectrophotometer described in Equipment was used. For the preparation of AMX RS pellets, an equivalent to 2.0 mg of AMX RS was accurately weighted and then homogenized with 148.0 mg of KBr, previously powdered, and dried at 105°C in an oven to constant weight, in order to obtain a 150 mg pellet (at 2 mg/pellet). This mixture was compressed by a mechanical press for 10 minutes to obtain a translucent pellet. The spectral region included in the analysis was from 4000 to 400 cm^{-1} (the mid-infrared region). The analysis was held in transmittance, and the spectrum was obtained with the aid of "IR Solution" software (Shimadzu, Kyoto, Japan). The same procedure was performed with amoxicillin in capsules. Finally, a comparison of the spectra obtained from AMX RS and AMX capsule was performed to verify the similarity between them.

In addition, three independent pellets were prepared, containing each adjuvant present in the pharmaceutical dosage form (croscarmellose sodium, sodium lauryl sulfate, and magnesium stearate), in a concentration of 1.0 mg/pellet. For this, 1.0 mg of each component was homogenized with 149.0 mg of KBr. Each mixture was compressed by a mechanical press for 10 minutes to obtain translucent pellets. The spectral region included in the analysis was from 4000 to 400 cm^{-1} (the mid-infrared region). This procedure was performed to define the best spectral region to be used in the quantitative analysis (a region without interference of adjuvants).

2.4. Quantitative Analysis

2.4.1. Obtaining the Calibration Curve. To perform the method, the spectrophotometer described in "Equipment" section was used. Samples were analyzed as KBr pellets containing the drug. After obtaining the spectrum in the mid-infrared range (4000–400 cm^{-1}), the spectral region from 1815.0 to 1736.0 cm^{-1} was selected, which corresponds to a characteristic band of the amoxicillin molecule (carbonyl), and its height was quantitatively analyzed in absorbance.

During preliminary tests, five concentrations of amoxicillin were selected for the analytical curve, as follows: 0.5, 0.75, 1.0, 1.25, and 1.50 mg/pellet.

For the preparation the pellets, the amounts equivalent to 0.50, 0.75, 1.00, 1.25, and 1.50 mg of AMX RS (previously diluted in potassium bromide of 1:10, w/w) were taken and diluted with sufficient amounts of KBr to obtain 150 mg pellets. The powders were mixed and ground to obtain a homogeneous mixture. Thereafter, this mixture was compressed with a mechanical press for 10 minutes to obtain translucent pellets.

2.4.2. Determination of Amoxicillin in the Pharmaceutical Dosage Form. Amoxicillin content in capsules was calculated by Equation (1), and its percentage content was calculated by Equation (2):

$$C_s = A_s \times \frac{C_{RS}}{A_{RS}}, \quad (1)$$

$$C_s\% = C_s \times \frac{100}{C_t}. \quad (2)$$

where C_s is the concentration of AMX in the sample (mg/pellet), A_s is the absorbance of the AMX sample pellet, C_{RS} is the concentration of AMX RS solution (mg/pellet), A_{RS} is the absorbance of the AMX RS pellet, $C_s\%$ is the percentage concentration of AMX in the sample, and C_t is the theoretical concentration of AMX in the sample (mg/pellet)

2.4.3. Method Validation. The method was validated by analyzing the following parameters: linearity, precision, accuracy, robustness, and selectivity, as recommended by ICH guidelines and Brazilian legislation [12, 13]. Limits of detection and quantification are not required for the validation of this class of the analytical method (Class I), according to ICH guidelines and Brazilian legislation [12, 13].

(1) Linearity. To check the linearity of the method, five concentrations of AMX RS (0.5 to 1.5 mg/pellet) were used and evaluated on three different days. Linearity was confirmed by linear regression of least squares and statistical analysis by ANOVA.

(2) Precision. Precision was evaluated in two different ways: by intermediate precision (interday and between analysts) and repeatability (intraday). Interday precision was performed by analysis of six pellets of AMX RS (at 1.0 mg/pellet) on three different days, at the same experimental conditions. Similarity between the absorbances obtained on different days was evaluated by analysis of variance (ANOVA). Between-analyst's precision was carried out by the analysis of six pellets of AMX RS (at 1.0 mg/pellet) performed by two different analysts. In this case, the absorbances were compared by *F*-test and *t*-test. Regarding intraday precision, it was evaluated by the analysis of six consecutive pellets of AMX RS (at 1.0 mg/pellet) in the same day and at the same working conditions. Thereafter, the percentage relative standard deviation value (RSD) between the absorbances was calculated [13].

(3) Accuracy. Accuracy was performed by the analysis of AMX RS recovery, in triplicate, at three levels (R1, R2, and R3), from 80 to 120% of the working concentration of the method (1.0 mg/pellet). The pellets for the recovery assay are prepared in accordance with the description in Table 1.

The recovery percentage was calculated by the equation determined by the Association of Official Analytical Chemists (AOAC) [16].

(4) Robustness. Robustness aims at showing the reliability of the method after small variations in the analytical parameters. The following parameters were varied: temperature of the workroom (21°C, with the air conditioning on, and 26 C, with the air conditioning off), compression time of the pellets (2 minutes above and below the working compression time), and KBr brand (Shimadzu). For this, six pellets of AMX RS (at 1.0 mg/pellet) were analyzed under each previously described condition. Similarity of the results was evaluated by the *F*-test and *t*-test, comparing the normal working conditions.

(5) Selectivity. Selectivity was analyzed in order to verify the ability of the method to quantify the drug in the presence of the adjuvants present in the pharmaceutical dosage form of the capsule [12, 13]. The used method has been previously described in Qualitative analysis.

3. Results

3.1. Qualitative Analysis. Figure 1 shows an overlay of the absorption spectra of the AMX RS and sample. Figure 2 displays the absorption spectra of each adjuvant present in the capsules. The spectral range chosen for quantitative analysis (from 1815 to 1736 cm^{-1}) is also identified in both figures.

3.2. Quantitative Analysis

3.2.1. Linearity. Three calibration curves, generated on three different days, were analyzed by a graphical representation. The average values of absorbance were plotted against their respective concentrations. A final analytical curve was obtained by averaging the three analytical curves, resulting in a correlation coefficient (*r*) of 0.9971, as shown in Figure 3.

ANOVA calculated for the data of the analytical curve showed an $F_{calculated}$ of 521.45 and $F_{critical}$ of 0.00445.

3.2.2. Precision. Precision was evaluated according to the repeatability (intraday precision) and intermediate precision (interday precision) and between analysts. Intraday precision shows a RSD of 3.80% among the absorbance values. Table 2 shows the absorbances obtained for the evaluation of interday precision and between analysts.

The absorbances obtained for interday precision (Table 2) were compared by ANOVA. $F_{calculated}$ and $F_{critical}$ were 1.94 and 3.68, respectively. To calculate the precision between analyst, *F*-test and *t*-test were used (Table 3). As the result, $t_{calculated}$ and $t_{critical}$ were 1.94 and 2.23, respectively, showing no significance.

3.2.3. Accuracy. Accuracy was evaluated by the recovery test and conducted at three different concentration levels, and the results are shown in Table 3.

3.2.4. Robustness. Robustness was evaluated by small variations in working parameters, such as the temperature of the

Table 1: Preparation of pellets for the recovery assay of the method of FT-IR spectrophotometry for amoxicillin.

	AMX sample (mg) (diluted 1:10 w/w in KBr)	AMX RS (mg) (diluted 1:10 w/w in KBr)	Amount of KBr (mg)[1]	Final theoretical concentration (mg/pellet)
Sample	5.0	—	145.0	0.50
R1	5.0	3.0	142.0	0.80
R2	5.0	5.0	140.0	1.00
R3	5.0	7.0	138.0	1.20
Reference standard	—	5.0	145.0	0.50

[1]Sufficient amount for the preparation of pellets with a total weight of 150 mg. AMX: amoxicillin; AMX RS: amoxicillin reference standard.

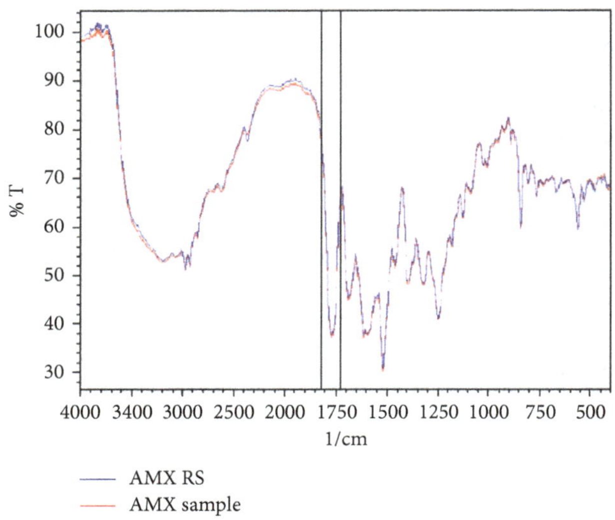

Figure 1: Overlap of the absorption spectra in the mid-infrared region of AMX RS and in the pharmaceutical dosage form of capsules. AMX RS: amoxicillin reference standard; AMX: amoxicillin. Spectral range chosen for quantitative analysis (from 1815 to 1736 cm^{-1}) is highlighted.

workroom, pellet compression time, and KBr brand. The obtained experimental values were evaluated by *F*-test and *t*-test, by comparing the absorbance values from the normal and varied conditions. The obtained absorbances are presented in Table 4.

For the temperature of the workroom, the values of $t_{calculated}$ and $t_{critical}$ were 4.96 and 2.36, respectively. This result is statistically significant.

Regarding the pellet compression time, the absorbances obtained for the normal working condition (10 minutes) were compared, individually, with each varied condition (8 and 12 minutes). The comparison between 8 and 10 minutes showed $t_{calculated}$ and $t_{critical}$ of 2.04 and 2.26, respectively. This result is statistically significant. In relation to the higher pellet compression times comparison (10 and 12 minutes), the $t_{calculated}$ was 1.97 and $t_{critical}$ 2.31, showing no significance.

Concerning the KBr brand, the absorbances obtained from two different brands were compared (Synth and Shimadzu). In this case, the values of $t_{calculated}$ and $t_{critical}$ were 1.21 and 2.23, respectively, showing no significance.

3.2.5. Selectivity. Results obtained from selectivity analysis have been previously described in Qualitative analysis. Figure 1 presents an overlay of the absorption spectra of the AMX RS and sample. Figure 2 presents the absorption spectra of each adjuvant present in the analyzed pharmaceutical dosage form. The spectral range chosen for quantitative analysis (from 1815 to 1736 cm^{-1}) is also identified in both figures.

3.3. Determination of Amoxicillin in the Pharmaceutical Dosage Form. Three quantifications of amoxicillin in the capsules were performed, according to Equations (1) and (2). Table 5 shows the results of these assays.

4. Discussion

Environmental preservation is an important issue nowadays and is increasingly a concern of chemical and pharmaceutical industries. In order to reduce the environmental damage caused by toxic waste generation, companies look for alternatives to reduce, prevent, and even eliminate the generation of chemical waste from their processes. One option is to replace methods that use organic solvents by others that do not use the environmentally friendly methods [15, 17].

It is in this context, the infrared spectrophotometry stands out because it is a method that allows to quantify substances without organic solvents. It is suitable for drugs with solubility problems, since they can be analyzed in the solid form [18]. Infrared spectrophotometry is based on the fact that the chemical bonds of the molecules have natural vibrational frequencies. Each molecule only absorbs selected frequencies of radiation in the infrared region, which are equivalent to its natural vibrational frequencies. This absorption increases the amplitude of vibrational motion of the chemical bonds. Thus, the frequency of vibration may be associated with a particular type of the band [19, 20].

In addition to the absence of organic solvents, the spectrophotometric method in the infrared region has other advantages, such as being a rapid technique that does not require a pretreatment of the sample and assisting in detecting impurities. Another advantage is the low cost of materials for the manufacture of pellets. On the other hand, this method requires high control of humidity and temperature of the working environment.

The development of the analytical methods by spectrophotometry in the mid-infrared region has been successfully used in the quantification of other drugs [18, 21–36].

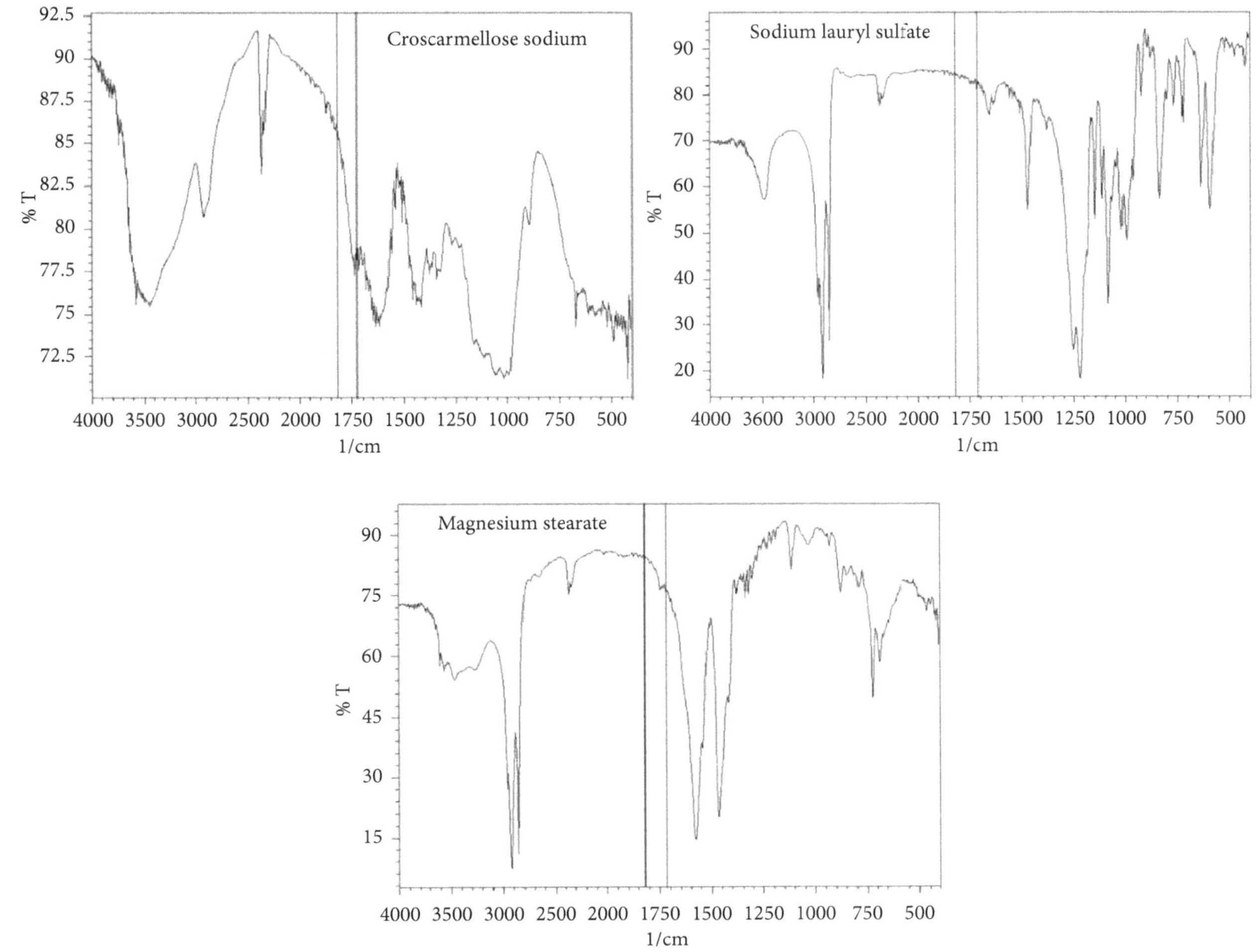

FIGURE 2: Absorption spectra in the mid-infrared region of the adjuvants present in the pharmaceutical dosage form of the capsule. Spectral range chosen for quantitative analysis (from 1815 to 1736 cm^{-1}) is highlighted.

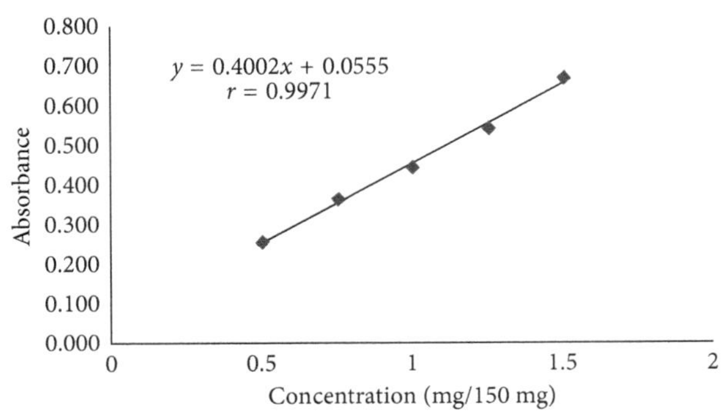

FIGURE 3: Graphical representation of the amoxicillin analytical curve by FT-IR spectrophotometry.

To develop the proposed method, the first step was a qualitative test. For this, an overlap of the spectra obtained with amoxicillin reference standard and amoxicillin in capsules was carried out (Figure 1). It was noticeable that both spectra have the same absorption bands, confirming the absence of adjuvants interference, impurities, or degradation products. The similarity between the spectra is a strong indicative of the identity of amoxicillin in the pharmaceutical dosage form.

The second step for the development of the quantitative method was to select a spectral range in the spectrum of amoxicillin in the mid-infrared region, to be quantitatively analyzed. For this, a comparison between the spectrum obtained with amoxicillin reference standard and each adjuvant present in the composition of amoxicillin in capsules was performed (Figure 2). It was observed that, between 1815 and 1736 cm^{-1}, which represents the amoxicillin carbonyl band, there was no adjuvants interference. Thus, this was the spectral range selected for the quantitative analysis.

After the selection of the spectral range for the quantitative analysis, the analytical curve was constructed. With this purpose, many concentrations of AMX RS were tested, and five of them were selected for the method, as follows: 0.5, 0.75, 1.0, 1.25, and 1.50 mg/pellet. Thereafter, the method was validated according to the ICH guidelines and Brazilian legislation [12, 13].

Three calibration curves, generated on three different days, were analyzed by a graphical representation. The average values of absorbance were plotted against their

TABLE 2: Absorbances obtained for the evaluation of interday and between analysts precision to validate the spectrophotometric method in the mid-infrared region for the analysis of amoxicillin capsules.

Interday			Between analysts	
Day 1	Day 2	Day 3	Analyst 1	Analyst 2
0.466	0.447	0.424	0.466	0.447
0.472	0.417	0.456	0.472	0.417
0.458	0.413	0.445	0.458	0.413
0.437	0.445	0.436	0.437	0.445
0.436	0.447	0.417	0.436	0.447
0.432	0.417	0.457	0.432	0.417

TABLE 3: Accuracy of the spectrophotometric method in the mid-infrared region for analysis of amoxicillin capsules.

	Theoretical concentration (mg/pellet)	Real concentration (mg/pellet)	Recovery (%)	RSD %	Average recovery (%)
R1	0.800	0.798	99.800	1.45	
R2	1.000	1.010	101.040	0.66	100.420
R3	1.200	1.205	100.420	0.80	

RSD: relative standard deviation.

TABLE 4: Absorbances obtained during the robustness evaluation of the spectrophotometric method in the mid-infrared region for the analysis of amoxicillin capsules.

Temperature of workroom		Pellet compression time			KBr brand	
21°C*	26°C	8 min	10 min*	12 min	Synth*	Shimadzu
0.472	0.495	0.415	0.428	0.435	0.430	0.437
0.447	0.494	0.420	0.424	0.434	0.433	0.433
0.460	0.504	0.418	0.427	0.432	0.435	0.431
0.458	0.508	0.417	0.417	0.413	0.434	0.432
0.447	0.491	0.422	0.421	0.434	0.439	0.430
0.491	0.503	0.413	0.417	0.433	0.432	0.426

*Normal working condition.

TABLE 5: Determination of amoxicillin in the capsules.

Assay	AMX content (mg/capsule)	AMX content (%)	Average content (%)	RSD (%)
1	501.535	100.307		
2	501.150	100.230	100.340	0.122
3	502.350	100.470		

AMX: amoxicillin; RSD: relative standard deviation.

respective concentrations. A final analytical curve was obtained by averaging the three analytical curves, resulting in a correlation coefficient (r) of 0.9971, as shown in Figure 3. The r value is situated in accordance with the required r values determined by the validation guidelines in the area of acceptance. Statistical analysis (ANOVA) of the results showed adequate regression of the calibration curve, as the $F_{calculated}$ (521.45) was higher than $F_{critical}$ (0.00445).

Precision was evaluated according to the repeatability (intraday precision) and intermediate precision (interday) and between analysts. The intraday precision showed adequate response, as the calculated RSD between the absorbances obtained from six samples, analyzed at the same day and same experimental conditions, was 3.80%.

For interday and between analysts precision (Table 2), the absorbances obtained in three different days and by a second analyst were compared by ANOVA and F-test and t-test, respectively. Regarding the interday precision, the $F_{calculated}$ was lower than $F_{critical}$, and the precision between analysts $t_{calculated}$ was lower than $t_{critical}$. These data show the appropriate precision (interday and between analysists) of the method, since the absorbances obtained were statistically equivalent.

Accuracy of the method was also proved, since the recovery assay provided an average value of 100.42% (Table 3). This value is in accordance with the range recommended by the Horwitz Trumpet [37].

Robustness was evaluated by small variations in the following analytical conditions: temperature of the workroom, pellet compression time, and KBr brand. The obtained experimental values were evaluated by the F-test and t-test, by comparing the absorbance values from the normal and varied conditions (Table 4). For the temperature of workroom, $t_{calculated}$ was higher than $t_{critical}$. It means the method is not robust for this aspect. In this way, the temperature of workroom must be strictly controlled for performing the method. When the air conditioning is turned off, the humidity of the environment increases, and this can be the main cause of the lack of robustness for this parameter. Regarding the other varied parameters (pellet compression time and KBr brand), all of them showed to be robust, since $t_{calculated}$ were lower than $t_{critical}$.

Selectivity of the method was carried out in order to verify whether the adjuvants present in the pharmaceutical dosage form were able to interfere with the analysis of amoxicillin. Figures 1 and 2 show that there are no interferences in the spectral range selected for quantitative analysis, so the selectivity was evidenced. It is important to highlight that, in Figure 2, each adjuvant was prepared in a concentration higher than that they are in the capsules, in order to better analyze the absorption bands. For this reason, the absorption bands presented in the excipient spectra are more prominent than that in Figure 1.

Three quantifications of amoxicillin in the capsules were performed, and the results are presented in Table 5. The average amoxicillin content was 100.34%, which is in accordance with the range recommended by the Brazilian Pharmacopoeia (90 to 120%) [20].

In this way, the method met all the requirements of the ICH guidelines [13] and Brazilian legislation [12] for validation of the analytical methods and can be used in quality control laboratories for the analysis of this drug. In addition, it presents low cost maintenance and is an environmentally friendly method.

5. Conclusion

The method has fulfilled all the requirements of the ICH guidelines and Brazilian legislation for the validation of the analytical methods. Spectrophotometry in the mid-infrared region for the quantification of amoxicillin in capsules showed adequate linearity, accuracy, precision, selectivity, and robustness for the pellet's compression time and KBr brand. On the other hand, the temperature of the workroom must be strictly controlled for performing the method. Therefore, this method can be used as an innovative alternative in quality control laboratories, since it is fast, clean (for the analysts and environment), presents low cost maintenance, and does not generate toxic chemical waste.

Conflicts of Interest

The authors declare that they have no conflicts of interest.

Acknowledgments

This work was supported by PACD-FCFAr-UNESP (Araraquara, Brazil), FAPESP (São Paulo, Brazil), and CNPq (Brasília, Brazil). EG Tótoli was funded by FAPESP (São Paulo, Brazil), and HRN Salgado was funded by CNPq (Brasília, Brazil).

References

[1] Brazilian National Health Surveillance Agency (ANVISA), "Online directions for the use of amoxicillin," 2016, http://www.anvisa.gov.br.

[2] H. P. Rang, M. M. Dale, J. M. Ritter, and R. J. Flower, *Rang & Dale: Farmacologia*, Elsevier, Rio de Janeiro, Brazil, 6th edition, 2007.

[3] B. G. Katzung, *Basic and Clinical Pharmacology*, AMGH, Porto Alegre, Brazil, 10th edition, 2010.

[4] M. C. Verdier, O. Tribut, P. Tattevin, Y. Le Tulzo, C. Michelet, and D. Bentué-Ferrer, "Simultaneous determination of 12 beta-lactam antibiotics in human plasma by high-performance liquid chromatography with UV detection: application to therapeutic drug monitoring," *Antimicrobial Agents and Chemotherapy*, vol. 55, no. 10, pp. 4873–4879, 2011.

[5] K. Xie, M. Zhao, H. Guo et al., "Determination and depletion of amoxicillin residues in eggs," *Food Additives & Contaminants: Part A*, vol. 30, no. 4, pp. 670–677, 2013.

[6] M. Szultka, R. Krzeminski, J. Szeliga, M. Jackowski, and B. Buszewski, "A new approach for antibiotic drugs determination in human plasma by liquid chromatography-mass spectrometry," *Journal of Chromatography A*, vol. 1272, pp. 41–49, 2013.

[7] M. Szultka, R. Krzeminski, J. Szeliga, M. Jackowski, and B. Buszewski, "Simultaneous determination of selected chemotherapeutics in human whole blood by molecularly imprinted polymers coated solid phase microextraction fibers and liquid chromatography-tandem mass spectrometry," *Journal of Chromatography B*, vol. 940, pp. 66–76, 2013.

[8] R. R. Carzola, G. R. Romero, A. G. Frenich, M. A. Rodrigues, and J. L. Martínez, "A análise simultânea de antibióticos em amostras biológicas por espectrometria líquida de alta eficiência de ultra-massa em tandem cromatografia," *Journal of Pharmaceutical and Biomedical Analysis*, vol. 89, pp. 203–212, 2014.

[9] M. Carlier, V. Fogão, J. A. Roberts, E. Van De Velde, J. J. De Waele, and A. G. Verstraete, "Quantificação de sete antibióticos β-lactâmicos e dois inibidores β-lactamase em plasma humano, utilizando um método de UPLC-MS/MS validado," *International Journal of Antimicrobial Agents*, vol. 40, no. 5, pp. 416–422, 2012.

[10] N. E. Blazhevskiĭ, S. P. Karpova, and V. I. Kabachyĭ, "Quantitative determination of penicillins by iodometry using potassium hydrogen peroxymonosulfate," *Antibiot Khimioter*, vol. 58, no. 11-12, pp. 3–7, 2013.

[11] World Health Organization, *Quality Assurance of Pharmaceuticals: A Compendium of Guidelines and Related materials, Good Manufacturing Practices and Inspection*, WHO, Geneva, Switzerland, 2nd edition, 2007.

[12] Brazilian Ministry of Health. Resolution of the Executive Board No. 899, *Approves the Guide for Validation of Analytical and Bioanalytical Methods*, Diário Oficial da União, Brasília, Brazil, 2003.

[13] ICH Q2 (R1), "Validation of analytical procedures Text and Methodology," in *Proceedings of International Conference on Harmonization of Technical Requirements for the Registration of Pharmaceutical for Human Use*, Geneva, Switzerland, 2005.

[14] The United States Pharmacopeia, *The National Formulary (NF 32)*, United States Pharmacopeia Convention, Rockville, MD, USA, 37th edition, 2014.
[15] A. M. Sanseverino, "Síntese orgânica limpa," *Química Nova*, vol. 23, no. 1, pp. 102–107, 2000.
[16] Association of Official Analytical Chemists, *Official Methods of Analysis*, AOAC, Gaithesburg, MD, USA, 17th edition, 2002.
[17] F. R. Nolasco, G. A. Tavares, and J. A. Bendassolli, "Implantação de programas de gerenciamento de resíduos químicos laboratoriais em universidades: análise crítica e recomendações," *Engenharia Sanitaria e Ambiental*, vol. 11, no. 2, pp. 118–124, 2006.
[18] A. H. Moreno and H. R. N. Salgado, "Development and validation of the quantitative analysis of ceftazidime in powder for injection by infrared spectroscopy," *Physical Chemistry*, vol. 2, no. 1, pp. 6–11, 2012.
[19] D. L. Pavia, G. M. Lampman, G. S. Kriz, and J. R. Vyvyan, *Introdução à Espectroscopia*, Cengage Learning, São Paulo, Brazil, 4th edition, 2010.
[20] ANVISA, *Brazilian Pharmacopeia*, Vol. 2, ANVISA, Brasília, Brazil, 5th edition, 2010.
[21] E. G. Tótoli and H. R. N. Salgado, "Development and validation of the quantitative analysis of ampicillin sodium in powder for injection by Fourier-transform infrared spectroscopy (FT-IR)," *Physical Chemistry*, vol. 2, no. 6, pp. 103–108, 2012.
[22] D. C. M. Vieira and H. R. N. Salgado, "Development and validation of the quantitative analysis of cefuroxime sodium in powder for injection by infrared spectroscopy," *Advances in Analytical Chemistry*, vol. 2, pp. 80–87, 2012.
[23] A. C. Kogawa and H. R. N. Salgado, "Development and validation of infrared spectroscopy method for the determination of darunavir in tablets," *Physical Chemistry*, vol. 33, pp. 1–6, 2013.
[24] M. S. Piantavini, R. Pontarolo, F. L. D. Pontes, C. P. Uber, D. P. Stremel, and M. M. Sena, "Chemometric quality inspection control of pyrantel pamoate, febantel and praziquantel in veterinary tablets by mid infrared spectroscopy," *Spectrochimica Acta Part A: Molecular and Biomolecular Spectroscopy*, vol. 125, pp. 393–403, 2014.
[25] M. A. Mallah, S. T. H. Sherazi, M. I. Bhanger, S. A. Mahesar, and M. A. Bajeer, "A rapid Fourier-transform infrared (FTIR) spectroscopic method for direct quantification of paracetamol content in solid pharmaceutical formulations," *Spectrochimica Acta Part A: Molecular and Biomolecular Spectroscopy*, vol. 141, pp. 64–70, 2015.
[26] L. P. Consortti and H. R. N. Salgado, "Green method for quantification of sodium cefotaxime in lyophilized powder by infrared spectroscopy," *Journal of Pharmaceutical Sciences & Emerging Drugs*, vol. 5, no. 1, pp. 1–6, 2017.
[27] J. C. R. Corrêa and H. R. N. Salgado, "A platform for designing quantitative infrared spectrophotometric method for drugs and pharmaceuticals analysis: a rediscover for an ecological and safer technique in the routine quality control laboratories," *World Journal of Pharmaceutical Sciences*, vol. 3, no. 6, pp. 2056–2059, 2014.
[28] A. C. Kogawa, F. A. Aguiar, C. M. Gaitani, and H. R. N. Salgado, "Validation of a stability indicating capillary electrophoresis method for the determination of darunavir in tablets and comparison with the of infrared absorption spectroscopic method," *World Journal of Pharmaceutical Sciences*, vol. 3, pp. 283–297, 2014.
[29] A. C. Kogawa, N. P. Mello, and H. R. N. Salgado, "Quantification of doxycycline in raw material by an eco-friendly method of infrared spectroscopy," *Pharmaceutica Analytica Acta*, vol. 7, no. 2, pp. 463–466, 2016.
[30] A. C. Kogawa and H. R. N. Salgado, "Spectrophotometry in infrared region: a new, low cost and green way to analyze tablets of rifaximin," *Current Pharmaceutical Analysis*, vol. 14, no. 2, pp. 108–111, 2018.
[31] A. C. Kogawa and H. R. N. Salgado, "Impact of rifaximin potency after treatment with different solvents," *Journal of Pharmaceutical and Biological Sciences*, vol. 5, pp. 77–80, 2017.
[32] A. C. Kogawa and H. R. N. Salgado, "Rifaximin stability: a look by UV, IR, HPLC and turbidimetry method," *Journal of AOAC International*, vol. 101, no. 2, pp. 410–413, 2018.
[33] B. A. Marco and H. R. N. Salgado, "Development and validation of an innovative method for the determination of cefadroxil monohydrate in capsules," *Physical Chemistry*, vol. 6, pp. 67–74, 2016.
[34] J. S. H. Natori, E. G. Tótoli, and H. R. N. Salgado, "Development and validation of a green analytical method for determination of norfloxacin in raw material by Fourier-Transform Infrared Spectrophotometry (FT-IR)," *Journal of AOAC International*, vol. 99, no. 6, pp. 1533–1536, 2016.
[35] E. G. Tótoli and H. R. N. Salgado, "Fourier-Transform Infrared (FT-IR) spectrophotometry: an eco-friendly method for the analysis of daptomycin injectable," *Journal of AOAC International*, vol. 100, no. 5, pp. 1569–1576, 2017.
[36] M. T. Trindade and H. R. N. Salgado, "Development and validation of a modern and stability-indicating method for the quantification of ceftriaxona sodium in powder for injection by infrared spectroscopy," *Physical Chemistry*, vol. 7, pp. 55–62, 2017.
[37] M. Valcárcel, *Principles of Analytical Chemistry: A Textbook*, Springer, Berlin Heidelberg, Germany, 6th edition, 2000.

11

Synthesis and Biological Evaluation of New (−)-Gossypol-Derived Schiff Bases and Hydrazones

Vu Van Vu,[1] Trinh Thi Nhung,[1] Nguyen Thi Thanh,[1] Luu Van Chinh,[2] Vu Dinh Tien,[1] Vu Thu Thuy,[1] Do Thi Thao,[3] Nguyen Hai Nam,[4] Angela Koeckritz,[5] and Tran Khac Vu[1]

[1]*School of Chemical Engineering, Hanoi University of Science and Technology, No. 1 Dai Co Viet, Hai Ba Trung, Hanoi, Vietnam*
[2]*Institute of Natural Products Chemistry, Vietnam Academy of Science and Technology, 18 Hoang Quoc Viet, Cau Giay, Hanoi, Vietnam*
[3]*Institute of Biotechnology, Vietnam Academy of Science and Technology, 18 Hoang Quoc Viet, Cau Giay, Hanoi, Vietnam*
[4]*Hanoi University of Pharmacy (HUP), 13-15 Le Thanh Tong, Hanoi, Vietnam*
[5]*Leibniz Institute for Catalysis at the University of Rostock, Albert-Einstein-Str. 29A, 18059 Rostock, Germany*

Correspondence should be addressed to Tran Khac Vu; vu.trankhac@hust.edu.vn

Academic Editor: Mire Zloh

A series of 14 new (−)-gossypol Schiff bases and hydrazones have been synthesized via an in situ procedure in high yields. Structural data showed that all target compounds exist as the enamine tautomer. Bioassays showed that several compounds exhibited cytotoxic effects against three human cancer cell lines. Compound **8a** showed the greatest cytotoxic effect against hepatocellular carcinoma (HepG2), lung carcinoma (LU-1), and breast cancer (MCF-7) cell lines with IC_{50} values of 20.93, 13.58, and 9.40 μM, respectively. However, in an antibacterial test, compounds **8a** and **8b** inhibited *Staphylococcus aureus* and *Bacillus cereus* and compound **8e** inhibited only *Staphylococcus aureus* at the same MIC values of 1024 μg/ml.

1. Introduction

Cancer is characterized by uncontrolled cell growth, metastasis, and invasion and is responsible for approximately 13% of all human deaths throughout the world [1]. The three most common and fatal types are lung, liver, and breast cancer. Currently, although there has been a lot of success in both cancer chemotherapy and anticancer drug development research, cancer remains a significant challenge in the future because of the drug resistance and adverse side effects of chemotherapy [2]. Novel antitumor agents based on natural products are becoming increasingly more popular in order to overcome these limitations and develop more effective anticancer agents [3].

Gossypol **(1)** (Figure 1), a polyphenolic dialdehyde found in high concentrations in the pigment glands of the cotton plant *Gossypium*, has recently received increased attention due to its wide range of biological activities, especially anticancer [4–7], contraceptive [8], antiviral [9–11], and antimicrobial [12] activities. However, the application of gossypol as a therapeutic agent has been limited because of a number of serious side effects [13] that have been shown to be associated with the aldehyde groups. More recently, significant attention has been focused on the potential therapeutic value of gossypol as a promising starting point for the development of antitumor or antiviral derivatives for medicinal applications with enhanced bioactivity and reduced side effects [14–18], in which the aldehyde groups are altered to give gossypol derivatives (e.g., Schiff bases, esters, and ethers). Many of these derivatives exhibit a variety of unusual disease-inhibiting activities, especially anticancer activity [19]. Research regarding the biological effects of gossypol both *in vitro* and *in vivo* showed that the (−)-enantiomer is more potent than the (+)-enantiomer or the racemic mixture **(1)** [20]. It has also been hypothesized that while low doses of (−) gossypol are selective, higher doses of either enantiomer result in nonselective action [21, 22]. Also related to the mechanism of action, (−)-gossypol has been reported to target Bcl-2, Bcl-xL, and Mcl-1 proteins with high affinities and is now in clinical trials as an orally administered

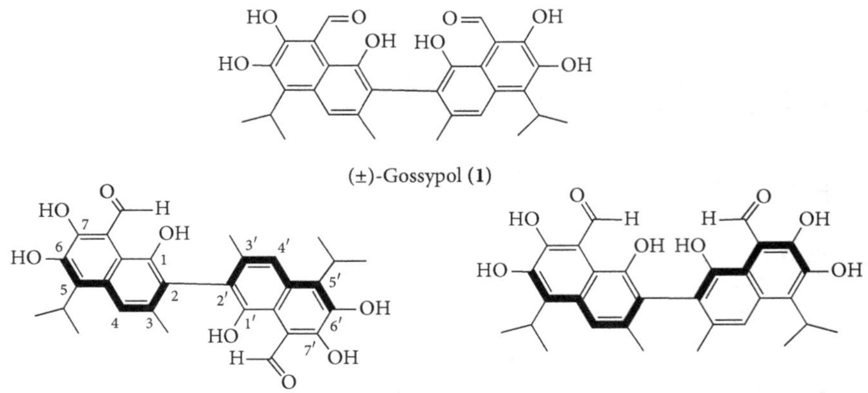

(±)-Gossypol (**1**)

R-(−)-Gossypol (**2**) *S*-(+)-Gossypol (**3**)

FIGURE 1: Racemic gossypol and its enantiomers.

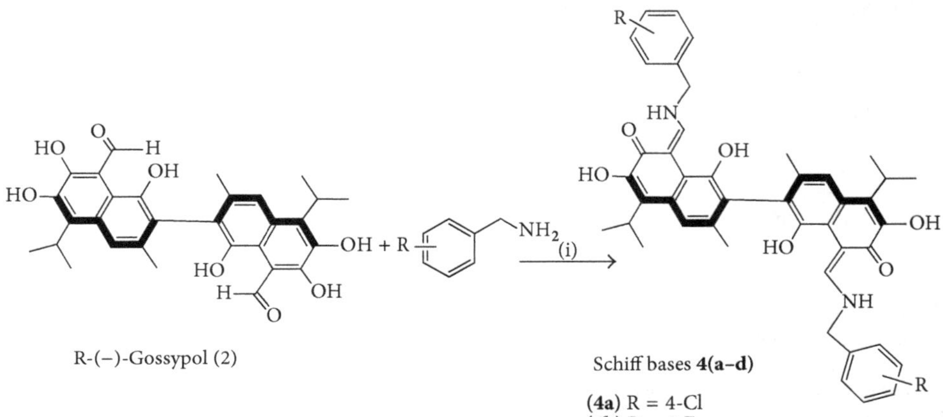

R-(−)-Gossypol (2)

Schiff bases **4(a–d)**

(**4a**) R = 4-Cl
(**4b**) R = 4-F
(**4c**) R = 4-CH_3
(**4d**) R = 4-OCH_3

SCHEME 1: Synthesis of (−)-gossypol Schiff bases **4a–d**; (i) EtOH, 8 h, r.t, 60.7–81.8%.

agent for the treatment of several types of human cancer [21, 22]. Intrigued by this observation, we synthesized and evaluated new (−)-gossypol hydrazones for their cytotoxic effects against three human cancer cell lines. This paper reports the results of this study.

2. Results and Discussion

2.1. Chemistry. New Schiff bases and hydrazones of (−)-gossypol were synthesized as outlined (Schemes 1–3). The starting material (−)-gossypol was first prepared according to a known procedure with modification [23], using L-tryptophan methyl ester as an effective agent for the resolution of racemic gossypol. The yield of (−)-gossypol (**2**) obtained was comparable to that reported in the literature [23] (see Supplementary Material available online at https://doi.org/10.1155/2017/3687182).

Schiff bases **4a–d** were synthesized (Scheme 1). The reactions of (−)-gossypol (**2**) with primary amines including 4-chlorobenzylamine, 4-fluorobenzylamine, 4-methylbenzylamine, and 4-methoxybenzylamine were conducted in ethanol at room temperature for 8 h. The resulting Schiff bases **4a–d** were obtained by simple filtering and washing with methanol and dichloromethane. The NMR data showed the enamine-enamine tautomer of these Schiff bases via characteristic chemical shifts of NH protons resulting from hydrogen bonding formation ranging from 13.34 to 13.49 ppm in the ^{1}H NMR spectrum [24]. In the ^{13}C NMR spectrum, the chemical shifts of ketone C-7 were observed at the lowest field from 171.9 to 172.3 ppm, followed by the chemical shifts of C-6 around 162 ppm. The chemical shifts of C-1 were easily observed at around 149 ppm.

Novel hydrazones **8a–e** were synthesized in a three-step procedure as outlined (Scheme 2). First, benzoic acid derivatives **5a–e** were converted to the corresponding esters **6a–e** in quantitative yields by Fisher esterification. These esters were then refluxed with hydrazine hydrate (80%) in ethanol to give the corresponding hydrazides (**7a–e**) in high yields, which were then used for the next step without further purification. Unlike the case of Schiff bases **4a–d**, hydrazides (**7a–e**) were less active than the primary amines. Initially, hydrazides (**7a–e**) were reacted with (−)-gossypol (**2**) in ethanol at room

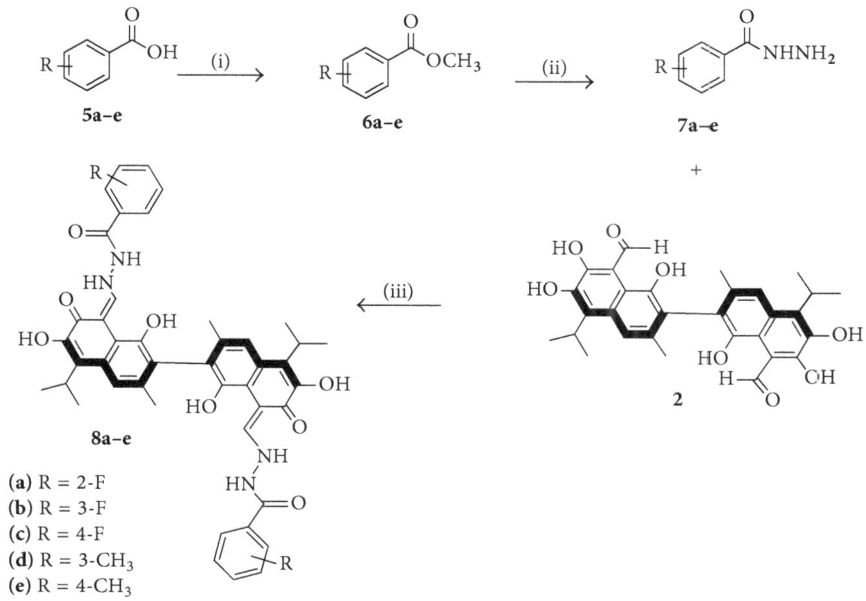

SCHEME 2: Synthesis of (–)-gossypol hydrazone **8a–e**; (i) MeOH, H_2SO_4, MeOH, reflux, 12 h; (ii) EtOH, NH_2NH_2 (80%), r.t, 10 h; (iii) EtOH, $ZnCl_2$, r.t, 16 h, 50.4–63.0%.

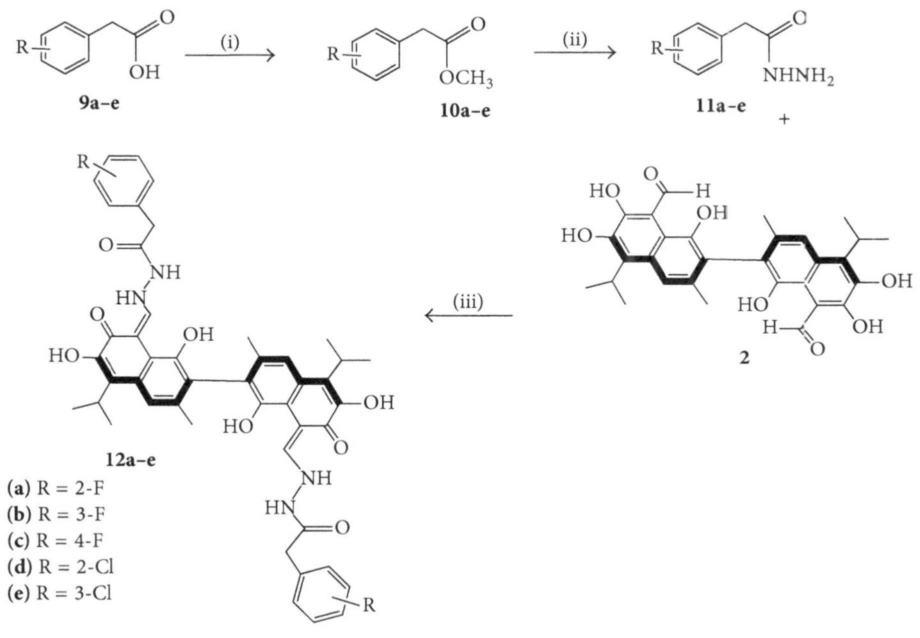

SCHEME 3: Synthesis of hydrazones **12a–e**; (i) MeOH, H_2SO_4, MeOH, reflux, 12 h; (ii) EtOH, NH_2NH_2 (80%), r.t, 10 h; (iii) EtOH, $ZnCl_2$, r.t, 12 h, 48.7–53.0%.

temperature for 16 h with very poor yields. The reaction was next carried out at reflux but failed to give the desired products owing to the decomposition of gossypol. Hydrazones are easily prepared in the presence of acetic acid as a catalyst. However, the reaction of **7a–e** with (–)-gossypol did not proceed as expected in the presence of acetic acid. In this study, $ZnCl_2$ was examined and found to be an excellent catalyst for the hydrazone formation. The reaction of hydrazides **(7a–e)** with (–)-gossypol **(2)** in the presence of a small amount of $ZnCl_2$ led to the formation of a series of hydrazones **(8a–e)** in high yields. To the best of our knowledge, this is the first time that $ZnCl_2$ has been used as an effective catalyst for the synthesis of new gossypol hydrazones **(8a–e)**. An analysis of ^1H and ^{13}C NMR spectra data showed that hydrazones **8a–e** exist as the enamine-enamine tautomer [24].

Compound **8c** was used as an example for the structural elucidation of hydrazones. The ^1H NMR spectrum of compound **8c** showed the characteristic resonance signal of

NH at the lowest field δ 14.76 ppm due to hydrogen bonding with the adjacent ketone group. The chemical shift at δ 12.39 ppm is attributed to the other NH group. Additionally, the resonance signal at δ 10.19 ppm as a sharp singlet is also attributed to 6-OH. The vinylic proton in the enamine is easily observed as a single peak at 8.81 ppm; signals at higher field belonging to the protons H-4 and 1-OH at δ 8.16 and 7.63 ppm, respectively, are coupled. Besides, the presence of the aromatic ring protons can be observed as doublets (J = 8.0 Hz) at δ 8.02 and 7.37 ppm. The ^{13}C NMR showed all the signals of the gossypol skeleton and aromatic rings in which the signals at 165.2 and 163.3 ppm are attributed to C-7 and C-6, respectively. The carbon connected to Fluorine in the aromatic ring resonates at δ 161.7 ppm, based on the characteristic splitting of Fluorine and the hydrazide carbonyl at δ 153.2 ppm.

Novel hydrazones **12a–e** were synthesized in the same way as mentioned above, starting from different phenylacetic acid derivatives **(9a–e)** (Scheme 3). The structure of synthesized hydrazones was confirmed similarly, based on NMR and mass spectroscopy data.

3. Biological Activity

The *in vitro* cytotoxic evaluation of the synthesized compounds was carried out according to the described protocol [25]. Briefly, a stock solution of the target compound was prepared in dimethylsulfoxide (DMSO) at a concentration of 1 mg/mL, followed by dilution to obtain a solution at a concentration of 100 μg/mL, which was serially diluted further for bioassays in 96-well plates. The determination of IC_{50} was carried out using three cancer cell lines, HepG2, LU-1, and MCF-7, with ellipticine and (−)-gossypol as positive controls. The IC_{50} values were determined from a dose-dependent curve plotted from five different concentration regimens (0–100 μM). At each regimen, mean of triplicate experiment was used for a point in the curve. The bioassay results are described in Table 1.

The bioassay results (Table 1) showed that only half of the target compounds exhibited cytotoxic effects against cancer cell lines tested, ranging from 9.40 to 99.55 μM. Unexpectedly, Schiff bases **4a–c**, bearing both electron withdrawing groups (Cl and F) and an electron donating group (CH_3) at the *para* position of the aromatic ring, showed no cytotoxicity against all three cell lines except compound **4d**, which contains an electron donating group (OCH_3). **4d** revealed a weak inhibitory effect against only the LU-1 cell line. For hydrazones **12a–e**, compounds **12c-e** containing electron withdrawing groups at *ortho*, *meta*, and *para* positions were inactive against all cell lines. However, compounds **12a** and **12b** containing electron withdrawing groups at *ortho* and *para* positions exhibited weak-to-moderate cytotoxic effects. The observation also shows that cytotoxic effects of hydrazones **8a–e** derived from benzoic acid derivatives were better than those of phenyl acetic acid derivatives, excluding compound **8e**, and that series **8a–d** were more cytotoxic towards MCF-7 and LU-1 than towards HepG2 cell line. It seems that the electron withdrawing group (F) at *ortho* position contributes towards greater toxicity compared to the *meta* position. In terms of IC_{50} values, all synthesized hydrazones exhibited less potency than that of the control, ellipticine, or the parent (−)-gossypol and some other gossypol derivatives reported [16, 17]. Regarding mechanism of action, most of gossypol derivatives were reported to target Bcl-2 family proteins [20]. In this study, it is early and hard to come up with a mechanism of action at this stage without predicted binding models of designed hydrazones to such target protein as Bcl-2 or Mcl-1 [20–22]. However, it can be predicted that such possible target proteins as Bcl-2 or Mcl-1 participate in the mechanism of action.

Among synthesized hydrazones, compound **8a** showed the best cytotoxic effect against HepG2, LU-1, and MCF-7 with IC_{50} values of 20.93, 13.58, and 9.40 μM, respectively. Accordingly, this compound could be considered as a lead for future design of gossypol hydrazones in which bioisosteric replacements in *ortho* position of the phenyl ring could be performed to improve the cytotoxic activity.

In order to expand the biological spectrum of gossypol derivatives, synthesized hydrazones were also evaluated for antibacterial activity against human pathogenic bacteria, including *Staphylococcus aureus* (ATCC 6538), *Bacillus cereus* (ATCC 21768), *Bacillus subtilis* (ATCC 6633), *Escherichia coli* (ATCC 25931), and *Pseudomonas aeruginosa* (ATCC 9027), according to described protocol [26]. Briefly, five bacterial species including two Gram-negative strains (*Escherichia coli* ATCC 25931 and *Pseudomonas aeruginosa* ATCC 9027) and three Gram-positive strains (*Bacillus subtilis* ATCC 6633, *Bacillus subtilis* ATCC 6633, and *Staphylococcus aureus* ATCC 6538) were obtained from the ATCC (Manassas, VA, USA). The bacterial strains were cultured aerobically on nutrient agar (NA) plates at 37°C for 24 h. For the antibacterial activity test, the bacteria were aerobically cultured in nutrient broth (NB) at 37°C for 24 h and then suspended in sterile saline at a density equivalent to that of the 0.5 McFarland standards. Bacterial suspensions with a concentration of 105 cfu/ml were used for *in vitro* antibacterial activity test.

The results (Table 2) showed that most of the synthesized hydrazones were inactive against the bacteria tested. Compounds **8a** and **8b** inhibited *Staphylococcus aureus* and *Bacillus cereus* and **8e** inhibited only *Staphylococcus aureus* at the same MIC values of 1024 μg/ml and showed less bacterial activity than that of the parent compound, (−)-gossypol and some aza-derivatives of gossypol [24]. From the structure activity relationship, inactivity or weak activity of new hydrazones can be a result of the presence in its structure of the moieties attached to gossypol, which probably do not enable better interaction of these derivatives with the lipid bilayer of bacteria.

4. Conclusions

We have presented the synthesis of new Schiff bases and hydrazones of (−)-gossypol and screened for *in vitro* cytotoxic activity against several human cancer cell lines and antibacterial activity against human pathogenic bacteria. In this research, for the first time, $ZnCl_2$ has been effectively used for the synthesis of gossypol hydrazones. Although no compounds were as effective as positive controls or gossypol

TABLE 1: Cytotoxic activity of target compounds.

Number	Compounds	IC$_{50}$ (μM)		
		HepG2	LU-1	MCF-7
(1)	**4a**	>100	>100	>100
(2)	**4b**	>100	>100	>100
(3)	**4c**	>100	>100	>100
(4)	**4d**	>100	66.83	>100
(5)	**8a**	20.93	13.58	9.40
(6)	**8b**	55.82	56.82	48.48
(7)	**8c**	89.72	79.83	86.40
(8)	**8d**	86.46	64.56	65.72
(9)	**8e**	>100	>100	>100
(10)	**12a**	55.68	54.70	63.33
(11)	**12b**	95.30	99.55	>100
(12)	**12c**	>100	>100	>100
(13)	**12d**	>100	>100	>100
(14)	**12e**	>100	>100	>100
(15)	Ellipticine	1.71	1.87	1.91
(16)	(−)-Gossypol	4.5	4.3	4.0

Note. The reference substance, ellipticine, exhibited cytotoxic activity against HepG2 (ATCC-HB 8065), LU-1 (ATCC-HTB-57), and MCF-7 (*ATCC*- HTB22) with IC$_{50}$ values of 1.71, 1.87, and 1.91 μM, respectively. The values shown for these compounds are the average of three determinations.

TABLE 2: Antibacterial activity of compounds against human pathogenic bacteria.

Number	Compounds	MIC (μg/ml)				
		Staphylococcus aureus (ATCC 6538)	*Bacillus cereus* (ATCC 21768)	*Bacillus subtilis* (ATCC 6633)	*Escherichia coli* (ATCC 25931)	*Pseudomonas aeruginosa* (ATCC 9027)
(1)	**4a**	—	—	—	—	—
(2)	**4b**	—	—	—	—	—
(3)	**4c**	—	—	—	—	—
(4)	**4d**	—	—	—	—	—
(5)	**8a**	1024	1024		—	—
(6)	**8b**	1024	—	1024	—	—
(7)	**8c**	—	—	—	—	—
(8)	**8d**	—	—	—	—	—
(9)	**8e**	1024	—	—	—	—
(10)	**12a**	—	—	—	—	—
(11)	**12b**	—	—	—	—	—
(12)	**12c**	—	—	—	—	—
(13)	**12d**	—	—	—	—	—
(14)	**12e**	—	—	—	—	—
(15)	Streptomycin sulfate	20	20	5	20	20
(16)	(−)-Gossypol	15	15	16	Nd*	Nd*

**Not determined.*

in terms of biological effects, the research could make contribution to the discovery of new gossypol derivatives based on research published herein.

5. Experimental

All chemicals and reaction solvents were purchased from Merck and Sigma-Aldrich. Melting points were determined in open capillaries on a Shimadzu Electrothermal IA 9200 apparatus and uncorrected. IR spectra were recorded on a Impact 410 FTIR using KBr discs. 1D-NMR and 2D-NMR Spectra were recorded on a Bruker AVANCE 500 MHz spectrometer in $CDCl_3$ and DMSO-d_6. Chemical shifts (δ) are in ppm relative to TMS; multiplicities are shown as follows: s (singlet), d (doublet), t (triplet), and m (multiplet); and coupling constants (J) are expressed in hertz (Hz). HRMS was recorded by using a FT-ICR MS and ESI-TOF-MS Agilent 6230 spectrophotometer (Varian). Progress of the reaction was monitored by thin-layer chromatography (TLC) using precoated TLC sheets with ultraviolet (UV) fluorescent silica gel (Merck 60F254) and spots were visualized by UV lamp at 254 nm. Column chromatography was carried out using silica gel (40–230 mesh).

5.1. Separation of (−)-Gossypol from (±)-Gossypol. A solution of L-tryptophan methyl ester hydrochloride (9.824 g, 0.038 mol) and NaOH (1.52 g, 0.038 mol) in ethanol (60 mL) was stirred at 45°C, and the racemic gossypol (9.85 g, 0.019 mol) was added. The mixture was stirred for 2 h. The reaction was monitored by TLC and then allowed to warm to room temperature. The resulting precipitate from the reaction was then filtered, washed with cold ethanol, and dried under reduced pressure to obtain Schiff base intermediate (8.28 g, 95%). Schiff base intermediate (8.808 g, 9.59 mmol) was dissolved in a mixture of acetic acid and ether (4 : 1) (50 mL). The reaction mixture was kept at −5–0°C under nitrogen atmosphere, and concentrated hydrochloric acid (1.5 mL) was slowly added for 10 minutes. The reaction was then stirred at room temperature for 72 and monitored by TLC. After the completion of reaction, the resulting D-tryptophan methyl ester was filtered and washed by an amount of dichloromethane. The filtrate was concentrated to dryness and was dissolved in 30 mL of mixture of dichloromethane and *n*-hexane (2 : 1) and was kept at 4°C for crystallization. (−)-Gossypol was obtained by filtering and washing with cold acetone in 85.2% yield (4.23 g), 94.4% ee by HPLC.

5.2. General Procedure for the Synthesis of Schiff Bases ***(4a–d).*** A mixture of (−)-gossypol (100 mg, 1 eq, and 0.193 mmol) and corresponding amines (2.2 eq, 0.42 mmol) in EtOH (10 mL) was stirred at room temperature for 8 h. The resulting precipitate was then filtered and washed with cold methanol and dichloromethane to afford desired Schiff bases **4a–d**.

(R,8Z,8′Z)-8,8′-Bis(((4-chlorobenzyl)amino)methylene)-1,1′,6,6′-tetrahydroxy-5,5′-diisopropyl-3,3′-dimethyl-[2,2′-binaphthalene]-7,7′(8H,8′H)-dione ***(4a).*** Yield 77.8%; mp: 227–229°C; ^{1}H NMR (500 MHz, DMSO-d6): 13.45 (t, J = 5.5 Hz, 2H, NH), 9.85 (d, J = 12.5 Hz, 2H), 8.37 (s, 2H), 7.83 (s, 2H), 7.44 (m, 3H), 7.39 (d, J = 8.5 Hz, 4H), 4.75 (m, 4H, $ArCH_2$), 3.69 (m, 2H, $CH(CH_3)_2$), 1.94 (s, 6H, $2CH_3$), 1.43 (t, J = 7.0 Hz, 12H, $4CH_3$). ^{13}C NMR (125 MHz, DMSO-d6): 172.2, 162.4, 149.6, 146.2, 136.7, 132.4, 131.4, 129.4, 128.7, 126.9, 126.7, 120.3, 116.6, 115.8, 103.5, 52.3, 26.5, 20.3, 20.14. HR-ESIMS $[M + H]^+$ found 765.24945; calculated $C_{44}H_{43}Cl_2N_2O_6$: 765.24982.

(R,8Z,8′Z)-8,8′-Bis(((4-fluorobenzyl)amino)methylene)-1,1′,6,6′-tetrahydroxy-5,5′-diisopropyl-3,3′-dimethyl-[2,2′-binaphthalene]-7,7′(8H,8′H)-dione ***(4b).*** Yield 81.8%; mp: 223–225°C; ^{1}H NMR (500 MHz, DMSO-d6): 13.47 (t, J = 5.0 Hz, 2H, NH), 9.87 (d, J = 12.5 Hz, 2H), 8.37 (s, 2H), 7.84 (s, 2H), 7.45–7.37 (m, 6H), 7.21 (m, 4H), 4.74 (brs, 4H, $ArCH_2$), 3.70 (m, 2H, $CH(CH_3)_2$), 1.94 (s, 6H, $2CH_3$), 1.43 (t, J = 6.5 Hz, 12H, $4CH_3$). ^{13}C NMR (125 MHz, DMSO-d6): 172.1, 162.6, 162.3, 160.7, 149.6, 146.2, 133.8, 131.3, 129.6, 126.9, 126.7, 120.3, 116.6, 115.6, 103.4, 52.3, 26.5, 20.3, 20.1. HR-ESIMS $[M + H]^+$ found 733.30890; calculated $C_{44}H_{43}F_2N_2O_6$: 733.30892.

(R,8Z,8′Z)-8,8′-Bis(((4-methylbenzyl)amino)methylene)-1,1′,6,6′-tetrahydroxy-5,5′-diisopropyl-3,3′-dimethyl-[2,2′-binaphthalene]-7,7′(8H,8′H)-dione ***(4c).*** Yield 60.7%; mp: 211–213°C; ^{1}H NMR (500 MHz, DMSO-d6): 13.49 (t, J = 6.0 Hz, 2H, NH), 9.88 (d, J = 12.5 Hz, 2H), 8.36 (s, 2H), 7.86 (s, 2H), 7.45 (s, 2H), 7.26 (d, J = 8.0 Hz, 4H, Ar), 7.19 (d, J = 8.0 Hz, 4H, Ar), 4.70 (m, 4H, ArCH2), 3.69 (m, 2H, $CH(CH_3)_2$), 2.28 (s, CH_3), 1.93 (s, 6H, $2CH_3$), 1.43 (t, J = 7.0 Hz, 12H, $4CH_3$). ^{13}C NMR (125 MHz, DMSO-d6): 171.9, 162.4, 149.7, 146.2, 136.9, 134.5, 131.3, 129.3, 128.7, 127.5, 126.6, 126.6, 120.2, 115.8, 103.4, 52.9, 26.5, 20.6, 20.2, 20.1. HR-ESIMS $[M + H]^+$ found 725.35891; calculated $C_{46}H_{49}N_2O_6$: 725.35906.

(R,8Z,8′Z)-8,8′-Bis(((4-methoxybenzyl)amino)methylene)-1,1′,6,6′-tetrahydroxy-5,5′-diisopropyl-3,3′-dimethyl-[2,2′-binaphthalene]-7,7′(8H,8′H)-dione ***(4d).*** Yield 65.0%; mp: 230–232°C; ^{1}H NMR (500 MHz, DMSO-d6): 13.34 (s, 2H, NH), 9.89 (d, J = 7.0 Hz, 2H), 7.42 (s, 2H), 7.36–7.29 (m, 6H), 6.95–6.92 (m, 6H), 4.68 (br, 4H, $2CH_2$-Ph), 3.75 (m, 2H, $CH(CH_3)_2$), 3.72 (s, 6H, OCH_3), 1.98 (s, 6H, $2CH_3$), 1.45 (t, J = 7.0 Hz, 12H, $4CH_3$). ^{13}C NMR (125 MHz, DMSO-d6): 172.3, 162.5, 159.2, 158.9, 149.7, 131.2, 130.2, 129.3, 129.1, 126.6, 116.3, 114.2, 113.9, 103.4, 55.2, 52.7, 52.9, 26.5, 20.6, 20.2. HR-ESIMS $[M + H]^+$ found 757.34880; calculated $C_{46}H_{49}N_2O_8$: 757.34889.

5.3. Representative Procedure for the Synthesis of Hydrazones ***(8a–e*** *and* ***12a–e).*** A mixture of 2-fluorobenzoic acid (**5a**) (1 g, 7.13 mmol) in MeOH (20 mL) was stirred for a few minutes. Concentrated H_2SO_4 (1 mL) was added, and the mixture was refluxed for 12 h. The reaction was monitored by TLC using *n*-hexane, with ethyl acetate as a developing system. CH_2Cl_2 (30 mL) was then added, and the mixture was extracted with distilled H_2O (3 × 20 mL). The organic phase was separated, dried on anhydrous Na_2SO_4, and evaporated under vacuum to afford corresponding ester **6a** in almost quantitative yield in oil form (1.09 g). The oil was then dissolved in EtOH, and $NH_2NH_2{\cdot}H_2O$ (3 eq) was added. The mixture was refluxed for 10 h. CH_2Cl_2 (30 mL) was then added and extracted with distilled H_2O (3 × 20 mL). The organic phase was separated, dried on anhydrous Na_2SO_4, and evaporated under vacuum to afford corresponding hydrazide **7a** in very good yield (white solid, 1.02 g, 93%) and used for next step. (−)-Gossypol (200 mg, 0.38 mmol) and $ZnCl_2$ (5 mg) were added to the solution of hydrazide **7a** (122 g, 0.80 mmol, and 2.1 eq) in ethanol (10 mL). The reaction was stirred at room temperature for 16 h. The resulting precipitate was then filtered and washed with cold methanol and dichloromethane to give hydrazone **8a**.

N′,N‴-((1Z,1′Z)-((R)-1,1′6,6′-Tetrahydroxy-5,5′-diisopropyl-3,3′-dimethyl-7,7′-dioxo-[2,2′-binaphthalene]-8,8′(7H,7′H)-diylidene)bis(methanylylidene))bis(2-fluorobenzohydrazide) ***(8a).*** Yield 50.4%; mp: 236–238°C; ^{1}H NMR (500 MHz, DMSO-d6): 14.53 (s, 2H, NH), 12.32 (s, 2H, NH), 10.09 (s,

2H), 8.82 (s, 2H), 8.17 (s, 2H), 7.69 (m, 2H), 7.61–7.56 (m, 4H), 7.36–7.32 (m, 4H), 3.91 (m, 2H, $CH(CH_3)_2$), 1.99 (s, 6H, $2CH_3$), 1.48 (t, J = 7.0 Hz, 12H, $4CH_3$). ^{13}C NMR (125 MHz, DMSO-d6): 160.2, 160.0, 158.3, 153.5, 150.8, 149.9, 133.1, 133.0, 132.8, 130.3, 128.7, 127.9, 124.7, 122.5, 122.4, 118.8, 116.3, 116.1, 115.9, 107.3, 56.0, 20.9, 20.5. HR-ESIMS $[M + H]^+$ found 791.28920; calculated $C_{44}H_{41}F_2N_4O_8$: 791.28925.

N′,N‴-((1Z,1′Z)-((R)-1,1′6,6′-Tetrahydroxy-5,5′-diisopropyl-3,3′-dimethyl-7,7′-dioxo-[2,2′-binaphthalene]-8,8′(7H,7′H)-diylidene)bis(methanylylidene))bis(3-fluorobenzohydrazide) ***(8b)***. Yield 63.0%; mp: 230–233°C; ^{1}H NMR (500 MHz, DMSO-d6): 14.69 (s, 2H, NH), 12.43 (s, 2H, NH), 10.20 (s, 2H), 8.83 (s, 2H), 8.19 (s, 2H), 7.79 (dd, J = 7.5 Hz, 4H), 7.63 (s, 2H), 7.58 (m, 2H), 7.46–7.42 (m, 2H), 7.69 (m, 2H), 7.61–7.56 (m, 4H), 7.36–7.32 (m, 4H), 3.91 (m, 2H, $CH(CH_3)_2$), 1.99 (s, 6H, $2CH_3$), 1.48 (t, J = 7.0 Hz, 12H, $4CH_3$). ^{13}C NMR (125 MHz, DMSO-d6): 162.9, 161.4, 160.9, 153.6, 150.7, 150.1, 143.9, 134.9, 132.7, 130.7, 128.7, 127.9, 123.9, 118.9, 118.7, 116.2, 114.6, 114.4, 107.3, 56.0, 26.0, 20.4.

N′,N‴-((1Z,1′Z)-((R)-1,1′6,6′-Tetrahydroxy-5,5′-diisopropyl-3,3′-dimethyl-7,7′-dioxo-[2,2′binaphthalene]-8,8′(7H,7′H)-diylidene)bis(methanylylidene))bis(4-fluorobenzohydrazide) ***(8c)***. Yield 57.9%; mp: 235–237°C; ^{1}H NMR (500 MHz, DMSO-d6): 14.76 (s, 2H, NH), 12.39 (s, 2H, NH), 10.19 (s, 2H), 8.81 (s, 2H), 8.16 (s, 2H), 8.02 (d, J = 8.0 Hz, 4H), 7.63 (s, 2H), 7.37 (d, J = 8.0 Hz, 4H), 3.92 (m, 2H, $CH(CH_3)_2$), 2.03 (s, 6H, $2CH_3$), 1.50 (t, J = 6.0 Hz, 12H, $4CH_3$). ^{13}C NMR (125 MHz, DMSO-d6): 165.2, 163.3, 161.7, 153.2, 150.8, 150.1, 143.9, 132.7, 130.4, 129.1, 128.64, 127.73, 118.87, 116.15, 115.90, 115.59, 115.42, 107.37, 56.0, 26.0, 20.4.

N′,N‴-((1Z,1′Z)-((R)-1,1′6,6′-Tetrahydroxy-5,5′-diisopropyl-3,3′-dimethyl-7,7′-dioxo-[2,2′-binaphthalene]-8,8′(7H,7′H)-diylidene)bis(methanylylidene))bis(3-methylbenzohydrazide) ***(8d)***. Yield 57.8%; mp: 204–206°C; ^{1}H NMR (500 MHz, DMSO-d6): 14.77 (s, 2H, NH), 12.29 (s, 2H, NH), 10.16 (s, 2H), 8.81 (s, 2H), 8.80 (s, 2H), 8.12 (s, 2H), 7.75–7.72 (m, 4H), 7.62 (s, 2H), 7.40 (m, 4H), 3.91 (m, 2H, $CH(CH_3)_2$), 2.37 (s, 6H, 2 CH_3), 2.02 (s, 6H, $2CH_3$), 1.47 (t, J = 5.5 Hz, 12H, $4CH_3$). ^{13}C NMR (125 MHz, DMSO-d6): 165.79, 153.07, 150.72, 150.08, 137.83, 132.59, 132.51, 128.63, 128.39, 128.14, 127.65, 124.80, 118.84, 116.18, 107.39, 26.60, 20.91, 20.56.

N′,N‴-((1Z,1′Z)-((R)-1,1′6,6′-Tetrahydroxy-5,5′-diisopropyl-3,3′-dimethyl-7,7′-dioxo-[2,2′-binaphthalene]-8,8′(7H,7′H)-diylidene)bis(methanylylidene))bis(4-methylbenzohydrazide) ***(8e)***. Yield 60.4%; mp: 213–215°C; ^{1}H NMR (500 MHz, DMSO-d6): 14.81 (s, 2H, NH), 12.28 (s, 2H, NH), 10.18 (s, 2H), 8.81 (s, 2H), 8.80 (s, 2H), 8.13 (s, 2H), 7.86 (d, J = 7.5 Hz, 2H), 7.45 (d, J = 7.5 Hz, 2H), 7.63 (s, 2H), 7.33 (d, J = 8.0 Hz, 2H), 7.28 (d, J = 7.0 Hz, 2H), 3.92 (m, 2H, $CH(CH_3)_2$), 2.37 (d, J = 10.5 Hz, 6H, 2 CH_3), 2.02 (s, 6H, $2CH_3$), 1.47 (t, J = 5.5. Hz, 12H, $4CH_3$). ^{13}C NMR (125 MHz, DMSO-d6): 162.56, 152.91, 150.72, 150.09, 142.06, 141.68, 132.61, 129.73, 129.41, 129.02, 128.97, 127.66, 127.16, 118.81, 116.16, 107.41, 21.01, 20.96, 20.58. HR-ESIMS $[M + H]^+$ found 783.33931; calculated $C_{46}H_{47}N_4O_8$: 783.33939.

N′,N‴-((1Z,1′Z)-((R)-1,1′6,6′-Tetrahydroxy-5,5′-diisopropyl-3,3′-dimethyl-7,7′-dioxo-[2,2′-binaphthalene]-8,8′(7H,7′H)-diylidene)bis(methanylylidene))bis(2-(2-fluorophenyl)acetohydrazide) ***(12a)***. Yield 51.7%; mp: 202–204°C; ^{1}H NMR (500 MHz, DMSO-d6): 15.26 (s, 2H, NH), 12.89 (s, 2H, NH), 10.83 (s, 2H), 9.61 (s, 2H), 9.03 (s, 2H), 8.44 (s, 2H), 8.23 (m, 2H), 8.16 (m, 2H), 8.01 (m, 4H), 4.73 (brs, 2H, $CH(CH_3)_2$), 4.48 (s, 4H, CH_2-Ph), 2.83 (s, 6H, $2CH_3$), 2.31 (t, J = 5.0 Hz, 12H, $4CH_3$). ^{13}C NMR (125 MHz, DMSO-d6): 165.21, 161.55, 159.61, 152.13, 150.81, 149.72, 143.83, 132.85, 131.95, 128.98, 127.65, 124.29, 122.53, 118.63, 116.02, 115.15, 114.98, 107.25, 33.84, 20.56, 20.51. HR-ESIMS $[M + H]^+$ found 819.32005; calculated $C_{46}H_{45}F_2N_4O_8$: 819.32055.

N′,N‴-((1Z,1′Z)-((R)-1,1′6,6′-Tetrahydroxy-5,5′-diisopropyl-3,3′-dimethyl-7,7′-dioxo-[2,2′-binaphthalene]-8,8′(7H,7′H)-diylidene)bis(methanylylidene))bis(2-(3-fluorophenyl)acetohydrazide) ***(12b)***. Yield 48.7%; mp: 188–190°C; ^{1}H NMR (500 MHz, DMSO-d6): 14.41 (s, 2H, NH), 12.04 (s, 2H, NH), 9.96 (s, 2H), 8.76 (s, 2H), 8.17 (s, 2H), 7.59 (s, 2H), 7.37 (m, 2H), 7.17 (d, J = 7.5 Hz, 4H), 7.08 (m, 2H), 3.90 (brs, 2H, $CH(CH_3)_2$), 3.62 (s, 4H, CH_2-Ph), 1.99 (s, 6H, $2CH_3$), 1.48 (t, J = 7.0 Hz, 12H, $4CH_3$). ^{13}C NMR (125 MHz, DMSO-d6): 165.60, 163.01, 161.07, 152.21, 150.77, 149.73, 143.84, 138.21, 132.65, 130.21, 128.66, 127.68, 125.26, 118.62, 115.99, 115.81, 113.55, 107.21, 40.20, 26.55, 20.55.

N′,N‴-((1Z,1′Z)-((R)-1,1′6,6′-Tetrahydroxy-5,5′-diisopropyl-3,3′-dimethyl-7,7′-dioxo-[2,2′-binaphthalene]-8,8′(7H,7′H)-diylidene)bis(methanylylidene))bis(2-(4-fluorophenyl)acetohydrazide) ***(12c)***. Yield 54.8%; mp: 195–197°C; ^{1}H NMR (500 MHz, DMSO-d6): 14.42 (s, 2H, NH), 12.00 (s, 2H, NH), 9.98 (s, 2H), 8.76 (s, 2H), 8.15 (s, 2H), 7.60 (s, 2H), 7.36–7.30 (m, 4H), 7.17–7.10 (m, 4H), 3.89 (brs, 2H, 2H, $CH(CH_3)_2$), 3.56 (s, 4H, CH_2-Ph), 1.98 (s, 6H, $2CH_3$), 1.47 (t, J = 5.0 Hz, 12H, $4CH_3$). ^{13}C NMR (125 MHz, DMSO-d6): 168.83, 166.03, 162.12, 160.19, 152.11, 150.78, 149.73, 143.83, 132.63, 128.85, 127.63, 118.62, 116.00, 115.11, 114.97, 107.22, 40.20, 26.50, 20.51.

N′,N‴-((1Z,1′Z)-((R)-1,1′6,6′-Tetrahydroxy-5,5′-diisopropyl-3,3′-dimethyl-7,7′-dioxo-[2,2′-binaphthalene]-8,8′(7H,7′H)-diylidene)bis(methanylylidene))bis(2-(2-chlorophenyl)acetohydrazide) ***(12d)***. Yield 52.7%; mp: 206–208°C; ^{1}H NMR (500 MHz, DMSO-d6): 14.43 (s, 2H, NH), 12.07 (s, 2H, NH), 9.99 (s, 2H), 8.78 (s, 2H), 8.19 (s, 2H), 7.61 (s, 2H), 7.44 (m, 4H), 7.31 (m, 4H), 3.91 (brs, 2H, 2H, $CH(CH_3)_2$), 3.76 (s, 4H, CH_2-Ph), 2.00 (s, 6H, $2CH_3$), 1.48 (t, J = 5.5 Hz, 12H, $4CH_3$). ^{13}C NMR (125 MHz, DMSO-d6): 165.07, 152.03, 150.79, 149.71, 143.85, 133.55, 133.41, 132.64, 132.43, 132.15, 129.04, 128.91, 128.68, 127.63, 127.10, 118.62, 116.02, 107.25, 38.31, 20.55, 20.5.

N′,N‴-((1Z,1′Z)-((R)-1,1′6,6′-Tetrahydroxy-5,5′-diisopropyl-3,3′-dimethyl-7,7′-dioxo-[2,2′-binaphthalene]-8,8′(7H,7′H)-diylidene)bis(methanylylidene))bis(2-(3-chlorophenyl)acetohydrazide) ***(12e)***. Yield 53.0%; mp: 194–196°C; ^{1}H NMR (500 MHz, DMSO-d6): 14.39 (s, 2H, NH), 12.04 (s, 2H, NH), 9.97 (s, 2H), 8.77 (s, 2H), 8.19 (s, 2H), 7.60 (s, 2H), 7.39–7.28 (m, 8H), 3.89 (brs, 2H, 2H, $CH(CH_3)_2$), 3.60 (s, 4H, CH_2-Ph),

1.99 (s, 6H, $2CH_3$), 1.47 (t, $J = 5.5$ Hz, 12H, $4CH_3$). ^{13}C NMR (125 MHz, DMSO-d6): 165.58, 152.24, 150.77, 149.72, 143.82, 137.89, 132.86, 130.15, 128.98, 128.65, 127.87, 127.69, 126.65, 118.63, 116.00, 107.21, 40.1, 20.55, 20.50. HR-ESIMS $[M + H]^+$ found 851.26141; calculated $C_{46}H_{45}Cl_2N_4O_8$: 851.26144.

Conflicts of Interest

The authors confirm that there are no conflicts of interest related to the contents of this article.

Acknowledgments

The authors acknowledge the financial support via Project 08/2015/HĐ-NĐT.

References

[1] N. Howlader, A. M. Noone, M. Krapcho et al., *SEER Cancer Statistics Review, 1975–2012*, National Cancer Institute, Bethesda, MD, USA, Based on November 2014 SEER data submission, posted to the SEER web site, April 2015 http://seer.cancer.gov/csr/1975_2012/.

[2] I. K. Mellinghoff and C. L. Sawyers, "The emergence of resistance to targeted cancer therapeutics," *Pharmacogenomics*, vol. 3, no. 5, pp. 603–623, 2002.

[3] M. E. Wall and M. C. Wani, "Camptothecin and taxol: discovery to clinic—thirteenth Bruce F. Cain Memorial Award Lecture," *Cancer Research*, vol. 55, no. 4, pp. 753–760, 1995.

[4] C. L. Oliver, J. A. Bauer, K. G. Wolter et al., "In vitro effects of the BH3 mimetic, (-)-gossypol, on head and neck squamous cell carcinoma cells," *Clinical Cancer Research*, vol. 10, no. 22, pp. 7757–7763, 2004.

[5] M. Zhang, H. Liu, Z. Tian, B. N. Griffith, M. Ji, and Q. Q. Li, "Gossypol induces apoptosis in human PC-3 prostate cancer cells by modulating caspase-dependent and caspase-independent cell death pathways," *Life Sciences*, vol. 80, no. 8, pp. 767–774, 2007.

[6] R. M. Mohammad, S. Wang, S. Banerjee, X. Wu, J. Chen, and F. H. Sarkar, "Nonpeptidic small-molecule inhibitor of Bcl-2 and Bcl-X_L, (−)-gossypol, enhances biological effect of genistein against BxPC-3 human pancreatic cancer cell line," *Pancreas*, vol. 31, no. 4, pp. 317–324, 2005.

[7] M. Zhang, H. Liu, R. Guo et al., "Molecular mechanism of gossypol-induced cell growth inhibition and cell death of HT-29 human colon carcinoma cells," *Biochemical Pharmacology*, vol. 66, no. 1, pp. 93–103, 2003.

[8] D. W. Hahn, C. Rusticus, A. Probst, R. Homm, and A. N. Johnson, "Antifertility and endocrine activities of gossypol in rodents," *Contraception*, vol. 24, no. 1, pp. 97–105, 1981.

[9] J. Yang, G. Chen, L. L. Li et al., "Synthesis and anti-H5N1 activity of chiral gossypol derivatives and its analogs implicated by a viral entry blocking mechanism," *Bioorganic & Medicinal Chemistry Letters*, vol. 23, no. 9, pp. 2619–2623, 2013.

[10] J. Yang, F. Zhang, J. Li et al., "Synthesis and antiviral activities of novel gossypol derivatives," *Bioorganic & Medicinal Chemistry Letters*, vol. 22, no. 3, pp. 1415–1420, 2012.

[11] R. J. Radloff, L. M. Deck, R. E. Royer, and D. L. Vander Jagt, "Antiviral activities of gossypol and its derivatives against herpes simplex virus type II," *Pharmacological Research Communications*, vol. 18, no. 11, pp. 1063–1073, 1986.

[12] G. Tegos, F. R. Stermitz, O. Lomovskaya, and K. Lewis, "Multidrug pump inhibitors uncover remarkable activity of plant antimicrobials," *Antimicrobial Agents and Chemotherapy*, vol. 46, no. 10, pp. 3133–3141, 2002.

[13] X. S. Liang, A. J. Rogers, C. L. Webber et al., "Developing gossypol derivatives with enhanced antitumor activity," *Investigational New Drugs*, vol. 13, no. 3, pp. 181–186, 1995.

[14] R. E. Royer, R. G. Mills, L. M. Deck, G. J. Mertz, and D. L. Vander Jagt, "Inhibition of human immunodeficiency virus type I replication by derivatives of gossypol," *Pharmacological Research*, vol. 24, no. 4, pp. 407–412, 1991.

[15] L. Tai-Shun, R. F. Schinazi, J. Zhu et al., "Anti-HIV-1 activity and cellular pharmacology of various analogs of gossypol," *Biochemical Pharmacology*, vol. 46, no. 2, pp. 251–255, 1993.

[16] Y. Lu, S. Wu, Y. Yue et al., "Gossypol with hydrophobic linear esters exhibits enhanced antitumor activity as an inhibitor of antiapoptotic proteins," *ACS Medicinal Chemistry Letters*, vol. 7, no. 12, pp. 1185–1190, 2016.

[17] L. Li, Z. Li, K. Wang et al., "Design, synthesis, and biological activities of aromatic gossypol schiff base derivatives," *Journal of Agricultural and Food Chemistry*, vol. 62, no. 46, pp. 11080–11088, 2014.

[18] Z. Yonghua, C. Xu, L. Yingchao, C. Xueli, and H. Xiaofeng, "Synthesis and biological evaluation of a novel apogossypolone derivative," *Letters in Drug Design & Discovery*, vol. 14, pp. 96–101, 2017.

[19] Y. Lu, J. Li, C. Dong, J. Huang, H. Zhou, and W. Wang, "Recent advances in gossypol derivatives and analogs: a chemistry and biology view," *Future Medicinal Chemistry*, vol. 9, no. 11, pp. 1243–1275, 2017.

[20] J. Qiu, L. R. Levin, J. Buck, and M. M. Reidenberg, "Different pathways of cell killing by gossypol enantiomers," *Experimental Biology and Medicine*, vol. 227, no. 6, pp. 398–401, 2002.

[21] G. Tang, C.-Y. Yang, Z. Nikolovska-Coleska et al., "Pyrogallol-based molecules as potent inhibitors of the antiapoptotic Bcl-2 proteins," *Journal of Medicinal Chemistry*, vol. 50, no. 8, pp. 1723–1726, 2007.

[22] G. Wang, Z. Nikolovska-Coleska, C. Yang et al., "Structure-based design of potent small-molecule inhibitors of anti-apoptotic Bcl-2 proteins," *Journal of Medicinal Chemistry*, vol. 49, no. 21, pp. 6139–6142, 2006.

[23] H.-X. Jiang, X.-X. Cao, H. Huang, and B. Jiang, "An expedient route for the practical preparation of optically active (-)-gossypol," *Tetrahedron: Asymmetry*, vol. 18, no. 20, pp. 2437–2441, 2007.

[24] P. Przybylski, K. Pyta, J. Stefańska et al., "Synthesis, crystal structures and antibacterial activity studies of aza-derivatives of phytoalexin from cotton plant - gossypol," *European Journal of Medicinal Chemistry*, vol. 44, no. 11, pp. 4393–4403, 2009.

[25] K. Likhitwitayawuid, C. K. Angerhofer, G. A. Cordell, J. M. Pezzuto, and N. Ruangrungsi, "Cytotoxic and antimalarial bisbenzylisoquinoline alkaloids from Stephania erecta," *Journal of Natural Products*, vol. 56, no. 1, pp. 30–38, 1993.

[26] C. Kurekci, S. L. Bishop-Hurley, P. E. Vercoe, Z. Durmic, R. A. M. Al Jassim, and C. S. McSweeney, "Screening of Australian plants for antimicrobial activity against Campylobacter jejuni," *Phytotherapy Research*, vol. 26, no. 2, pp. 186–190, 2012.

12

New Silver Complexes with Mixed Thiazolidine and Phosphine Ligands as Highly Potent Antimalarial and Anticancer Agents

Nur Rahimah Fitrah Mohd Sofyan,[1] Fariza Juliana Nordin,[2] Mohd Ridzuan Mohd Abd Razak,[3] Syahrina Nur 'Ain Abdul Halim,[1] Nur Adila Fatin Mohd Khir,[1] Amirrudin Muhammad,[3] Nor Fadilah Rajab,[2] and Rozie Sarip[1]

[1]Department of Chemistry, Faculty of Science, University of Malaya, Lembah Pantai, 50603 Kuala Lumpur, Malaysia
[2]Faculty of Health Sciences, Universiti Kebangsaan Malaysia, Kuala Lumpur Campus, Jalan Raja Muda Abdul Aziz, 50300 Kuala Lumpur, Malaysia
[3]Bioassay Unit, Herbal Medicine Research Centre, Institute for Medical Research, Jalan Pahang, 50588 Kuala Lumpur, Malaysia

Correspondence should be addressed to Rozie Sarip; rozie@um.edu.my

Academic Editor: Kokhwa Lim

Five silver(I) complexes containing a mixed ligand system of phosphine and thiazolidine were successfully synthesized. The structural information of the complexes was assembled using various spectroscopic techniques such as CHN elemental analysis, Fourier transformed infrared (FTIR), ^{1}H, ^{13}C, and ^{31}P{^{1}H} NMR spectroscopy, and thermogravimetric analysis (TGA). A bidentate phosphine ligand acted as a chelating agent which bond to the silver in 1 : 2 molar ratios. Meanwhile, thiazolidine was attached to the silver in a 1 : 1 molar ratio. The antiplasmodial properties of all synthesized complexes were investigated on chloroquine-resistant *P. falciparum* parasite via HRP2 assays and cytotoxicity tests on Vero cells. Of all the synthesized complexes, complex **2** showed the highest SI value (more than 12.4) followed by complex **5** (6.6). The potent properties of compounds **2** and **5** were also noted in the *in vitro* antiproliferative assays involving MDA-MB-231 and MCF-7 breast cancer cell lines as well as HT-29 colon cancer cell line. Complex **2** was selective for MDA-MB-231 cells ($GI_{50} = 1.9 \pm 0.3\,\mu M$), while complex **5** acted predominantly on breast carcinoma cells (GI_{50} MDA-MB-231 $= 4.7 \pm 1.1\,\mu M$; MCF-7 $= 2.9 \pm 0.9\,\mu M$) instead of colon carcinoma (HT-29) cells ($GI_{50} = 15.1 \pm 1.9\,\mu M$).

1. Introduction

Over the years, the synthesis of complexes with transition metals and mixed ligand that offers few points of donating ability has attracted much attention due to their special structures [1–3], catalytic abilities [4, 5], and especially their potential applications in biomedical field [6]. In particular, silver(I) complexes containing mono- or bidentate phosphine ligands and thiol or thione moieties have well-established applications in the form of anticancer [7, 8] and antimalarial [9, 10] agents. Numerous ligands derived from mono- and diphosphines have been explored for their ability to coordinate with silver, hence enabling their development as antiproliferative agents [11]. Thiazolidine is one of the compounds which show good potential in various biological activities [12, 13]. Based on the analyses and databases from the National Cancer Institute (NCI, USA), approximately about 42,247 compounds which consist of 734 nonfused and 146 fused thiazolidine derivatives are active in three tumor cell line assays. On the contrary, malaria is a parasitic disease that occurs predominantly in tropical countries. In 2015 alone, 212 million malaria cases have been reported worldwide [14]. An increasing number of research on the application of silver(I) complexes has been conducted as the complexes are a leading candidate for the curing of various infectious diseases [15, 16].

The development of a new metallotherapeutic drug containing silver coordination compound provides many benefits to human body because of its low toxicity [17]. Although cisplatin has a high cure rate, its uses are limited owing to side effects and toxicity issues [18, 19]. On the contrary, the commercial antimalarial drug chloroquine showed some degree of ineffectiveness in light of parasite resistance [20]. As such, to overcome the limitations of cisplatin and chloroquine, we studied another form of metal-based drug which was less harmful and more effective by using silver complexes with a mixed ligand system of bioactive thiazolidine and phosphine. Even though malaria and cancer are totally different diseases with different symptoms, it is surprisingly possible to cure both diseases using the same drug since they have similar pathophysiology and mechanistic treatment pathway [21, 22]. Based on recent researches, there are also a number of anticancer drugs that display potent antimalarial properties [23]. Hence, we attempted to uncover compounds that can fulfil both roles, as an anticancer and antimalaria agent. In this research, we report the preparation of the silver(I) complexes with thiazolidine and different phosphine ligands along with their *in vitro* antiplasmodial and antiproliferative activities.

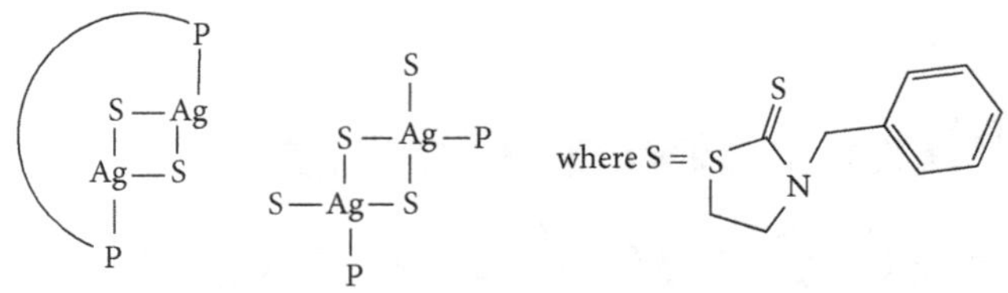

Figure 1: Proposed structure of silver complexes—corresponding P-ligand as listed in Table 1.

2. Materials and Methods

2.1. Experimental and Instrumentation. All the solvents and reagents were of analytical grade and purchased commercially from Sigma-Aldrich Ltd. The silver nitrate and 1,2-bis(diphenylphosphino)methane (dppm), 1,1-bis(diphenylphosphino) ethane (dppe), 1,1-bis(diphenylphosphino)ferrocene (dppf), triphenylphosphine (PPh_3), tri(o-tolyl)phosphine, ethanol, methanol, and acetonitrile were used as supplied without further purification unless stated otherwise. The CHN analyses were performed by PerkinElmer CHNS/O 2400 Series II. The infrared (IR) spectra were determined using a PerkinElmer Spectrum One FT-IR spectrophotometer (ATR) at a frequency range 450–4000 cm^{-1}. JEOL FT-NMR ECX 400 (ECX 400) was employed to measure the NMR spectra of ^1H, ^{13}C, and $^{31}P\{^1H\}$ at 400 MHz in deuterated solvents without internal reference. The presence of metals and other elements was detected by energy-dispersive X-ray spectroscopy (EDX), powder X-ray diffraction (PXRD) was recorded on an X-ray diffractometer (PANalytical, Netherlands) with Cu Kα characteristic radiation (wavelength $\lambda = 0.154$ nm) at the voltage of 40 kV and current of 40 mA, the scanning rate was 4.25°/min, and the scanning scope of 2θ was from 0 to 90° at room temperature (25°C), while the thermogravimetric analysis were carried out on a PerkinElmer TGA 4000 thermogravimetric analyzer at a heating rate of 10°C/min.

2.1.1. Synthesis of 3-Benzyl-1,3-thiazolidine-2-thione. The thiazolidine ligand was synthesized in accordance to a method reported in the literature, with slight modifications [24]. Generally, benzylaminoethanol (10 mmol, 1.4 mL) was added dropwise into a solution of potassium hydroxide (50 mmol, 2.81 mg) in ethanol (50 mL). A clear solution was formed. Carbon disulfide (50 mmol, 3 mL) was added dropwise to the reaction mixture and then refluxed (90°C) for 18 h, resulting in an orange precipitate. This precipitate was filtered out and left to dry in an oven overnight.

Yield, 88%, m.p. 132–133°C. Anal. Calc. for $C_{10}H_{11}NS_2$: C, 57.38; H, 5.30; N, 6.69; and S, 30.64. Found: C, 57.06; H, 4.96; N, 6.34; and S, 30.61. IR: ν(C-N) 1149 and ν(C=S) 1242. ^1H NMR (ppm, CD_3CN): 7.37–7.25 (*m*, 5.41H, Ar-H); 4.93 (*s*, 2.07H, N-CH_2); 3.95 (*t*, 2.01H, N-CH_2); and 3.23 (*t*, 2H, S-CH_2). ^{13}C NMR (400 MHz, CD_3CN, δ ppm): 196.9 (C=S); 135–127 (4C, C-Ar); 56.3 (C-N); 51.9 (C-N); and 26.8 (C-S).

2.2. Synthesis of Silver Complexes. The complexes were prepared in accordance with the previously reported procedure [25] with slight modifications (Figure 1; Table 1). Different mono- and bidentate phosphines were utilized. For the bidentate phosphine, a suspension of silver nitrate (2.00 mmol, 0.17 mg) and bis-(diphenylphosphino)methane (1.00 mmol, 0.19 mg), 1,2-bis(diphenylphosphino)ethane (1.00 mmol, 0.20 mg), or 1,1-bis(diphenylphosphino)ferrocene (1 mmol, 0.50 mg) in acetonitrile (10 mL) was stirred at 40°C. A solution of 3-benzyl-thiazolidine-2-thione (2 mmol, 0.21 mg) in methanol (10 mL) was then added. The resulting solution was filtered, and the clear solution was reduced to dryness.

While for monodentate phosphine [26], a solution of silver nitrate (0.14 mmol, 0.02 mg) in acetonitrile (10 mL) was added to a solution of 3-benzyl-thiazolidine-2-thione (0.28 mmol, 0.06 mg) in methanol (10 mL) followed by stirring at room temperature for 4 h. Then, a solution of triphenylphosphine (0.14 mmol, 0.04 mg) or tri(*o*-tolyl) phosphine (0.14 mmol, 0.04 mg) in acetonitrile (5 mL) and methanol (5 mL) was added. The resulting solution was then filtered and reduced to dryness.

2.2.1. [Ag_2(dppm)(3-benzyl-1,3-thiazolidine-2-thione)$_2$].(NO_3)$_2$, 1. Yield, 54%, m.p. 190–191°C. Anal. Calc: C, 54.56; H, 4.24; N, 2.36, and S; 10.79 Found: C, 54.24; H, 3.92; N, 2.24, and S; 10.43. IR data (cm^{-1}): ν (NO_3^-) 1310, ν (C-N) 1223, ν (C=S) 1152, and ν (P-C_{ph}) 1094. ^1H NMR (400 MHz, CD_3CN, δ ppm): 7.50–7.18 (*m*, 30H, Ar-H); 4.89 (*s*, 4.00H, N-CH_2); 3.91 (*t*, 4.36H, $J = 8$ Hz, N-CH_2); 3.67 (*s*, 2.09H, CH_2-P); and 3.13 (*t*, 4.24H, $J = 8$ Hz, S-CH_2). ^{13}C NMR (400 MHz, CD_3CN, δ ppm): 196.7 (C=S); 136–128 (8C, Ar-C); 56.6 (C-N); 52.3 (C-N); 26.8 (C-S); and 25 (CH_2-P). $^{31}P\{^1H\}$ NMR (400 MHz, CD_3CN, δ ppm): 5.2, 8.0 (s)

Table 1: The complexes with its phosphine ligand.

Complex	P-ligand	
1	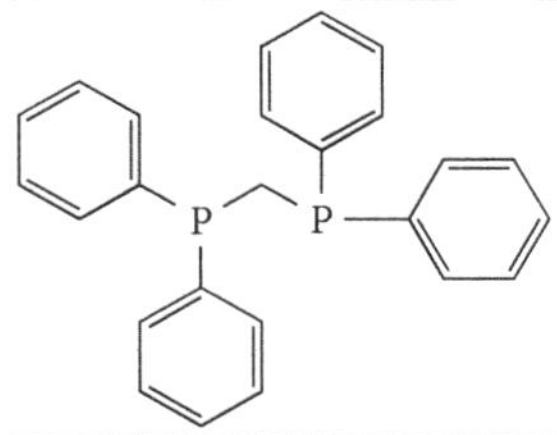	dppm
2	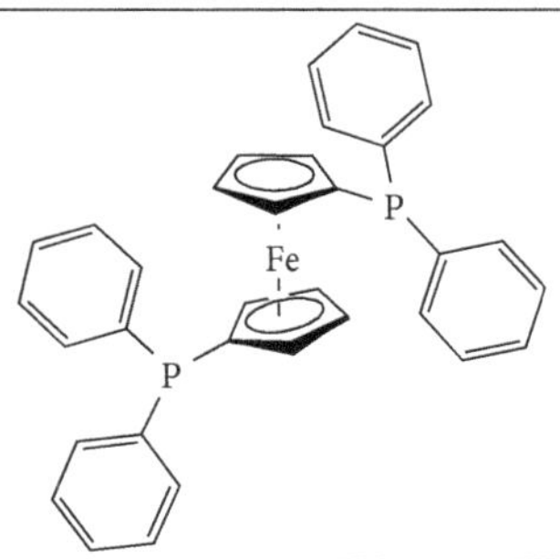	dppf
3	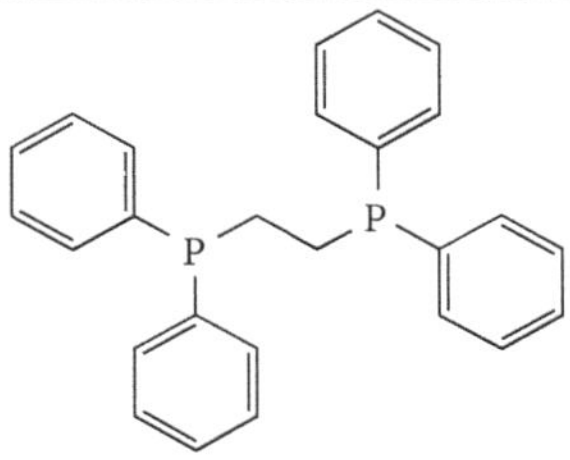	dppe
4	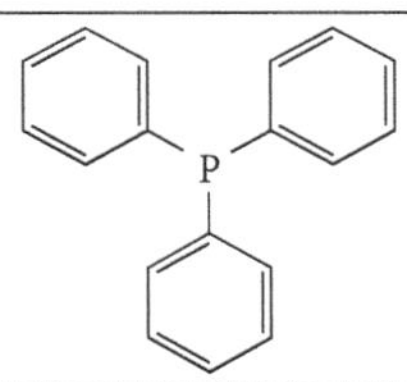	PPh_3
5	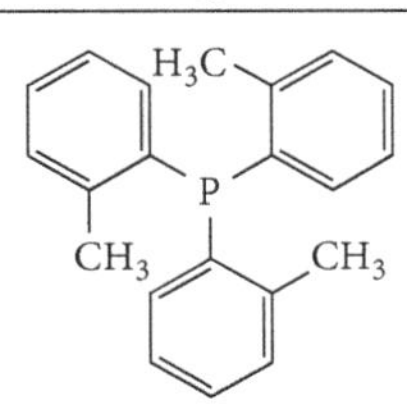	tri(*o*-tolyl)

2.2.2. $[Ag_2(dppf)(3\text{-benzyl-}1,3\text{-thiazolidine-}2\text{-thione})_2].(NO_3)_2$, **2**. Yield: 54%, Mp: 190–191°C. Anal. Calc: C, 54.56; H, 4.24; N, 2.36, and S; 10.79 Found: C, 54.24; H, 3.92; N, 2.24, and S; 10.43. IR data (cm^{-1}): ν (NO_3^-) 1307, ν (C-N) 1223, ν (C=S) 1163, and ν (P-C_{ph}) 1095. ^1H NMR (400 MHz, CD_3CN, δ ppm): 7.50–7.28 (*m*, 30H, Ar-H); 4.93 (*s*, 4.37H, N-CH_2); 4.35 (*s*, 4H, C_5H_4); 4.17 (*s*, 4.28H, C_5H_4); 3.95 (*t*, 4H, $J = 8$ Hz, N-CH_2); and 3.11 (*t*, 4.32H, $J = 8$ Hz, S-CH_2). ^{13}C NMR (400 MHz, CD_3CN, δ ppm): 197.1 (C=S); 135–127 (8C, Ar-C); 74 (C_5H_4); 72 (C_5H_4); 57.1 (C-N); 52.5 (C-N); and 27.3 (C-S). $^{31}P\{^1H\}$ NMR (400 MHz, CD_3CN, δ ppm): −1.4 (s)

2.2.3. $[Ag_2(dppe)(3\text{-benzyl-}1,3\text{-thiazolidine-}2\text{-thione})_2].(NO_3)_2$, **3**. Yield, 55%, m.p. 156–157°C. Anal. Calc: C, 53.49; H, 4.49; N, 2.70, and S; 12.42 Found: C, 53.12; H, 4.27; N, 2.61, and S; 12.08. IR data (cm^{-1}): ν (NO_3^-) 1307, ν (C-N) 1223, ν (C=S) 1153, and ν (P-C_{ph}) 1097. ^1H NMR (400 MHz, CD_3CN, δ ppm): 7.42–7.22 (*m*, 30H, Ar-H); 4.88 (*s*, 4.01H, N-CH_2); 3.92 (*t*, 4H, $J = 8$ Hz, N-CH_2); 3.12 (*t*, 4.12H, $J = 8$ Hz, S-CH_2); and 2.46 (*s*, 4.15H, CH_2-CH_2-P). ^{13}C NMR (400 MHz, CD_3CN, δ ppm): 196.9 (C=S); 135–127 (8C, C-Ar); 56.9 (C-N); 52.4 (C-N); 27.3 (C-S); and 24 (CH_2-CH_2-P). $^{31}P\{^1H\}$ NMR (400 MHz, CD_3CN, δ ppm): 4.3 (s)

2.2.4. $[Ag_2(PPh_3)_2(3\text{-benzyl-}1,3\text{-thiazolidine-}2\text{-thione})_4].(NO_3)_2$, **4**. Yield, 54%, m.p. 135–136°C. Anal. Calc: C, 57.86; H, 4.73; N, 3.55, and S; 16.26 Found: C, 57.72; H, 4.46; N, 3.38, and S; 16.04. IR data (cm^{-1}): ν (NO_3^-) 1314, ν (C-N) 1223, ν (C=S) 1154, ν (P-C_{ph}) 1092. ^1H NMR (400 MHz, CD_3CN, δ ppm): 7.33–7.15 (*m*, 50.37H, Ar-H); 4.91 (*s*, 7.80H, N-CH_2); 3.94 (*t*, 8H, $J = 8$ Hz, N-CH_2); and 3.18 (*t*, 7.88H, $J = 8$ Hz, S-CH_2). ^{13}C NMR (400 MHz, CD_3CN, δ ppm): 196.9 (C=S); 136–127 (8C, C-Ar); 56.4 (C-N); 52.1 (C-N); and 26.9 (C-S). $^{31}P\{^1H\}$ NMR (400 MHz, CD_3CN, δ ppm): 8.3 (s)

2.2.5. $[Ag_2(Tri(o\text{-tolyl})phosphine)_2(3\text{-benzyl-}1,3\text{-thiazolidine-}2\text{-thione})_4].(NO_3)_2$, **5**. Yield, 37%, m.p. 144–145°C. Anal. Calc: C, 59.27; H, 5.22; N, 3.37, and S; 15.54 Found: C, 59.12; H, 4.97; N, 3.06, and S; 15.36. IR data (cm^{-1}): ν (NO_3^-) 1315, ν (C-N) 1266, ν (C=S) 1160, and ν (P-C_{ph}) 1129. ^1H NMR (400 MHz, CD_3CN, δ ppm): 7.40–6.62 (*m*, 44.24H, Ar-H); 4.92 (*s*, 8H, N-CH_2); 3.95 (*t*, 7.52H, $J = 8$ Hz, N-CH_2); 3.21 (*t*, 8.35H, $J =$ 8 Hz, S-CH_2); and 2.31 (*s*, 17.86H, CH_3-Ar-P). ^{13}C NMR (400 MHz, CD_3CN, δ ppm): 197.1 (C=S); 143–126 (10C, C-Ar); 56.4 (C-N); 52.1 (C-N); 26.9 (C-S); and 20 (C-Ar-P). $^{31}P\{^1H\}$ NMR (400 MHz, CD_3CN, δ ppm): −27.8, 37 (s)

2.3. In Vitro Antiplasmodial Assay

2.3.1. In Vitro Culture and Synchronization of P. falciparum. The chloroquine-resistant *P. falciparum* was grown in an incubator at 5% CO_2. The culture was prepared in a 25 cm^3 culture flask with filtered vent and maintained in a complete RPMI 1640 culture medium (Invitrogen, USA). Fresh red cells of blood group O were used as a host to grow the *P. falciparum*, with the initial culture containing 1% parasitemia at a hematocrit of 2.5%. The parasite density was monitored daily by means of making thin blood smears stained with 10% Giemsa solution, after which they were observed under the microscope at magnification of 1000x. The parasites were synchronized using a 5% sorbitol [27] and then cultured for one complete cycle. The *P. falciparum* infected red blood cell culture with a parasitemia level of approximately 5 to 7% was used *in vitro* for the histidine-rich protein (HRP2) assay.

2.3.2. P. falciparum Histidine-Rich Protein 2 (HRP2) Assay. The HRP2 is a specific protein secreted by the *P. falciparum* parasites during their blood stage cycle [28]. The HRP2 assay was carried out according to the procedure specified in the literature [29, 30] with some modifications. Briefly, the substance was dissolved in 100% dimethyl sulfoxide (DMSO) to obtain 5 mg/ml stock solutions. For each compound stock plate, the compounds (5 mg/ml) were serially diluted (2-fold dilution) to give 7-point concentrations (ranging from 0.08 mg/ml to 5 mg/ml) in DMSO in wells A1 to A7 of a 96-well plate. 15 μl of each serially diluted stock was transferred into watery plates containing 225 μl of sterile H_2O. An aliquot of the mixture was used in the HRP2 assay.

Ring-infected RBCs with 5% parasitemia were adjusted to 0.05% parasitemia and 1.5% hematocrit. A total of 190 μl of parasitized RBCs at 1.5% hematocrit were added into each well of the test plate. 10 μl of serially diluted compounds from the preprepared plates were transferred to the test plates containing parasitized RBCs after which they were incubated in a candle jar at 37°C for 72 h. The final concentrations of the compounds ranged from 0.25 μg/ml to 15.7 μg/ml, with the concentration of DMSO being 0.3%.

Chloroquine (CQ) (Sigma, USA), quinine (Q) (Sigma, USA), mefloquine (Mef) (Sigma, USA), and artemisinin (Art) (Sigma, USA) were used as standard control to validate the test. The ranges of the standards concentration were as follows: (1) 27.7–1772.6 nM for CQ, (2) 54.6–3495 nM for Q, (3) 9.4–601.3 nM for Mef, and (4) 0.8–51.2 nM for Art. Sterile H_2O and infected RBCs without the tested compounds were the negative controls in this study.

After incubating for 72 h, the test plates were kept at −80°C overnight. After being thawed at room temperature, 100 μl of the *P. falciparum*-infected RBC lysates was transferred from the test plates into ELISA plates coated with immunoglobulin M (IgM) capture antibodies (MPFM-55A, ICL, Inc, Newberg, OR, USA) specific for *P. falciparum* HRP2 (1 μg/ml in phosphate-buffered saline (PBS)). Subsequently, the ELISA plates were incubated in a humidity chamber for 1 h at room temperature. The plates were washed three times with 0.05% PBS-Tween 20 (PBST), following which 100 μl of the horseradish peroxidase-conjugated (0.2 μg/ml in PBS) detector antibodies (MPFG-55P, ICL, Inc., Newberg, OR, USA) was added to each well. Incubation was done in a humidity chamber for 1 h at room temperature. A subsequent washing step as mentioned above was followed by the addition of 100 μl of 3,3', 5,5-tetramethylbenzidine (TMB) chromogen (Zymed Lab., Inc., San Francisco, CA, USA) into each well. After being incubated for 10 min in the absence of light, 50 μl of 1 M sulfuric acid was added to the wells. The absorbance was determined by an ELISA plate reader at a wavelength of 450 nm (FLUOstar Omega, Germany). Finally, the collected data were keyed into to the HN-nonLin software (http://malaria.farch.net) to obtain the 50% effective concentration (EC_{50}) values directly from the graphs. All tests were performed in triplicates.

2.4. In Vitro Cytotoxicity Assay. The Vero cells, the kidney epithelial cells isolated from African green monkey, were used as a representative of normal cell lines in cytotoxicity assay for determination of cytotoxic activity of each compound. The Vero cells were maintained in complete DMEM culture medium containing 25 mM HEPES, 0.4% sodium bicarbonate ($NaHCO_3$), 100 U of Penstrep (100 U of each penicillin and streptomycin) supplemented with 10% fetal bovine serum (FBS). The cytotoxicities of the synthesized compounds were measured via 3-(4, 5-dimethylthiazol-2-yl)-2, 5-diphenyltetrazolium bromide (MTT) assay [31]. Prior to the day of testing, the stock plates were prepared by serially diluting (2-fold dilution) the compounds (2.5 mg) to 7-point concentrations (ranging from 0.08 mg/ml to 5 mg/ml) with DMSO. Then, a 6 μl of serially diluted stocks was transferred into a 96-well plate containing 294 μl of complete DMEM media (medium plates). Subsequently, 100 μl of the compounds was taken from the medium plate (as prepared) and added to the test plate containing 100 μl of complete DMEM media with 1×10^3 Vero cells. The final concentration ranged between 0.78 μg/ml and 25 μg/ml. The final concentration of DMSO in all the tests was less than 1%. All tests were performed in triplicates.

The positive control for the cell growth was the cell suspension without the test substance, while the negative control cell suspension with 0.05% Triton × 100. The cultures were incubated at 37°C in 5% CO_2 incubator for 72 h. Then, 50 μl of MTT solution (5 mg MTT in 1 mL PBS and 2.5 mL of DMEM media) was added to each well. Following further incubation for 4 h at 37°C and 5% CO_2 incubator, the medium was removed and replaced with 200 μl of DMSO to solubilize the MTT formazan product. The solution was mixed for 15 min and once for 30 s before the absorbance and was measured using a microplate reader (FLUOstar Omega, Germany) at a wavelength 540 nm. The 50% effective concentration (EC_{50}) of antiplasmodial activity and 50% cytotoxic concentration (CC_{50}) were determined based on a dose response curve. A selectivity index (SI)—which corresponded to the ratio between the antiplasmodial and cytotoxic activities—was calculated according to the following formula:

$$SI_{\text{Plasmodium}} = \frac{CC_{50\ \text{normal cell lines}}}{EC_{50\ \text{Plasmodium}}}. \quad (1)$$

2.5. Antiproliferative Assay. The three human cancer cell lines used in the present study, human breast cancer cell lines, MDA-MB-231 and MCF-7, and human colon cancer cell line, and HT-29 were purchased from American Type Culture Collection (ATCC, Rockville, MD, USA). Cells were grown as monolayer culture in Dulbecco's modified Eagle medium (DMEM) (Gibco BRL, Grand Island, NY, USA) supplemented with 10% fetal bovine serum (Gibco BRL), 25 mM HEPES (Sigma), and 1% antibiotic (Gibco BRL) in tissue culture flasks in a humidified incubator at 37°C at an atmosphere of 95% air and 5% carbon dioxide. Cells were kept in the logarithmic growth phase by routine passage every 2-3 days using 0.025% trypsin-EDTA treatment.

Sulforhodamine B (SRB) assay was carried out as described in the previous studies to determine the 50% growth inhibition (GI_{50}) values of all the compounds [32]. Initially, the cells were seeded in 96-well plates at a density of either 1×10^5 cells/ml (MCF-7) or 2×10^5 cells/ml (MDA-MB-231 and HT-29), after which incubation was done overnight to allow the cells to adhere to the bottom of the plates. The next day, some of the plates were processed to determine the density at time zero (t_0). According to the National Cancer Institute (NCI), the use of t_0 control allows the determination of the amount of cells killed as well as the net inhibition of growth [32]. The cells in the remaining plates were treated with compounds **1**, **2**, **3**, **4**, or **5** at concentrations ranging from 0.383–12.269 μM, 0.378–12.103 μM, 0.657–21.031 μM, 0.496–15.846 μM, and 1.881–60.176 μM, respectively. After 48 h, the cells were fixed in the plates using 50 μl of 50% (w/v) trichloroacetic acid (TCA) solution, following which they were further incubated at 4°C for 1 h. The plates were then washed five times with tap water and air-dried prior to staining with 100 μl of 0.4% (w/v) SRB staining solution. Further incubation was done for 10 min at room temperature. Subsequently, the plates were washed three times with 1% (v/v) acetic acid to remove the unbound stains. After air-drying, the wells were added with 200 μl of 10 mM Trizma base and shaken well for 10 min. Henceforth, the absorbance was measured using a microplate reader at a wavelength of 490 nm. All experiments were carried out in triplicates. GI_{50} was calculated using the following formula:

$$GI_{50} = OD_{sample} - \frac{OD_{t_0}}{OD_{control} - OD_{t_0}} \times 100. \quad (2)$$

3. Results and Discussion

The synthesized 3-benzyl-1,3-thiazolidine-2-thione ligands were reacted with silver nitrate and one of the following: (**a**) 1,2-bis(diphenylphosphino)methane (dppm), (**b**) 1,1-bis (diphenylphosphino)ferrocene (dppf), (**c**) 1,1-bis(diphenylphosphino)ethane (dppe), (**d**) triphenylphosphine (PPh_3), or (**e**) tri(*o*-tolyl)phosphine. The molar ratio (Ag : S : P) of the three reactants was 2 : 2 : 1 if phosphines (**a**)–(**c**) were used, or 1:2:1 if (**d**) or (**e**) were used. The solvent was a mixture of acetonitrile/methanol. Based on the reaction between silver, thiazolidine, mono-, or bidentate phosphine ligands, complexes **1**, **3**, **4**, and **5** produced a clear black solution, while complex **2**, a clear orange solution. According to the ^{1}H NMR data, the synthesized silver(I) complexes were found to be nonhygroscopic and thus stable (free from decomposition) for at least a year of monitoring. Besides, all the complexes were found to have good solubility in certain organic solvents such as acetonitrile, dimethyl sulfoxide, diethyl ether, and dimethylformamide.

3.1. Spectroscopic Data Analysis. The 3-benzyl-1,3-thiazolidine-2-thione ligand offers three types of donor atoms which are thiocarbonyl sulfur atom, the nitrogen atom, and the endocyclic sulfur atom. ν (C=S) in complexes **1** to **5** were assigned at 1275–1030 cm^{-1} that observed to be shifted to a lower energy as compared to the free 3-benzyl-1,3-thiazolidine-2-thione ligand. The displacement to lower energies of the thioamide band shows the coordination to silver metal through the C=S sulfur atom. Meanwhile, the ν (C-N) at 1360–1180 cm^{-1} in the spectra was observed to shift to higher energies; this did not occur in the free ligand. The nitrogen and sulfur within the ring containing electron pairs which were in resonance with the thiocarbonyl group lead to higher delocalization of electrons and lower the ability of coordination [33]. The presence of phosphine ligand was confirmed by its characteristic ν (P-C_{ph}) band in the range of 1130–1090 cm^{-1} and a sharp band at about 1300 cm^{-1}, indicating the existence of noncoordinated NO_3^- in all complexes.

In the ^{1}H NMR spectra of the thiazolidine ligand, there were signals observed in the δ 7.29–δ 7.39 ppm region which corresponded to aromatic protons. As for the complexes, the additional aromatic protons observed at the δ 6.68–7.50 ppm region indicated the existence of aromatic groups from the phosphine ligand. Also, the peaks for the N-CH_2 and S-CH_2 signals in all the complexes were shifted slightly upfield relative to the free thiazolidine ligand. For complex **1**, a singlet signal was seen at δ 3.67 ppm in light of the -CH_2 protons in the dppm ligand. As for complex **3**, there was a peak at δ 2.46 ppm corresponding to the four protons of the ethylene group in the dppe ligand. Meanwhile, the two broad signals at δ 4.35 ppm and δ 4.17 ppm for complex **2** were attributable to the cyclopentyl protons of the dppf ligand. For complex **5**, there was a singlet peak at δ 2.31 ppm owing to the -CH_3 protons in the orthoposition of the tri(*o*-tolyl)phosphine ligands.

The FTIR and ^{1}H NMR data strongly suggested the coordination between the thiazolidine ligand and silver centre via the thione sulfur. In the ^{13}C NMR spectrum, the proof of the bonding between the aforementioned molecules was also reflected by the C=S signal, in which a change in the chemical shift value (ca. δ 0.2 ppm) was present; again, this was not seen as in free thiazolidine ligands [33]. A downfield shift in δ C-N and δ C-S relative to the free thiazolidine ligands was also observed. Furthermore, there were additional carbon peaks in the spectra at δ 25.3 ppm for complex **1**, δ 74.8 ppm and δ72.8 ppm for complex **2**, δ 24.4 ppm for complex **3**, and δ 20.5 ppm for complex **5**, all of which denote the carbon signals from the corresponding phosphine ligands.

It was observed that in comparison to free phosphine ligands, the $^{31}P\{^1H\}$ NMR resonances in all the complexes were shifted downfield in light of the formation of σ bonds between P and Ag. As per the $^{31}P\{^1H\}$ NMR data of complexes **2**, **3**, and **4**, the singlet peaks can be ascribed to two chemically-equivalent P atoms each in dppf and dppe, as well as one in triphenylphosphine. However, the spectrum of compound **1** surprisingly showed two different singlet peaks at δ 5.2 and δ 8.0 ppm. In contrast, its corresponding free dppm ligand had only one singlet peak. This phenomenon could presumably be attributed to the fact that there were nonequivalent phosphorus atoms in the complex probably due to the ^{31}P-$^{109/107}Ag$ coupling [34–36]. For complex **5**, the $^{31}P\{^1H\}$ NMR spectrum revealed two singlet peaks at δ −27.8 and 38 ppm. The peak which was at an unusually low frequency (δ −27.8 ppm) was also reported by Rizatto et. al. that reflected the indication of the formation of an unexpected

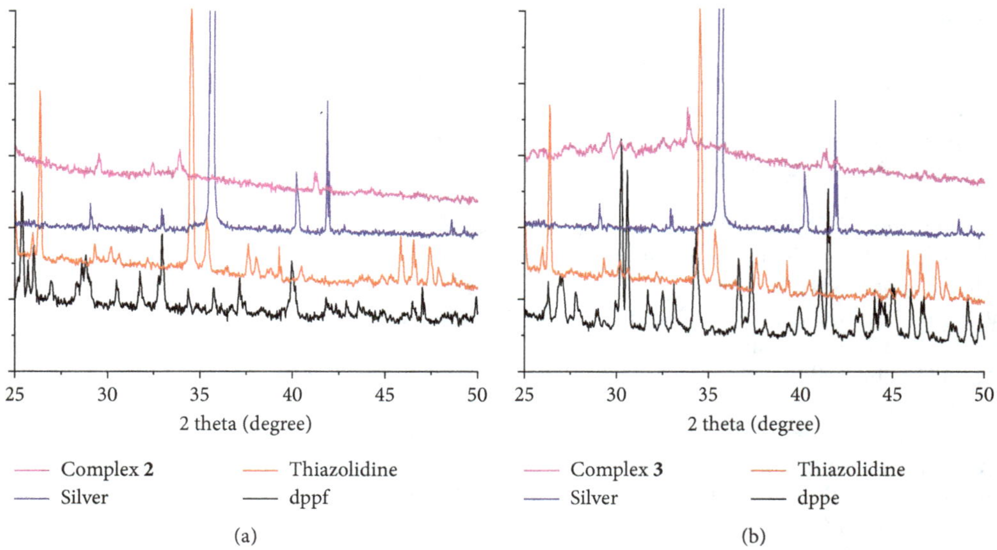

Figure 2: PXRD diffraction patterns of complexes **2** and **3**.

product with the presence of an uncoordinated P-atom from the tri(*o*-tolyl)phosphine ligand [37, 38]. Since we are yet to obtain the X-ray crystallography, it is worth noting that the affirmations of complex **5** structures are still uncertain.

The obtained complexes were subjected to EDX analysis (refer Supplementary Materials (available here)) to confirm the presence of silver metal in each compound. The analysis revealed the presence of silver metal, which was in agreement with the molar ratio used as well as all the other components that were expected to be present.

3.2. Powder X-Ray Diffraction Analysis (PXRD). The growth of crystals sufficient for X-ray single crystal analysis of all the synthesized complexes was attempted several times by utilising various techniques. However, it is difficult to crystallize the silver complexes especially containing sulfur and nitrogen as donor atoms due to its tendency of intramolecular interactions and formation into a polymeric state [39, 40]. Thus, we probed the complexes for powder X-ray diffraction analysis to obtain its phase identification, determination of unit cell dimensions, as well as measurement of sample purity. These data complement with our findings from other spectroscopic methods that suggest the structural rearrangement of the complexes.

The PXRD diffractograms were obtained within the 2θ range of 25–50° (refer Supplementary Materials (available here)). The diffractogram of each complex was compared to their corresponding thiazolidine and phosphine ligands based on the diffraction pattern of $AgNO_3$ as a standard reference (and starting material). The diffractogram patterns demonstrated the silver characteristics of face-centred cubic. Nonetheless, the broader peaks observed were due to the noncrystalline sample with relatively smaller sizes (ca. <20 Å) [25, 36]. On Comparison of peaks of the complexes **2** and **3**, $AgNO_3$, 3-benzyl-1,3-thiazolidine-2-thione and its phosphine ligand (Figure 2) indicated the successful complex formation due to the presence of matched peaks which show the insertion of ligands to the metal center. The prominent peak for the (111) unit cell dimensions observed at 2θ value ca. 34° slightly shifted from the $AgNO_3$ peak (ca. 36°) and confirmed the purity of the synthesized compounds which was also justified by the CHN, FTIR, and NMR results.

3.3. Thermogravimetric Analysis (TGA). The thermal decompositions of complexes **1–5** were carried out to study its decomposition behaviour (refer Supplementary Materials (available here)). The TGA curved for complex **1** showed the decomposition of the silver metal to Ag_2O and the phosphine molecules to its oxide form which was 34.99% (theoretical: 35%) starting from 450°C to 900°C and for complex **3** was 32.41% (theoretical: 34.54%) starting from 490°C to 900°C. Other than that, the percentage decomposition of Ag, phosphine, and Fe molecules to the oxide forms for complex **2** was 40.16% (theoretical: 39.32%) starting from 710°C to 886°C. Meanwhile, for complex **4**, the percentage of decomposition of Ag was 25.73% (theoretical: 29.38%) between 500°C and 880°C, while that for complex **5** was 25.27% (theoretical: 27.89%) from 300°C to 900°C. Thus, with respect to the data, it was shown that the experimental mass losses were in agreement to the theoretical values.

3.4. Antiplasmodial Activity. Thiazolidine derivatives have been extensively studied for their antimicrobial activities against pathogenic bacteria, fungi, anti-HIV, and *P. falciparum* [41, 42]. In addition, the introduction of an amide bond with a heterocyclic ring system (4-thiazolidine) to the lateral side chain of 4-aminoquinoline (an antimalarial agent) has been shown to improve the antimalarial activity of this compound [43]. In the present study, five Ag complexes with thiazolidine ligand showed promising *in vitro* antiplasmodial activities against chloroquine-resistant *P. falciparum* of K1 strain. The EC_{50} values of the complexes

Table 2: Antiplasmodial and cytotoxicity activities of silver(I) complexes.

Complexes	*P. falciparum* K1 $EC_{50} \pm SD$ (μM)	Vero cell line $CC_{50} \pm SD$ (μM)	SI
1	2.5 ± 0.1	2.9 ± 0.6	1.1
2	1.7 ± 0.2	>21	>12.4
3	1.5 ± 0.1	5.2 ± 0.9	3.5
4	1.04 ± 0.02	1.6 ± 0.2	1.6
5	1.8 ± 0.1	11.8 ± 2.5	6.6

Table 3: GI_{50} (μM) of different cancer cell lines after 48 h of exposure to compounds.

Complexes	$GI_{50} \pm SD$ (μM)		
	MDA-MB-231	MCF-7	HT-29
1	1.7 ± 1.3	0.5 ± 0.2	1.6 ± 0.7
2	1.9 ± 0.3	>21.03	>21.03
3	1.0 ± 0.1	0.2 ± 0.1	0.4 ± 0.3
4	1.5 ± 0.3	1.2 ± 0.2	1.5 ± 0.9
5	4.7 ± 1.1	2.9 ± 0.9	15.1 ± 1.9

fell within the acceptable cutoff values, which were more than 1–5 μM for further *in vivo* preclinical antimalarial studies [44]. The cytotoxic effects of each complex on Vero cells were assessed to determine the selectivity index or ratio, of cytotoxicity to biological activity (SI). The antiplasmodial activities of the complexes were considered to be specific and safe when the SI was more than 10 [44, 45]. Of all the synthesized Ag complexes, complex **2** exhibited highest SI followed by complex **5**, as shown in Table 2.

3.5. Antiproliferative Activity. Breast and colon cancer are the most common cancer worldwide [46, 47]. According to the National Cancer Registry 2007 of Malaysia, breast and colorectal cancer also are the most common cancer in Malaysia [48]. One of the main treatments for cancer is chemotherapy [49, 50]; however, the development of drug resistance [51, 52] and drug toxicity [53] results in significant relapse as well as decreased overall survival rates in cancer patients [54]. Thus, searching for potential drug with high efficacy and low drug toxicity remains a huge challenge in the anticancer drug discovery research and development.

Clinical successes of cisplatin, carboplatin, and oxaliplatin have resulted in the usage of metal complexes in the treatment of malignant tumors [55]. The development of anticancer drugs from coinage metals such as Ag is currently a very active field [56]. Previous studies have suggested that Ag-mixed ligand complexes have antiproliferative activities [57]. Hence, the antiproliferative potential of our newly synthesized Ag complexes with thiazolidine and phosphine ligand was verified via an experiment on different human carcinomas.

The evaluation of new anticancer drug agents through preclinical testing using cell culture is important to eliminate unsuitable candidates before pursuing into clinical research. In the present study, sulforhodamine B (SRB) was used to evaluate the anticancer properties of our drug candidates. Although MTT has been the gold standard for cytotoxicity assays, it showed interactions with many compounds and thus may yield inaccurate results [58, 59]. On the contrary, SRB assay is highly reproducible, and this assay is dependent on the protein content, thus test compound interference can be avoided [60].

The antiproliferative activities were evaluated on three human cancer cell lines, that is, metastatic breast carcinoma (MDA-MB-231), breast adenocarcinoma (MCF-7), and colon carcinoma (HT-29). Dose-response curves were constructed to calculate the GI_{50} (μM) values, which corresponded to the concentrations required to inhibit the growth of 50% of the cells. Table 3 shows the GI_{50} values of the synthesized compounds against the tested human carcinomas.

The selectivity of Ag complexes towards the tumor cells was ligand-dependent, which could probably be attributable to the stability and hydrophilicity-lipophilicity of the complexes formed by the type of the ligand [61]. Interestingly, compound **2** was selective to inhibit the 50% of MDA-MB-231 cell growth ($GI_{50} = 1.9 \pm 0.3\,\mu$M), while compound **5** acted more potent to inhibit breast carcinoma growth (GI_{50}: MDA-MB-231 = $4.7 \pm 1.1\,\mu$M; MCF-7 = $2.9 \pm 0.9\,\mu$M) instead of colon carcinoma and HT-29 ($GI_{50} = 15.1 \pm 1.9\,\mu$M). Fichtner et al. 2012 reported that silver-carbene complexes were a potent cytotoxic and resistant-breaking anticancer agent, but unfortunately, their efficacy was at the expense of high toxic effect and low selectivity in *in vivo* setting [62]. However, the type of ligands that attached to the metal can contribute to its anticancer properties as they can be involved in target recognition and interfere in biochemical pathways [63]. The presence of phosphine ligands increases the lipophilicity and membrane permeability of metal-based complexes that make them active [64]. On the contrary, thiazolidine was known to exert anticancer activity mainly via PPARγ-independent mechanism of actions [65, 66]. Our synthesized Ag complexes with bioactive thiazolidine and phosphine ligands were able to halt the proliferation of breast and colon cancer cells, thus warranting further investigations for its mechanism of action *in vivo*.

4. Conclusions

In conclusion, a series of five silver(I) complexes with phosphines and 3-benzyl-1,3-thiazolidine-2-thione have been successfully prepared and characterized by spectroscopic methods. The antimalarial activities of all the complexes have been investigated, whereby it was found that complex **2** had the highest SI value (more than 12.4) followed by complex **5** (6.6). The relatively high SI value makes complex **2** promising for further investigations towards its development as an antimalarial drug. The anticancer applications of all the synthesized compounds have also been explored and noted to concur with the results for antiplasmodial activity, whereby complex **2** was the most potent. Complex **2** was selective for MDA-MB-231 cells while complex **5** acted predominantly on breast carcinoma cells rather than those of colon carcinoma. Thus, with these findings, we have provided a preliminary insight into the potential agents that have the ability to act as dual purpose (anticancer and antimalarial) drugs, and in light of that, more in-depth studies on their molecular mechanisms of action are warranted.

Conflicts of Interest

The authors declare that there are no conflicts of interest regarding the publication of this paper.

Acknowledgments

This work was supported by the Fundamental Research Grant Scheme (FRGS) (Grant no. FP047-2014A) and University of Malaya IPPP Grant (Grant no. PG001-2015A).

References

[1] T.-H. Huang, J. Yan, H. Yang, C. Tan, and Y. Yang, "Synthesis, structures, and properties of polynuclear silver complexes containing tetra-phosphine ligand with Ag ·C interactions," *Australian Journal of Chemistry*, vol. 69, no. 3, pp. 336–342, 2016.

[2] J. M. Antelo, L. Adrio, B. Bermúdez et al., "Spectroscopic and solid state characterization of bimetallic terdentate *[C,N,S]* thiosemicarbazone palladium(II) metallacycles with bridging and chelating *[P,P]* diphosphine ligands," *Journal of Organometallic Chemistry*, vol. 740, pp. 83–91, 2013.

[3] R. Sharma, T. S. Lobana, M. Kaur et al., "Variable coordinating activity of sulfur in silver(I) complexes with thiophene based N1-substituted thiosemicarbazones: first case of thiopheneyl-thione sulfur bridging in a dinuclear complex," *Journal of Chemical Sciences*, vol. 128, no. 7, pp. 1103–1112, 2016.

[4] M. Aufiero, F. Proutiere, and F. Schoenebeck, "Redox reactions in palladium catalysis: on the accelerating and/or inhibiting effects of copper and silver salt additives in cross-coupling chemistry involving electron-rich phosphine ligands," *Angewandte Chemie International Edition*, vol. 51, no. 29, pp. 7226–7230, 2012.

[5] R. Ramachandran, G. Prakash, P. Vijayan, P. Viswanathamurthi, and J. G. Malecki, "Synthesis of heteroleptic copper(I) complexes with phosphine-functionalized thiosemicarbazones: an efficient catalyst for regioselective *N*-alkylation reactions," *Inorganica Chimica Acta*, vol. 464, pp. 88–93, 2017.

[6] M. Porchia, F. Tisato, M. Zancato, V. Gandin, and C. Marzano, "In vitro antitumor activity of water-soluble copper(I) complexes with diimine and monodentate phosphine ligands," *Arabian Journal of Chemistry*, 2017, In press.

[7] R. S. Corrêa, M. M. da Silva, A. E. Graminha et al., "Ruthenium(II) complexes of 1,3-thiazolidine-2-thione: cytotoxicity against tumor cells and anti-*Trypanosoma cruzi* activity enhanced upon combination with benznidazole," *Journal of Inorganic Biochemistry*, vol. 156, pp. 153–163, 2016.

[8] K. H. Lam, R. Gambari, M. C. W. Yuen et al., "The preparation of 2,6-disubstituted pyridinyl phosphine oxides as novel anticancer agents," *Bioorganic and Medicinal Chemistry Letters*, vol. 19, no. 8, pp. 2266–2269, 2009.

[9] K. Fourmy, M. Gouygou, O. Dechy-Cabaret, and F. Benoit-Vical, "Gold(I) complexes bearing phosphole ligands: synthesis and antimalarial activity," *Comptes Rendus Chimie*, vol. 20, no. 4, pp. 333–338, 2017.

[10] M. Navarro, F. Vásquez, R. A. Sánchez-Delgado, H. Pérez, V. Sinou, and J. Schrével, "Toward a novel metal-based chemotherapy against tropical diseases. 7. Synthesis and in vitro antimalarial activity of new gold-chloroquine complexes," *Journal of Medicinal Chemistry*, vol. 47, no. 21, pp. 5204–5209, 2004.

[11] S. Gao, Z. F. Li, M. Liu et al., "A series of silver(I) complexes with bis[2-(diphenylphosphino)phenyl]ether (DPEphos) and bis(diphenylphosphino)ethane (dppe): synthesis, structural characterization and photophysical properties," *Polyhedron*, vol. 83, pp. 10–15, 2014.

[12] A. Verma and S. K. Saraf, "4-Thiazolidinone–a biologically active scaffold," *European Journal of Medicinal Chemistry*, vol. 43, no. 5, pp. 897–905, 2008.

[13] N. C. Desai, A. Dodiya, and N. Shihory, "Synthesis and antimicrobial activity of novel quinazolinone–thiazolidine–quinoline compounds," *Journal of Saudi Chemical Society*, vol. 17, no. 3, pp. 259–267, 2013.

[14] WHO, *World Health Statistics 2015*, World Health Organization, Geneva, Switzerland, 2015.

[15] A. Kascatan-Nebioglu, M. J. Panzner, C. A. Tessier, C. L. Cannon, and W. J. Youngs, "*N*-heterocyclic carbene–silver complexes: a new class of antibiotics," *Coordination Chemistry Reviews*, vol. 251, no. 5-6, pp. 884–895, 2007.

[16] M. J. Carter, K. Tingley-Kelley, and R. A. Warriner Iii, "Silver treatments and silver-impregnated dressings for the healing of leg wounds and ulcers: a systematic review and meta-analysis," *Journal of the American Academy of Dermatology*, vol. 63, no. 4, pp. 668–679, 2010.

[17] R. Ahmad Khan, K. Al-Farhan, A. de Almeida et al., "Light-stable bis(norharmane)silver(I) compounds: synthesis, characterization and antiproliferative effects in cancer cells," *Journal of Inorganic Biochemistry*, vol. 140, pp. 1–5, 2014.

[18] E. R. Jamieson and S. J. Lippard, "Structure, recognition, and processing of cisplatin-DNA adducts," *Chemical Reviews*, vol. 99, no. 9, pp. 2467–2498, 1999.

[19] L. P. Rybak, D. Mukherjea, S. Jajoo, and V. Ramkumar, "Cisplatin ototoxicity and protection: clinical and experimental studies," *Tohoku Journal of Experimental Medicine*, vol. 219, no. 3, pp. 177–186, 2009.

[20] C. Hemmert, A. Fabié, A. Fabre, F. Benoit-Vical, and H. Gornitzka, "Synthesis, structures, and antimalarial activities of some silver(I), gold(I) and gold(III) complexes involving *N*-heterocyclic carbene ligands," *European Journal of Medicinal Chemistry*, vol. 60, pp. 64–75, 2013.

[21] R. Hooft van Huijsduijnen, R. K. Guy, K. Chibale et al., "Anticancer properties of distinct antimalarial drug classes," *PLoS One*, vol. 8, no. 12, Article ID e82962, 2014.

[22] C. N. Kundu, S. Das, A. Nayak, S. R. Satapathy, D. Das, and S. Siddharth, "Anti-malarials are anti-cancers and vice versa–one arrow two sparrows," *Acta Tropica*, vol. 149, pp. 113–127, 2015.

[23] A. Nzila, J. Okombo, R. P. Becker, R. Chilengi, T. Lang, and T. Niehues, "Anticancer agents against malaria: time to revisit?," *Trends in Parasitology*, vol. 26, no. 3, pp. 125–129, 2010.

[24] M. M. Alhamadsheh, N. C. Waters, D. P. Huddler, M. Kreishman-Deitrick, G. Florova, and K. A. Reynolds, "Synthesis and biological evaluation of thiazolidine-2-one 1,1-dioxide as inhibitors of *Escherichia coli* β-ketoacyl-ACP-synthase III (FabH)," *Bioorganic and Medicinal Chemistry Letters*, vol. 17, no. 4, pp. 879–883, 2007.

[25] P. Aslanidis, P. J. Cox, S. Divanidis, and P. Karagiannidis, "Diphosphines as bridging ligands in polymeric and dimeric thione-S-ligated Ag(I) nitrate complexes," *Inorganica Chimica Acta*, vol. 357, no. 9, pp. 2677–2686, 2004.

[26] R. Sultana, T. S. Lobana, R. Sharma et al., "Heterocyclic thioamide derivatives of coinage metals (Cu, Ag): synthesis, structures and spectroscopy," *Inorganica Chimica Acta*, vol. 363, no. 13, pp. 3432–3441, 2010.

[27] C. Lambros and J. P. Vanderberg, "Synchronization of *Plasmodium falciparum* erythrocytic stages in culture," *Journal of Parasitology*, vol. 65, no. 3, pp. 418–420, 1979.

[28] H. Noedl, W. H. Wernsdorfer, R. S. Miller, and C. Wongsrichanalai, "Histidine-rich protein II: a novel approach to malaria drug sensitivity testing," *Antimicrobial Agents and Chemotherapy*, vol. 46, no. 6, pp. 1658–1664, 2002.

[29] H. Noedl, J. Bronnert, K. Yingyuen, B. Attlmayr, H. Kollaritsch, and M. Fukuda, "Simple histidine-rich protein 2 double-site sandwich enzyme-linked immunosorbent assay for use in malaria drug sensitivity testing," *Antimicrobial Agents and Chemotherapy*, vol. 49, no. 8, pp. 3575–3577, 2005.

[30] M. R. Mohd Abd Razak, A. Afzan, R. Ali et al., "Effect of selected local medicinal plants on the asexual blood stage of chloroquine resistant *Plasmodium falciparum*," *BMC Complementary and Alternative Medicine*, vol. 14, no. 1, p. 492, 2014.

[31] T. Mosmann, "Rapid colorimetric assay for cellular growth and survival: application to proliferation and cytotoxicity assays," *Journal of Immunological Methods*, vol. 65, no. 1-2, pp. 55–63, 1983.

[32] S. L. Holbeck, J. M. Collins, and J. H. Doroshow, "Analysis of food and drug administration-approved anticancer agents in the NCI60 panel of human tumor cell lines," *Molecular Cancer Therapeutics*, vol. 9, no. 5, pp. 1451–1460, 2010.

[33] J. D. S. Chaves, F. Neumann, T. M. Francisco et al., "Synthesis and cytotoxic activity of gold(I) complexes containing phosphines and 3-benzyl-1,3-thiazolidine-2-thione or 5-phenyl-1,3,4-oxadiazole-2-thione as ligands," *Inorganica Chimica Acta*, vol. 414, pp. 85–90, 2014.

[34] T. S. Lobana, S. Khanna, and A. Castineiras, "Sulfur bridging by acetophenone thiosemicarbazone in [Ag(μ-dppm)2(μ-SR) Ag(ONO2)](NO3) dimer with a new {Ag2(μ-P,P)2(μ-SR)} core," *Inorganic Chemistry Communications*, vol. 10, no. 11, pp. 1307–1310, 2007.

[35] C. Pettinari, J. Ngoune, A. Marinelli, B. W. Skelton, and A. H. White, "Silver(I) methanesulfonate complexes containing diphosphine ligands: spectroscopic and structural characterization," *Inorganica Chimica Acta*, vol. 362, no. 9, pp. 3225–3230, 2009.

[36] S. Nawaz, A. A. Isab, K. Merz et al., "Synthesis, characterization and antimicrobial studies of mixed ligand silver(I) complexes of triphenylphosphine and heterocyclic thiones: crystal structure of bis[{(μ2-diazinane-2-thione)(diazinane-2-thione)(triphenylphosphine)silver(I) nitrate}]," *Polyhedron*, vol. 30, no. 9, pp. 1502–1506, 2011.

[37] P. S. Pregosin, "31P and 13C NMR studies on metal complexes of phosphorus-donors: recognizing surprises," *Coordination Chemistry Reviews*, vol. 252, no. 21-22, pp. 2156–2170, 2008.

[38] H. Caldwell, S. Isseponi, P. S. Pregosin, A. Albinati, and S. Rizzato, "Reactions of Ru(Cp∗) complexes with P(o-tolyl) 3," *Journal of Organometallic Chemistry*, vol. 692, no. 19, pp. 4043–4051, 2007.

[39] M. Altaf, H. Stoeckli-Evans, A. Cuin et al., "Synthesis, crystal structures, antimicrobial, antifungal and antituberculosis activities of mixed ligand silver(I) complexes," *Polyhedron*, vol. 62, pp. 138–147, 2013.

[40] K. Nomiya, S. Takahashi, and R. Noguchi, "Synthesis and crystal structure of a hexanuclear silver(I) cluster [Ag(Hmna)] 6·4H2O (H2mna = 2-mercaptonicotinic acid) and a supramolecular gold(I) complex H[Au(Hmna)2] in the solid state, and their antimicrobial activities," *Journal of the Chemical Society, Dalton Transactions*, no. 13, pp. 2091–2097, 2000.

[41] A. Mital, D. Murugesan, M. Kaiser, C. Yeates, and I. H. Gilbert, "Discovery and optimisation studies of antimalarial phenotypic hits," *European Journal of Medicinal Chemistry*, vol. 103, pp. 530–538, 2015.

[42] V. R. Solomon, W. Haq, M. Smilkstein et al., "Synthesis and antimalarial activity of novel side chain modified antimalarial agents derived from 4-aminoquinoline," *Medicinal Chemistry*, vol. 4, no. 5, pp. 446–456, 2008.

[43] V. R. Solomon, W. Haq, K. Srivastava, S. K. Puri, and S. B. Katti, "Design, synthesis of 4-aminoquinoline-derived thiazolidines and their antimalarial activity and heme polymerization inhibition studies," *Journal of Enzyme Inhibition and Medicinal Chemistry*, vol. 28, no. 3, pp. 619–626, 2013.

[44] K. Katsuno, J. N. Burrows, K. Duncan et al., "Hit and lead criteria in drug discovery for infectious diseases of the developing world," *Nature Reviews Drug Discovery*, vol. 14, no. 11, pp. 751–758, 2015.

[45] B. Weniger, S. Robledo, G. J. Arango et al., "Antiprotozoal activities of Colombian plants," *Journal of Ethnopharmacology*, vol. 78, no. 2-3, pp. 193–200, 2001.

[46] J. Ferlay, H. R. Shin, F. Bray, D. Forman, C. Mathers, and D. M. Parkin, "Estimates of worldwide burden of cancer in 2008: GLOBOCAN 2008," *International Journal of Cancer*, vol. 127, no. 12, pp. 2893–2917, 2010.

[47] K. I. Deen, H. Silva, R. Deen, and P. C. Chandrasinghe, "Colorectal cancer in the young, many questions, few answers," *World Journal of Gastrointestinal Oncology*, vol. 8, no. 6, pp. 481–488, 2016.

[48] O. Zainal Ariffin and I. T. N. Saleha, *Malaysia-National Cancer Registry Report, Malaysia Cancer Statistics–Data and Figure 2007*, National Cancer Registry, Ministry of Health (Malaysia), Putrajaya, Malaysia, 2011.

[49] M. H. Redden and G. M. Fuhrman, "Neoadjuvant chemotherapy in the treatment of breast cancer," *Surgical Clinics of North America*, vol. 93, no. 2, pp. 493–499, 2013.

[50] N. Matsuda, X. Wang, B. Lim et al., "Safety and efficacy of panitumumab plus neoadjuvant chemotherapy in patients with primary HER2-negative inflammatory breast cancer," *JAMA Oncology*, 2018.

[51] S. Crawford, "Is it time for a new paradigm for systemic cancer treatment? Lessons from a century of cancer chemotherapy," *Frontiers in Pharmacology*, vol. 4, p. 68, 2013.
[52] K. O. Alfarouk, C. M. Stock, S. Taylor et al., "Resistance to cancer chemotherapy: failure in drug response from ADME to P-gp," *Cancer Cell International*, vol. 15, no. 1, p. 71, 2015.
[53] X. Han, Y. Zhou, and W. Liu, "Precision cardio-oncology: understanding the cardiotoxicity of cancer therapy," *npj Precision Oncology*, vol. 1, no. 1, p. 31, 2017.
[54] M. Morris, C. Platell, K. McCaul et al., "Survival rates for stage II colon cancer patients treated with or without chemotherapy in a population-based setting," *International Journal of Colorectal Disease*, vol. 22, no. 8, pp. 887–895, 2007.
[55] B. Desoize, "Metals and metal compounds in cancer treatment," *Anticancer Research*, vol. 24, no. 3a, pp. 1529–1544, 2004.
[56] S. J. Tan, Y. K. Yan, P. P. F. Lee, and K. H. Lim, "Copper, gold and silver compounds as potential new anti-tumor metallodrugs," *Future Medicinal Chemistry*, vol. 2, no. 10, pp. 1591–1608, 2010.
[57] S. Shukla and A. P. Mishra, "Synthesis, structure, and anticancerous properties of silver complexes," *Journal of Chemistry*, vol. 2013, Article ID 527123, 6 pages, 2013.
[58] S. Wang, H. Yu, and J. K. Wickliffe, "Limitation of the MTT and XTT assays for measuring cell viability due to superoxide formation induced by nano-scale TiO_2," *Toxicology in Vitro*, vol. 25, no. 8, pp. 2147–2151, 2011.
[59] Y. J. Wang, S. M. Zhou, G. Xu, and Y. Q. Gao, "Interference of phenylethanoid glycosides from *Cistanche tubulosa* with the MTT assay," *Molecules*, vol. 20, no. 5, pp. 8060–8071, 2015.
[60] A. van Tonder, A. M. Joubert, and A. D. Cromarty, "Limitations of the 3-(4,5-dimethylthiazol-2-yl)-2,5-diphenyl-2H-tetrazolium bromide (MTT) assay when compared to three commonly used cell enumeration assays," *BMC Research Notes*, vol. 8, p. 47, 2015.
[61] U. Kalinowska-Lis, A. Felczak, L. Chęcińska et al., "Antibacterial activity and cytotoxicity of silver(I) complexes of pyridine and (benz)imidazole derivatives. X-ray crystal structure of [Ag(2,6-di(CH_2OH)py)2]NO_3," *Molecules*, vol. 21, no. 2, p. 87, 2016.
[62] I. Fichtner, J. Cinatl, M. Michaelis et al., "In vitro and in vivo investigations into the carbene silver acetate anticancer drug candidate SBC1," *Letters in Drug Design and Discovery*, vol. 9, no. 9, pp. 815–822, 2012.
[63] P. Zhang and P. J. Sadler, "Redox-active metal complexes for anticancer therapy," *European Journal of Inorganic Chemistry*, vol. 2017, no. 12, pp. 1541–1548, 2017.
[64] F. K. Keter, I. A. Guzei, M. Nell, W. E. van Zyl, and J. Darkwa, "Phosphinogold(I) dithiocarbamate complexes: effect of the nature of phosphine ligand on anticancer properties," *Inorganic Chemistry*, vol. 53, no. 4, pp. 2058–2067, 2014.
[65] M. S. A. El-Gaby, Z. H. Ismail, S. M. Abdel-Gawad, H. M. Aly, and M. M. Ghorab, "Synthesis of thiazolidine and thiophene derivatives for evaluation as anticancer agents," *Phosphorus, Sulfur, and Silicon and the Related Elements*, vol. 184, no. 10, pp. 2645–2654, 2009.
[66] V. Asati, D. K. Mahapatra, and S. K. Bharti, "Thiazolidine-2,4-diones as multi-targeted scaffold in medicinal chemistry: potential anticancer agents," *European Journal of Medicinal Chemistry*, vol. 87, pp. 814–833, 2014.

Synthesis, Characterization, and Cytotoxic Evaluation of Some Newly Substituted Diazene Candidates

Mohamed El-Naggar,[1] Abdel-Nasser El-Shorbagi,[2,3] Dina H. Elnaggar,[4] Abd El-Galil E. Amr,[5,6] Mohamed A. Al-Omar,[5] and Elsayed A. Elsayed[7,8]

[1]*Chemistry Department, Faculty of Sciences, University of Sharjah, Sharjah 27272, UAE*
[2]*College of Pharmacy, University of Sharjah, Sharjah 27272, UAE*
[3]*Sharjah Institute for Medical Research, University of Sharjah, Sharjah 27272, UAE*
[4]*Applied Organic Chemistry Department, National Research Centre, Dokki, Giza 12622, Egypt*
[5]*Drug Exploration & Development Chair (DEDC), Pharmaceutical Chemistry Department, College of Pharmacy, King Saud University, Riyadh 11451, Saudi Arabia*
[6]*Department of Applied Organic Chemistry, Industrial Chemical Division, National Research Center, Cairo, Dokki 12622, Egypt*
[7]*Bioproducts Research Department, Zoology Department, Faculty of Science, King Saud University, Riyadh, Saudi Arabia*
[8]*Chemistry of Natural and Microbial Products Department, National Research Centre, Dokki 12622, Cairo, Egypt*

Correspondence should be addressed to Abd El-Galil E. Amr; aeamr1963@yahoo.com

Academic Editor: Artur M. S. Silva

A series of azocompounds containing methyl salicylate **4a–k** and 1-naphthyl moiety **6–8** was synthesized and tested as anticancer agents. Nitrosation of methyl 5-amino-2-hydroxybenzoate or 1-aminonaphthalene by using $NaNO_2$ in the presence of HCl afforded diazonium salt derivatives **2** and **5**, which were treated with substituted imino or substituted amino derivatives, to give the corresponding substituted amino-pent-2-en-3-yl-diazenylbenzoate **4a–k** or 2-substituted-1-(naphthalen-1-yl)diazene derivatives **6a–h**, **7a,b**, and **8a,b**. All the synthesized compounds were elucidated by elemental analysis and spectroscopic evidence.

1. Introduction

In a previous work, the authors reported that certain synthesized substituted heterocyclic aromatic derivatives showed antiviral activities [1] and alpha reductase inhibition [2]. Also, some of new compounds containing the nitrogen atom have been synthesized and used as antihypertensive α-blocking [3], antiparkinsonian [4], antialzheimer [5], antimicrobial [6–8], and anti-inflammatory [9, 10] agents. There is an increasing and critical demand for more powerful anticancer agents, due to increased detection of cancer resistance [11]. Azocompounds are one of the largest types of organically synthesized compounds, which are effective both as drugs and cosmetics [12]. Also, they are extensively used in the field of dyeing textiles, biomedical studies, advanced applications in organic synthesis, and remarkable biological activities including antibacterial activity [13–20]. A number of azocompounds were synthesized comprising a drug moiety in their skeleton [21]. Salicylates are a class of chemicals with appreciative biological activities. Several biomedical studies have used methyl salicylate derivatives with promising results [22–24]. Aromatic amines such as benzidine, 1-naphthylamine, and 4-methylaniline have been recognized as potential carcinogens [25, 26]. From these observations, we decided to synthesize two series of azocompounds. One of these series contains methyl salicylate moiety, and the other involves 1-naphthyl moiety. All the synthesized compounds were screened as anticancer agents.

2. Experimental

2.1. Chemistry. All reagents and solvents were obtained from the commercial supplier and used without further purification. Melting point is uncorrected and was determined on an electrothermal melting point apparatus (Stuart Scientific, England, UK). Precoated silica gel plates

(Kieselgel 0.25 mm, 60G F_{254}, Merck, Germany) were used for thin layer chromatography (TLC) for reaction monitoring, and UV light was used for detection. The separation was done using column chromatography with silica gel 60 (Merck) eluted with appropriate solvent. Nuclear magnetic resonance spectra (NMR) were recorded using Bruker 500 MHz instrument. Chemical shifts were reported in ppm, and spectra were corrected using the residual solvent as a reference. Mass spectra were recorded on a Bruker Daltonics spectrometer.

2.1.1. General Procedure for the Synthesis of Compounds **4a–k**. A mixture of methyl 5-aminosalicylate **1** (0.56 g, 3.4 mmol) in concentrated hydrochloric acid (5 mL) was stirred with gentle heating until complete dissolving. Then, the mixture was cooled at a temperature from zero to −4°C, sodium nitrite (0.28 g, 4 mmol) in 2 mL of water was added portionwise at a temperature of −4°C to form the corresponding diazonium salt **2**, and the diazonium salt formed was added to a mixture of 4-(substituted imino)pentan-2-one derivatives **3** (2 mmol) [27] in ethanol (15 mL) slowly under the same temperature allowing the mixture to stir at room temperature for an additional 3 hours. The solid formed was collected by filtration, washed with cold water several times, and then left to dry. The obtained product was recrystallized from ethanol to give target compounds **4a–k**, respectively.

(1) Methyl 5-(4-(benzylamino)-2-oxopent-3-en-3-yl) diazenyl)-2-hydroxybenzoate (**4a**). Yellow crystalline solid, m.p. 167-168°C, yield = 62%, and ^{1}H-NMR (DMSO-d_6) δ (ppm) 2.42 (s, 3H, CH_3-CNH), 2.57 (s, 3H, CO-CH_3), 3.93 (s, 3H, OCH_3), 4.82 (s, 2H, Ar-CH_2), 7.05 (d, 1H, J = 8.8 Hz, Ar-H, benzoate ring), 7.36–7.42 (m, 5H, ArH), 7.72 (d, 1H, J = 2.5 Hz, Ar-H, benzoate ring), 7.78 (s, 1H, Ar-H, benzoate ring), 10.50 (br s, 1H, NH), and 14.40 (s, 1H, OH); ^{13}C-NMR (DMSO-d_6) δ (ppm) 16.8, 29.0, 47.6, 53.0, 113.8, 118.9, 123.0, 125.6, 128.1, 128.2, 128.6, 129.4, 137.3, 144.5,159.1, 159.4, 169.3, and 196.4 (20C). ESI-MS: m/z, 368.16 (M^+ + H). Anal. calcd for $C_{20}H_{21}N_3O_4$ (367.40): C, 65.38; H, 5.76; and N, 11.44. Found: C, 65.35; H, 5.78; and N, 11.40.

(2) Methyl 5-(2-oxo-4-(phenethylamino)pent-3-en-3-yl) diazenyl)-2-hydroxybenzoate (**4b**). Yellow crystalline solid, m.p. 131-132°C, yield = 60%, and ^{1}H-NMR (DMSO-d_6) δ (ppm) 2.39 (s, 3H, CH_3-CNH), 2.52 (s, 3H, CO-CH_3), 2.98 (t, 2H, J = 6.8 Hz, $PhCH_2CH_2N$), 3.85 (t, 2H, J = 6.7 Hz, $ArCH_2CH_2N$), 3.95 (s, 3H, OCH_3), 7.03 (d, 1H, J = 8.8 Hz, Ar-H, benzoate ring), 7.34–7.30 (m, 5H, ArH), 7.59 (d, 1H, J = 2.5 Hz, Ar-H, benzoate ring), 7.76 (s, 1H, Ar-H, benzoate ring), 10.50 (br s, 1H, NH), and 13.84 (s, 1H, OH); ^{13}C-NMR (DMSO-d_6) δ (ppm) 16.3, 29.1, 35.0, 44.6, 53.1, 113.8, 118.7, 122.4, 126.6, 127.0, 128.3, 129.0, 129.2, 138.7, 144.9, 159.1, 159.3, 169.5, and 196.3 (21C). ESI-MS: m/z, 382.18 (M^+ + H). Anal. calcd for $C_{21}H_{23}N_3O_4$ (381.43): C, 66.13; H, 6.08; and N, 11.02. Found: C, 66.45; H, 6.02; and N, 11.00.

(3) Methyl 5-(2-oxo-4-(1-phenylethylamino)pent-3-en-3-yl)diazenyl)-2-hydroxybenzoate (**4c**). Yellow crystalline solid, m.p. 110-111°C, yield = 66%, and ^{1}H-NMR (DMSO-d_6) δ (ppm) 1.60 (d, 3H, J = 7.6 Hz, $ArCH(CH_3)N$), 2.41 (s, 3H, CH_3-C), 2.49 (s, 3H, CO-CH_3), 3.94 (s, 3H, OCH_3), 5.17 (q, 1H, J = 6.7 Hz, $Ar\underline{C}H(CH_3)N$), 7.06 (d, 1H, J = 8.8 Hz, Ar-H, benzoate ring), 7.40–7.45 (m, 5H, ArH), 7.76 (d, 1H, J = 2.5 Hz, Ar-H, benzoate ring), 7.86 (s, 1H, Ar-H, benzoate ring), 10.50 (br s, 1H, NH), and 14.63 (s, 1H, OH); ^{13}C-NMR (DMSO-d_6) δ (ppm) 16.8, 24.0, 29.0, 53.1, 53.7, 114.0, 119.0, 122.2, 126.4, 126.6, 128.2, 128.5, 129.4, 143.0, 144.5, 159.2, 159.4, 169.3, and 196.4 (21C). ESI-MS: m/z, 382.18 (M^+ + H). Anal. calcd for $C_{21}H_{23}N_3O_4$ (381.43): C, 66.13; H, 6.08; and N, 11.02. Found: C, 66.04; H, 6.03; and N, 10.98.

(4) Methyl 5-(4-(2-hydroxyethylamino)-2-oxopent-3-en-3-yl)diazenyl)-2-hydroxybenzoate (**4d**). Yellow crystalline solid, m.p. 187-188°C, yield = 70%, and ^{1}H-NMR (DMSO-d_6) δ (ppm) 2.42 (s, 3H, CH_3-C), 2.54 (s, 3H, CO-CH_3), 3.59 (t, 2H, J = 5.0 Hz, NCH_2CH_2OH), 3.66 (t, 2H, J = 4.2 Hz, NCH_2CH_2OH), 3.92 (s, 3H, OCH_3), 5.13 (br s, 1H, NCH_2CH_2OH) 7.07 (d, 1H, J = 8.8 Hz, Ar-H, benzoate ring), 7.79 (d, 1H, J = 2.5 Hz, Ar-H, benzoate ring), 8.01 (s, 1H, Ar-H, benzoate ring), and 10.5 (br s, 1H, NH), 14.21 (s, 1H, OH); ^{13}C-NMR (DMSO-d_6) δ (ppm) 16.7, 29.1, 46.0, 53.1, 59.7, 113.9, 118.7, 122.1, 127.0, 128.4, 134.5, 154.1, 159.3, 169.5, and 196.4 (15C). ESI-MS: 322.27 (M^+ + H). Anal. calcd for $C_{15}H_{19}N_3O_5$ (321.33): C, 56.07; H, 5.96; and N, 13.08. Found: C, 56.00; H, 5.92; and N, 13.00.

(5) Methyl 5-(4-(isobutylamino)-2-oxopent-3-en-3-yl) diazenyl)-2-hydroxybenzoate (**4e**). Yellow crystalline solid, m.p. 94–95°C, yield = 75%, and ^{1}H-NMR (DMSO-d_6) δ (ppm) 1.00 (d, 6H, J = 6.7 Hz, $NHCH_2CH(CH_3)_2$), 1.95–1.90 (m, 1H, $NHCH_2CH(CH_3)_2$), 2.42 (s, 3H, CH_3-C), 2.54 (s, 3H, CO-CH_3), 3.39 (dd, 2H, J = 5.4 Hz, $NHCH_2CH(CH_3)_2$), 3.92 (s, 3H, OCH_3), 7.07 (d, 1H, J = 8.8 Hz, Ar-H, benzoate ring), 7.76 (d, 1H, J = 2.5 Hz, Ar-H, benzoate ring), 7.95 (s, 1H, Ar-H, benzoate ring), 10.50 (br s, 1H, NH), and 14.09 (s, 1H, OH); ^{13}C-NMR (DMSO-d_6) δ (ppm) 16.45, 2.36, 28.3, 29.1, 50.7, 53.1, 114.0, 118.9, 122.0, 126.8, 128.4, 134.5, 145.2, 159.3, 169.4, and 196.2 (17C). ESI-MS: m/z, 334.18 (M^+ + H). Anal. calcd for $C_{17}H_{23}N_3O_4$ (333.38): C, 61.25; H, 6.95; and N, 12.60. Found: C, 61.12; H, 6.90; and N, 12.54.

(6) Methyl 5-(4-(isopropylamino)-2-oxopent-3-en-3-yl) diazenyl)-2-hydroxybenzoate (**4f**). Yellow crystalline solid, m.p. 104-105°C, yield = 61%, and ^{1}H-NMR (DMSO-d_6) δ (ppm) 1.30 (d, 6H, J = 6.4 Hz, $NHCH(CH_3)_2$), 2.41 (s, 3H, CH_3-C), 2.56 (s, 3H, $COCH_3$), 3.92 (s, 3H, OCH_3), 4.15–4.12 (m, 1H, J = 6.4 Hz, $NHCH(CH_3)_2$), 7.08 (d, 1H, J = 8.8 Hz, Ar-H, benzoate ring), 7.78 (d, 1H, J = 2.5 Hz, Ar-H, benzoate ring), 7.92 (s, 1H, Ar-H, benzoate ring), 10.50 (br s, 1H, NH), and 14.10 (s, 1H, OH); ^{13}C-NMR (DMSO-d_6) δ (ppm) 16.3, 23.2, 29.1, 45.6, 53.1, 114.0, 118.8, 121.8, 127.1, 127.9, 134.5, 145.1, 159.3, 169.4, and 196.1 (16C). ESI-MS: m/z, 320.16 (M^+ + H). Anal. calcd for $C_{16}H_{21}N_3O_4$ (319.36): C, 60.17; H, 6.63; and N, 13.16. Found: C, 60.10; H, 6.58; and N, 13.10.

(7) Methyl 5-(4-(butylamino)-2-oxopent-3-en-3-yl) diazenyl)-2-hydroxybenzoate (**4g**). Yellow crystalline solid, m.p. 131-132°C, yield = 65%, and ^{1}H-NMR (DMSO-d_6) δ (ppm) 0.97 (t, 3H, J = 7.3 Hz, $NHCH_2CH_2CH_2CH_3$), 1.43-1.40 (m, 2H, J = 7.6 Hz, $NHCH_2CH_2CH_2CH_3$), 1.62–1.66 (m, 2H, J = 6.9 Hz, $NHCH_2CH_2CH_2CH_3$), 2.41 (s, 3H, CH_3-C), 2.53 (s, 3H, $COCH_3$), 3.52 (t, 2H, J = 6.7 Hz, $NHCH_2CH_2CH_2CH_3$), 3.91 (s, 3H, OCH_3), 7.07 (d, 1H,

J = 8.8 Hz, Ar-H, benzoate ring), 7.72 (d, 1H, J = 2.5 Hz, Ar-H, benzoate ring), 7.93 (s, 1H, Ar-H, benzoate ring), 10.50 (br s, 1H, NH), and 14.05 (s, 1H, OH); ^{13}C-NMR (DMSO-d$_6$) δ (ppm) 14.0, 16.4, 20.2, 29.0, 31.0, 43.1, 53.1, 113.9, 118.9, 122.6, 126.5, 128.2, 134.5, 145.1, 159.3, 169.4, and 196.2 (17C). ESI-MS: m/z, 334.18 (M$^+$ + H). Anal. calcd for $C_{17}H_{23}N_3O_4$ (333.38): C, 61.25; H, 6.95; and N, 12.60. Found: C, 61.12; H, 6.86; and N, 12.52.

(8) Methyl 5-(4-(3-hydroxypropylamino)-2-oxopent-3-en-3-yl)diazenyl)-2-hydroxybenzoate **(4h)**. Yellow crystalline solid, m.p. 135-136°C, yield = 63%, and ^{1}H-NMR (DMSO-d$_6$) δ (ppm) 1.80–1.75 (m, 2H, J = 6.3 Hz, $NHCH_2CH_2CH_2OH$), 2.41 (s, 3H, CH_3-C), 2.54 (s, 3H, CO-CH_3), 3.54 (br t, 2H, J = 5.7 Hz, $NHCH_2CH_2CH_2OH$), 3.59 (br t, 2H, J = 6.7 Hz, $NHCH_2CH_2CH_2OH$), 3.91 (s, 3H, OCH_3), 4.70 (br s, 2H, $NHCH_2CH_2CH_2OH$), 7.07 (d, 1H, J = 8.8 Hz, Ar-H, benzoate ring), 7.77 (d, 1H, J = 2.5 Hz, Ar-H, benzoate ring), 7.98 (s, 1H, Ar-H, benzoate ring), and 10.50 (br s, 1H, NH), 13.95 (s, 1H, OH); ^{13}C-NMR (DMSO-d$_6$) δ (ppm) 16.3, 29.1, 32.2, 40.0, 53.1, 58.4, 113.9, 118.7, 122.0, 127.1, 128.3, 134.5, 145.2, 159.3, 169.5, and 196.2 (16C). ESI-MS): m/z, 336.16 (M$^+$ + H). Anal. calcd for $C_{16}H_{21}N_3O_5$ (335.36): C, 57.30; H, 6.31; and N, 12.53. Found: C, 57.10; H, 6.22; and N, 12.42.

(9) Methyl 5-(2-oxo-4-(pentylamino)pent-3-en-3-yl) diazenyl)-2-hydroxybenzoate **(4i)**. Yellow crystalline solid, m.p. 127-128°C, yield = 72%, and ^{1}H-NMR (DMSO-d$_6$) δ (ppm) 0.92 (t, 3H, J = 6.9 Hz, $NHCH_2CH_2CH_2CH_2CH_3$), 1.40–1.45 (m, 4H, $NHCH_2CH_2CH_2CH_2CH_3$), 1.66–1.62 (m, 2H, $NHCH_2CH_2CH_2CH_2CH_3$), 2.41 (s, 3H, CH_3-C), 2.54 (s, 3H, CO-CH_3), 3.53 (t, 2H, J = 6.7 Hz, $NHCH_2CH_2CH_2CH_2CH_3$), 3.91 (s, 3H, OCH_3), 7.06 (d, 1H, J = 8.8 Hz, Ar-H, benzoate ring), 7.78 (d, 1H, J = 2.5, Ar-H, benzoate ring), 7.93 (s, 1H, Ar-H, benzoate ring), 10.50 (br s, 1H, NH), and 14.05 (s, 1H, OH); ^{13}C-NMR (DMSO-d$_6$) δ (ppm) 14.3, 16.4, 22.3, 28.6, 29.2, 30.7, 43.4, 53.0, 113.8, 118.9, 122.7, 126.3, 128.2, 134.5, 145.1, 159.4, 169.4, and 196.2 (18C). ESI-MS: m/z, 348.19 (M$^+$ + H). Anal. calcd for $C_{18}H_{25}N_3O_4$ (347.41): C, 62.23; H, 7.25; and N, 12.10. Found: C, 62.05; H, 7.16; and N, 12.05.

(10) Methyl 5-(4-(ethylamino)-2-oxopent-3-en-3-yl) diazenyl)-2-hydroxybenzoate **(4j)**. Yellow crystalline solid, m.p. 106-107°C, yield = 75%, and ^{1}H-NMR (DMSO-d$_6$) δ (ppm) 1.27 (t, 3H, J = 7.2 Hz, $NHCH_2CH_3$), 2.41 (s, 3H, CH_3-C), 2.54 (s, 3H, CO-CH_3), 3.55 (q, 2H, $NHCH_2CH_3$), 3.91 (s, 3H, OCH_3), 7.07 (d, 1H, J = 8.8 Hz, Ar-H, benzoate ring), 7.78 (d, 1H, J = 2.5 Hz, Ar-H, benzoate ring), 7.93 (s, 1H, Ar-H, benzoate ring), 10.50 (br s, 1H, NH), and 13.88 (s, 1H, OH); ^{13}C-NMR (DMSO-d$_6$) δ (ppm) 14.8, 16.3, 29.1, 38.3, 53.1, 113.9, 118.8, 122.3, 126.9, 128.1, 134.5, 145.3, 159.4, 169.5, and 196.1 (15C). ESI-MS): m/z, 306.15 (M$^+$ + H). Anal. calcd for $C_{15}H_{19}N_3O_4$ (305.33): C, 59.01; H, 6.27; and N, 13.76. Found: C, 58.86; H, 6.20; and N, 13.70.

(11) Methyl 5-(2-oxo-4-(propylamino)pent-3-en-3-yl) diazenyl)-2-hydroxybenzoate **(4k)**. Yellow crystalline solid, m.p. 137-138°C, yield = 69%, and ^{1}H-NMR (DMSO-d$_6$) δ (ppm) 0.99 (t, 3H, J = 7.3 Hz, $NHCH_2CH_2CH_3$), 1.68–1.64 (m, 3H, J = 7.2 Hz, $NHCH_2CH_2CH_3$), 2.41 (s, 3H, CH_3-C), 2.54 (s, 3H, CO-CH_3), 3.50 (q, 2H, J = 6.9 Hz, $NHCH_2CH_3$), 3.92 (s, 3H, OCH_3), 7.08 (d, 1H, J = 8.8 Hz, Ar-H, benzoate ring), 7.77 (d, 1H, J = 2.5 Hz, Ar-H, benzoate ring), 7.96 (s, 1H, Ar-H, benzoate ring), 10.52 (br s, 1H, NH), and 13.98 (s, 1H, OH); ^{13}C-NMR (DMSO-d$_6$) δ (ppm) 11.8, 16.4, 22.5, 29.1, 45.1, 53.1, 114, 118.8, 122.2, 126.9, 128.3, 134.5, 145.2, 159.3, 169.4, and 196.2 (16C). ESI-MS: m/z, 320.14 (M$^+$ + H). Anal. calcd for $C_{16}H_{21}N_3O_4$ (319.36): C, 60.17; H, 6.63; and N, 13.16. Found: C, 60.00; H, 6.56; and N, 13.10.

2.1.2. General Procedure for the Synthesis of Compounds **6–8**. A mixture of 1-aminonaphthalene (0.48 g, 3.4 mmol) and conc. HCl (5 mL) was stirred with gentle heating until complete dissolving. The reaction mixture was cooled at a temperature from zero to −4°C in an ice bath, and then sodium nitrite (0.28 g, 4 mmol) in water (2 mL) was added portionwise to the reaction mixture at the same temperature to form the diazonium salt **5**. To the diazonium salt 5, a solution of appropriate amide or iminoderivatives [27], namely, *N*-substituted 3-oxo-butanamide, ethyl 3-(substituted imino) butanoate, or *N*-alkyl-3-(alkyl imino)-butanamide in absolute ethanol (10 mL), was added while stirring at a temperature of −4°C, and the mixture was stirred at room temperature for an additional 3 hours. The solid formed was collected by filtration, washed with cold water several times, dried, and crystallized from dimethylformamide/ethanol to give the corresponding **6a–h**, **7a,b**, and **8a,b**, respectively.

(1) 2-(Naphthalen-1-yl-diazenyl)-3-hydroxy-N-phenethylbut-2-enamide **(6a)**. Brownish crystalline solid, m.p. 141-142°C, yield = 70%, and ^{1}H-NMR (DMSO-d$_6$) δ (ppm) 2.50 (s, 3H, C-CH_3), 2.88 (t, 2H, J = 7.2 Hz, $ArCH_2CH_2NH$), 3.62 (q, 2H, J = 7.0 Hz, $ArCH_2CH_2NH$), 7.23–7.20 (m, 1H, Ar), 7.33–7.30 (m, 4H, ArH), 7.63–7.60 (m, 2H, naphthyl-H), 7.74–7.70 (m, 1H, naphthyl-H), 7.78 (d, 1H, J = 8.1 Hz, naphthyl-H), 7.83 (d, 1H, J = 6.7 Hz, naphthyl-H), 7.91 (d, 1H, J = 8.4 Hz, naphthyl-H), 8.04 (d, 1H, J = 7.9 Hz, naphthyl-H), 9.40 (br s, 1H, NHCO), and 15.67 (s, 1H, OH); ^{13}C-NMR (DMSO-d$_6$) δ (ppm) 26.4, 35.2, 40.7, 110.9, 119.8, 123.1, 125.1, 126.8, 126.9, 127.2, 127.5, 128.1, 128.9, 129.2, 129.3, 134.2, 136.9, 139.5, 164.9, and 198.7 (22C). ESI- MS: m/z, 360.17 (M$^+$ + H). Anal. calcd for $C_{22}H_{21}N_3O_2$ (359.42): C, 73.52; H, 5.89; and N, 11.69. Found: C, 73.45; H, 5.80; and N, 11.60.

(2) 2-(Naphthalen-1-yl-diazenyl)-3-hydroxy-N-methylbut-2-enamide **(6b)**. Brownish crystalline solid, m.p. 175-176°C, yield = 71%, and ^{1}H-NMR (DMSO-d$_6$) δ (ppm) 2.52 (s, 3H, C-CH_3), 2.88 (d, 3H, J = 4.9 Hz, $CONHCH_3$), 7.63–7.60 (m, 2H, naphthyl-H), 7.72–7.68 (m, 1H, naphthyl-H), 7.77 (d, 1H, J = 8.2 Hz, naphthyl-H), 7.83 (d, 1H, J = 7.6 Hz, naphthyl-H), 7.92 (d, 1H, J = 8.0 Hz, naphthyl-H), 8.02 (d, 1H, J = 8.1 Hz, naphthyl-H), 9.23 (br s, 1H, NHCO), and 15.72 (s, 1H, OH); ^{13}C-NMR (DMSO-d$_6$) δ (ppm) 25.8, 26.4, 110.8, 119.8, 123.0, 124.9, 126.9, 127.1, 127.5, 128.3, 129.3, 134.2, 136.9, 165.5, and 198.5 (15C). ESI-MS: m/z, 270.12 (M$^+$ + H). Anal. calcd for $C_{15}H_{15}N_3O_2$ (269.30): C, 66.90; H, 5.61; and N, 15.60. Found: C, 65.80; H, 5.50; and N, 15.52.

(3) 2-(Naphthalen-1-yl-diazenyl)-N-tert-butyl-3-hydroxybut-2-enamide **(6c)**. Brownish crystalline solid, m.p. 126-127°C, yield = 59%, and ^{1}H-NMR (DMSO-d$_6$) δ (ppm) 1.43 (s, 9H, $C(CH_3)_3$), 2.52 (s, 3H, C-CH_3), 7.63–7.60 (m, 2H, naphthyl-H), 7.74–7.40 (m, 1H, naphthyl-

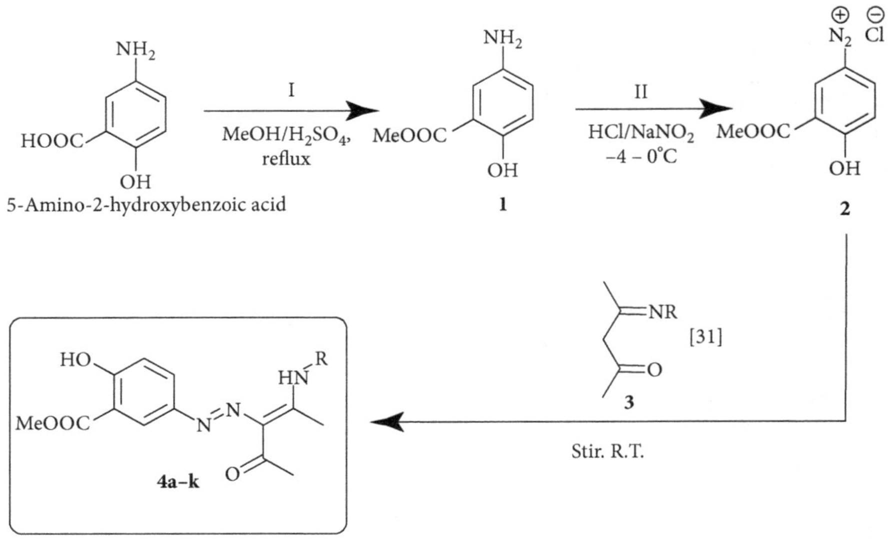

R		R	
a,	Ph-CH_2–	g,	$CH_3(CH_2)_2CH_2$–
b,	$PhCH_2CH_2$–	h,	$HOCH_2CH_2CH_2$–
c,	Ph(CH_3)CH–	i,	$CH_3(CH_2)_3CH_2$–
d,	$HOCH_2CH_2$–	j,	CH_3CH_2–
e,	$(CH_3)_2CHCH_2$–	k,	$CH_3CH_2CH_2$–
f,	$(CH_3)_2CH$–		

SCHEME 1: Synthetic route of compounds **4a–k**.

H), 7.78 (d, 1H, J = 8.1 Hz, naphthyl-H), 7.83 (d, 1H, J = 7.6 Hz, naphthyl-H), 7.90 (d, 1H, J = 8.4 Hz, naphthyl-H), 8.02 (d, 1H, J = 8.1 Hz, naphthyl-H), 9.42 (s, 1H, NHCO), and 15.69 (s, 1H, OH); ^{13}C-NMR (DMSO-d_6) δ (ppm) 26.5, 28.8, 51.3, 111.0, 119.7, 123.1, 125.1, 126.9, 127.1, 127.6, 128.4, 129.2, 134.2, 136.9, 164.7, and 199.3 (18C). ESI-MS: m/z, 312.17 (M^+ + H). Anal. calcd for $C_{18}H_{21}N_3O_2$ (311.38): C, 69.43; H, 6.80; and N, 13.49. Found: C, 69.30; H, 6.75; and N, 13.40.

(4) 2-(Naphthalen-1-yl-diazenyl)-3-hydroxy-N-isopropylbut-2-enamide **(6d)**. Brownish crystalline solid, m.p. 124-125°C, yield = 62%, and ^{1}H-NMR (DMSO-d_6) δ (ppm) 1.23 (d, 6H, J = 6.6 Hz, $CH(CH_3)_2$), 2.52 (s, 3H, C-CH_3), 4.11–4.00 (m, 1H, $CH(CH_3)_2$), 7.65–7.60 (m, 2H, naphthyl-H), 7.74–7.70 (m, 1H, naphthyl-H), 7.78 (d, 1H, J = 8.1 Hz, naphthyl-H), 7.83 (d, 1H, J = 6.8 Hz, naphthyl-H), 7.90 (d, 1H, J = 8.3 Hz, naphthyl-H), 8.02 (d, 1H, J = 8.1 Hz, naphthyl-H), 9.26 (br s, 1H, NHCO), and 15.71 (s, 1H, OH); ^{13}C-NMR (DMSO-d_6) δ (ppm) 22.6, 40.8, 49.1, 110.9, 119.7, 123.1, 125.1, 127.0, 127.1, 127.5, 128.1, 129.3, 134.2, 136.9, 164.2, and 199.0 (17C). ESI-MS: m/z, 280.15 (M^+ + H). Anal. calcd for $C_{17}H_{19}N_3O_2$ (297.35): C, 68.67; H, 6.44; and N, 14.13. Found: C, 68.60; H, 6.34; and N, 14.00.

(5) 2-(Naphthalen-1-yl-diazenyl)-N-ethyl-3-hydroxybut-2-enamide **(6e)**. Brownish crystalline solid, m.p. 130–131°C, yield = 75%, and ^{1}H-NMR (DMSO-d_6) δ (ppm) 1.17 (t, 3H, J = 7.2 Hz, $NHCH_2CH_3$), 2.52 (s, 3H, C-CH_3), 3.38–3.34 (m, 2H, $NHCH_2CH_3$), 7.64–7.60 (m, 2H, naphthyl-H), 7.74–7.70 (m, 1H, naphthyl-H), 7.77 (d, 1H, J = 8.1 Hz, naphthyl-H), 7.82 (d, 1H, J = 7.6 Hz, naphthyl-H), 7.91 (d, 1H, J = 8.4 Hz, naphthyl-H), 8.02 (d, 1H, J = 8.1 Hz, naphthyl-H), 9.33 (br s, 1H, NHCO), and 15.72 (s, 1H, OH); ^{13}C-NMR (DMSO-d_6) δ (ppm) 14.9, 26.5, 33.60, 110.9, 119.8, 123.0, 125.0, 126.9, 127.1, 127.5, 128.2, 129.3, 134.2, 136.9, 164.8, and 198.7 (16C). ESI-MS: m/z, 284.14 (M^+ + H). Anal. calcd for $C_{16}H_{17}N_3O_2$ (283.33): C, 67.83; H, 6.05; and N, 14.83. Found: C, 67.74; H, 6.00; and N, 14.78.

(6) 2-(Naphthalen-1-yl-diazenyl)-N-butyl-3-hydroxybut-2-enamide **(6f)**. Brownish crystalline solid, m.p. 142-143°C, yield = 71%, and ^{1}H-NMR (DMSO-d_6) δ (ppm) 0.93 (t, 3H, J = 7.3 Hz, $NHCH_2CH_2CH_2CH_3$), 1.38–1.35 (m, 2H, J = 7.6 Hz, $NHCH_2CH_2CH_2CH_3$), 1.57–1.54 (m, 2H, J = 7.3 Hz, $NHCH_2CH_2CH_2CH_3$) 2.53 (s, 3H, C-CH_3), 3.36 (q, 2H, J = 7.0 Hz, $NHCH_2CH_2CH_2CH_3$), 7.65–1.62 (m, 2H, naphthyl-H), 7.75–7.71 (m, 1H, naphthyl-H), 7.77 (d, 1H, J = 8.1 Hz, naphthyl-H), 7.82 (d, 1H, J = 7.5 Hz, naphthyl-H), 7.91 (d, 1H, J = 8.4 Hz, naphthyl-H), 8.02 (d, 1H, J = 8.1 Hz, naphthyl-H), 9.36 (br s, 1H, NHCO), and 15.72 (s, 1H, OH); ^{13}C-NMR (DMSO-d_6) δ (ppm) 14.1, 20.2, 26.5, 31.3, 38.3, 110.9, 119.8, 123.1, 125.1, 126.9, 127.1, 127.5, 128.2, 129.3, 134.2, and 136.9, 164.9, 198.8 (18C). ESI-MS: m/z, 312.17 (M^+ + H). Anal. calcd for $C_{18}H_{21}N_3O_2$ (311.38): C, 69.43; H, 6.80; and N, 13.49. Found: C, 69.35; H, 6.72; and N, 13.40.

(7) 2-(Naphthalen-1-yl-diazenyl)-N-sec-butyl-3-hydroxybut-2-enamide **(6g)**. Yellow crystalline solid, m.p. 105-106°C, yield = 55%, and ^{1}H-NMR ($CDCl_3$) δ (ppm) 1.01 (t, 3H, J = 7.4 Hz, $CH_3CH_2CH(NH)CH_3$), 1.28 (d, 3H, J = 6.6 Hz, $CH_3CH_2CH(NH)CH_3$), 1.66–1.60 (m, 2H, $CH_3CH_2CH(NH)CH_3$), 2.62 (s, 3H, C-CH_3), 4.12–4.08 (m, 1H, $CH_3CH_2CH(NH)CH_3$) 7.56–7.52 (m, 3H, naphthyl-H),

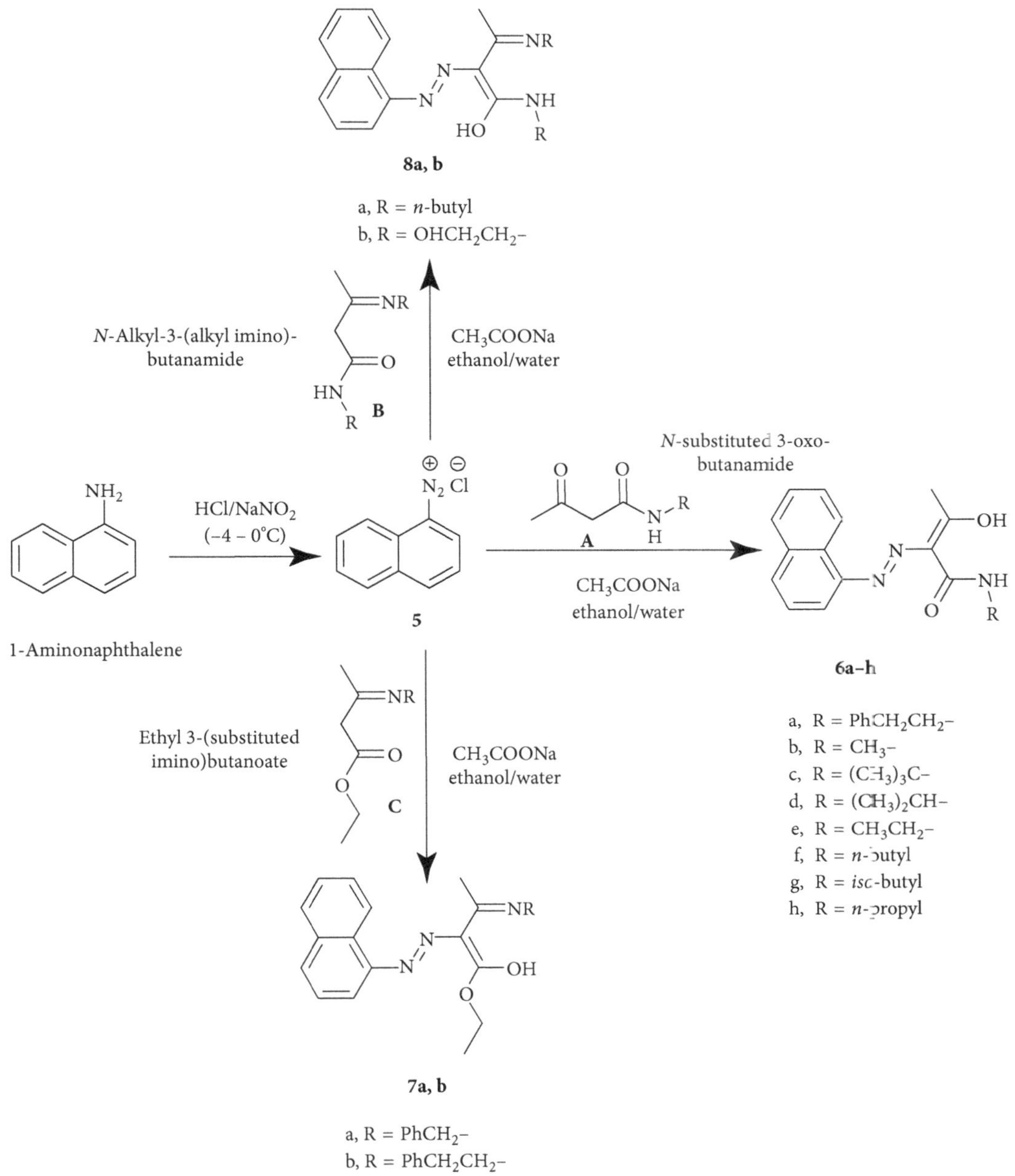

SCHEME 2: Synthetic route of compounds **6–8**.

7.68 (d, 1H, $J = 8.0$ Hz, naphthyl-H), 7.82 (d, 1H, $J = 8.0$ Hz, naphthyl-H), 7.90 (d, 1H, $J = 7.6$ Hz naphthyl-H), 8.07 (d, 1H, $J = 7.9$ Hz naphthyl-H), 9.40 (br s, 1H, NHCO), and 15.81 (s, 1H, OH); ^{13}C-NMR ($CDCl_3$) δ (ppm). 10.4, 20.3, 26.2, 29.4, 46.0, 110.7, 120.2, 123.5, 124.8, 126.1, 126.4, 126.7, 127.6, 128.6, 134.1, 137.2, 164.7, and 199.4 (18C). ESI-MS: *m/z*, 312.16 (M^+ + H). Anal. calcd for $C_{18}N_{21}N_3O_2$ (311.38): C, 69.43; H, 6.80; and N, 13.49. Found: C, 69.35; H, 6.72; and N, 13.40.

(8) 2-(Naphthalen-1-yl-diazenyl)-3-hydroxy-N-propylbut-2-enamide **(6h)**. Brownish crystalline solid, m.p. 157-158°C, yield = 65%, and ^{1}H-NMR ($CDCl_3$) δ (ppm) 1.04 (t, 3H, $J = 7.4$ Hz, $CONHCH_2CH_2CH_3$), 1.68–1.63 (m, 2H, $J = 7.3$ Hz, $CONHCH_2CH_2CH_3$), 2.62 (s, 3H, C-CH_3), 3.42 (q, 2H, $J = 7.0$ Hz, $CONHCH_2CH_2CH_3$), 7.57–7.53 (m, 3H, naphthyl-H), 7.68 (d, 1H, $J = 8.1$ Hz, naphthyl-H), 7.82 (d, 1H, $J = 7.5$ Hz, naphthyl-H), 7.90 (d, 1H, $J = 7.5$ Hz, naphthyl-H), 8.07 (br d, 1H, $J = 8.5$ Hz, naphthyl-H), 9.54 (br s, 1H, NHCO), and 15.79 (s, 1H, OH); ^{13}C-NMR ($CDCl_3$) δ (ppm) 11.6, 22.6, 26.2, 40.4, 110.7, 120.2, 123.5, 124.9, 126.1, 126.4, 126.7, 127.6, 128.6, 134.1, 137.1, 165.3, and 199.4 (17C). ESI-MS: *m/z*, 298.15 (M^+ + H). Anal. calcd for $C_{17}H_{19}N_3O_2$ (297.35): C, 68.67; H, 6.44; and N, 14.13. Found: C, 68.55; H, 6.36; and N, 14.04.

(9) Ethoxy 3-(benzylimino)-2-(naphthalen-1-yl-diazenyl) but-1-en-1-ol **(7a)**. Brownish crystalline solid, m.p. 110-111°C, yield = 77%, and ^{1}H-NMR (DMSO-d_6) δ (ppm) 1.32 (t, 3H, $J = 7.1$ Hz, OCH_2CH_3), 2.59 (s, 3H, C-CH_3), 4.27 (q, 2H, $J = 7.1$ Hz, OCH_2CH_3), 4.83 (s, 2H, $J = 3.4$ Hz, $ArCH_2N$), 7.50–4.46 (m, 9H, naphthyl-H and ArH), 7.65 (d, 1H, $J =$

TABLE 1: Inhibition percentages of different synthesized compounds against MCF-7, A549, and HT-29 cell line.

Compound	Inhibition percentage ± SD (%)		
	MCF-7	A549	HT-29
4a	23 ± 3	34 ± 4	0.0 ± 8
4b	4 ± 5	6 ± 3	0.0 ± 8
4c	35 ± 3	10 ± 8	0.0 ± 3
4d	10 ± 5	9 ± 8	12 ± 8
4e	10 ± 2	3 ± 9	0.0 ± 4
4f	7 ± 3	0.0 ± 8	0.0 ± 6
4g	13 ± 7	0.0 ± 6	13 ± 7
4h	10 ± 5	11 ± 5	8 ± 6
4i	16 ± 5	15 ± 7	13 ± 4
4j	11 ± 4	13 ± 6	9 ± 6
4k	13 ± 7	10 ± 5	6 ± 6
6a	12 ± 6	7 ± 4	0.0 ± 7
6b	15 ± 3	15 ± 4	6 ± 7
6c	8 ± 4	5 ± 5	0.0 ± 4
6d	11 ± 5	3 ± 6	9 ± 7
6e	6 ± 5	2 ± 8	0.0 ± 7
6f	20 ± 2	0.0 ± 7	0.0 ± 8
6g	1 ± 6	0.0 ± 8	0.0 ± 8
6h	10 ± 5	0.0 ± 6	2 ± 8
7a	11 ± 4	17 ± 2	11 ± 6
7b	10 ± 3	0.0 ± 6	39 ± 7
8a	22 ± 5	0.0 ± 6	0.0 ± 8
8b	2 ± 4	0.0 ± 3	0.0 ± 8

Data are represented as mean ± SD.

7.5 Hz, naphthyl-H), 7.72 (d, 1H, J = 8.1 Hz, naphthyl-H), 7.88 (d, 1H, J = 8.2 Hz, naphthyl-H), and 14.90 (s, 1H, OH); ^{13}C-NMR (DMSO-d$_6$) δ (ppm) 14.9, 17.2, 50.2, 60.3, 111.2, 121.9, 124.1, 126.1, 126.4, 126.4, 126.6, 127.2, 128.1, 128.5, 129.3, 129.5, 134.3, 136.9, 137.9, 166.7, and 198.7 (23C). ESI-MS: *m/z*, 374.19 (M^+ + H). Anal. calcd for $C_{23}H_{23}N_3O_2$ (373.45): C, 73.97; H, 6.21; and N, 11.25. Found: C, 73.85; H, 6.12; and N, 11.20.

(10) Ethoxy 2-(naphthalen-1-yl-diazenyl)-3-(phenethylimino)but-1-en-1-ol (**7b**). Brownish crystalline solid, m.p. 142-143°C, yield = 72%, and ^{1}H-NMR (DMSO-d$_6$) δ (ppm) 1.30 (t, 3H, J = 7.1 Hz, OCH_2CH_3), 2.48 (s, 3H, C-CH_3), 3.08 (t, 2H, J = 7.1 Hz, $ArCH_2CH_2N$), 3.89 (t, 2H, J = 6.9 Hz, $ArCH_2CH_2N$), 4.23 (q, 3H, J = 7.1 Hz, OCH_2CH_3), 7.20–7.15 (m, 1H, naphthyl-H), 7.28–7.24 (m, 2H, naphthyl-H), 7.32–7.28 (m, 2H, naphthyl-H), 7.58–7.55 (m, 4H, ArH), 7.78–7.74 (m, 1H, naphthyl-H), 7.96–7.92 (m, 1H, naphthyl-H), and 14.59 (s, 1H, OH); ^{13}C-NMR (DMSO-d$_6$) δ (ppm). 14.9, 16.5, 35.9, 46.4, 60.1, 111.9, 122.6, 125.1, 126.3, 126.5, 126.7, 126.9, 127.5, 128.6, 128.9, 129.2, 129.2, 134.4, 136.9, 138.9, 166.7, and 198.7 (24C). ESI-MS: *m/z*, 388.19 (M^+ + H). Anal. calcd for $C_{24}H_{25}N_3O_2$ (387.47): C, 74.39; H, 6.50; and N, 10.84. Found: C, 74.18; H, 6.42; and N, 10.76.

(11) 1-(Butylamino)-3-(butylimino)-2-[(1-(naphthalen-1-yl)diazene]but-1-en-1-ol (**8a**). Brownish crystalline solid, m. p. 91-92°C, yield = 71%, and ^{1}H-NMR (DMSO-d$_6$) δ (ppm). 0.93 (t, 3H, J = 6.4 Hz, $NHCH_2CH_2CH_2CH_3$), 0.96 (t, 3H, J = 7.3 Hz, $NCH_2CH_2CH_2CH_3$), 1.46–1.40 (m, 4H, J = 7.7 Hz, $NHCH_2CH_2CH_2CH_3$ and $NCH_2CH_2CH_2CH_3$), 1.61 (q, 2H, $NHCH_2CH_2CH_2CH_3$), 1.67–1.64 (m, 2H, $NCH_2CH_2CH_2CH_3$), 2.62 (s, 3H, C-CH_3), 3.39 (q, 2H, J = 6.4 Hz, $NHCH_2CH_2CH_2CH_3$), 3.65 (q, 2H, J = 6.8 Hz, $NCH_2CH_2CH_2CH_3$), 7.56–7.52 (m, 4H, naphthyl-H), 7.76 (d, 1H, J = 8.2 Hz, naphthyl-H), 7.95–7.90 (m, 1H, naphthyl-H), 8.4 (d, 1H, J = 7.4 Hz, naphthyl-H), 11.3 (br s, 1H, NH), and 15.37 (s, 1H, OH); ^{13}C-NMR (DMSO-d$_6$) δ (ppm) 14.1, 14.2, 14.7, 20.0, 20.4, 31.4, 31.7, 37.9, 44.0, 111.3, 121.3, 122.8, 126.0, 126.3, 126.8, 126.8, 128.6, 129.0, 134.5, 148.7, 165.9, and 172.7 (22C). ESI-MS: *m/z*, 367.25 (M^+ + H). Anal. calcd for $C_{22}H_{30}N_4O$ (366.50): C, 72.10; H, 8.25; and N, 15.29. Found: C, 72.00; H, 8.18; and N, 15.20.

(12) N-(2-hydroxyethyl imino)-3-(2-hydroxyethylamino)-2-(naphthalen-1-yldiazenyl)but-2-en-2-ol (**8b**). Brownish crystalline solid, m.p. 157-158°C, yield = 70%, and ^{1}H-NMR (DMSO-d$_6$) δ (ppm) 2.64 (s, 3H, C-CH_3), 3.47 (q, 3H, J = 5.3 Hz, $CONHCH_2CH_2OH$), 3.62–3.66 (m, 6H, $CONHCH_2CH_2OH$ and $CNCH_2CH_2OH$), 4.95 (t, 1H, J = 4.6 Hz, $CONHCH_2CH_2OH$), 5.08 (t, 1H, J = 4.4 Hz, $CNCH_2CH_2OH$), 7.54–7.50 (m, 4H, naphthyl-H), 7.74 (d, 1H, J = 7.8 Hz, naphthyl-H), 7.94–7.88 (m, 1H, naphthyl-H), 8.74–8.70 (m, 1H, naphthyl-H), 11.40 (br s, 1H, NH), and 15.37 (s, 1H, OH); ^{13}C-NMR (DMSO-d$_6$) δ (ppm).15.0, 41.1, 47.1, 60.0, 60.5, 111.1, 121.3, 123.9, 126.3, 126.4, 126.7, 126.8, 128.2, 129.2, 134.5, 148.8, 165.8, and 172.8 (18C). ESI-MS: *m/z*, 343.18 (M^+ + H). Anal. calcd for $C_{18}H_{22}N_4O_3$ (342.39): C, 63.14; H, 6.48; and N, 16.36. Found: C, 63.00; H, 6.40; and N, 16.30.

2.2. Cytotoxicity Assay. Three cancer cell lines were used in the current study: MCF-7 (human breast adenocarcinoma), A549 (lung carcinoma), and HT-29 (human colorectal adenocarcinoma). Cancer cell lines were obtained from Sigma-Aldrich Chemical Company, USA, and were seeded in Dulbecco's Modified Eagle Media (DMEM) supplemented with 10% heat-inactivated fetal bovine serum (FBS), 100 IU/mL penicillin, 100 mg/mL streptomycin, and 2 mM glutamine. The cells were maintained in a humidified atmosphere with 5% CO_2 at 37°C.

2.2.1. Antiproliferative Activity. The antiproliferative activity of the tested compounds was measured using the 3-(4,5-dimethylthiazol-2-yl)-2,5-diphenyltetrazolium bromide (MTT) assay as mentioned before [28–30]. In brief, cells were seeded in 96-well plates 24 h before different treatments. Test compounds were dissolved in dimethyl sulfoxide (DMSO) and further diluted with the cultivation medium. All target compounds (**1–23**) were preliminary evaluated in vitro at the single concentration of 5 μM against different cancer cell lines. Doxorubicin was used in parallel as a positive control drug at the single concentration of 2 μM. Different tested samples were added to the wells, and the cells were incubated at 37°C in a 5% CO_2-humidified incubator for 48 h. After treatment, 10 μL of the MTT solution (5 mg/mL) was added to each well, and the plates were further incubated for 2 h. The media along with the MTT solution was removed and replaced by 100 μL of the DMSO solution to dissolve the precipitating formazan crystals. The intensity of the developed color was measured at 570 nm with an automatic multiwell plate reader (Varioskan™ LUX multimode microplate reader, Thermo Fisher Scientific,

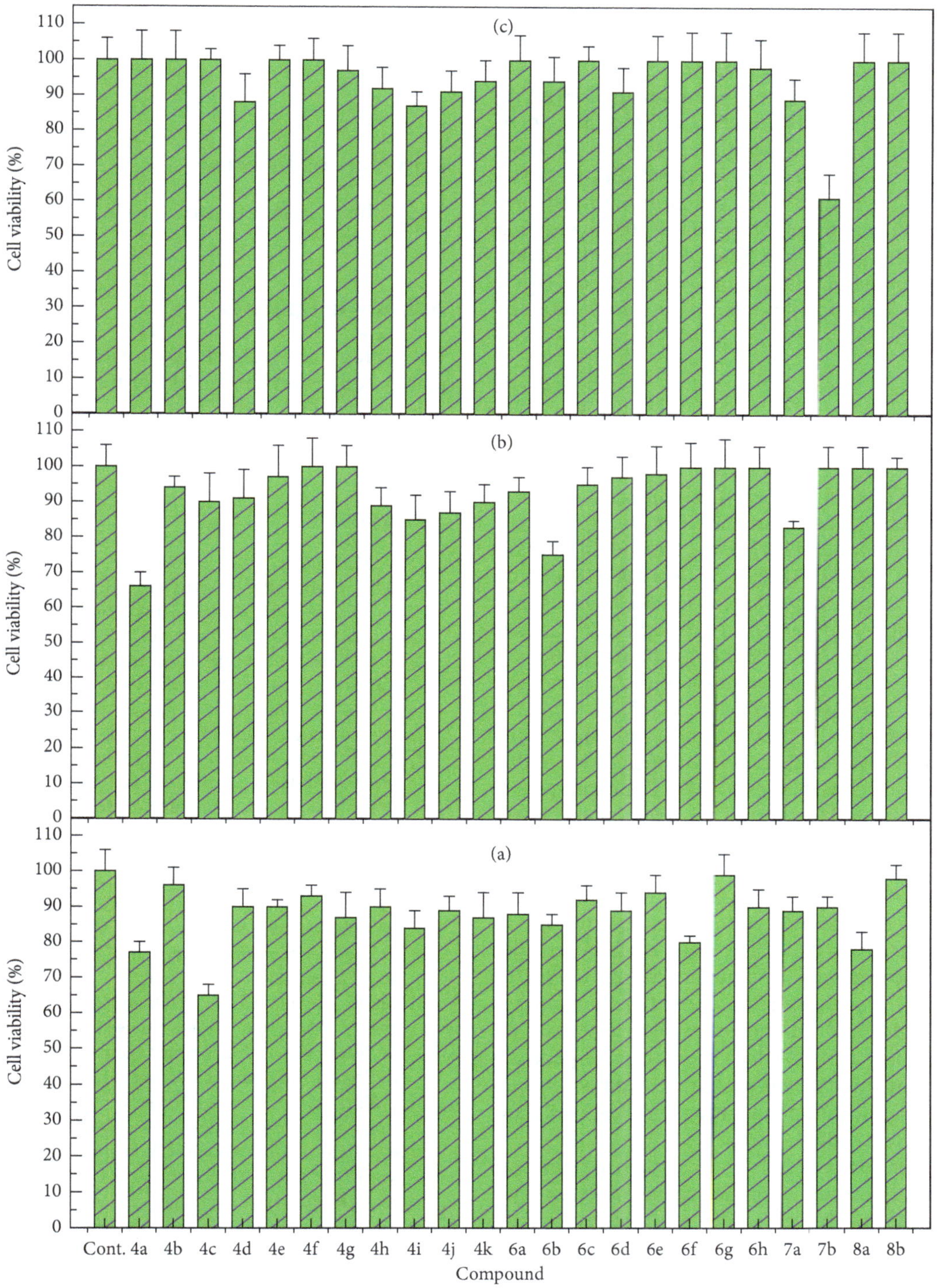

FIGURE 1: Effect of different compounds on the viability of MCF-7, A549 and HT-29 cell line. Data are represented as mean ± SD. (a) MCF-7; (b) A549; (c) HT-29.

USA). Percentage inhibition of proliferation was calculated as a fraction of the vehicle control.

3. Results and Discussion

3.1. Chemistry. In the present work, a series of methyl 2-hydroxy-5-4-oxo-2-(substituted amino)pent-2-en-3-yl) diazenyl)benzoate **4a–k** were synthesized via the diazotization of methyl 5-aminc-2-hydroxybenzoate **2** as the starting material. Nitrosation of methyl 5-amino-2-hydroxybenzoate **1** by using $NaNO_2$ in the presence of HCl afforded the corresponding diazonium salt derivative **2**. Treatment of **2** with 4-(substituted amino)pent-3-en-2-one derivatives **3** [31] in absolute ethanol with stirring at room temperature gave the corresponding methyl 2-hydroxy-5-(4-oxo-2-(substituted amino)pent-2-en-3-yl)diazenyl) benzoate **4a-k**, respectively (Scheme 1).

Additionally, nitrosation of 1-aminonaphthalene by using $NaNO_2$ in the presence of HCl afforded the corresponding diazonium salt of naphthalen-1-amine **5**. Treatment of diazonium compound **5** with *N*-substituted 3-oxo-butanamides **(A)** or ethyl 3-(substituted imino)butanoates **(B)** gave the corresponding azocompounds **6a–h** and **7a,b**, respectively. Finally, compound **5** was reacted with *N*-alkyl-3-(alkyl imino)butanamides **(C)** in the presence of sodium acetate to afford the corresponding compounds **8a,b**, respectively (Scheme 2).

3.2. Anticancer Activity. The results of the antiproliferative activity of the synthesized compounds (Table 1, Figure 1) clearly showed that the synthesized compounds greatly varied in their effects on the tested cell lines. All the evaluated 23 compounds showed variable effects on MCF-7 cells. On the contrary, compounds **4f, 6f, 6g, 8a,** and **8b** were not effective against A549 and HT-29 cells. For MCF-7 cells, the most effective compounds are **4c**, **4a,** and **8a**, whereas their toxicities were moderate on cells and reached 35, 23, and 22%, respectively. Compounds **6f, 4i, 6b, 4k, 6a, 4j, 6d,** and **7a** showed lower anticancer activities, and their toxicities obtained at the tested dose recorded 20, 16, 15, 13, 12, 11, 11, and 11%, respectively. The rest of the synthesized compounds showed weak anticancer activities, where they maximally reduced cell viabilities by 10%. Concerning A549 cells, compounds **4a** and **6b** showed a maximal potent effect, and their percentages of inhibition of cancer cells were obtained as 34 and 25%, respectively. On the contrary, compounds **7a, 4i, 4j,** and **4h** showed lower anticancer activities, and their maximal cytotoxicities reached 17, 15, 13, and 11%, respectively. For HT-29 cells, it can be seen that the most potent compound (compound **7b**) against HT-29 cells reduced cell growth by about 39%. On the contrary, compounds **4f, 4d,** and **7a** showed weaker cytotoxic activities ranging only from 13 to 11%. It can be generally concluded that the most potent synthesized compounds are **7b** against HT-29 cells, compound **4a** against A549 cells, and compound **4c** against MCF-7 cells.

The abovementioned results showed that different cell lines varied greatly in their response against different synthesized compounds. This correlates well with previously reported results [32, 33], where this can be attributed to the inherent differences between different cells in their specific membrane structure and organization. For the synthesized derivatives, our results are in good agreement with those previously reported by Jakopec et al. [34], who reported on the cytotoxic activities of diazenes on several tumor cell lines. They attributed their results to both (1) the attachment of unsubstituted benzene on the amide nitrogen and (2) the extent of compound basicity. Moreover, the length of the conjugation plays a crucial role in the activity of the synthesized derivatives [34].

Generally, it can be observed that the obtained anticancer activities of the newly synthesized diazene candidates are not promising as expected; therefore, the future work is planned to investigate further structural modifications of the prepared derivatives.

Conflicts of Interest

The authors declare that they have no conflicts of interest.

Acknowledgments

The authors are grateful to the Deanship of Scientific Research, King Saud University, for funding this work through research group project "RGP-1435-047."

References

[1] N. M. Khalifa, M. A. Al-Omar, A. E. Amr, and M. E. Haiba, "HIV-1 and HSV-1 virus activities of some new polycyclic nucleoside pyrene candidates," *International Journal of Biological Macromoleculesd*, vol. 54, pp. 51–56, 2013.

[2] A. E. Amr, N. A. Abdel-Latif, and M. M. Abdalla, "Synthesis of some new testosterone derivatives fused with substituted pyrazoline ring as 5α-reductase inhibitors," *Acta Pharmaceutica (Zagreb, Croatia)*, vol. 56, no. 2, pp. 203–218, 2006.

[3] B. F. Abdel Wahab, S. F. Mohamed, A. E. Amr, and M. M. Abdalla, "Synthesis and reactions of some newly synthesized thiosemicarbazides, triazoles and Schiff bases as antihypertensive α-blocking agents," *Monatshefte für Chemie - Chemical Monthly*, vol. 139, no. 9, pp. 1083–1090, 2008.

[4] N. O. Al-Harbi, S. A. Bahashwan, A. A. Fayed, M. S. Aboonq, and A. E. Amr, "Anti-parkinsonism, hypoglycemic and antimicrobial activities of some new poly ring heterocyclic candidates," *International Journal of Biological Macromolecules*, vol. 57, pp. 165–173, 2013.

[5] M. M. Abdalla, M. A. Al-Omar, R. A. Al-Salahi, A. E. Amr, and N. M. Sabry, "A new investigation for some steroidal derivatives as anti-alzheimer agents," *International Journal of Biological Macromolecules*, vol. 51, no. 2, pp. 56–63, 2012.

[6] L. Yurttaş, Y. Özkay, H. K. Gençer, and U. Acar, "Synthesis of some new thiazole derivatives and their biological activity evaluation," *Journal of Chemistry*, vol. 2015, Article ID 464379, 7 pages, 2015.

[7] K. A. Khan, H. M. Faidallah, and A. M. Asiri, "Synthesis and biological evaluation of new N,N′-Bis(1-substituted-ethylidene)-ethane1,2-diamine and their acetyl and trifluoroacetyl derivatives as cytotoxic and antimicrobial agents," *Journal of Chemistry*, vol. 2013, Article ID 478635, 9 pages, 2013.

[8] Z. A. Kaplancıklı, M. D. Altıntop, B. Sever, Z. Cantürk, and A. Özdemir, "Synthesis and in vitro evaluation of new thiosemicarbazone derivatives as potential antimicrobial agents," *Journal of Chemistry*, vol. 2016, Article ID 1692540, 7 pages, 2016.

[9] A. E. Amr and M. M. Abdulla, "Synthesis and anti-inflammatory activities of new cynopyrane derivatives fused with steroidal nuclei," *Archiv der Pharmazie*, vol. 339, no. 2, pp. 88–95, 2006.

[10] L. Pan, N. Hang, C. Zhang et al., "Synthesis and biological evaluation of novel benzimidazole derivatives and analogs targeting the NLRP3 inflammasome," *Molecules*, vol. 22, no. 2, p. 213, 2017.

[11] C. Holohan, S. Van Schaeybroeck, D. B. Longley, and P. G. Johnston, "Cancer drug resistance: an evolving paradigm," *Nature Reviews Cancer*, vol. 13, no. 10, pp. 714–726, 2013.

[12] D. M. Marmion, *Hand Book of Colorant*, Wiley, Hoboken, NJ, USA, 1999.

[13] S. Piotto, S. Concilio, L. Sessa et al., "Synthesis and antimicrobial studies of new antibacterial azo-compounds active against *Staphylococcus aureus* and *Listeria monocytogenes*," *Molecules*, vol. 22, p. 1372, 2017.

[14] S. Concilio, P. Iannelli, L. Sessa et al., "Biodegradable antimicrobial films based on poly (lactic acid) matrices and active azo compounds," *Journal of Applied Polymer Science*, vol. 132, no. 33, article 42357, 2015.

[15] S. Piotto, S. Concilio, L. Sessa et al., "Novel antimicrobial polymer films active against bacteria and fungi," *Polymer Composites*, vol. 34, no. 9, pp. 1489–1492, 2013.

[16] S. Piotto, S. Concilio, L. Sessa et al., "Small azobenzene derivatives active against bacteria and fungi," *European Journal of Medicinal Chemistry*, vol. 68, pp. 178–184, 2013.

[17] L. Sessa, S. Concilio, P. Iannelli, F. De Santis, A. Porta, and S. Piotto, "Antimicrobial azobenzene compounds and their potential use in biomaterials," in *Proceedings of the AIP Conference*, pp. 16–19, Oudeniz, Turkey, April 2015.

[18] I. M. Awad, A. A. Aly, A. M. Abdel Alim, R. A. Abdel, and S. H. Ahmed, "Synthesis of some 5-azo(4'-substituted benzene-sulphamoyl)-8-hydroxyquinolines with antidotal and antibacterial activities," *Journal of Inorganic Biochemistry*, vol. 33, no. 2, pp. 77–89, 1998.

[19] S. A. Ibrahim, M. A. Gahami, Z. A. Khafagi, and S. A. Gyar, "Structure and antimicrobial activity of some new azopyrazolone chelates of Ni(II) and Cu(II) acetates, sulfates, and nitrates," *Journal of Inorganic Biochemistry*, vol. 43, no. 1, pp. 1–7, 1991.

[20] A. A. Jarahpour, M. Motamedifar, K. Pakshir, N. Hadi, and Z. Zarei, "Synthesis of novel azo Schiff bases and their antibacterial and antifungal activities," *Molecules*, vol. 9, no. 10, pp. 815–824, 2004.

[21] K. M. Rathod and N. S. Thakre, "Synthesis and antimicrobial activity of azo compounds containing m-cresol moiety," *Chemical Science Transactions*, vol. 2, no. 1, pp. 25–28, 2013.

[22] S. Samadhiya and H. Halve, "Synthetic utility of schiff bases as potential herbicidal agents," *Oriental Journal of Chemistry*, vol. 17, pp. 119–122, 2001.

[23] P. Mao, Z. Liu, M. Xie et al., "Naturally occurring methyl salicylate glycosides," *Mini-Reviews in Medicinal Chemistry*, vol. 14, no. 1, pp. 56–63, 2014.

[24] W. Xin, C. Huang, X. Zhang et al., "Methyl salicylate lactoside inhibits inflammatory response of fibroblast-like synoviocytes and joint destruction in collagen-induced arthritis in mice," *British Journal of Pharmacology*, vol. 171, no. 14, pp. 3526–3538, 2014.

[25] X. Zhang, J. Sun, W. Xin et al., "Anti-inflammation effect of methyl salicylate 2-O-β-D-lactoside on adjuvant induced-arthritis rats and lipopolysaccharide (LPS)-treated murine macrophages RAW264.7 cells," *International Immunopharmacology*, vol. 25, no. 1, pp. 88–95, 2015.

[26] A. L. Walpole and M. H. C. Williams, "Aromatic amines as carcinogens in industry," *British Medical Bulletin*, vol. 14, no. 2, pp. 141–145, 1958.

[27] R. Sirgamalla, A. Kommakula, S. Banoth et al., "Synthesis of amides from aliphatic acids and amines by using of I2/TBHP at room temperature," *Chemistry Select*, vol. 3, no. 4, pp. 1062–1065, 2018.

[28] H. A. Omar, S. A. Arafa, I. A. Maghrabi, and J. R. Weng, "Sensitization of hepatocellular carcinoma cells to Apo2L/TRAIL by a novel Akt/NF-kappaB signalling inhibitor," *Basic & Clinical Pharmacology & Toxicology*, vol. 114, no. 6, pp. 464–471, 2014.

[29] H. A. Omar, W. R. Mohamed, S. A. Arafa et al., "Hesperidin alleviates cisplatininduced hepatotoxicity in rats without inhibiting its antitumor activity," *Pharmacological Reports*, vol. 68, no. 2, pp. 349–356, 2016.

[30] J. J. Wu, H. A. Omar, Y. R. Lee et al., "6-Shogaol induces cell cycle arrest and apoptosis in human hepatoma cells through pleiotropic mechanisms," *European Journal of Pharmacology*, vol. 762, pp. 449–458, 2015.

[31] A. Dib, "Synthesis and characterization of schiff bases derived from acetylacetone and their theoretical study," *International Journal of ChemTech Research*, vol. 5, no. 1, pp. 204–211, 2013.

[32] E. A. Elsayed, M. A. Sharaf-Eldin, and M. Wadaan, "*In vitro* evaluation of cytotoxic activities of essential oil from *Moringa oleifera* seeds on HeLa, HepG2, MCF-7, CACO-2 and L929 cell lines," *Asian Pacific Journal of Cancer Prevention*, vol. 16, no. 11, pp. 4671–4675, 2015

[33] E. A. Elsayed, M. Farooq, D. Dailin et al., "*In vitro* and *in vivo* biological screening of kefiran polysaccharide produced by *Lactobacillus kefiranofaciens*," *Biomedical Research*, vol. 28, no. 2, pp. 594–600, 2017.

[34] S. Jakopec, K. Dubravcic, A. Brozovic, S. Polanc, and M. Osmak, "Structurally similar diazenes exhibit significantly different biological activity," *Cell Biology and Toxicology*, vol. 22, no. 1, pp. 61–71, 2006.

14

Synthesis of Kojic Ester Derivatives as Potential Antibacterial Agent

Carolynne Zie Wei Sie, Zainab Ngaini, Nurashikin Suhaili, and Eswaran Madiahlagan

Faculty of Resource Science and Technology, Universiti Malaysia Sarawak, 94300 Kota Samarahan, Sarawak, Malaysia

Correspondence should be addressed to Zainab Ngaini; nzainab@unimas.my

Academic Editor: Gabriel Navarrete-Vazquez

The search for lead product with beneficial pharmacological properties has become a great challenge and costly. Extraction and synthetic modification of bioactive compounds from natural resources has gained great attention and is cost effective. In this study, kojic acid was produced from fungal fermentation, using sago waste as substrate, and chemically incorporated with chalcones and azobenzene to form a series of kojic ester derivatives and evaluated for antibacterial activities. Kojic ester bearing halogenated chalcone demonstrated active inhibition against *Staphylococcus aureus* compared to that of standard ampicillin. The inhibition increased as the electronegativity of halogens decreased, while incorporation of azobenzene derivatives on kojic acid backbone demonstrated fair antibacterial activity against *Escherichia coli* with minimum inhibitory concentration (MIC) of 190–330 ppm. The presence of C=C and N=N reactive moieties in both chalcone and azo molecules contributed to the potential biological activities of the kojic acid ester.

1. Introduction

Kojic acid is a natural pyrone produced from fungal fermentation from various types of carbon substrates in the presence of fungi [1, 2]. Kojic acid and its derivatives have drawn attention and were studied intensively due to its significant biological activities in medicine and pharmacological field such as antifungal, antibiotic [3, 4], anti-inflammatory and analgesic [5], and antibacterial [6] properties.

Many recent studies reported on chemical modification of kojic acid for various applications such as antibacterial activity [7] and dye sensitized solar cell (DSSC) [8]. Synthesis of bioactive molecules employing kojic acid as a precursor via esterification of available hydroxyl groups has been widely reported to overcome the drawbacks of its hydrophilic properties in various applications [9–12]. Esterification of kojic acid, for instance, is able to enhance the hydrophobic properties which in turn improved its performance as tyrosinase inhibitor and antioxidant [13, 14].

Incorporation of kojic acid into other biological active compounds has also been reported with enhanced biological activities [15, 16]. Chalcones are an example of biological active compound which is derived from natural sources such as fruits, vegetables, spices, tea, and soy-based foodstuff [17, 18] and used as an intermediate precursor of flavonoids and isoflavonoids [19]. Chalcone is also commonly prepared via Claisen Schmidt condensation. The synthesis of chalcone has drawn much attentions due to the presence of its α,β-unsaturated ketone moieties which contributes to various biological activities such as antimalaria [20], anticancer [21], antiprotozoal, anti-inflammatory, antibacterial, antifilarial, anticonvulsant [22], antifungal, insect antifeedant, antimutagenic [23], and antioxidant [24, 25]. The presence of halogen substituents particularly at the *para* position of chalcone backbone has been reported with excellent biological activities such as antibacterial [26] and anti-inflammatory [27], while azobenzene bearing aryl/alkyl and N=N reactive moieties have been extensively reported due to its excellent antibacterial [28, 29], antioxidant [30], anti-inflammatory [31], antifungal [32], and antitumor [33] properties. Azobenzene is also recognized for its medicinal properties such as antineoplastic [34], antidiabetic [35], and antiseptic [36].

Herein, we report on the isolation of kojic acid from sago waste and utilized it as a precursor for the synthesis of kojic acid ester bearing chalcone **3a–c** and azobenzene **5a–d** moieties. These compounds were screened for potential antibacterial activities against *E. coli* and *S. aureus*, and the effect of different moieties towards biological activities of kojic acid was studied.

2. Experimental

All the reagents were used as received without any further purification. Melting points were determined in open capillary tubes and are uncorrected. The IR spectra were recorded with KBr on Perkin Elmer 1605 FTIR spectrophotometer. The ^{1}H NMR were recorded in $CDCl_3$ or DMSO-d_6 on JEOL ECA 500 spectrometer at 500 MHz using TMS as internal standard. The elemental carbon hydrogen nitrogen sulphur (CHNS) elemental analyses were performed using Thermo Scientific™ FLASH 2000 CHNS/O Analyzer.

*2.1. Fermentation of Kojic Acid (**1**).* Sago waste was used as the substrate for kojic acid fermentation. Prior to fermentation, sago waste (5 g) was dried, ground, and sieved. The medium was supplemented with 3% (w/v) urea and 10% (w/v) mineral salts solution containing KH_2PO_4, (w/v) $MgSO_4{\cdot}7H_2O$, and yeast extract. Mixed strains of *Aspergillus flavus NSH9* and *A. flavus Link 44-1* were added for fermentation process. The culture was incubated at 30 ± 2°C in static condition for 20 days. The slurry suspension culture (40 mL) was extracted with ethyl acetate (2 × 15 mL). The organic layer was evaporated in vacuo to form crude brown solid and recrystallized from ethanol to afford **1** (0.15 g, 30%) as a yellowish needle like solid; m.p. 149-150°C [Reported 151–154°C [37]]; ν_{max} (KBr/cm^{-1}) 3173 (OH), 2919 (C=C), 1661 (C=O), 1228 (C-O); ^{1}H NMR (500 MHz, DMSO-d_6): δ 9.04 (1H, s, OH-5), 8.02 (s, 1H, H-6), 6.34 (1H, s, H-3), 5.67 (1H, s, OH-7), 4.29 (1H, s, H-7); ^{13}C NMR (125 MHz, DMSO-d_6): δ 174.4 (C-4), 168.6 (C-2), 146.2 (C-5), 139.8 (C-6), 110.4 (C-3), 60.0 (C-7).

*2.2. General Method for Synthesis of 4-Carboxyl Chalcone (**2a–c**).* Acetophenone (1.2 g, 0.1 mmol) was added to 4-carboxybenzaldehyde derivatives (0.15 g, 0.1 mmol) in methanol (50 mL). KOH (1.7 g, 0.3 mmol) in methanol was added and the mixture was stirred at room temperature for 4 h under nitrogen atmosphere. The precipitate formed was filtered and recrystalized from ethanol to afford title compound.

*4-[(E)-3-Oxo-3-phenyl-prop-1-enyl]benzoic Acid (**2a**).* Compound **2a** (2.32 g, 92%) as a yellowish powder. m.p. 223-224°C; (Found: C, 76.61; H, 4.37%. $C_{16}H_{12}O_3$ Requires C, 76.2; H, 4.8%); ν_{max} (KBr/cm^{-1}) 3059 (C=CH), 1677 (C=O), 1608 (COO), 1578 (aromatic); ^{1}H NMR (500 MHz, DMSO-d_6): δ 8.17 (2H, d, J = 7.45 Hz, H-3), 8.05 (1H, d, J = 16.05 Hz, H-β), 8.02 (2H, d, J = 8.06 Hz, H-8), 7.99 (2H, d, J = 8.00 Hz, H-2), 7.78 (1H, d, J = 15.45 Hz, H-α); 7.69 (1H, t, J = 7.73 Hz, H-10), 7.59 (2H, t, J = 7.75 Hz, H-9).

*4-[(E)-3-(4-Bromophenyl)-3-oxo-prop-1-enyl]benzoic Acid (**2b**).* Compound **2b** (2.65 g, 88%) as a yellowish powder; m.p. 246-247°C; (Found: C, 58.44; H, 3.65%. $C_{16}H_{11}BrO_3$ Requires C, 58.0; H, 3.3%); ν_{max} (KBr/cm^{-1}) 3064 (C=CH), 1663 (C=O), 1607 (COO), 1583 (aromatic); ^{1}H NMR (500 MHz, DMSO-d_6): δ 8.13 (2H, d, J = 8.00 Hz, H-8), 8.05 (1H, d, J = 15.45 Hz, H-β), 8.02 (2H, d, J = 8.00 Hz, H-3), 7.98 (2H, d, J = 8.60 Hz, H-9), 7.79 (1H, d, J = 15.45 Hz, H-α), 7.79 (2H, d, J = 8.00 Hz, H-2).

*4-[(E)-3-(4-Chlorophenyl)-3-oxo-prop-1-enyl]benzoic Acid (**2c**).* Compound **2c** (2.12 g, 74%) as a yellowish powder. m.p. 259-260°C; (Found: C, 67.09; H, 3.27%. $C_{16}H_{11}ClO_3$ Requires C, 67.0; H, 3.9%); ν_{max} (KBr/cm^{-1}) 2943 (C=CH), 1678 (C=O), 1612 (COO), 1577 (aromatic); ^{1}H NMR (500 MHz, DMSO-d_6): δ 8.20 (2H, d, J = 8.40 Hz, H-8), 8.00 (1H, d, J = 16.05 Hz, H-β), 7.93 (2H, d, J = 8.40 Hz, H-3), 7.90 (2H, d, J = 8.40 Hz, H-9), 7.78 (1H, d, J = 16.05 Hz, H-α), 7.65 (2H, d, J = 7.65 Hz, H-2).

*2.3. General Method for Synthesis of Kojic Acid-Chalcone Derivatives (**3a–c**).* Chalcone **2a–c** (1.26 g, 0.05 mmol) was added to a solution of 1 (0.71 g, 0.05 mmol), DMAP (0.12 g, 1 mmol), and DCC (1.03 g, 0.05 mmol) in DCM (30 mL). The reaction mixture was stirred at 0°C for 4 h, washed with acetic acid (1 M, 2 × 20 mL), followed by Na_2CO_3 solution (1 M, 2 × 20 mL), and dried over magnesium sulphate. The crude product obtained was purified by column chromatography over silica gel (eluting with 1 : 2 EA/Hex) to afford title compound.

*[6-(Hydroxymethyl)-4-oxo-pyran-3-yl] 4-[(E)-3-oxo-3-phenyl-prop-1-enyl] Benzoate (**3a**).* Compound **3a** (0.4 g, 23%) as a white powder. m.p. 162-163°C; (Found: C, 69.56; H, 4.25%. $C_{22}H_{16}O_6$ Requires C, 70.2; H, 4.3%); ν_{max} (KBr/cm^{-1}) 2987 (C=CH), 1730 (C=O ester), 1665 (C=O), 1644 (C=O), 1605 (aromatic), 1201 (C-O); ^{1}H NMR (500 MHz, DMSO-d_6): δ 8.69 (1H, s, H-6), 8.11 (7H, m, H-9, H-10, H-15, H-β), 7.79 (2H, d, J = 16.05 Hz, H-α), 7.67 (1H, m, H-17), 7.59 (2H, m, H-16), 6.65 (1H, s, H-3), 5.02 (2H, s, H-7).

*[6-(Hydroxymethyl)-4-oxo-pyran-3-yl]4-[(E)-3-(4-bromophenyl)-3-oxo-prop-1-enyl] Benzoate (**3b**).* Compound **3b** (0.4 g, 19%) as a white powder. m.p. 194-195°C; (Found: C, 58.37; H, 3.24%. $C_{22}H_{15}BrO_6$ Requires C, 58.0; H, 3.3%); ν_{max} (KBr/cm^{-1}) 2926 (C=CH), 1697 (C=O ester), 1651 (C=O), 1623 (C=O), 1570 (aromatic), 1242 (C-O); ^{1}H NMR (500 MHz, DMSO-d_6): δ 8.67 (1H, s, H-6), 8.13 (2H, d, J = 8.6 Hz, H-15), 8.12 (2H, d, J 8.00 Hz, H-10), 8.04 (1H, d, J = 15.45 Hz, H-β), 7.99 (2H, d, J = 8.00 Hz, H-16), 8.00 (1H, d, J = 12.6 Hz, H-α), 7.80 (2H, d, J = 8.55 Hz, H-9), 6.50 (1H, s, H-3), 4.40 (2H, s, H-7).

*[6-(Hydroxymethyl)-4-oxo-pyran-3-yl]4-[(E)-3-(4-chlorophenyl)-3-oxo-prop-1-enyl]benzoate (**3c**).* Compound **3c** (0.5 g, 25%) as a white powder. m.p. 202-203°C; (Found: C, 63.54; H, 3.31%. $C_{22}H_{15}ClO_6$ Requires C, 64.3; H, 3.7%); ν_{max} (KBr/cm^{-1}) 2927 (C=CH), 1731 (C=O ester), 1666 (C=O), 1643 (C=O), 1607 (aromatic), 1239 (C-O); ^{1}H NMR (500 MHz, DMSO-d_6): δ 8.67 (1H, s, H-6), 8.13 (4H, t, J = 7.73 Hz, H-15, H-10), 8.04 (1H, d, J = 14.9 Hz, H-β), 7.99 (2H,

d, $J = 8.0$ Hz, H-16), 7.86 (1H, d, $J = 11.45$ Hz, H-α), 7.79 (2H, d, $J = 7.45$ Hz, H-9), 6.50 (1H, s, H-3), 4.39 (2H, s, H-7).

*2.4. Synthesis of Azobenzene Derivatives (**4a–d**).* HCl (8 M, 100 mL) was added drop wise to a solution of 4-aminoethylbenzoate (5 g, 30.0 mmol) in methanol (50 mL) in an ice bath at 0–5°C. Sodium nitrite (3.1 g, 45.0 mmol) in distilled water (10 mL) and phenol (3.4 g, 36 mmol) in methanol (50 mL) were added drop wise to the reaction mixture. NaOH solution (1 M, 100 mL) was then added drop wise to pH 8.5–9.5. The reaction mixture was stirred for 4 hr under nitrogen atmosphere. Methanol (50 mL) was added and acidified with HCl (8 M, 50 mL) to form precipitate. The solid was filtered, washed, and recrystallized from ethanol. KOH (0.13 g, 2.4 mmol) was added to the solid (0.2 g, 0.8 mmol) in methanol (50 mL) and heated under reflux for 4 h under nitrogen atmosphere. Cold water (30 mL) was added to the reaction mixture, acidified with acetic acid (8 M), and filtered and recrystallized from ethanol to give azobenzene intermediate. Azobenzene (1.5 g, 5 mmol) was added to a solution of bromoalkene derivatives, t- BuOK (0.62 g, 5.0 mmol) and KI (20 mg) in acetone (100 mL), and heated at reflux for 48 h under nitrogen atmosphere. DCM (50 mL) and water (30 mL) were added and the layers were separated. The aqueous layer was extracted with DCM (2 × 30 mL) and the combined organic layers were dried and evaporated under reduced pressure and recrystallized from ethanol to give title compound **4a–d** as colorless crystals. FTIR and NMR data were consistent with the reported literature [8].

*2.5. Synthesis of Kojic Acid-Azobenzene (**5a–d**).* Thionyl chloride (0.036 g, 0.3 mmol) was added to a solution of **1** (0.014 g, 0.1 mmol) in DMF (20 mL) and heated under reflux for 4 h. The reaction mixture was extracted with DCM (2 × 15 mL). The organic layer was dried, filtered, and concentrated under reduced pressure to give kojyl chloride as yellowish needle solid (0.012 g, 70%). Azo intermediate was added to a solution of kojyl chloride prepared (0.1 g, 0.6 mmol) in DMF (30 mL) with TEA (0.015 g, 0.15 mmol) in dried acetone (20 mL) and heated under refluxed for 24 h. The reaction mixture was washed with dilute HCl (1 M, 2 × 20 mL). The organic layer was dried, filtered, and concentrated under reduced pressure. Crude product was recrystallized from ethanol to afford **5b–e** as colorless crystals. FTIR and NMR data were consistent with the reported literature [8].

2.6. Antibacterial Screening

2.6.1. Turbidimetric Kinetic Method. The antibacterial activities of synthesized **4a–d** and **5a–d** were screened against *Escherichia coli* using turbidimetric kinetic method. *Escherichia coli* were cultured on LB plate agar for 24 h at 37°C and a single colony was transferred to the media containing nutrient broth prepared. The inoculum was allowed to grow at 37°C with 250 rpm stirring overnight. Inoculum (0.2 mL) was added to 10 mL of culture medium which had been added with different concentrations (50 ppm, 80 ppm, and 100 ppm) of respective compounds in DMSO. Three replicates of each concentration were prepared. The control used was the medium broth of inoculums added with DMSO with no active compound. The mixture was shaken at 230 rpm at 37°C. The transmittances (T) of the aliquots were taken every 1 h interval for six continuous hours using UV-Visible Spectrophotometer Optima SP-300. The antibacterial activity was determined from the graph of In Nt versus time plotted. The In Nt values represent the number of colony forming units (units/mL) which follows the equation of In Nt = 27.1 – 8.56 T [38, 39]

2.6.2. Via Kirby Bauer Disc Diffusion Method. The antibacterial activities of the synthesized **2a–c** and **3a–c** were evaluated against *Escherichia coli* and *Staphylococcus aureus* using Kirby Bauer Disc Diffusion method. *E. coli* and *S. aureus* were used as inoculums and cultured in Mueller-Hinton Broth (MHB) and incubated at 37.5°C with constant shaking at 150 rpm for 18 h. The bacteria suspension prepared was inoculated onto the entire surface of a Mueller-Hinton Agar (MHA) plate with sterilised cotton tipped swab to form an even lawn. Sterilised filter paper disc impregnated with 10 μL of the compound in DMSO was placed on the bacteria surface of MHA plate using a sterilised forcep. The plates were incubated at 37.5°C for 24 h. The zone of inhibition was measured in millimetres (mm) to estimate the potency of the tested compound.

3. Results and Discussion

3.1. Preparation of Kojic Acid via Fermentation. Kojic acid **1** was prepared via fermentation of sago hampas employing mixed strains of *Aspergillus flavus NSH9* and *Aspergillus flavus Link 44-1. Aspergillus flavus* was used for a higher yield of **1**[15]. The optimum parameters were from a combination of 3% (w/v) urea and 10% mineral salts solution containing KH_2PO_4, (w/v) $MgSO_4{\cdot}7H_2O$ and yeast. KH_2PO_4 and $MgSO_4$ supported the growth of bacteria strain and boosted the production of kojic acid [40]. Yeast is a nitrogen source for metabolic activation to enhance kojic acid production. Its high levels of essential components such as vitamins and oligoelements support growth of bacteria and fermentation process [3].

3.2. Synthesis of Kojic Acid Ester Derivatives. Kojic ester **2a–c** was prepared via Claisen Schmidt condensation followed by Steglich esterification with isolated kojic acid to yield **3a–c**, whereas kojic ester **(5a–d)** was prepared via kojyl chloride [11] followed by addition of **4a–d** to afford **5a–d** (70–87%). The overall synthesis is shown in Scheme 1. Structures of the compounds synthesized were confirmed by elemental analysis, IR spectra, and ^{1}H NMR spectra.

The IR spectra for **2a–c** showed absorptions band at 3064–2943 cm^{-1} and 1678–1663 cm^{-1} attributed to C=C and C=O group of chalcone linkage. A sharp band at 1612–1607 cm^{-1} indicated the presence of carboxyl group. Absorption peaks at 1583–1577 cm^{-1} revealed the presence of aromatic group. The ^{1}H NMR showed aromatic protons of **2a** at δ 8.17–7.59 ppm as doublet and triplet while aromatic protons of **2b-c** appeared at δ 8.20–7.65 ppm as four doublets with coupling constant 8–8.6 Hz. The peaks of $CH_\alpha = CH_\beta$ were observed at δ 7.79–7.78 and 8.05–8.00 ppm as

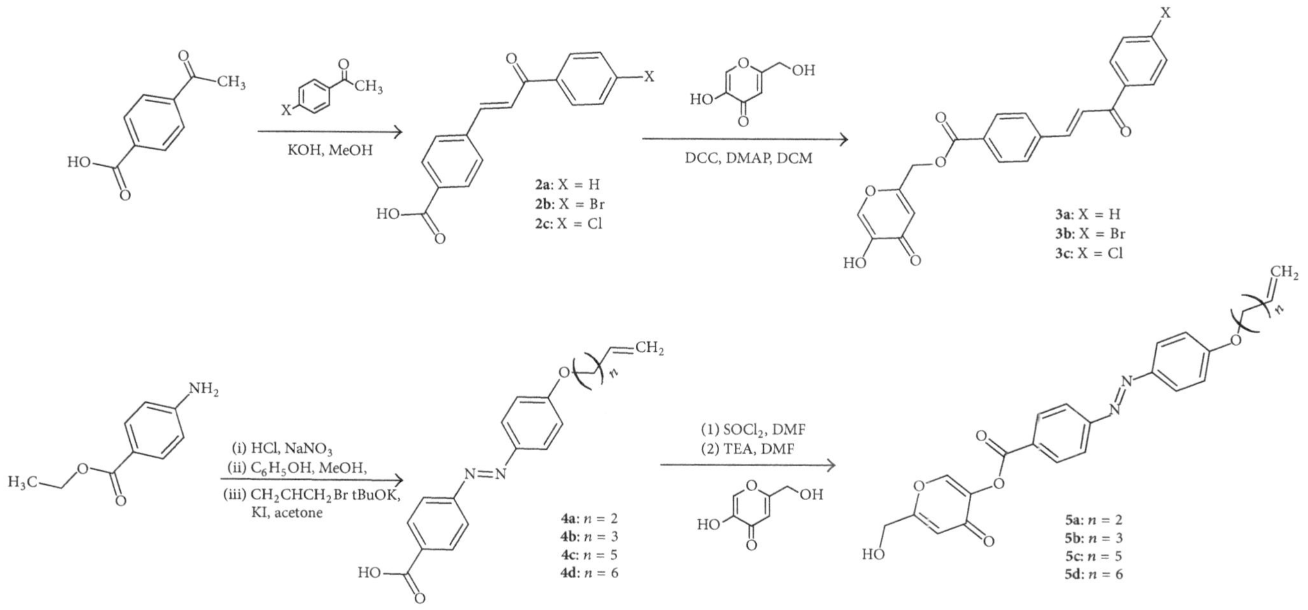

SCHEME 1: Synthesis of kojic esters (**3a–c**) and (**5a–d**).

two doublets with J_{ab} 15.5–16.1 Hz, which indicated *trans* configuration.

Formation of **3a–c** was confirmed by FTIR and ^{1}H NMR spectroscopies. The IR spectrum for **3a–c** showed appearance of C=O ester absorption band at 1731–1697 cm^{-1}. The IR spectrum revealed two strong bands at 1666–1651 cm^{-1} and 1644–1623 cm^{-1} attributed to ν(C=O) of chalcone and kojic acid, respectively. The absorption frequency at 1605–1570 cm^{-1} was attributed to aromatic group while ν(C-O) was represented by absorption band at 1242–1201 cm^{-1}. The ^{1}H NMR showed that all corresponding peaks indicated the formation of **3a–c**. The existence of kojic acid moieties in **3a–c** was proven by appearance of two kojic acid olefinic proton at δ 8.69–8.67 and 6.65–6.50 ppm as singlet. The oxymethyl protons of kojic acid were represented by peaks at δ 5.02–4.39 ppm.

The NMR spectra of kojic ester (**5a–d**) via formation of kojyl chloride revealed the presence of peak attributed to oxymethyl proton (OCH_2) at δ 5.26–5.24 ppm as singlet, which indicated 7-*O*-substituent **1** [41]. In comparison to 5-*O*-substituted **3a–c**, 7-*O*-substituted **5a–d** showed shifted H-6 olefinic proton signals to upfield and oxymethyl proton signals to downfield. Carbonyl group of ketone is more deshielded than carbonyl group of ester due to deshielding effect of pyrone ring [42].

3.3. Antibacterial Activities of Kojic Ester Derivatives. The kojic ester derivatives **3a–c** and **5a–d** were evaluated for antibacterial activities against gram negative strain *E. coli* at 37°C via turbidimetric method at three different concentrations, 50 ppm, 80 ppm, and 100 ppm, and compared to chalcones (**2a–c**) and azobenzene (**4a–d**) derivatives. Due to solubility limitation of **2a–c** and **3a–d** in organic solvent, turbidimetric method was only applied on **4a–d** and **5a–d**. Negative control, which is comprised of solvent and inoculums, showed significant increase in growth of *E. coli*. The antibacterial assay was expressed by plotting graph of In N_t versus time. The In N_t value indicated number of colony forming (growth of *E. coli*) units/mL which followed the expression of In N_t = 27.1 − 8.56 T [39].

TABLE 1: MIC for **4a–d** and **5a–d**.

Compounds	MIC (ppm)
4a	>400
4b	>400
4c	>400
4d	>400
5a	200
5b	190
5c	330
5d	330

Kojic ester **4a–d** exhibited weak inhibition while **5a–d** exhibited good inhibition against *E. coli*. The inhibition of **5a–d** is concentration dependent where 100 ppm > 80 ppm > 50 ppm. The minimum inhibitory concentration (MIC) of **4a–d** and **5a–d** was determined by extrapolating the concentration to zero growth rate of *E. coli*. The MIC of **4a–d** and **5a–d** is summarized in Table 1. Kojic ester moieties in azobenzene are envisaged to increase the lipophilic properties in **5a–d** to interact and destruct the bacterial cell membrane [43]. The presence of C=O and N=N moieties in kojic ester offers more reactive sites that easily protonated under acidic condition and reacted with the bacterial surfaces which further enhance the biological activity [44]. Compound **5a-b** showed better activity than **5c-d** due to the increase of

TABLE 2: Inhibition zones of **2a–c** and **3a–c**.

Compounds	Zone of inhibition (mm)	
	E. coli	*S. aureus*
2a	-	16.0
2b	-	13.0
2c	-	8.0
3a	-	8.0
3b	-	11.0
3c	-	13.0
DMSO	-	-
Kojic acid	-	-
Ampicillin	16.0	11.0

chain length after the cutoff effect associated with decrease in solubility.

Antibacterial activity of kojic esters **2a–c** and **3a–c** bearing chalcone moieties was conducted by measuring the zone of inhibition on agar plates (Kirby Bauer Disc Diffusion method) against *S. aureus* and *E. coli* bacterial strain due to solubility limitation in DMSO. Ampicillin was used as reference standard and DMSO as negative control.

The zone inhibition of **2a–c** and **3a–c** is shown in Table 2. Kojic acid showed no inhibition against both *S. aureus* and *E. coli,* while both series of **2a–c** and **3a–c** showed no inhibition against *E. coli* compared to standard ampicillin. Compounds **2a–c** and **3a–c**, however, showed active inhibition against *S. aureus* with **2c** and **3a-b** being inhibited better than ampicillin. The selective antibacterial activity of **2a–c** and **3a–c** against gram positive bacterial strain is due to the significance different in the membrane composition and architecture of gram positive and gram negative organism [45]. *S. aureus* has simpler mesh like cell membrane compared to *E. coli* which eases the penetration of compound [46].

Antibacterial activities of **2a–c** increased with the decrease in electronegativity of substituent on chalcone. Electronegativity in decreasing order is as follows: Cl (8 mm) > Br (13 mm) > H (16 mm). Polarized ability of antibacterial agent increases with the decreased in electronegativity which reasoned to enhance activity of acting antibacterial agent [47].

The antibacterial activities of **3a–c**, however, showed an opposite trend. The inhibition was increased with the increased in electronegativity. The incorporation of kojic acid transformed the lipophilic properties of the compound. A good balance between hydrophilic and hydrophobic properties of compound makes good antibacterial agent [48]. Compound **3c** with chlorine substituents showed the biggest zone of inhibition with 13 mm while **3a** with hydrogen showed the least zone of inhibition (8 mm).

4. Conclusions

A series of kojic esters **2a–c** and **3a–c** has been successfully synthesized and exhibited good antibacterial activity towards gram positive *S. aureus* bacteria strain. Antibacterial activities of **2a–c** increased with the decrease in electronegativity of halogen substituent on chalcone with **2c** showing the highest inhibition, better than the ampicillin reference standard. Lipophilic properties of kojic ester bearing chalcone derivatives **3a-b** inhibited better antibacterial activity. Kojic ester **5a–d** bearing azobenzene derivatives showed good inhibition against *E. Coli* compared to **4a–d**. The presence of C=O ester, N=N, and alkene chains has increased the lipophilic properties of of **5a–d** and disrupts the bacteria cell walls.

Conflicts of Interest

The authors declare that there are no conflicts of interest regarding the publication of this paper.

Acknowledgments

This work was supported by Universiti Malaysia Sarawak and Ministry of Education for Research Funds F07(DPP17)/1173/2014(17) and FRGS/ST01(02)/968/2013(09). The authors would also like to express special thanks to Professor Dr. Arbakariya Ariff from UPM for providing the *A. flavus* Link 44-1 strain.

References

[1] J. Chaudhary, A. N. Pathak, and S. Lakhawat, "Production technology and applications of kojic acid," *Annual Research & Review in Biology*, vol. 4, no. 21, pp. 3165–3196, 2014.

[2] A. F. B. Lajis, M. Basri, R. Mohamad et al., "Enzymatic synthesis of kojic acid esters and their potential industrial applications," *Chemical Papers*, vol. 67, no. 6, pp. 573–585, 2013.

[3] S. A. El-Aasar, "Cultural conditions studies on kojic acid production by *Aspergillus parasiticus*," *nternational Journal of Agriculture & Biology*, vol. 8, no. 4, pp. 468–473, 2006.

[4] M. D. Aytemir, D. D. Erol, R. C. Hider, and M. Özalp, "Synthesis and evaluation of antimicrobial activity of new 3-Hydroxy-6-methyl-4-oxo-4H-pyran-2-carboxamide derivatives," *Turkish Journal of Chemistry*, vol. 26, no. 7, pp. 757–764, 2003.

[5] H. Kayahara, N. Shibata, K. Tadasa, H. Maeda, T. Kotani, and I. Ichimoto, "Amino acid and peptide derivatives of kojic acid and their antifungal properties," *Agricultural and Biological Chemistry*, vol. 54, no. 9, pp. 2441-2442, 1990.

[6] B. V. S. Reddy, M. R. Reddy, C. Madan, K. P. Kumar, and M. S. Rao, "Indium(III) chloride catalyzed three-component coupling reaction: a novel synthesis of 2-substituted aryl(indolyl)kojic acid derivatives as potent antifungal and antibacterial agents," *Bioorganic & Medicinal Chemistry Letters*, vol. 20, no. 24, pp. 7507–7511, 2010.

[7] S. Emami, E. Ghafouri, M. A. Faramarzi, N. Samadi, H. Irannejad, and A. Foroumadi, "Mannich bases of 7-piperazinylquinolones and kojic acid derivatives: synthesis, in vitro antibacterial activity and in silico study," *European Journal of Medicinal Chemistry*, vol. 68, pp. 185–191, 2013.

[8] C. Z. Sie and Z. Ngaini, "Incorporation of kojic acid-azo dyes on TiO_2 thin films for dye sensitized solar cells applications," *Journal of Solar Energy*, vol. 2017, Article ID 2760301, pp. 1–10, 2017.

[9] K.-J. Liu and J.-F. Shaw, "Lipase-catalyzed synthesis of kojic acid esters in organic solvents," *Journal of the American Oil Chemists' Society*, vol. 75, no. 11, pp. 1507–1511, 1998.

[10] N. El-Boulifi, S. E. Ashari, M. Serrano, J. Aracil, and M. Martínez, "Solvent-free lipase-catalyzed synthesis of a novel hydroxyl-fatty acid derivative of kojic acid," *Enzyme and Microbial Technology*, vol. 55, pp. 128–132, 2014.

[11] S. R. Ho, S. B. Heung, M. A. Soo, H. K. Duck, and S. C. Ih, "Synthesis of new anti-melanogenic compounds containing two molecules of kojic acid," *Bulletin of the Korean Chemical Society*, vol. 29, no. 8, pp. 1569–1571, 2008.

[12] T. Kobayashi, S. Adachi, K. Nakanishi, and R. Matsuno, "Semi-continuous production of lauroyl kojic acid through lipase-catalyzed condensation in acetonitrile," *Biochemical Engineering Journal*, vol. 9, no. 2, pp. 85–89, 2001.

[13] H. S. Rho, M. Goh, J. Lee et al., "Ester derivatives of kojic acid and polyphenols containing adamantane moiety with tyrosinase inhibitory and anti-inflammatory properties," *Bulletin of the Korean Chemical Society*, vol. 32, no. 4, pp. 1411–1414, 2011.

[14] C. M. Yang, J. Y. Hong, K. W. Lee, B. G. Lee, and D. I. Chang, "Kojic Acid Derivative," United States Patent, Patent number: 5486624, 1996.

[15] C. Balakrishna, N. Payili, S. Yennam, P. Uma Devi, and M. Behera, "Synthesis of new kojic acid based unnatural α-amino acid derivatives," *Bioorganic & Medicinal Chemistry Letters*, vol. 25, no. 21, pp. 4753–4756, 2015.

[16] K. Morisaki and S. Ozaki, "Design of novel hybrid vitamin C derivatives: Thermal stability and biological activity," *Chemical & Pharmaceutical Bulletin*, vol. 44, no. 9, pp. 1647–1655, 1996.

[17] D. E. Okwu and N. Ukanwa, "Isolationand characterization of flavonoids chalcones and anthocynidines from bridelia ferruginea benth," *Der Chemica Sinica*, vol. 1, no. 2, pp. 21–28, 2010.

[18] G. Di Carlo, N. Mascolo, A. A. Izzo, and F. Capasso, "Flavonoids: old and new aspects of a class of natural therapeutic drugs," *Life Sciences*, vol. 65, no. 4, pp. 337–353, 1999.

[19] G. Achanta, A. Modzelewska, L. Feng, S. R. Khan, and P. Huang, "A boronic-chalcone derivative exhibits potent anticancer activity through inhibition of the proteasome," *Molecular Pharmacology*, vol. 70, no. 1, pp. 426–433, 2006.

[20] R. Ferrer, G. Lobo, N. Gamboa et al., "Synthesis of [(7-chloroquinolin-4-yl)amino] chalcones: potential antimalarial and anticancer agents," *Scientia Pharmaceutica*, vol. 77, no. 4, pp. 725–741, 2009.

[21] A. Modzelewska, C. Pettit, G. Achanta, N. E. Davidson, P. Huang, and S. R. Khan, "Anticancer activities of novel chalcone and bis-chalcone derivatives," *Bioorganic & Medicinal Chemistry*, vol. 14, no. 10, pp. 3491–3495, 2006.

[22] M. Rahman, "Chalcone: a valuable insight into the recent advances and potential pharmacological activities," *Chemical Sciences Journal*, vol. 29, 2011.

[23] A. H. Shridhar, J. Keshavayya, S. K. Peethambar, and H. Joy Hoskeri, "Synthesis and biological activities of Bis alkyl 1,3,4-oxadiazole incorporated azo dye derivatives," *Arabian Journal of Chemistry*, vol. 9, pp. S1643–S1648, 2016.

[24] A. Suksamrarn, P. Poomsing, N. Aroonrerk, T. Punjanon, S. Suksamrarn, and S. Kongkun, "Antimycobacterial and antioxidant flavones from *Limnophila geoffrayi*," *Archives of Pharmacal Research*, vol. 26, no. 10, pp. 816–820, 2003.

[25] K. L. Ameta, N. S. Rathore, and B. Kumar, "Synthesis of some novel chalcones and their facile one-pot conversion to 2-aminobenzene-1, 3-dicarbonitriles using malononitrile," *Analele Universității din Bucuresti – Chimie (serie nouă)*, vol. 20, no. 1, pp. 15–24, 2011.

[26] A. N. Chaudhary and V. Juyal, "Synthesis of chalcone and their derivatives as antimicrobial agents," *International Journal of Pharmacy and Pharmaceutical Sciences*, vol. 3, no. 3, pp. 125–128, 2011.

[27] M. A. El-Hashash, S. A. Rizk, and S. R. Atta-Allah, "Synthesis and regioselective reaction of some unsymmetrical heterocyclic chalcone derivatives and spiro heterocyclic compounds as antibacterial agents," *Molecules*, vol. 20, no. 12, pp. 22069–22083, 2015.

[28] A. M. Khedr, M. Gaber, and E. H. Abd El-Zaher, "Synthesis, structural characterization, and antimicrobial activities of Mn(II), Co(II), Ni(II), Cu(II) and Zn(II) complexes of triazole-based azodyes," *Chinese Journal of Chemistry*, vol. 29, no. 6, pp. 1124–1132, 2011.

[29] P. K. Nikhil, M. P. Pratik, and R. Manoj, "Study on antibacterial activity for multidrug resistance stain by using phenyl pyrazolones substituted 3-amino 1H-pyrazolon(3, 4-b) quinoline derivative in vitro condition," *International Journal of Pharmtech Research*, vol. 3, pp. 540–548, 2011.

[30] H. Yoo, H. K. Seung, J. Lee et al., "Synthesis and antioxidant activity of 3-methoxyflavones," *Bulletin of the Korean Chemical Society*, vol. 26, no. 12, pp. 2057–2060, 2005.

[31] J.-S. Bae, H. S. Freeman, and A. El-Shafei, "Metallization of non-genotoxic direct dyes," *Dyes and Pigments*, vol. 57, no. 2, pp. 121–129, 2003.

[32] H. Xu and X. Zeng, "Synthesis of diaryl-azo derivatives as potential antifungal agents," *Bioorganic & Medicinal Chemistry Letters*, vol. 20, no. 14, pp. 4193–4195, 2010.

[33] T. A. Farghaly and Z. A. Abdallah, "Synthesis, azo-hydrazone tautomerism and antitumor screening of N-(3-ethoxycarbonyl-4,5,6,7-tetrahydro-benzo[b]thien-2-yl)-2-arylhydrazono-3-oxobutanamide derivatives," *Arkivoc*, vol. 2008, no. 17, pp. 295–305, 2008.

[34] R. G. Child, R. G. Wilkinson, and A. Tomcu-Fucik, "Effect of substrate orientation of the adhesion of polymer joints," *Chemical Abstracts*, vol. 87, article 6031, 1977.

[35] H. G. Garg and C. Praksh, "Preparation of 4-arylazo-3,5-disubstituted-(2H)-1,2,6-thiadiazine 1,1-dioxides," *Journal of Medicinal Chemistry*, vol. 15, no. 4, pp. 435–439, 1972.

[36] M. R. S. P. Piste, D. P. Indalkar, N. Z. Dnyandev, and S. M. Pankaj, "Synthesis and antimicrobial activity of substituted p-amino azobenzene with thymol moiety- a green protocol," *International Journal of Chemistry Research*, vol. 3, no. 2, pp. 25–29, 2012.

[37] J.-M. Noh, S.-Y. Kwak, H.-S. Seo, J.-H. Seo, B.-G. Kim, and Y.-S. Lee, "Kojic acid-amino acid conjugates as tyrosinase inhibitors," *Bioorganic & Medicinal Chemistry Letters*, vol. 19, no. 19, pp. 5586–5589, 2009.

[38] Z. Ngaini, S. M. H. Fadzillah, and H. Hussain, "Synthesis and antimicrobial studies of hydroxylated chalcone derivatives with variable chain length," *Natural Product Research*, vol. 26, no. 10, pp. 892–902, 2012.

[39] N. B. Pappano, O. P. Centorbi, and F. H. Ferretti, "Determination of minimum concentration inhibitory chalcone derivatives," *Revise Microbiology*, vol. 2, no. 1, pp. 183–188, 1990.

[40] R. Mohamad, M. S. Mohameh, N. Suhaili, M. M. Salleh, and A. B. Ariff, "Kojic acid: applications and development of fermentation process for production," *Biotechnology and Molecular Biology Reviews*, vol. 5, no. 2, pp. 24–37, 2010.

[41] H. Kaatz, K. Streffer, U. Wollenberger, and M. G. Peter, "Inhibition of mushroom tyrosinase by kojic acid octanoates," *Zeitschrift für Naturforschung*, vol. 54c, pp. 70–74, 1999.

[42] D. L. Pavia, G. M. Lampman, G. S. Kriz, and J. A. Vyvyan, *Introduction to Spectroscopy*, Cengage Learning, Belmont, Calif, USA, 2008.

[43] R. Bartzatt and C. Malesa, "Synthesis, structural analysis and antibacterial activity of a butyl ester derivative of ampicillin," *Chemotherapy*, vol. 49, no. 5, pp. 213–221, 2003.

[44] Z. Zhong, R. Xing, S. Liu, L. Wang, S. Cai, and P. Li, "Synthesis of acyl thiourea derivatives of chitosan and their antimicrobial activities in vitro," *Carbohydrate Research*, vol. 343, no. 3, pp. 566–570, 2008.

[45] G. M. Morris, H. Ruth, W. Lindstrom et al., "Software news and updates AutoDock4 and AutoDockTools4: automated docking with selective receptor flexibility," *Journal of Computational Chemistry*, vol. 30, no. 16, pp. 2785–2791, 2009.

[46] M. J. Alves, H. J. C. Froufe, A. F. T. Costa et al., "Docking studies in target proteins involved in antibacterial action mechanisms: Extending the knowledge on standard antibiotics to antimicrobial mushroom compounds," *Molecules*, vol. 19, no. 2, pp. 1672–1684, 2014.

[47] I. A. El-Kady, A. N. Zohri, and S. R. Hamed, "Kojic acid production from agro-industrial by-products using fungi," *Biotechnology Research International*, vol. 2014, Article ID 642385, pp. 1–10, 2014.

[48] C. I. Wei, T. S. Huang, S. Y. Fernando, and K. T. Chung, "Mutagenicity studies of kojic acid," *Toxicology Letters*, vol. 59, no. 1-3, pp. 213–220, 1991.

15

Synthesis, Characterization, X-Ray Crystallography, and Antileishmanial Activities of *N*-Linked and *O*-Linked Glycopyranosides

Haroon ur Rashid,[1] Sher Wali Khan,[2] Momin Khan,[3,4] Akhtar Nadhman,[5] Noor Rehman,[2] Muhammad Tariq,[2] and Sammer Yousuf[6]

[1]*Department of Chemistry, Sarhad University of Science and Information Technology, Peshawar, Khyber Pakhtunkhwa, Pakistan*
[2]*Department of Chemistry, Shaheed BB University, Sheringal, Upper Dir, Khyber Pakhtunkhwa, Pakistan*
[3]*Department of Biotechnology, Quaid-I-Azam University, Islamabad 45320, Pakistan*
[4]*Department of Microbiology, Institute of Basic Medical Sciences, Khyber Medical University, Peshawar, Pakistan*
[5]*Institute of Integrative Biosciences (IIB), CECOS University of Information Technology and Emerging Sciences, Peshawar, Khyber Pakhtunkhwa, Pakistan*
[6]*H. E. J. Research Institute of Chemistry, International Center for Chemical and Biological Sciences, University of Karachi, Karachi 75270, Pakistan*

Correspondence should be addressed to Sher Wali Khan; sherwalikhn@yahoo.co.uk

Academic Editor: Vinod Kumar Tiwari

Novel *N*-linked **5a–e** and *O*-linked glycopyranosides **7a–e** were synthesized in high yield from commercially available L-tartaric acid containing two asymmetric centers and C_2 axis of symmetry. The compound L-tartaric acid was completely protected and then partially hydrolyzed to get the monoester, which upon treatment with different amino and hydroxyl derivatives of glycopyranoses gave the desired amides and esters. The synthesized derivatives were purified by chromatography and characterized by spectroanalytical techniques. The structure of compound **7c** in the series was supported by X-ray analysis. Leishmanicidal activities of compounds **5a–e** and **7a–e** were investigated which showed moderate to good activities.

1. Introduction

Leishmania has long been known to human beings especially in African and Indian populations since the mid eighteenth century [1–3]. In 1903, Leishman and Donovan discovered the protozoan in spleen tissues of patients in India, which is now known as *Leishmania donovani* [4, 5]. Nicolle and Comte reported infection by *Leishmania* in dogs in 1908 [6]. Leishmaniasis is a vector-borne disease carried by sand fly and caused by protozoa of the genus *Leishmania*. Recent reports confirm that people in about 88 countries are affected by *Leishmania* and 350 million are at risk, and more or less 2 million new cases are reported each year [7]. In spite of the socioeconomical significance of this tropical infectious disease, the discovery of new potential drugs against it is underway [8, 9].

Glycolipid and glycoprotein oligosaccharides present on cell surfaces are known to play an important role in various biological processes, that is, cellular recognition, adhesion, tumor metastasis, and cell growth regulation [10]. One major class of glycoprotein oligosaccharides consists of *N*-oligosaccharides linked to asparagine by an amide bond [11, 12]. Amides are usually prepared by thermolysis of carboxylic acids with amines, by coupling of carboxylic acids and amines in the presence of a coupling reagent [13] or by prior conversion of the carboxylic acid into a more electrophilic derivative [14]. The concept to add a chiral aglycone moiety such as tartaric acid to prepare glycosides is an active area of research. L-tartaric acid is cheap and readily available starting material; it has two chiral centers with two carboxylic acid groups which can be easily derivatized. Derivatives of tartaric acid containing ester and amide linkages have promisingly useful biological applications, for example, anti-inflammatory, antifungal, thrombin inhibitors, antimicrobial, and β-secretase inhibitory activities [15–20].

Hence, glyco derivatives of tartaric acid have potential as curative agents for various infectious diseases. Herein, we report an efficient and scalable synthesis of N-linked (**5a–e**) and O-linked glycopyranosides (**7a–e**) through anomeric N- and O-acylation, respectively, using commercially available L-tartaric acid as a starting material.

2. Experimental

All chemicals were of highest purity available and used as supplied. D-glucose, D-mannose, D-galactose, D-glucosamine, and D-galactosamine hydrochloride were purchased from Fluka and Sigma Aldrich. Dry solvents like methanol, dichloromethane, chloroform, and n-hexane were obtained by distillation using standard procedures or by passage through a column of anhydrous alumina using equipment from anhydrous engineering (University of Bristol, UK) based on Grubbs' design. Reactions under anhydrous conditions were carried out under nitrogen gas using three-way stopcock and rubber septa.

Liquid reagents, solutions, or solvents were added via syringe or cannula through rubber septa. Solid reagents were added via Schlenk type adapters. All reactions were monitored by TLC on Kieselgel 60 F254 (Merck); ethyl acetate/n-hexane and methanol/chloroform were used as eluent. Chromatograms were detected under UV light (λ_{max} 254 and 365 nm) and by charring with 10% sulfuric acid in ethanol, ninhydrin, and vanillin, respectively. Column chromatography was performed using silica gel [Merck, 230–400 mesh (40–63 μm)]. Extracts were concentrated under reduced pressure using both rotary evaporator (bath temperature up to 40°C) at a pressure of either 15 mmHg (diaphragm pump) or 0.1 mmHg (oil pump), as appropriate and a high vacuum line at room temperature. Melting points were determined in degree Celsius (°C) using Gallenkamp digital melting point apparatus and are uncorrected. IR spectra were recorded on a Bruker IFS 66 (FT-IR), Nicolet 205 FT-IR; Nicolet 360 smart orbit (ATR); Thermo Scientific Nicolet 6700 FT-IR and Shimadzu Fourier Transform Infrared Spectrophotometer Model 270. Solid samples were taken in KBr pellets and oils were used in NaCl cells for recording their spectra. ^{1}HNMR spectra were recorded on an NMR Bruker apparatus at 300 MHz and Varian 400 MHz INOVA instrument. ^{13}C NMR spectra were recorded on NMR Bruker apparatus at 75 MHz and Varian 100 MHz INOVA instrument. Chemical shifts are quoted in parts per million from $SiMe_4$ or residual solvent proton signals and coupling constants (J) given in Hertz. Multiplicities are abbreviated as b (broad), s (singlet), d (doublet), t (triplet), q (quartet), m (multiplet), or combinations thereof. Positive ion Matrix Assisted Laser Desorption Ionization Time-of-Flight (MALDI-TOF) mass spectra were recorded using an HP-MALDI instrument using gentisic acid as matrix material. Optical rotations were measured on an ATAGO, AP-100 Automatic polarimeter. Single crystal X-ray diffraction data were collected on a Bruker Smart APEX II, CCD detector diffractometer [21]. Data reductions were performed using the SAINT program. The structures were solved by direct methods [22] and refined by full-matrix least squares on F2 by using the SHELXTL-PC package [23]. The figures were plotted with the aid of ORTEP program [24].

*2.1. General Procedure for the Synthesis of N-Linked Glycopyranosides (**5a–e**).* In a round bottom flask (100 mL), monoester **1** (0.61 g, 3 mmol), EDC (0.74 g, 3 mmol × 1.3 eq.), and catalytic amount of DMAP were placed under nitrogen. CH_2Cl_2 (30 mL) was added as a solvent. After 30 min glycopyranosyl amine **4a–e** (3 mmol) was added and the mixture was stirred for 18 h. The byproduct urea was removed by extraction of the reaction mixture with ethyl acetate or chloroform and water (30 mL × 2). The crude product was purified by column chromatography using ethyl acetate : hexane (3 : 7) as eluent.

*2.2. (2R,3R,5R,6R)-2-(Acetoxymethyl)-6-((4R,5R)-5-(methoxycarbonyl)-2,2-dimethyl-1,3-dioxolane-4-carboxamido)tetrahydro-2H-pyran-3,4,5-triyl Triacetate **5a**.* Yield: 72%. Colorless oil $[\alpha]_D^{25}$ = 12.5. (IR υ cm^{-1}): 3347 (NH), 2955 (CH), 1742 ($COOCH_3$), 1701 (CONH), 1299 (C-O-C). ^{1}HNMR (400 MHz, $CDCl_3$): δ (ppm): 7.18 (d, J = 9.3 Hz, NH), 5.23, 5.18, 5.05, 4.96 (pseudo t, J ~ 9.2 Hz in each, CH), 4.65 (d, J = 5.4 Hz, CH), 4.59 (d, J = 5.4 Hz, CH), 4.30 (dd, J = 12.5, 3.6 Hz, CH), 4.21 (ddd, J = 10.4, 3.6, 2.1 Hz, CH), 3.95 (dd, J = 12.5, 2.1 Hz, CH), 3.75 (s, OCH_3), 1.94, 1.92, 1.91, 1.89 (s, CH_3), 1.56 (s, CH_3), 1.50 (s, CH_3). ^{13}C NMR (100 MHz, $CDCl_3$): δ (ppm): 171.2 (CONH), 170.7 ($COOCH_3$), 170.0, 169.9, 169.5, 169.2 (CO), 113.3 (qt C), 81.8 (CH), 76.3 (CH), 75.2 (CH), 73.3, 72.5, 70.0, 67.8 (CH), 61.2 (CH), 52.4 (OCH_3), 25.9, (CH_3), 25.2 (CH_3), 20.2, 20.1 (3), (CH_3). HRMS-ESI for $C_{22}H_{31}NO_{14}$ Na: $[M + Na]^+$ calcd: 556.1745, found: 556.1621. Anal. Calc. For $C_{22}H_{31}NO_{14}$: C, 49.53; H, 5.86; N, 2.63; O, 41.99 Found C, 49.43; H, 5.56; N, 2.78; O, 42.10.

*2.3. (2R,3R,5R,6R)-2-(Acetoxymethyl)-6-((4R,5R)-5-(methoxycarbonyl)-2,2-dimethyl-1,3-dioxolane-4-carboxamido)tetrahydro-2H-pyran-3,4,5-triyl Triacetate **5b**.* Yield: 69%. Colorless oil $[\alpha]_D^{25}$ = 13.3. (IR υ cm^{-1}): 3337 (NH), 2925 (CH), 1732 ($COOCH_3$), 1703 (CONH), 1292 (C-O-C). ^{1}HNMR (400 MHz, $CDCl_3$): δ (ppm): 7.47 (d, J = 9.3 Hz, NH), 5.34, 5.31, 5.24 (pseudo t, J = 10.0, Hz in each, CH), 4.95 (t, J = 3.9 Hz, CH), 4.93 (d, J = 5.5 Hz, CH), 4.80 (d, J = 5.5 Hz, CH), 4.28 (dd, J = 12.4, 3.5 Hz, CH), 4.10 (ddd, J = 10.3, 3.6, 2.1 Hz, CH), 3.95 (dd, J = 12.4, 3.5 Hz, CH), 3.77 (s, OCH_3), 2.34, 2.23, 2.13, 1.99 (s, CH_3), 1.52 (s, CH_3), 1.46 (s, CH_3). ^{13}C NMR (100 MHz, $CDCl_3$): δ (ppm): 172.2 (CONH), 170.9 ($COOCH_3$), 170.1, 169.6, 169.3, 169.1 (CO), 114.3 (qt C), 81.4 (CH), 76.5 (CH), 75.1 (CH), 73.1, 72.3, 70.1, 67.9 (CH), 61.8 (CH), 52.9 (OCH_3), 25.5 (CH_3), 25.1 (CH_3), 20.1, 20.0 (3) (CH_3). HRMS-ESI for $C_{22}H_{31}NO_{14}$ Na: $[M + Na]^+$ calcd: 556.1745, found: 556.1627. Anal. Calc. For $C_{22}H_{31}NO_{14}$: C,

49.53; H, 5.86; N, 2.63; O, 41.99 Found C, 49.43; H, 5.56; N, 2.78; O, 42.10.

2.4. (2R,3R,5R,6R)-2-(Acetoxymethyl)-6-((4R,5R)-5-(methoxycarbonyl)-2,2-dimethyl-1,3-dioxolane-4-carboxamido)tetrahydro-2H-pyran-3,4,5-triyl Triacetate **5c**. Yield: 65%. Colorless oil $[\alpha]_D^{25}$ = 27.1. (IR υ cm^{-1}): 3339 (NH), 2928 (CH), 1736 ($COOCH_3$), 1700 (CONH), 1290 (C-O-C). ^{1}HNMR (400 MHz, $CDCl_3$): δ (ppm): 7.37 (d, J = 9.3 Hz, NH), 5.42, 5.40, 5.36 (pseudo t, J ~ 6.5 Hz in each, CH), 4.98 (t, J = 11.9 Hz, CH), 4.97 (d, J = 5.6 Hz, CH), 4.90 (d, J = 5.6 Hz, CH), 4.48 (dd, J = 12.7, 3.4 Hz, CH), 4.12 (ddd, J = 10.4, 3.6, 2.1 Hz, CH), 3.92 (dd, J = 12.4, 3.5 Hz, CH), 3.84 (s, OCH_3), 2.11, 2.10, 2.09, 1.99 (s, CH_3), 1.54 (s, CH_3), 1.48 (s, CH_3). ^{13}C NMR (100 MHz, $CDCl_3$): δ (ppm): 171.8 (CONH), 170.2 ($COOCH_3$), 170.3, 169.5, 169.4, 169.2 (CO), 114.2 (qt C), 81.5 (CH), 76.1 (CH), 75.2 (CH), 73.3, 72.2, 70.1, 67.5 (CH), 61.9 (CH), 52.5 (OCH_3), 26.5 (CH_3), 25.7 (CH_3), 20.4, 19.5 (3) (CH_3). HRMS-ESI for $C_{22}H_{31}NO_{14}$ Na: $[M + Na]^+$ calcd: 556.1745, found: 556.1643. Anal. Calc. For $C_{22}H_{31}NO_{14}$: C, 49.53; H, 5.86; N, 2.63; O, 41.99 Found C, 49.43; H, 5.56; N, 2.78; O, 42.10.

2.5. (2R,3S,5R,6R)-5-Acetamido-2-(acetoxymethyl)-6-((4R, 5R)-5-(methoxycarbonyl)-2,2-dimethyl-1,3-dioxolane-4-carboxamido)tetrahydro-2H-pyran-3,4-diyl Diacetate **5d**. Yield: 67%. Gel like. $[\alpha]_D^{25}$ = 43.3. (IR υ cm^{-1}): 3343 (NH), 2948 (CH), 1746 ($COOCH_3$), 1710 (CONH), 1293 (C-O-C). ^{1}HNMR (400 MHz, $CDCl_3$): δ (ppm): 7.69 (d, J = 9.5 Hz, NH), 7.57 (d, J = 9.3 Hz, NH), 5.43 (t, J = 10.0 Hz, 1H), 5.41-5.32 (m, CH), 5.37, 4.99 (pseudo t, J ~ 10.0 Hz in each, CH), 4.96 (d, J = 5.5 Hz, CH), 4.94 (d, J = 5.5 Hz, CH), 4.44 (dd, J = 12.7, 3.4 Hz, CH), 4.22 (ddd, J = 10.4, 3.6, 2.1 Hz, CH), 3.91 (dd, J = 12.4, 3.5 Hz, CH), 3.86 (s, OCH_3), 2.21, 2.20, 2.19, 2.13 (s, CH_3), 1.57 (s, CH_3), 1.49 (s, CH_3). ^{13}C NMR (100 MHz, $CDCl_3$): δ (ppm): 172.8 (CONH), 171.2 ($COOCH_3$), 170.2, 169.8, 169.5, 169.1 (CO), 114.1 (qt C), 81.8 (CH), 76.1 (CH), 75.3 (CH), 73.2, 72.1, 70.0, 67.8 (CH), 61.4 (CH), 52.2 (OCH_3), 26.2 (CH_3), 25.8 (CH_3), 20.3, 19.3 (3) (CH_3). HRMS-ESI for $C_{22}H_{32}N_2O_{13}$ Na: $[M + Na]^+$ calcd: 555.1904, found: 555.1619. Anal. Calc. For $C_{22}H_{32}N_2O_{13}$: C, 49.62; H, 6.06; N, 5.26; O, 39.06 Found C, 49.32; H, 6.16; N, 5.21; O, 39.16.

2.6. (2R,3S,5R,6R)-5-Acetamido-2-(acetoxymethyl)-6-((4R, 5R)-5-(methoxycarbonyl)-2,2-dimethyl-1,3-dioxolane-4-carboxamido)tetrahydro-2H-pyran-3,4-diyl Diacetate **5e**. Yield: 68%. Gel like. $[\alpha]_D^{25}$ = 47.1. (IR υ cm^{-1}): 3341 (NH), 2942 (CH), 1745 ($COOCH_3$), 1711 (CONH), 1290 (C-O-C). ^{1}HNMR (400 MHz, $CDCl_3$): δ (ppm): 7.67 (d, J = 9.6 Hz, NH), 7.55 (d, J = 9.3 Hz, NH), 5.41 (t, J = 10.8 Hz, CH), 5.40 (m, CH), 5.38, 4.97 (pseudo t, J ~ 6.9 Hz in each, CH), 4.93 (d, J = 5.6 Hz, CH), 4.91 (d, J = 5.5 Hz, CH), 4.41 (dd, J = 12.7, 3.4 Hz, CH), 4.21 (ddd, J = 10.4, 3.6, 2.1 Hz, CH), 3.90 (dd, J = 12.4, 3.5 Hz, CH), 3.83 (s, OCH_3), 2.23, 2.22, 2.20, 2.16 (s, CH_3), 1.56 (s, CH_3), 1.46 (s, CH_3). ^{13}C NMR (100 MHz, $CDCl_3$): δ (ppm): 172.2 (CONH), 171.3 ($COOCH_3$), 170.5, 169.5, 169.2, 169.1 (CO), 113.4 (qt C), 81.6 (CH), 76.4 (CH), 75.2 (CH), 73.4, 72.3, 70.1, 67.3, (CH), 62.4 (CH), 53.2 (OCH_3), 26.2 (CH_3), 25.7 (CH_3), 20.7, 19.5 (3) (CH_3). HRMS-ESI for $C_{22}H_{31}NO_{14}$ Na: $[M + Na]^+$ calcd: 555.1904, found: 555.1719. Anal. Calc. For $C_{22}H_{32}N_2O_{13}$: C, 49.62; H, 6.06; N, 5.26; O, 39.06 Found C, 49.32; H, 6.16; N, 5.21; O, 39.16.

*2.7. General Procedure for the Synthesis of O-Linked Glycopyranosides (**7a–e**).* Hydrazinium acetate (1.10 g, 12 mmol) was added to a stirred solution of anomeric mixture of acetylated monosaccharides **1a–e** (3.12 g, 8 mmol) in DMF (40 mL). The reaction mixture was heated at 55°C for 30 min under a nitrogen atmosphere. After completion, the reaction mixture was diluted with H_2O (100 mL) and extracted with EtOAc (40 mL × 3). The combined organic layers were washed with water and brine (40 × 2), dried over anhydrous $MgSO_4$. The solvent was removed under reduced pressure and the crude product was used without further purification. To the resulting hemiacetal (1.04 g, 3 mmol), monoester **1** (0.61 g, 3 mmol), EDC (0.74 g, 3 mmol × 1.3 eq.), and catalytic amount of DMAP were added under nitrogen with CH_2Cl_2 (30 mL). The reaction mixture was stirred for 18 h; the byproduct urea was removed by extraction of the reaction mixture with ethyl acetate or chloroform and water (30 mL × 3). The crude product was purified by column chromatography using ethyl acetate: n-hexane (3 : 7) as eluent.

2.8. (4R,5R)-4-Methyl-5-((2S,3R,5R,6R)-3,4,5-triacetoxy-6-(acetoxymethyl)tetrahydro-2H-pyran-2-yl)-2,2-dimethyl-1,3-dioxolane-4,5-dicarboxylate **7a**. Yield: 72%. Colorless oil $[\alpha]_D^{25}$ = 26.5. (IR υ cm^{-1}): 2991 (CH), 1742 (CO), 1366 (C-O-C). ^{1}HNMR (400 MHz, $CDCl_3$): δ (ppm): 6.45 (d, J = 4.0 Hz, 1H, H-1$^\alpha$), 5.56 (d, J = 8.0 Hz, 1H, H-1$^\beta$), 5.48, 5.45, 4.93 (pseudo t, J ~ 9.8 Hz in each, CH), 4.91 (d, J = 6.3 Hz, CH), 4.88 (d, J = 6.3 Hz, CH), 4.32 (dd, J = 12.5, 3.6 Hz, CH), 4.12 (ddd, J = 10.4, 3.6, 2.1 Hz, CH), 3.93 (dd, J = 12.5, 2.1 Hz, CH), 3.86 (s, OCH_3), 2.19, 2.17, 2.12, 2.04 (s, CH_3), 1.55 (s, CH_3), 1.46 (s, CH_3). ^{13}C NMR (100 MHz, $CDCl_3$): δ (ppm): 172.6 ($COOCH_3$), 171.2, 169.5, 169.3, 169.0 (CO), 113.4 (qt C), 92.3 (CH), 92.1 (CH), 76.3 (CH), 75.4 (CH), 74.2, 73.3, 72.1, 70.3 (CH), 56.6 (CH), 53.4 (OCH_3), 26.6 (CH_3), 25.9 (CH_3), 20.3 (2), 19.8, 19.3 (CH_3). HRMS-ESI for $C_{22}H_{30}O_{15}$ Na: $[M + Na]^+$ calcd: 557.1585, found: 557.1452. Anal. Calc. For $C_{22}H_{30}O_{15}$: C, 49.44; H, 5.66; O, 44.90 Found C, 49.39; H, 5.70; O, 44.81.

2.9. (4R,5R)-4-Methyl-5-((2S,3R,5R,6R)-3,4,5-triacetoxy-6-(acetoxymethyl)tetrahydro-2H-pyran-2-yl)-2,2-dimethyl-1,3-dioxolane-4,5-dicarboxylate **7b**. Yield: 74%. Colorless gel $[\alpha]_D^{25}$ = 21.6. (IR υ cm^{-1}): 2993 (CH), 1744 (CO), 1361 (C-O-C). ^{1}HNMR (400 MHz, $CDCl_3$): δ (ppm): 6.47 (d, J = 3.9 Hz, 1H, H-1$^\alpha$), 5.53 (d, J = 8.3 Hz, 1H, H-1$^\beta$), 5.47, 4.93 (pseudo t, J ~ 6.9 Hz in each, CH), 5.50 (t, J = 8.9 Hz, CH), 4.94 (d, J = 6.3 Hz, CH), 4.88 (d, J = 6.3 Hz, CH), 4.27 (dd, J = 12.5, 3.6 Hz, CH), 4.14 (ddd, J = 10.4, 3.6, 2.1 Hz, CH), 3.96 (dd, J = 12.5, 2.1 Hz, CH), 3.86 (s, OCH_3), 2.19, 2.17, 2.12, 2.04 (s, CH_3), 1.55 (s, CH_3), 1.45 (s, CH_3). ^{13}C NMR (100 MHz, $CDCl_3$): δ (ppm): 172.3 ($COOCH_3$), 171.7, 169.4, 169.3, 169.1 (CO), 113.3 (qt C), 92.7 (CH), 92.4 (CH), 76.2 (CH), 75.6 (CH), 74.4, 73.5,

72.2, 70.2, (CH), 56.5 (CH), 53.2 (OCH_3), 25.6 (CH_3), 24.9 (CH_3), 20.6 (2), 19.7, 19.5 (CH_3). HRMS-ESI for $C_{22}H_{30}O_{15}$ Na: $[M + Na]^+$ calcd: 557.1585, found: 557.1466. Anal. Calc. For $C_{22}H_{30}O_{15}$: C, 49.44; H, 5.66; O, 44.90 Found C, 49.39; H, 5.70; O, 44.81.

*2.10. (4R,5R)-4-Methyl-5-((2R,3R,5R,6R)-3,4,5-triacetoxy-6-(acetoxymethyl)tetrahydro-2H-pyran-2-yl)-2,2-dimethyl-1,3-dioxolane-4,5-dicarboxylate **7c**.* Yield: 70%. White foam m.p = 151–153°C. $[\alpha]_D^{25}$ = 17.3. (IR υ cm^{-1}): 2995 (CH), 1745 (CO), 1336 (C-O-C). ^{1}HNMR (400 MHz, $CDCl_3$): δ (ppm): 6.42 (d, J = 3.7 Hz, 1H, H-1$^\alpha$), 5.51 (d, J = 4.0 Hz, 1H, H-1$^\beta$), 5.46, 5.43 (pseudo t, J ~ 5.6 Hz in each, CH), 4.96 (t, J = 9.8 Hz, CH), 4.93 (d, J = 5.3 Hz, CH), 4.89 (d, J = 5.3 Hz, CH), 4.26 (dd, J = 12.5, 3.6 Hz, CH), 4.13 (ddd, J = 10.4, 3.6, 2.1 Hz, CH), 3.93 (dd, J = 12.5, 2.1 Hz, CH), 3.76 (s, OCH_3), 2.15, 2.14, 2.12, 2.11 (s, CH_3), 1.56 (s, CH_3), 1.46 (s, CH_3). ^{13}C NMR (100 MHz, $CDCl_3$): δ (ppm): 171.3 ($COOCH_3$, 170.7, 169.6, 169.3, 169.1 (CO), 113.4 (qt C), 92.4 (CH), 92.1 (CH), 76.1 (CH), 75.6 (CH), 74.2, 73.2, 72.1, 70.5 (CH), 56.7 (CH), 53.3 (OCH_3), 26.6 (CH_3), 25.8 (CH_3), 20.2 (2), 19.8, 19.4 (CH_3). HRMS-ESI for $C_{22}H_{30}O_{15}$ Na: $[M + Na]^+$ calcd: 557.1585, found: 557.1475. Anal. Calc. For $C_{22}H_{30}O_{15}$: C, 49.44; H, 5.66; O, 44.90 Found C, 49.39; H, 5.70; O, 44.88.

*2.11. (4R,5R)-4-((2S,3R,5S,6R)-3-Acetamido-4,5-diacetoxy-6-(acetoxymethyl)tetrahydro-2H-pyran-2-yl)-5-methyl-2,2-dimethyl-1,3-dioxolane-4,5-dicarboxylate **7d**.* Yield: 67%. Colorless oil $[\alpha]_D^{25}$ = 45.0. (IR υ cm^{-1}): 3342 (NH), 2981 (CH), 1739 (CO), 1360 (C-O-C). ^{1}HNMR (400 MHz, $CDCl_3$): δ (ppm): 7.87 (d, J = 9.5 Hz, NH), 6.49 (d, J = 3.7 Hz, 1H, H-1$^\alpha$), 5.57 (d, J = 8.3 Hz, 1H, H-1$^\beta$), 5.46 (m, CH) 5.43, 4.96 (pseudo t, J ~ 7.4 Hz in each, CH), 4.94 (d, J = 6.3 Hz, CH), 4.83 (d, J = 6.3 Hz, CH), 4.37 (dd, J = 12.5, 3.6 Hz, CH), 4.23 (ddd, J = 10.4, 3.6, 2.1 Hz, CH), 3.93 (dd, J = 12.5, 2.1 Hz, CH), 3.87 (s, OCH_3), 2.13, 2.11, 2.10, 2.04 (s, CH_3), 1.57 (s, CH_3), 1.49 (s, CH_3). ^{13}C NMR (100 MHz, $CDCl_3$): δ (ppm): 171.6 ($COOCH_3$), 170.3, 169.3, 169.2, 169.0 (CO), 113.6 (qt C), 92.6 (CH), 92.2 (CH), 76.2 (CH), 75.1 (CH), 74.5, 73.3, 72.6, 70.1 (CH), 57.9 (CH), 53.5 (OCH_3), 26.4 (CH_3), 25.6 (CH_3), 20.1 (2), 19.5, 19.2, (CH_3). HRMS-ESI for $C_{22}H_{31}NO_{14}$ Na: $[M + Na]^+$ calcd: 556.1745, found: 556.1614. Anal. Calc. For $C_{22}H_{31}NO_{14}$: C, 49.53; H, 5.86; N, 2.63; O, 41.99 Found C, 49.55; H, 5.76; N, 2.60; O, 41.90.

*2.12. (4R,5R)-4-((2S,3R,5S,6R)-3-Acetamido-4,5-diacetoxy-6-(acetoxymethyl)tetrahydro-2H-pyran-2-yl)-5-methyl-2,2-dimethyl-1,3-dioxolane-4,5-dicarboxylate **7e**.* Yield: 63%. Colorless oil $[\alpha]_D^{25}$ = 33.3. (IR υ cm^{-1}): 3348 (NH), 2990 (CH), 1741 (CO), 1356 (C-O-C). ^{1}HNMR (400 MHz, $CDCl_3$): δ (ppm): 7.86 (d, J = 9.7 Hz, NH), 6.52 (d, J = 3.9 Hz, 1H, H-1$^\alpha$), 5.67 (d, J = 8.7 Hz, 1H, H-1$^\beta$), 5.46 (m, CH), 5.43, 4.96 (pseudo t, J ~ 6.0 Hz in each, CH), 4.93 (d, J = 5.4 Hz, CH), 4.82 (d, J = 5.3 Hz, CH), 4.38 (dd, J = 12.5, 3.6 Hz, CH), 4.33 (ddd, J = 10.4, 3.6, 2.1 Hz, CH), 3.93 (dd, J = 12.5, 2.1 Hz, CH), 3.85 (s, OCH_3), 2.14, 2.12, 2.10, 2.06 (s, CH_3), 1.56 (s, CH_3), 1.46 (s, CH_3). ^{13}C NMR (100 MHz, $CDCl_3$): δ (ppm): 171.4 ($COOCH_3$), 170.3, 169.5, 169.4, 169.2 (CO), 114.4 (qt C), 93.4 (CH), 92.7 (CH), 76.3 (CH), 75.2 (CH), 74.5, 73.4, 72.6, 70.1 (CH), 58.7 (CH), 53.5 (OCH_3), 26.6 (CH_3), 25.6 (CH_3), 20.1 (2), 19.4, 19.1 (CH_3). HRMS-ESI for $C_{22}H_{31}NO_{14}$ Na: $[M + Na]^+$ calcd: 556.1745, found: 556.1616. Anal. Calc. For $C_{22}H_{31}NO_{14}$: C, 49.53; H, 5.86; N, 2.63; O, 41.99 Found C, 49.55; H, 5.81; N, 2.60; O, 41.91.

3. Results and Discussion

As outlined in synthetic Schemes 1 and 2, our proposed strategy was to transform five different glycopyranoses to glycopyranosyl amines **4a–e** which would serve as nucleophiles. In sugars, the anomeric position is comparatively more reactive due to the presence of the endocyclic oxygen atom. Owing to its higher reactivity, microwave assisted Kochetkov amination was tried. The Kochetkov amination is used to introduce amino functionality at the anomeric position in unprotected sugars [25]. However, when glucose was reacted under microwave with Kochetkov reagents, that is, ammonium carbonate in DMSO, a mixture of amino and untreated glucose was obtained after prolonged lipophilization. In the case of glucosamine and galactosamine hydrochlorides, which already contain amino functionality, to ensure chemoselectivity in coupling with tartaric acid, a relative long route was adopted involving protecting the entire hydroxyl and amino functionalities on these sugars. Thus per acetylation [26], followed by anomeric bromination [27], azide formation [28] and reduction of azides afforded the target acetylated glycopyranosyl amines **4a–e**.

Following the same coupling procedure with increased time (Table 2), O-linked glycopyranosides **7a–e** were also prepared in 63–74% yields, starting from acetylated monosaccharides **1a–e**, Scheme 3.

It was observed that during per acetylation, a mixture of both α and β anomers is formed in a 3 : 1 ratio. In the bromination step, the thermodynamically more stable α anomer was obtained exclusively. In the resultant bromo analogue when left for longer time (two days), slight decomposition was observed. Therefore the product was used in the next step immediately. The reaction of bromide **3a** with sodium azide in DMF proceeded in low yield; however under sonication, the azide analogue was produced in excellent yield. The substitution involves inversion of configuration at the anomeric site and thus the α-glycopyranosyl bromide gave the corresponding β-glycopyranosyl azide. The azide intermediates were reduced immediately under catalytic hydrogenation in the next step. The hydrogenation reaction furnished the desired amines in quantitative yield, which were stable when completely dried under vacuum. Once a series of glycopyranosyl amines **4a–e** were generated, the next step was to synthesize the monoester **1** starting from commercially available L-tartaric acid. Both the hydroxyl and carboxyl groups were protected followed by partial hydrolysis with one equivalent of lithium hydroxide; however the yield was variable. By the use of a stronger base sodium hydroxide, the yield was improved. It is important to mention that the formation of monoester **1** from the diester does not depend

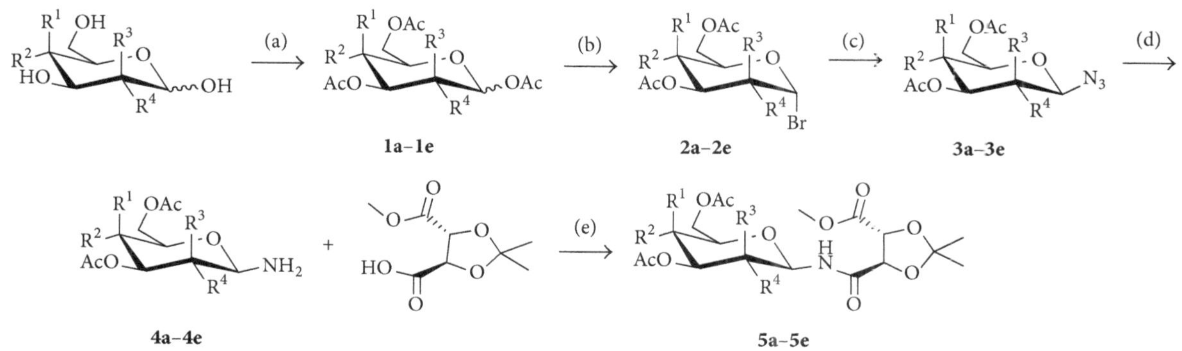

(**a**) D-Glucose $R^1, R^3 = H; R^2, R^4 = OH$
(**b**) D-Galactose $R^1, R^4 = OH; R^2, R^3 = H$
(**c**) D-Mannose $R^1, R^4 = H; R^2, R^3 = OH$
(**d**) D-Glucosamine·HCl $R^1, R^3 = H; R^2 = OH; R^4 = NH_2 \cdot HCl$
(**e**) D-Galactosamine·HCl $R^1 = OH; R^2, R^3 = H; R^4 = NH_2 \cdot HCl$

Scheme 1: Reagents and conditions: (a) Ac_2O, pyridine, cat. DMAP, N_2, r.t., overnight; (b) HBr/AcOH (30%), CH_2Cl_2, N_2, r.t., 2 h; (c) NaN_3, DMF, sonication, r.t., 15 min; (d) Pd/C (10%), H_2, MeOH, 30 min; (e) EDC, CH_2Cl_2, cat. DMAP, N_2, r.t., 18 h.

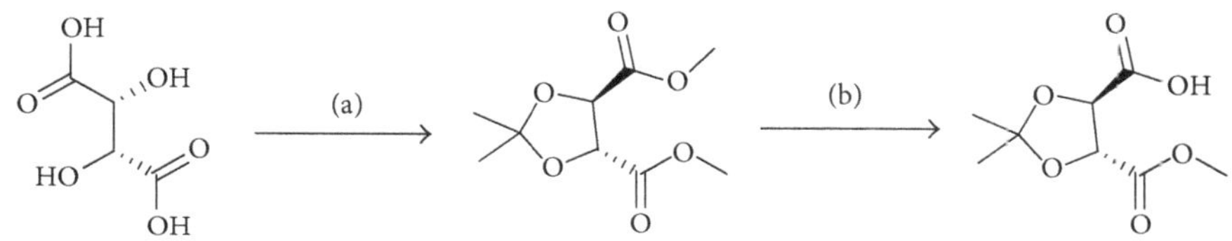

(1)

Scheme 2: Reagents and conditions: (a) MeOH, H^+, reflux, 2 h, 2-methoxypropene, CSA, DMF, N_2, r.t., overnight; (b) NaOH (1 eq), MeOH, r.t., 2 h.

on which ester group of the two is cleaved as both groups are homotopic and therefore monohydrolysis of any of these groups will lead to identical molecules. The same strategy was then used for all of the compounds in the series.

With glycopyranosyl amines and monoester in hand, it was decided to use coupling reagents to form peptides **5a–e**. Different coupling conditions were tried to generate the target amides, Table 1. DCC in the presence of a DMAP catalyst furnished the target amide in low yield. EDC when used in 1.3 equivalents for 12 h afforded the amides in good yield and the reaction was reproducible. Once coupling conditions were optimized, the same coupling conditions were used for the rest of the series to furnish the respective amides, Table 1.

The coupling reaction was high yielding and the compounds were purified by column chromatography. IR spectral data of the compounds **5a–e** exhibited the appearance of characteristic (NH) stretching bands in the range of 3347–3331 cm^{-1} and disappearance of NH_2 stretching. Two C=O stretching bands ranging from 1742 to 1736 and 1701 to 1692 cm^{-1} were assigned to ester and amide, respectively. ^{1}HNMR spectral data shows characteristic doublets for the NH protons in the range of 7.69–7.18 ppm with a coupling constant of 9.6–9.3 Hz. The anomeric protons appeared as pseudotriplets in the range of 5.42–5.36 ppm with coupling constant of 12.2–6.5 Hz which confirms the formation of the β-anomer. In ^{13}C NMR spectra, peaks ranging from 172.8–171.2 ppm were assigned to C=O (amide) and 170.9–170.2 ppm to C=O (ester), respectively. Elemental analysis of the glycopyranosyl amides **5a–e** confirmed the structures of the compounds. The exact mass of the compounds was confirmed by HRMS-ESI.

The anomeric acetyl group was hydrolyzed selectively using hydrazinium acetate, in DMF [29–31]; the resulting hemiacetal was washed with water and brine and used without further purification. To the crude hemiacetal, monoester was added using EDC as a coupling agent. When compound **6a** was coupled with monoester 1 using 1.3 eq. EDC, DMAP with increased reaction time afforded compound **7a** in 72% yield. Once the reaction conditions were optimized, the same conditions were used for the rest of the series to furnish the respective esters, Table 2.

The synthesized compounds were purified by column chromatography and characterized by spectroscopic techniques. IR spectral data of the compounds **7a–e** showed characteristic (C=O) stretching bands ranging from 1745 to 1739 cm^{-1} in addition to C-O-C stretching in 1366–1356. The NH stretching bands **7d-e** appeared in 3348–3342 cm^{-1}. ^{1}HNMR spectral data showed doublets **7d-e**, for the NH protons in the range of 7.87–7.36 ppm with coupling constant of 9.7–9.5 Hz. Both α- and β-anomers were found to be 1 : 3 except compound **7c** where α and β ratio was 1 : 1 as confirmed by ^{1}HNMR spectroscopy. The anomeric protons appeared as doublets in the range of 5.67–5.43 ppm with coupling constant of 8.3–8.0 Hz for the β-anomers

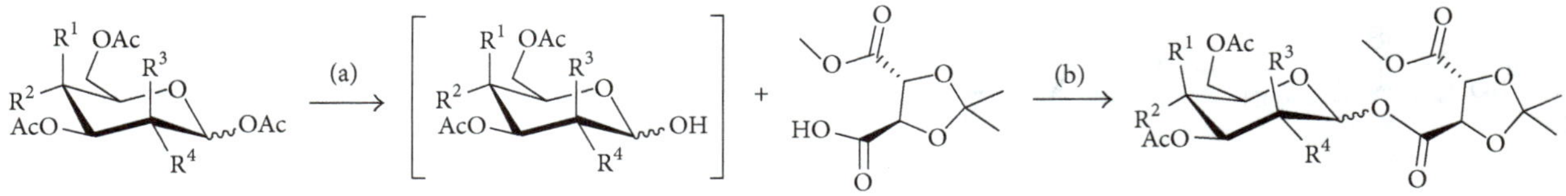

1a–1e **6a–6e** **7a–7e**

SCHEME 3: *Reagents and conditions*: (a) N_2H_4/AcOH (1.5 eq.), DMF, 55°C, N_2, 30 min; (b) EDC, CH_2Cl_2, cat. DMAP, N_2, rt, 18 h.

TABLE 1: Different coupling condition for amide synthesis.

Entry	Starting material		Reagents/conditions	Product (isolated yield%)
1	**4a**	HA	1.2 eq. DCC, cat. DMAP, CH_2Cl_2, r.t., 8 h	46
2	**4a**	HA	1 eq. EDC, cat. DMAP, CH_2Cl_2, r.t., 8 h	36
3	**4a**	HA	1.1 eq. EDC, cat. DMAP, CH_2Cl_2, r.t., 12 h	70
4	**4a**	HA	1.3 eq. EDC, cat. DMAP, CH_2Cl_2, r.t., 12 h	76
5	**5a**	HA	1.3 eq. EDC, cat. DMAP, CH_2Cl_2, r.t., 12 h	61
6	**6a**	HA	1.3 eq. EDC, cat. DMAP, CH_2Cl_2, r.t., 18 h	79

and in 6.52–6.42 ppm with coupling constant of 4.0–3.7 Hz for the α-anomers, respectively. The coupling constant of anomeric hydrogen for mannopyranosyl **7c** was 4.0 Hz in both the anomers. Finally the synthesis of glycopyranosides was supported by one of the crystal structure for α-anomer, *O-(2,3,4,6-tetra-O-acetyl-α-D-mannopyranosyl)-(4R,5R)-2,2-dimethyl-1,3-dioxolane-4,5-dicarboxylate* **7c** by X-ray crystallography (Figure 1).

Crystal system = monoclinic. Torsion angles for compound **7c** are C5-O1-C1-O13 = −62.9(2)°, C5-O1-C1-C2 = 54.9(2)°, C15-O13-C1-O1 = −76.4(2) (2)°, C15-O13-C1-C2 = 160.28(19)°, C13-O8-C2-C3 = −103.8(2)°, C13-O8-C2-C1 = 136.00(19)°, O1-C1-C2-O8 = 73.3(2)°, O13-C1-C2-O8 = −165.60(16)°, O1-C1-C2-C3 = −46.9(2)°, O13-C1-C2-C3 = 74.2(2)°, C11-O6-C3-C4 = −149.8(2)°, C11-O6-C3-C2 = 88.7(3)°, O8-C2-C3-O6 = 49.8(2)°, C1-C2-C3-O6 = 167.17(17)°, O8-C2-C3-C4 = −69.0(2)°, C1-C2-C3-C4 = 48.4(3)°, C9-O4-C4-C3 = −119.6(3)°, C9-O4-C4-C5 = 123.8(3)°, O6-C3-C4-O4 = 67.2(2)°, C2-C3-C4-O4 = −172.34(18)°, O6-C3-C4-C5 = −176.24(17)°, C2-C3-C4-C5 = −55.8(2)°, C1-O1-C5-C6 = 176.43(19)°, C1-O1-C5-C4 = −61.4(2)°, O4-C4-C5-O1 = 176.08(18)°, C3-C4-C5-O1 = 60.3(2)°, O4-C4-C5-C6 = −65.2(3)°, C3-C4-C5-C6 = 179.0(2)°, C7-O2-C6-C5 = −153.1(2)°, O1-C5-C6-O2 = −68.2(3)°, C4-C5-C6-O2 = 172.1(2)°, C6-O2-C7-O3 = 1.9(5)°, C6-O2-C7-C8 = −175.2(3)°, C4-O4-C9-O5 = −7.1(6)°, C4-O4-C9-C10 = 174.8(3)°, C3-O6-C11-O7 = 3.7(4)°, C3-O6-C11-C12 = −175.5(2)°, C2-O8-C13-O9 = 6.4(3)°, C2-O8-C13-C14 = −171.9(2)°, C1-O13-C15-O12 = −5.2(4)°, C1-O13-C15-C16 = 173.8(2)°, C19-O15-C16-C15 = 131.8(4)°, C19-O15-C16-C17 = 11.0(5)°, O12-C15-C16-O15 = −7.6(4)°, O13-C15-C16-O15 = 173.5(2)°, O12-C15-C16-C17 = 108.9(4)°, O13-C15-C16-C17 = −70.1(3)°, C19-O14-C17-C18 = 130.6(5)°, C19-O14-C17-C16 = 11.5(6)°, O15-C16-C17-O14 = −13.4(5)°, C15-C16-C17-O14 = −133.2(4)°, O15-C16-C17-C18 = −131.8(3)°, C15-C16-C17-C18 = 108.4(3)°, C22-O11-C18-O10 = −2.3(9)°, C22-O11-C18-C17 = 178.0(4)°, O14-C17-C18-O10 = −31.2(7)°, C16-C17-C18-O10 = 82.9(6)°, O14-C17-C18-O11 = 148.5(4)°, C16-C17-C18-O11 = −97.4(4)°, C17-O14-C19-O15 = −5.3(7)°, C17-O14-C19-C21 = 109.3(8)°, C17-O14-C19-C20 = −123.6(6)°, C16-O15-C19-O14 = −4.3(6)°, C16-O15-C19-C21 = −121.1(8)°, and C16-O15-C19-C20 = 113.7(6)°.

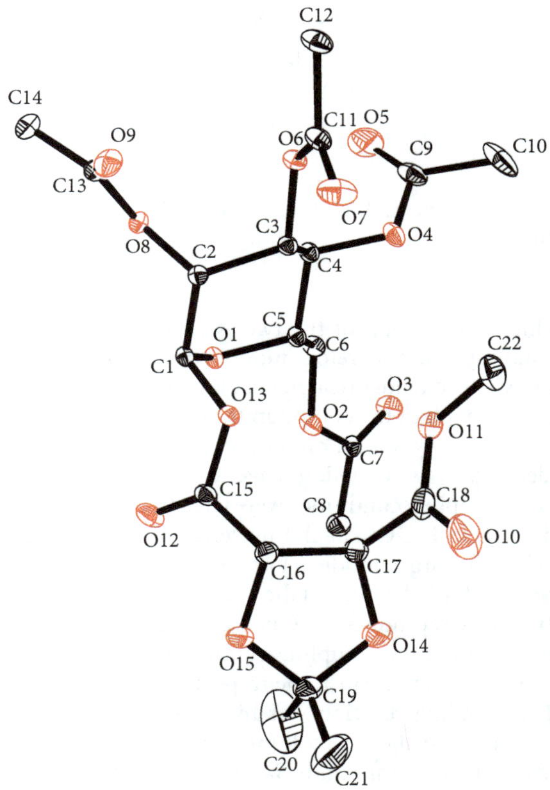

FIGURE 1: Crystal structure of α-anomer, *O-(2,3,4,6-tetra-O-acetyl-α-D-mannopyranosyl)-(4R,5R)-2,2-dimethyl-1,3-dioxolane-4,5-dicarboxylate* **7c**.

TABLE 2: Coupling conditions and isolated yield for **5a–e** and **7a–e**.

Entry	Starting material		Reagents/conditions	Product (isolated yield%)	$\alpha:\beta$ ratio	$[\alpha]_D^{25}$
1	**4a**	HA	1.3 eq. EDC, cat. DMAP, r.t., 12 h	72	0 : 1	12.5
2	**4b**	HA	1.3 eq. EDC, cat. DMAP, r.t., 12 h	69	0 : 1	13.3
3	**4c**	HA	1.3 eq. EDC, cat. DMAP, r.t., 12 h	65	0 : 1	27.1
4	**4d**	HA	1.3 eq. EDC, cat. DMAP, r.t., 12 h	67	0 : 1	43.3
5	**4e**	HA	1.3 eq. EDC, cat. DMAP, r.t., 12 h	68	0 : 1	47.1
6	**6a**	HA	1.3 eq. EDC, cat. DMAP, r.t., 18 h	72	1 : 3	26.5
7	**6b**	HA	1.3 eq. EDC, cat. DMAP, r.t., 18 h	74	1 : 3	21.6
8	**6c**	HA	1.3 eq. EDC, cat. DMAP, r.t., 18 h	70	1 : 1	17.3
9	**6d**	HA	1.3 eq. EDC, cat. DMAP, r.t., 18 h	67	1 : 3	45.0
10	**6e**	HA	1.3 eq. EDC, cat. DMAP, r.t., 18 h	63	1 : 3	33.3

$c = 24$ mg/2 mL in dichloromethane.

TABLE 3: % inhibition of compounds **5a–e** and **7a–e** against *L. tropica leishmania*.

Compound	*L. tropica*[a]	Compound	*L. tropica*[a]
5a	0.65 ± 0.01	**7a**	1.35 ± 0.04
5b	0.68 ± 0.09	**7b**	0.73 ± 0.12
5c	0.81 ± 0.16	**7c**	0.63 ± 0.18
5d	0.68 ± 0.27	**7d**	0.77 ± 0.08
5e	0.89 ± 0.11	**7e**	0.66 ± 0.18
	Standard drug IC_{50}	(μg/mL ± SD)	
	Amphotericin B	0.56 ± 0.20	

[a]Percentage inhibition activity: 100 = (nonsignificant; 0.95–0.80 = low; 0.79–0.70 = moderate; 0.69–0.60 = good; below 0.59–0.56 = significant activity).

4. Material and Methods

4.1. Antileishmanial Activity. Antileishmanial activities of compounds **5a–e** and **7a–e** were assayed by Zhai et al.'s method (1999) [32] using a preestablished culture of *Leishmania*. Triple-N media slants were overlaid with 199 media for *leishmanial* growth. NNN medium was prepared by mixing 4 g of agar in 100 mL of distilled water. The mixture was then dissolved and sterilized by autoclaving at 121°C and then allowed to cool to 55°C. 15–20 mL of defibrinated sheep blood was aseptically added to the mixture with gentle rolling with glass beads; 1 ampoule of gentamicin was mixed in blood and mixed with the agar mixture. Slopes of culture medium were prepared by dispensing 2-3 mL of the blood agar mixture into sterile tubes that were then set in a slant position until the agar completely solidified. To prepare 199 medium different constituents were mixed in 1000 mL of distilled water. The pH was adjusted to 7.4 and medium was filtered, sterilized, and kept at 37°C for 24 h to check sterility. Preestablished culture of *Leishmania tropica* KWH23 was inoculated in 199 medium in Triple-N medium slants and incubated at 24°C for 6-7 days and the results are summarized in Table 3.

5. Results

All the synthesized *N*-linked **5a–e** and *O*-linked glycopyranosides **7a–e** were tested for their antileishmanial activity using *Leishmania tropica* KWH23 promastigotes for *in vitro* screening. The results are shown in Table 3. Compounds **5a, 5b,** and **5d** showed good activity while the compounds **5c** and **5e** exhibited low activity.

Among the synthesized *O*-linked glycopyranosides **7a–7e**, compounds **7c** and **7e** showed good activity; **7b** and **7d** showed moderate activity. Compound **7a** showed nonsignificant activity.

6. Conclusion

Most of the synthesized compounds **5a–e** and **7a–e** showed moderate to good activities against *Leishmania tropica* KWH23 promastigotes. Compounds **5a, 5b,** and **5d** were the most active. Among the synthesized compounds, **7a–7e, 7c,** and **7e** were more active than **7a, 7b,** and **7d**. Drug resistance has been reported in various species of *Leishmania* against various antileishmanial drugs like antimonials, amphotericin B, pentamidine, miltefosine, and so forth. In this scenario these synthetic compounds may be further explored for research in medicinal chemistry and drug designing. These compounds may prove to be good candidates against leishmaniasis. A number of other new compounds can also be synthesized while following the same *N*- and *O*-acylation synthetic routes. Such compounds can act as valuable glycodrugs as many of the glycosides and their derivatives have been reported to be biologically active compounds.

Conflicts of Interest

The authors confirm that this article's content has no conflicts of interest.

Acknowledgments

This work was financially supported by Higher Education Commission (HEC), Pakistan, under International Research Support Initiative Program (IRSIP) and National Research Program (NRPU) for Universities.

Supplementary Materials

Supplementary data include NMR (^{1}H and ^{13}C), IR, and MS spectra of the synthesized compounds (**5a**– **5d** and **7a**-**7e**). ^{1}HNMR of compound **5a**. ^{13}CNMR of compound **5a**. ESI-MS of compound **5a**. FT-IR of compound **5a**. ^{1}HNMR of compound **5b**. ^{13}CNMR of compound **5b**. ESI-MS of compound **5b**. ^{1}HNMR of compound **5c**. ^{13}CNMR of compound **5c**. ^{1}HNMR of compound **5d**. ^{13}CNMR of compound **5d**. ESI-MS of compound **5d**. ^{1}HNMR of compound **7a**. ^{13}CNMR of compound **7a**. ESI-MS of compound **7a**. ^{1}HNMR of compound **7b**. ^{13}CNMR of compound **7b**. ESI-MS of compound **7b**. ESI-MS of compound **7b** backup. ^{1}HNMR of compound **7c**, new. ^{1}HNMR of compound **7c**. ^{13}CNMR of compound **7c**. ^{13}CNMR of compound **7b**, new. ESI-MS of compound **7c**. Mass analysis of compound **7c**. ^{1}HNMR of compound **7d**. ^{13}CNMR of compound **7d**. ESI-MS of compound **7d**. Mass analysis of compound **7d**. ^{1}HNMR of compound **7e**. ^{13}CNMR of compound **7e**. *(Supplementary Materials)*

References

[1] N. Singh, B. B. Mishra, S. Bajpai, R. K. Singh, and V. K. Tiwari, "Natural product based leads to fight against leishmaniasis," *Bioorganic & Medicinal Chemistry*, vol. 22, no. 1, pp. 18–45, 2014.

[2] P. Dwivedi, K. B. Mishra, B. B. Mishra, N. Singh, R. K. Singh, and V. K. Tiwari, "Click inspired synthesis of antileishmanial triazolyl O-benzylquercetin glycoconjugates," *Glycoconjugate Journal*, vol. 32, no. 3-4, pp. 127–140, 2015.

[3] B. B. Mishra, J. K. Gour, N. Kishore, R. K. Singh, V. Tripathi, and V. K. Tiwari, "An antileishmanial prenyloxy-naphthoquinone from roots of plumbago zeylanica," *Natural Product Research (Formerly Natural Product Letters)*, vol. 27, no. 4-5, pp. 480–485, 2013.

[4] W. B. Leishman, "On the possibility of the occurrence of trypanosomiasis in India," *British Medical Journal*, vol. 1, no. 2213, pp. 1252–1254, 1903.

[5] C. Donovan, "On the possibility of the occurrence of trypanosomia-sis in India," *British Medical Journal*, vol. 2, pp. 79–82, 1903.

[6] C. Nicolle and C. Comte, "Origine canine du Kala-azar," *Bulletin So-ciété Pathologie Exotique*, vol. 1, pp. 299–301, 1908.

[7] WHO (World Health Organisation), "Communicable disease sur-veillance and response," http://www.who.int/emc/diseases/leish/leisdis1.html.

[8] S. L. Croft, "The current status of antiparasite chemotherapy," *Parasitology*, vol. 114, no. 1, pp. S3–S15, 1997.

[9] H. D. Engers, R. Bergquist, and F. Moddaber, "Progress on vaccines against parasite," *Developments in Biological Standardization*, vol. 87, pp. 73–84, 1996.

[10] H. Hojo and N. Nakahara, "Recent Progress in the Solid-phase Synthesis of Glycopeptide," *Current Protein & Peptide Science*, vol. 1, no. 1, pp. 23–48, 2000.

[11] H. Lis and N. Sharon, "Protein glycosylation: structural and functional aspects," *European Journal of Biochemistry*, vol. 218, no. 1, pp. 1–27, 1993.

[12] A. Varki, "Biological roles of oligosaccharides: all of the theories are correct," *Glycobiology*, vol. 3, no. 2, pp. 97–130, 1993.

[13] S.-Y. Han and Y.-A. Kim, "Recent development of peptide coupling reagents in organic synthesis," *Tetrahedron*, vol. 60, no. 11, pp. 2447–2467, 2004.

[14] R. B. Greenwald, H. Zhao, J. Xia, and A. Martinez, "Poly(ethylene glycol) transport forms of vancomycin: a long-lived continuous release delivery system," *Journal of Medicinal Chemistry*, vol. 46, no. 23, pp. 5021–5030, 2003.

[15] C. Dai, D. Li, J. Popovici-Muller et al., "2-(2-Aminothiazol-4-yl)pyrrolidine-based tartrate diamides as potent, selective and orally bioavailable TACE inhibitors," *Bioorganic and Medicinal Chemistry Letters*, vol. 21, no. 10, pp. 3172–3176, 2011.

[16] J.-B. Behr, T. Gourlain, A. Helimi, and G. Guillerm, "Design, synthesis and biological evaluation of hetaryl-nucleoside derivatives as inhibitors of chitin synthase," *Bioorganic & Medicinal Chemistry Letters*, vol. 13, no. 10, pp. 1713–1716, 2003.

[17] A. Dahlgren, J. Branalt, I. Kvarnström, I. Nilsson, D. Musil, and B. Samuelsson, "Synthesis of potential thrombin inhibitors. Incorporation of tartaric acid templates as P2 proline mimetics," *Bioorganic and Medicinal Chemistry*, vol. 10, no. 5, pp. 1567–1580, 2002.

[18] H. B. Küçük, A. Yusufoğlu, E. Mataraci, and S. Dösler, "Synthesis and biological activity of new 1,3-dioxolanes as potential antibacterial and antifungal compounds," *Molecules*, vol. 16, no. 8, pp. 6806–6815, 2011.

[19] S. Li, G. Liao, J. Xiao et al., "Derivatives of substituted tartaric acid and usage for preparing beta-secretase inhibitors," US 8268884 B2, 2012.

[20] B. B. Mishra and V. K. Tiwari, "Natural products: an evolving role in future drug discovery," *European Journal of Medicinal Chemistry*, vol. 46, no. 10, pp. 4769–4807, 2011.

[21] Siemens, SMART and SAINT, Siemens Analytical X-Ray Instruments Inc., Madison, Wisconsin, USA, 1996.

[22] A. Altomare, G. Cascarano, C. Giacovazzo, and A. Guagliardi, "Completion and refinement of crystal structures with SIR92," *Journal of Applied Crystallography*, vol. 26, no. 3, pp. 343–350, 1993.

[23] G. M. Sheldrick, SHELXTL-PC (Version 5.1), Siemens analytical instruments, Inc., Madison, Wisconsin, Wis, 1997.

[24] C. K. Jhnson, *ORTEPII Report ORNL-5138*, Oak Ridge National Laboratory, Tennessee, Tenn, USA, 1976.

[25] L. M. Likhosherstov, O. S. Novikova, V. A. Derevitskaja, and N. K. Kochetkov, "A new simple synthesis of amino sugar β-d-glycosylamines," *Carbohydrate Research*, vol. 46, no. 1, pp. C1–C5, 1986.

[26] P. J. Garegg, "Saccharides of biological importance: challenges and opportunities for organic synthesis," *Accounts of Chemical Research*, vol. 25, no. 12, pp. 575–580, 1992.

[27] H. J. Berthold, S. Franke, J. Thiem, and T. Schotten, "Ex post glycoconjugation of phthalocyanines," *The Journal of Organic Chemistry*, vol. 75, no. 11, pp. 3859–3862, 2010.

[28] V. V. Sureshbabu, R. Venkataramanarao, and H. P. Hemantha, "A facile synthesis of C-terminal neoglycopeptides: incorporation of urea moiety between sugars and peptides employing curtius rearrangement," *International Journal of Peptide Research and Therapeutics*, vol. 14, no. 1, pp. 34–40, 2008.

[29] T. Ren and D. Liu, "Synthesis of targetable cationic amphiphiles," *Tetrahedron Letters*, vol. 40, no. 43, pp. 7621–7625, 1999.

[30] R. R. Schmidt, "New methods for the synthesis of glycosides and oligosaccharides—are there alternatives to the koenigs-knorr method?" *Angewandte Chemie International Edition*, vol. 25, no. 3, pp. 212–235, 1986.

[31] G. Excoffier, D. Gagnaire, and J.-P. Utille, "Coupure sélective par l'hydrazine des groupements acétyles anomères de résidus glycosyles acétylés," *Carbohydrate Research*, vol. 39, no. 2, pp. 368–373, 1975.

[32] L. Zhai, M. Chen, J. Blom, T. G. Theander, S. B. Christensen, and A. Kharazmi, "The antileishmanial activity of novel oxygenated chalcones and their mechanism of action," *Journal of Antimicrobial Chemotherapy*, vol. 43, no. 6, pp. 793–803, 1999.

16

Confirmed Mechanism for 1,8-Diaminonaphthalene and Ethyl Aroylpyrovate Derivatives Reaction, DFT/B3LYP, and Antimicrobial Activity of the Products

Sraa Abu-Melha

Department of Chemistry, Faculty of Science of Girls, King Khaled University, Abha, Saudi Arabia

Correspondence should be addressed to Sraa Abu-Melha; sabomlha@kku.edu.sa

Academic Editor: Xinyong Liu

Two new series of perimidine derivatives were prepared from the reaction of 1,8-diaminonaphthalene with ethyl aroylpyrovate followed by coupling for the products. The structures, the mechanism, and the tautomeric forms of the products have been discussed on the basis of spectral data aided with the computational study. Applying Guassian09 software, the reaction mechanism was assured. The energy contents were computed after optimization for all proposed structures. The likely extracted compound has a reduced heat during formation and internal energy which coincides with experimental outcomes. The antimicrobial activity of all products was screened, and their results indicated the presence of five derivatives more potent than the reference drugs used. The simulation procedure was executed by Autodock 4.2 tools over the expected compounds yielded (**12** and **21**). This computational technique asserts on the drug behavior inside the causative organism proteins. The organisms here match with that used in the antimicrobial and antifungal study.

1. Introduction

It was reported that the reaction of 1,2-diaminoaromatic compound **1** with 1,2-dicarbonyl compounds **2** or **3** afforded the quinoxaline or pyrazin-2(1*H*)-one derivatives **4** and **5** [1, 2]. However, condensation of 1,8-diaminonaphthalene with benzil **2** or alkyl carboxylate derivative of 3-furanone **7** furnished the perimidine derivatives **8** and **9**, respectively [3–5] (Schemes 1 and 2). Perimidine ring (peri-naptho-fused pyrimidine systems) is a heterocycle containing two nitrogen atoms which is the unusual system with an excess and deficiency of *p*-electrons simultaneously [6, 7]. This feature makes them susceptible to electrophilic [8, 9] and nucleophilic attack [9]. The biological activities of perimidines have attracted our attention due to their wide range of application, for example, antitumor, antimicrobial, antifungal, and antiulcer [10–12]. Also, some of them were applied as coupler substances for preparation of oxidation dyestuffs, hair colouring compositions [13, 14], and fluorescent molecular sensors [15].

On the other side, ethyl aroylpyrovates **10** contain three different carbonyl groups with a different polarity and reactivity which condensed with various reagents to afford new heterocyclic compounds [16, 17]. It was proven that the most reactive carbonyl is that adjacent to the ester group [1, 18] (Scheme 3).

From all the previous screening, the author was interested herein to investigate the reactivity of ethyl aroylpyrovate **10** in the reaction with 1,8-diaminonaphthalene and elucidate the product or products of such reaction. This due to the previous reaction can afford one or more of the seven products which will be illustrated (Scheme 4). Moreover, Gaussian09 computation software was considered to establish a well view about the reaction mechanism [19]. DFT/B3LYP is a suitable method for executing the optimization process [20] in the gas phase and ethanol solvent for comparative assertion. Essential parameters have been considered as heat during formation, frontier energy gap, dipole moment, and oscillator strength. Such parameters were intended due to their convinced impact in

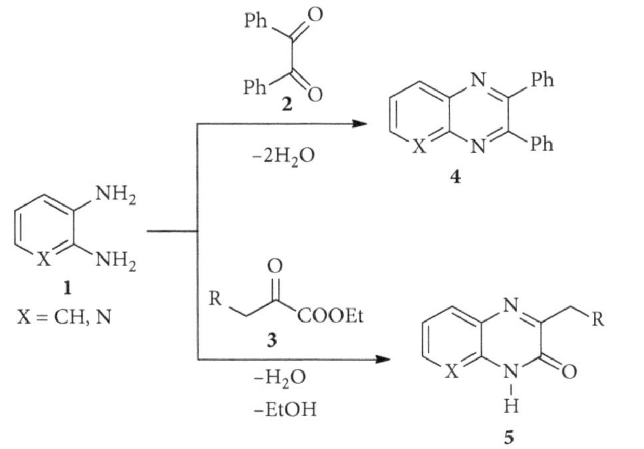

SCHEME 1: Reaction of 1,2-diaminoaromatic compound **1** with dicarbonyl compounds **2** and **3** [1, 2].

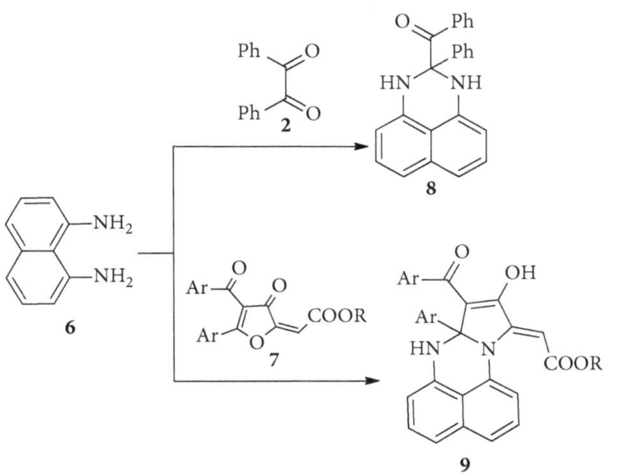

SCHEME 2: Condensation of 1,8-diaminonaphthalene with benzyl **2** or alkyl carboxylate derivative of 3-furanone **7** [3–5].

stability differentiation between all possible outputs [21]. As well as investigating the antimicrobial activity for novel synthesized derivatives was also intended. The simulation procedure for antimicrobial behavior was also considered to strengthen the results.

2. Experimental

2.1. Physical Techniques. All melting points were uncorrected and measured on the Gallenkamp electric melting point apparatus (capillary method, Gallenkamp Co., London, UK). Infrared spectra were determined on a Mattson 5000 FT-IR spectrometer (Shimadzu Co., Kyoto, Japan, not all frequencies are reported) using KBr discs. The nuclear magnetic resonance NMR spectra were determined using a WP 300 spectrometer (Bruker Co., Billerica, MA, USA) at 300 MHz for ^{1}H or 75.5 MHz for ^{13}C using TMS as the internal standard. The mass spectra were recorded on a Qp-2010 mass spectrometer (Shimadzu, Tokyo, Japan) at 70 eV (EI mode). Elemental analyses (C, H, and N) were determined on the PerkinElmer 2400 analyzer. The reactions were pursued by TLC (aluminum sheets, Merck), and the spots were disclosed by exposure to the UV lamp.

2.2. Synthesis of Ethyl 2-(2-Oxo-2-arylethyl)-2,3-dihydro-1H-perimidine-2-carboxylate (***12a–e***). In ethanolic solution, a mixture of 1,8-diaminonaphthalene **6** (0.40 g, 2.5 mmol) and ethyl aroylpyrovate **10** (2.5 mmol) was refluxed for 5 hrs. The reaction mixture was allowed to cool at room temperature, and then, the solid formed was collected by filtration, dried, and recrystallized from ethanol.

2.2.1. Ethyl 2-(2-Oxo-2-phenylethyl)-2,3-dihydro-1H-perimidine-2-carboxylate (***12a***). Dark red solid (yield, 70%), m.p. = 190–192°C. IR (KBr) ν_{max} 3435, 3317 (NH), 1725, 1628 (2C=O). ^{1}H NMR (DMSO-d_6) δ 1.33 (t, J = 7.2 Hz, 3H, CH_3), 3.89 (s, 2H, CH_2), 4.32 (q, J = 7.2 Hz, 2H, CH_2), 6.60–8.08 (m, 11H, Ar-H), 10.80 (brs, 2H, 2NH); MS (EI): m/z (%): 360 (M^+, 20), 347 (25), 256 (40), 170 (20), and 92 (100). Analysis for $C_{22}H_{20}N_2O_3$ (360.41): calcd: C, 73.32; H, 5.59; and N, 7.77%. Found: C, 73.15; H, 5.39; and N, 7.60%.

2.2.2. Ethyl 2-(2-Oxo-2-(p-tolyl)ethyl)-2,3-dihydro-1H-perimidine-2-carboxylate (***12b***). Dark red solid (yield, 74%), m.p. = 172–174°C. IR (KBr) ν_{max} 3443, 3316 (NH), 1717, 1610 (2C=O). ^{1}H NMR (DMSO-d_6) δ 1.33 (t, J = 7.0 Hz, 3H, CH_3), 2.37 (s, 3H, CH_3), 3.87 (s, 2H, CH_2), 4.32 (q, J = 7.0 Hz, 2H, CH_2), 6.60–7.92 (m, 10H, Ar-H), 10.89 (brs, 2H, 2NH); MS (EI): m/z (%): 374 (M^+, 10), 350 (60), 344 (100), 336 (63), 328 (38), 313 (52), 285 (28), 275 (25), 241 (18), and 134 (12). Analysis for $C_{23}H_{22}N_2O_3$ (374.44): calcd: C, 73.78; H, 5.92; and N, 7.48%. Found: C, 73.59; H, 5.81; and N, 7.26%.

2.2.3. Ethyl 2-(2-(4-Chlorophenyl)-2-oxoethyl)-2,3-dihydro-1H-perimidine-2-carboxylate (***12c***). Dark red solid (yield, 68%), m.p. = 180–182°C. IR (KBr) ν_{max} 3434, 3316 (NH), 1716, 1627 (2C=O). ^{1}H NMR (DMSO-d_6) δ1.32 (t, J = 7.0 Hz, 3H, CH_3), 3.86 (s, 2H, CH_2), 4.33 (q, J = 7.0 Hz, 2H, CH_2), 6.49–7.98 (m, 10H, Ar-H), 10.82 (brs, 2H, 2NH); MS (EI): m/z (%): 396 (M^++2, 10), 395 (M^++1, 11), 394 (M^+, 25), 361 (26), 347 (28), 319 (52), 277 (75), 257 (100), 111 (14), 105 (54), and 91 (32). Analysis for $C_{22}H_{19}ClN_2O_3$ (394.85): calcd: C, 66.92; H, 4.85; and N, 7.09%. Found: C, 66.78; H, 4.74; and N, 6.89%.

2.2.4. Ethyl 2-(2-(4-Fluorophenyl)-2-oxoethyl)-2,3-dihydro-1H-perimidine-2-carboxylate (***12d***). Dark red solid (yield, 73%), m.p. = 168–170°C. IR (KBr) ν_{max} 3429 (br. NH), 1671, 1610 (2C=O). ^{1}H NMR (DMSO-d_6) δ 1.22 (t, J = 7.0 Hz, 3H, CH_3), 3.13 (s, 2H, CH_2), 3.67 (q, J = 7.0 Hz, 2H, CH_2), 6.29–8.04 (m, 10H, Ar-H), 10.50 (brs, 2H, 2NH); ^{13}C NMR (DMSO-d_6) δ 11.91 (CH_3), 35.4 (CH_2), 42.69 (CH_2), 100.42 (C2), [113.85, 115.26, 115.55, 118.25, 119.33, 120.57, 122.49, 127.63, 128.41, 130.42, 133.46, 135.16, 148.63, 149.05 (Ar-C)], 163.20 (C=O), 196.23 (C=O). MS (EI): m/z (%): 378 (M^+, 12), 377 (M^+−1, 3.4), 367 (28), 360 (22), 348 (6), 336

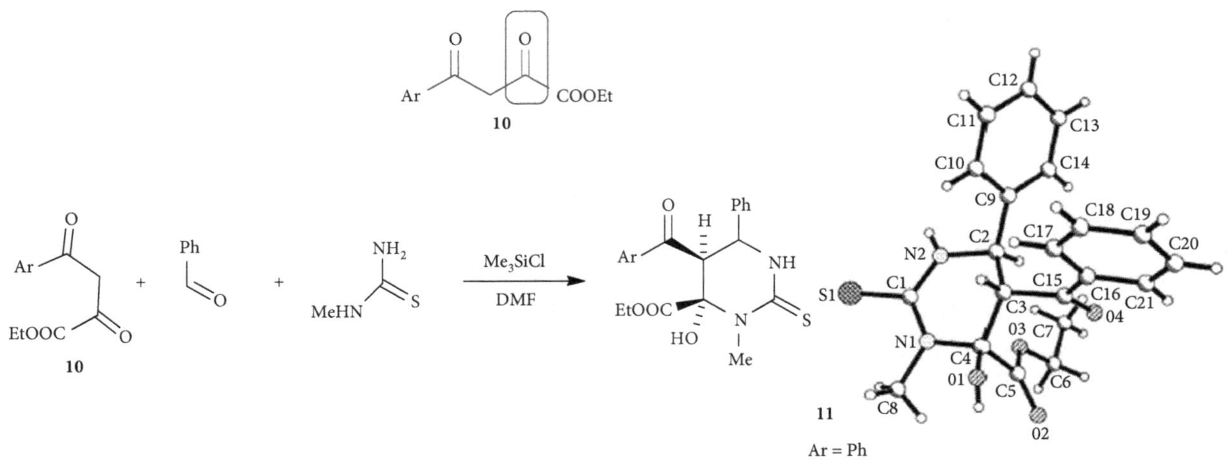

SCHEME 3: The X-ray analyses which prove the reactivity of carbonyl group-2 of ethyl aroylpyrovate **10** [1, 18].

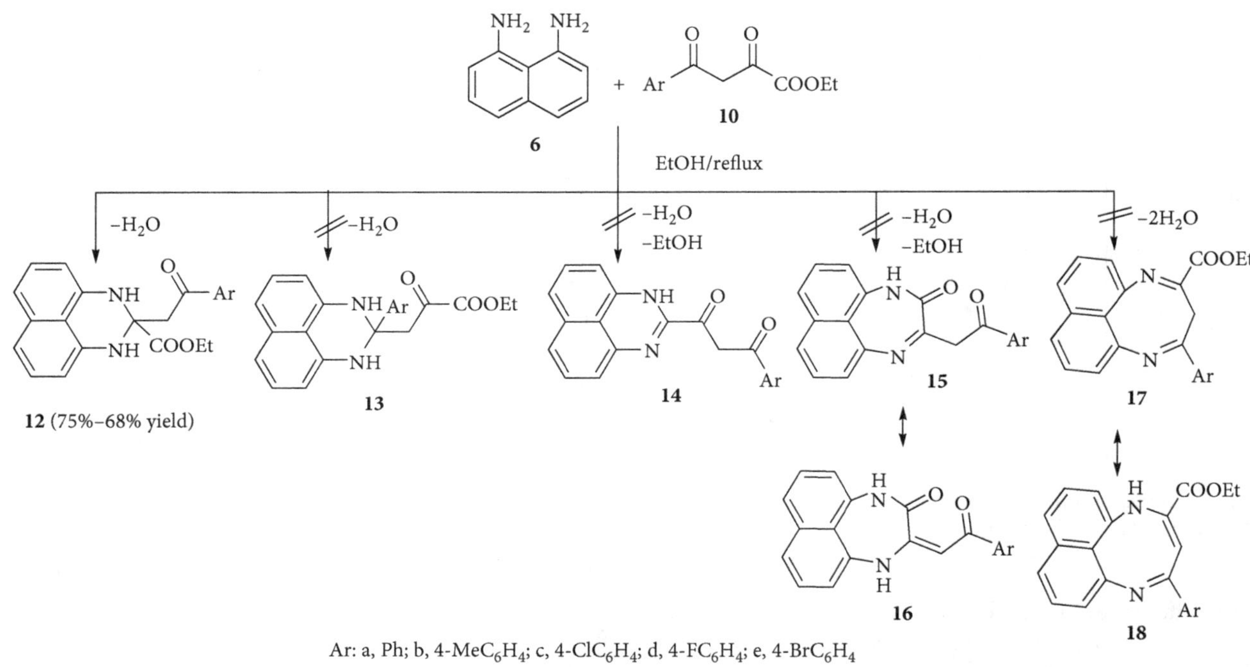

SCHEME 4: Reaction of 1,8-diaminonaphthalene **6** with ethyl aroylpyrovate derivatives **10a–e**.

(27), 332 (12), 304 (14), 168 (2.4), 156 (1.1), 122 (2.3), 93 (27), 92 (23), and 77 (45). Analysis for $C_{22}H_{19}FN_2O_3$ (378.40): calcd: C, 69.83; H, 5.06; and N, 7.40%. Found: C, 69.61; H, 4.94; and N, 7.52%.

*2.2.5. Ethyl 2-(2-(4-Bromophenyl)-2-oxoethyl)-2,3-dihydro-1H-perimidine-2-carboxylate (**12e**).* Dark red solid (yield, 75%), m.p. = 175–176°C. IR (KBr) ν_{max}3442, 3318 (NH), 1717, 1627 (2C=O). ^{1}H NMR (DMSO-d_6) δ 1.33 (*t*, J = 7.0 Hz, 3H, CH_3), 3.87 (*s*, 2H, CH_2), 4.34 (*q*, J = 7.0 Hz, 2H, CH_2), 6.47–7.99 (*m*, 10H, Ar-H), 10.91 (brs, 2H, 2NH); MS (EI): m/z (%): 441 (M^++2, 1.8), 440 (M^++1, 12), 439 (M^+, 3), 424 (34), 412 (23), 392 (14), 381 (100), 380 (23), and 154 (1.1). Analysis for $C_{22}H_{19}BrN_2O_3$ (439.31): calcd: C, 60.15; H, 4.36; and N, 6.38%. Found: C, 60.03; H, 4.27; and N, 6.19%.

*2.3. Coupling Reaction of Compounds **12a–e** to Synthesize Perimidine Derivatives **21a–e**.* To a stirred solution of ethyl 2-(2-aryl-2-oxoethyl)-2,3-dihydro-1H-perimidine-2-carboxylate derivatives **12a–e** (1 mmole) in ethanol, sodium hydroxide solution (0.04 g of NaOH in 15 mL ethanol with two drops of water), was added and the reaction mixture was cooled in an ice bath to 0–5°C. To the resulting solution, while being stirred, was added dropwise over a period of 10 min a solution of aniline diazonium chloride, prepared as usual by diazotizing aniline (1 mmole) in hydrochloric acid (6 M, 1 mL) with sodium nitrite (1 M, 1 mL). The whole mixture was then left in a refrigerator overnight. The precipitated solid was filtered, washed with water, and finally crystallized from ethanol/dioxane to give the respective hydrazone derivatives **21a–e**.

*2.3.1. 2-[N-Phenyl-2-oxo-2-phenyl-ethanehydrazonoyl]-1H-perimidine (**21a**).* Dark brown solid (yield 76%), m.p. >300°C [22], IR $\acute{\upsilon}$: br 3427 (2NH), 1623 (CO) cm^{-1},^{1}H NMR (DMSO-d_6) δ: 6.72–7.80 (*m*, 16H, Ar-H), 13.62 (*s*, 2H, 2NH); MS m/z (%) 390 (M^+, 5), 314 (M^+-Ph, 35), 298 (M^+-PhNH, 23), 285 (M^+-PhCO, 6), 167 ($C_{11}H_7N_2$, 5), 139 (12), 127 (5), 105 (5), 93 (5), and 84 (12). Analysis for $C_{25}H_{18}N_4O$ (390.45): calcd: C, 76.91; H, 4.65; and N, 14.35%. Found: C, 76.74; H, 4.46; and N, 14.09%.

*2.3.2. 2-[N-Phenyl-2-oxo-2-(4-methyl)phenyl-ethanehydrazonoyl]-1H-perimidine (**21b**).* Dark brown solid (yield 69%), m.p. = 162–164°C, ^{1}H NMR (DMSO-d_6) δ: 2.28 (s, 3H, CH_3), 7.40–8.35 (m, 15H, Ar-H), 10.90 (s, 2H, 2NH); MS m/z (%) 404 (M^+, 15), 403 (M^+−1, 21), 388 (M^+−CH_3, 39), 330 (M^+−$C_3H_5O_2$, 74), 310 (43), 298 (100), 266 (24), 165 (6), 148 (5), 121 (13), 118 (12), and 91 (4). Analysis for $C_{26}H_{20}N_4O$ (404.47): calcd: C, 77.21; H, 4.98; and N, 13.85%. Found: C, 76.94; H, 4.76; and N, 13.66%.

*2.3.3. 3-[N-Phenyl-2-oxo-2-(4-chloro)phenyl-ethanehydrazonoyl]-1H-perimidine (**21c**).* Dark brown solid (yield 68%), m.p. =150–152°C, IR $\acute{\upsilon}$: 3386, 3184 (2NH), 1644 (CO) cm^{-1},^{1}H NMR (DMSO-d_6) δ: 7.05–8.39 (*m*, 15H, Ar-H), 13.45 (*s*, 2H, 2NH); MS m/z (%) 424 (M^+, 10), 387 (21), 369 (100), 336 (65), 313 (32), 308 (45), 299 (63), 284 (64), 111 (7), 87 (22), and 71 (33). Analysis for $C_{25}H_{17}ClN_4O$ (424.89): calcd: C, 70.67; H, 4.03; and N, 13.19%. Found: C, 70.54; H, 3.89; and N, 13.36%.

*2.3.4. 2-[N-Phenyl-2-oxo-2-(4-fluoro)phenyl-ethanehydrazonoyl]-1H-perimidine (**21d**).* Dark brown solid (yield 70%), m.p. =157–159°C, IR $\acute{\upsilon}$: br 3436 (2NH), 1630 (CO) cm^{-1}, ^{1}H NMR (DMSO-d_6) δ: 6.80–8.37 (*m*, 15H, Ar-H),12.60 (*s*, 2H, 2NH); MS m/z (%) 408 (M^+, 22), 371 (62), 305 (24), 269 (17), 252 (12), 169 (8), 165 (6), 134 (12), 106 (11), 95 (2), 84 (33), and 65 (100). Analysis for $C_{25}H_{17}FN_4O$ (408.44): calcd: C, 73.52; H, 4.20; and N, 13.72%. Found: C, 73.33; H, 4.05; and N, 13.65%.

*2.3.5. 2-[N-Phenyl-2-oxo-2-(4-bromo)phenyl-ethanehydrazonoyl]-1H-perimidine (**21e**).* Dark brown solid (yield 63%), m.p. >300°C, IR $\acute{\upsilon}$: 3375, 3225 (2NH), 1620 (CO) cm^{-1},^{1}H NMR (DMSO-d_6) δ: 7.27–8.10 (m, 15H, Ar-H), 10.91 (s, 1H, 1NH), 12.95 (s, 1H, NH); MS m/z (%) 470 (M^++1, 12), 468 (M^+-1, 7), 450 (16), 426 (22), 384 (59), 365 (59), 155 (2.3), 76 (6), and 65 (13). Analysis for $C_{25}H_{17}BrN_4O$ (469.34): calcd: C, 63.98; H, 3.65; and N, 11.94%. Found: C, 63.78; H, 3.56; and N, 11.69%.

2.4. Theoretical Implementation

2.4.1. Geometrical Optimization. Gaussian09 program [19] is software used to investigate all the proposed structures **12–18** and **21** (Ar = Ph). DFT/B3LYP method is a suitable method used to investigate the physicochemical properties of proposed tautomeric forms. 3-21G is the base set used to accomplish the comparative investigation. Essential files yielded from this implementation (log and chk files). These files were visualized over the program screen to extract important parameters. HOMO and LUMO images were also obtained. Other additive energies according to optimized structures applying HyperChem 8.1 program were computed. Molecular mechanics (MM^+) method was followed by semiempirical AM1 one to extract log files. All parameters were obtained after reaching to the equilibrium state in the gas phase.

2.4.2. Molecular Docking. Utilization for AutoDock tools 4.2 was to execute the docking process between the proposed drug and protein receptors of variable bacteria and fungi. *Aspergillus niger* (1ukc), *Bacillis subtilis* (1ydo), *Escherichia coli* (5o22), *Geotrichum candidum* (1thg), *Klebsiella pneumoniae* (3gdz), *Pseudomonas aeruginosa* (2w7q), *Salmonella typhimurium* (1af7), *Staphylococcus aureus* (1bqb), *Staphylococcus epidermidis* (3kp3), and *Streptococcus pyogenes* (2esr) were the microorganism protein receptors used in this investigation. Such process was accomplished after the addition of Gasteiger charges over all compound atoms. The nonpolar atoms of hydrogen were incorporated to effect on the multiplication of rotatable bonds. The presence of hydrogen atoms with Kollman unite charge type is an essential preliminary feature [23]. The affinity maps by ×× Å grid points with 0.375 Å spacing were adapted using the Autogrid program [24]. Electrostatic terms and Vander Waals force were evaluated after adapting the Autodock parameter sets and the dielectric function of distance dependent, respectively. The docking simulation process to be the completed Lamarckian genetic algorithm (LGA) and the Solis and Wets local search method must be used [25]. The torsion of the investigated compound and occasionally the initial position as well as the orientation must be set. The rotatable torsions must be released completely through the docking process. Each docking process was accomplished through tens following runs and adjusted to be block after reaching 250000 energy considerations. Throughout the process, the size was settled to be 150, a translational step of 0.2 Å, and quaternion and torsion steps were of 5.

3. Results and Discussion

3.1. Chemistry. Scheme 4 illustrates all possible products **12–18** from the reaction of 1,8-diaminonaphthalene **6** and ethyl aroylpyrovate **10** in refluxing ethanol. Such possibilities were tested using Gaussian 09 software to optimize their structural forms (Figure 1) and computed essential parameters. These parameters are significant in asserting the reaction mechanism built on experimental aspects. The TLC experiment and the spectral data (^{1}H, ^{13}C NMR, mass, and IR) of the products indicate the isolation of only one product

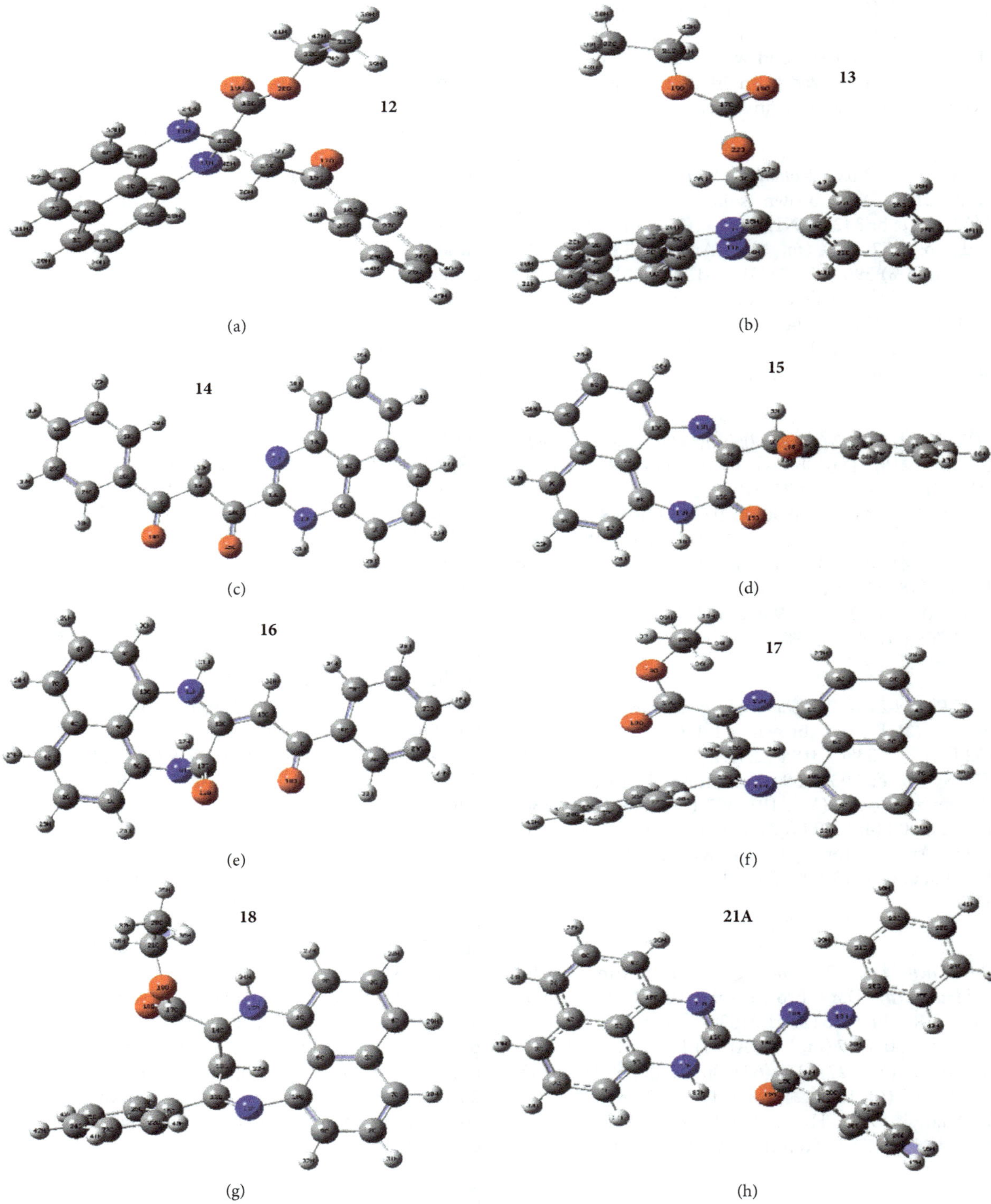

Figure 1: Continued.

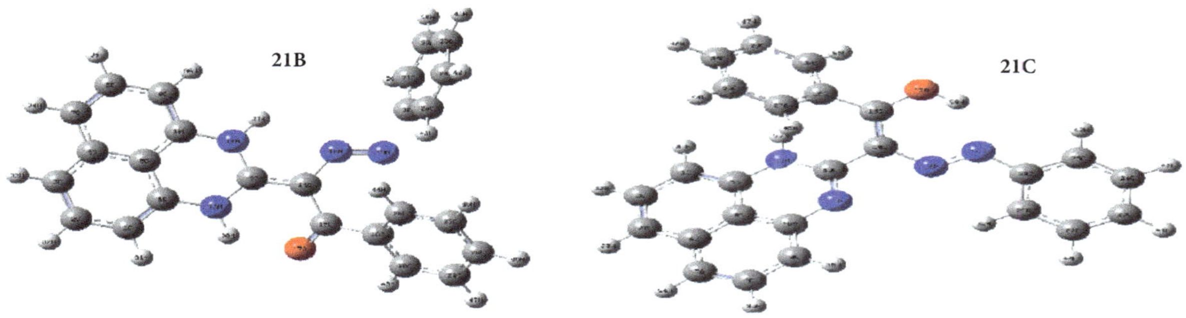

(i) (j)

FIGURE 1: The optimized geometries of all suggested derivatives (a–g) **12-18** and (h–j) **21A–C**.

in each case for such reaction. Their mass spectra show the molecular ion peak corresponding to elimination of only one water molecule which matches strongly with both structures **12** or **13**. Structure **13** was ruled out based on the data of IR and ^{13}C NMR spectra. This is due to the presence of two absorption bands for the two carbonyl groups characteristic for ester and aroyl groups at 1725–1716 and 1628–1610 cm^{-1}. The low value of the second carbonyl group can be attributed to the conjugation with aryl moiety and the hydrogen bond with the NH group. Moreover, all ^{1}H NMR spectra of the products **12a–e** revealed the presence of triplet and quartet signals for CH_3CH_2 integral ester groups near 1.33 and 4.32 ppm. As well as, the singlet signal for the CH_2 group was found near 3.87 ppm (see Experimental). Another evidence for the structures **12a–e** is the value of C2 concerning to the perimidine ring in the ^{13}C NMR spectrum of compound **12d** (as an example for series prepared). Such was appeared at down field (δ = 100.42 ppm) due to the presence of electron withdrawing ester group at the same carbon. However, the C2 in structure **13** appeared at the up-field value near δ = 81.8 ppm [5]. All previous concerns were assured from optimizing characteristic parameters executed from the DFT/B3LYP method (Table 1). 3-21G base set leads to produce best distribution for atomic contents of each compound in the gas phase and ethanol. The reduced formation energy values for compounds **12** and **13** point to the truth of the oriented mechanism proposed. The excellence for the formation energy value of compound **12** propose its priority as the only one product yielded from the condensation reaction. The high value of dipole moment of such compound (**12**) is suitable for the high difference in electronegativity inside the compound. Moreover, the reduced oscillator strength of such compound is suitable for facilitated electronic transition inside the compound (**12**) which coincides with its reactivity. This proposal was also confirmed by applying HyperChem 8.1 program which provides data (Table 2) and confirms the previous aspects. The reduced content energies (total energy, binding energy, and heat during formation) of compound **12** in comparing with others (**13–18**) confirm its high stability. This leads to its prior in production without conflict.

In continuation, the reactivity of the active methylene group in ethyl 2-(2-aryl-2-oxoethyl)-2,3-dihydro-1H-perimidine-2-carboxylate (**12a–e**) toward the coupling reaction with diazotized aniline in ethanol in the presence of NaOH has been investigated as shown in Scheme 5. Mass, IR, ^{1}H NMR, and elemental analyses confirmed the chemical structure of coupling products as the structure **21a–e**. The disappearance of the carbonyl absorption band of the ester group in all IR spectra of all derivatives (**21a–e**) proved that the coupling reaction proceeds via the *Japp-Klingemann* reaction [26] with elimination of the ethyl format molecule (Scheme 5). The actual tautomeric structure of the coupling products **21A–C** was studied and confirmed based on computational data extracted. The minimized internal energies (total energy and binding energy) beside the formation energy (Tables 1 and 2) of **21** tautomers suggest their high stability which outweigh their production from the coupling reaction and indicate that the most stable tautomer is **21A**. The HOMO and LUMO levels (Figures 2 and 3) were extracted over the optimized structures for comparison. The HOMO level appeared is very concentrated in the compounds over 1,8-diaminonaphthalene moiety of the perimidine ring in **21**. However, the LUMO level appeared is concentrated in the Ar-CO moiety which considered a good receptor for electrons as appeared in the coupling reaction. The two levels have the same positions in the coupling products (**21a–e**) which verify the same electron density attitude after reaction.

3.2. Biological Activity. The agar diffusion technique was used in screening the antimicrobial activity of all derivatives **12a–e** and **21a–e** [27] against two fungi, namely, *A. Niger* and *G. candidum* in comparison with *Amphotericin B* as a reference drug. Four species of each of Gram-positive and Gram-negative bacteria were tested against standard antibacterial agents as *ampicillin* and *gentamicin*, respectively (Tables 3 and 4). A suspension solution of the organisms was added to the sterile nutrient agar media at 45°C, and the mixture was transferred to sterile Petri dishes and allowed to solidify. Holes of 6 mm in diameter were made using a cork borer. A solution of each perimidines **12a–e** or **21a–e** or the reference drugs (5 mg/mL) in dimethylsulfoxide (DMSO) was prepared. The amount tested of either the synthesized compounds or reference drugs was 100 μL. Dimethylsulfoxide (DMSO) was used as a negative control. The plates were left for 1 h at room temperature as a period of pre-incubation diffusion to minimize the effects of variation in time between the applications of different solutions.

TABLE 1: DFT/B3LYP (3-21G) parameters over all proposed tautomers (**12–18** and **21**A–C).

The indexes	Compounds proposed									
	12	**13**	**14**	**15**	**16**	**17**	**18**	**21A**	**21B**	**21C**
DFT (in the gas phase)	Method used									
E (A.U.)	−1179.0538	−1179.0176	−1024.8445	−1024.8229	−1024.7594	−1102.9729	−1102.8973	−1250.7728	−1250.7425	−1250.7516
HOMO (ev)	−0.16943	−0.17047	−0.18788	−0.20387	−0.20432	−0.20184	−0.19332	−0.16835	−0.18688	−0.17179
LUMO (ev)	−0.06024	−0.04643	−0.07797	−0.07446	−0.07475	−0.07489	−0.07923	−0.08974	−0.05425	−0.08767
$\Delta E_{L\text{-}H}$ (ev)	0.10919	0.12404	0.10991	0.12941	0.12957	0.12695	0.11409	0.07861	0.13263	0.08412
Dipole moment (debye)	5.0333	2.9408	2.4273	2.3373	6.1572	3.3955	4.8856	6.4436	3.3095	3.0171
Oscillator strength	0.0008	0.0001	0.0312	0.0250	0.1241	0.0032	0.0250	0.0009	0.0027	0.0190
DFT (in ethanol)	Method used									
E (A.U.)	−1179.0712	−1179.0348	−1024.8601	−1024.8385	−1024.7798	−1102.9871	−1102.9147	−1250.7924	−1250.7603	−1250.7632
HOMO (ev)	−0.18124	−0.18105	−0.19124	−0.20708	−0.20530	−0.20738	−0.20142	−0.18347	−0.19622	−0.18512
LUMO (ev)	−0.06190	−0.05161	−0.08301	−0.08130	−0.08213	−0.08088	−0.08783	−0.08816	−0.06388	−0.08547
$\Delta E_{L\text{-}H}$ (ev)	0.11934	0.12944	0.108139	0.12578	0.12317	0.1265	0.11359	0.09531	0.13234	0.09965
Dipole moment (debye)	6.2687	3.9833	2.9013	3.0273	8.6666	4.7047	6.8986	8.3543	4.2647	4.5073
Oscillator strength	0.0012	0.0002	0.0369	0.0211	0.5074	0.004	0.0377	0.0038	0.0032	0.0098

Table 2: MM^+ internal energies computed over all proposed tautomers (**12–18** and **21A–C**).

Function	Compounds proposed									
	12	**13**	**14**	**15**	**16**	**17**	**18**	**21A**	**21B**	**21C**
Total energy	−102992.175	−102982.874	−87772.794	−87765.524	−87763.048	−94928.863	−94899.331	−105923.799	−105925.533	−105917.534
Total energy (kcal/mol)	−164.1285	−164.1136	−139.8748	−139.8632	−139.8593	−151.2787	−151.2317	−168.8003	−168.8031	−168.7903
Binding energy (a.u.)	−5236.8731	−5227.5718	−4456.1687	−4448.8985	−4446.4225	−4988.7493	−4959.2172	−5579.3923	−5581.125	−5573.1271
Heat during formation (ΔH_f) (kcal/mol)	−30.5761	−21.2748	36.1772	43.4475	45.9235	53.7847	83.3168	142.2527	142.5192	148.5179
$\Delta\Delta H_f$ (kcal/mol)	0.0000	9.3013	66.753	74.0236	76.4996	84.3608	113.8929	0.0000	0.2665	6.2652

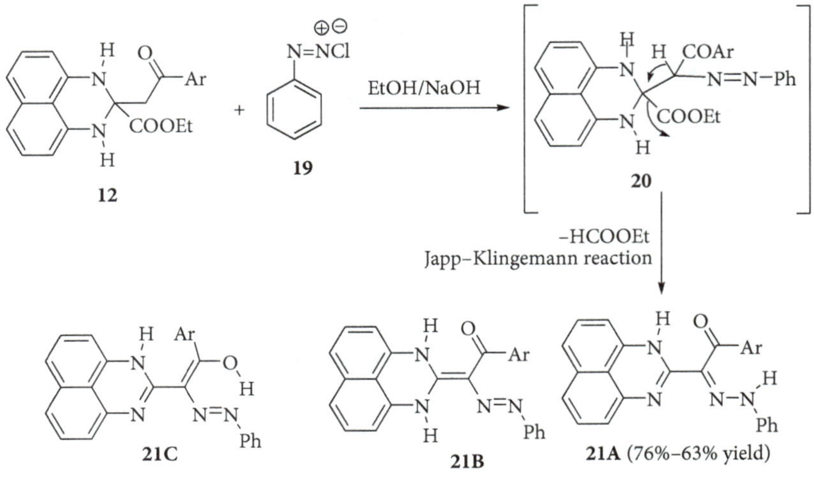

Ar: a, Ph; b, 4-MeC$_6$H$_4$; c, 4-ClC$_6$H$_4$; d, 4-FC$_6$H$_4$; e, 4-BrC$_6$H$_4$

Scheme 5: Coupling reaction of compounds **12a–e**.

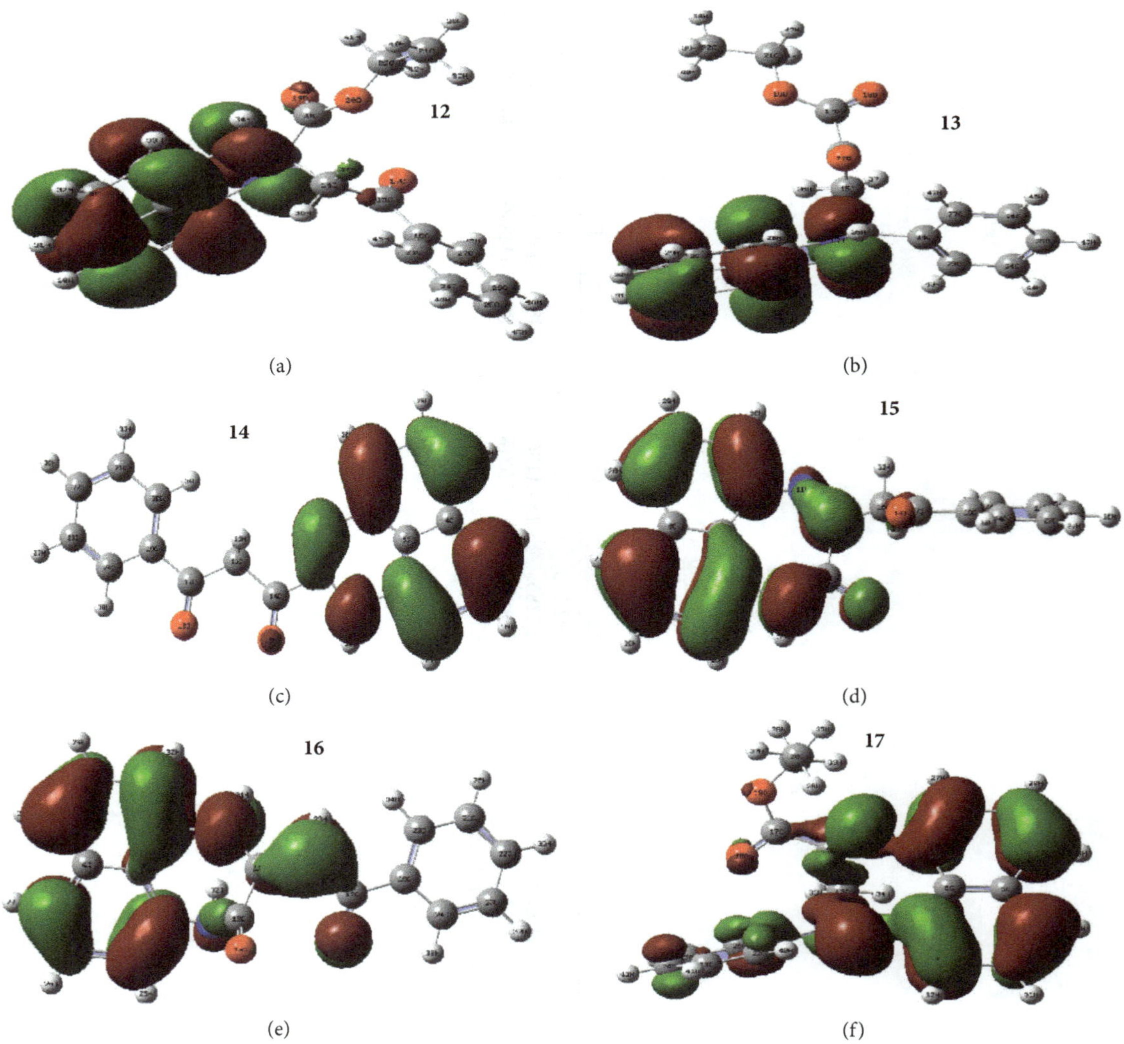

Figure 2: Continued.

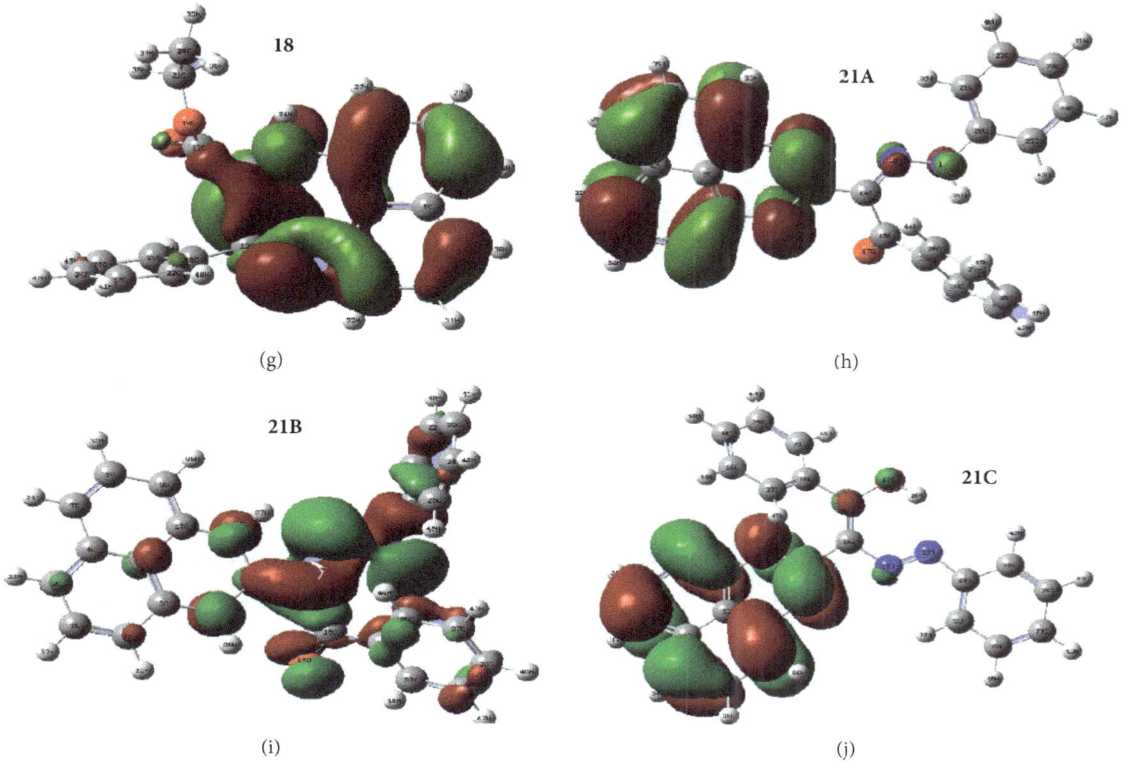

FIGURE 2: (a–j) Frontier orbitals regarding the HOMO level for optimized structures.

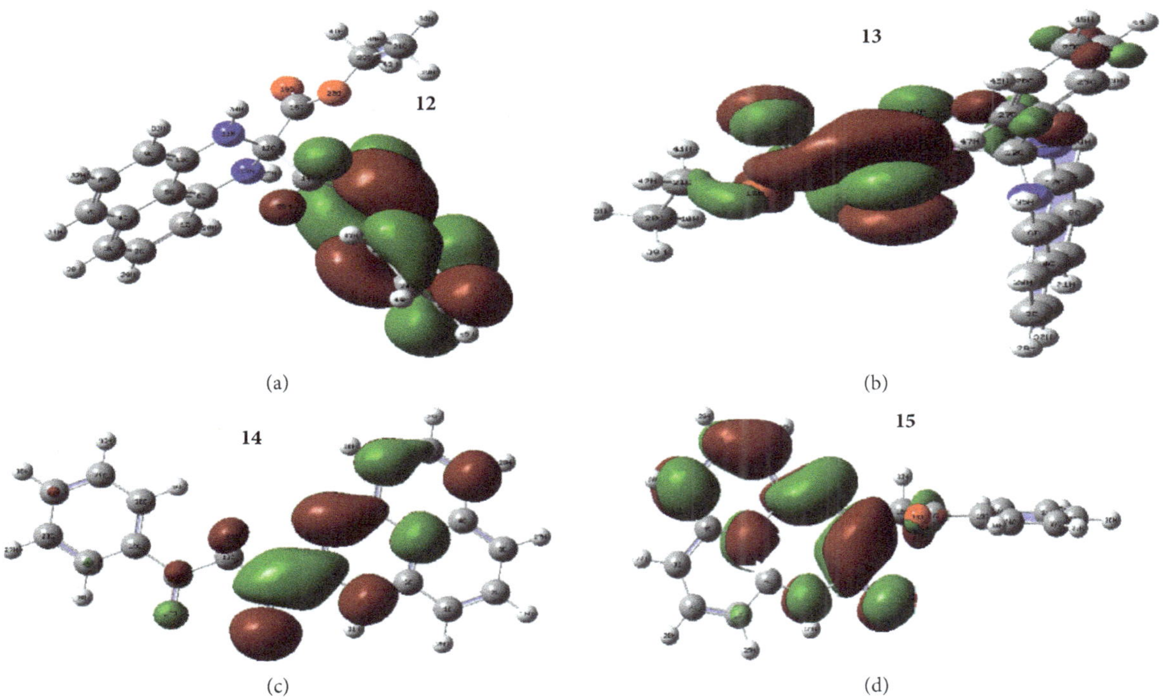

FIGURE 3: Continued.

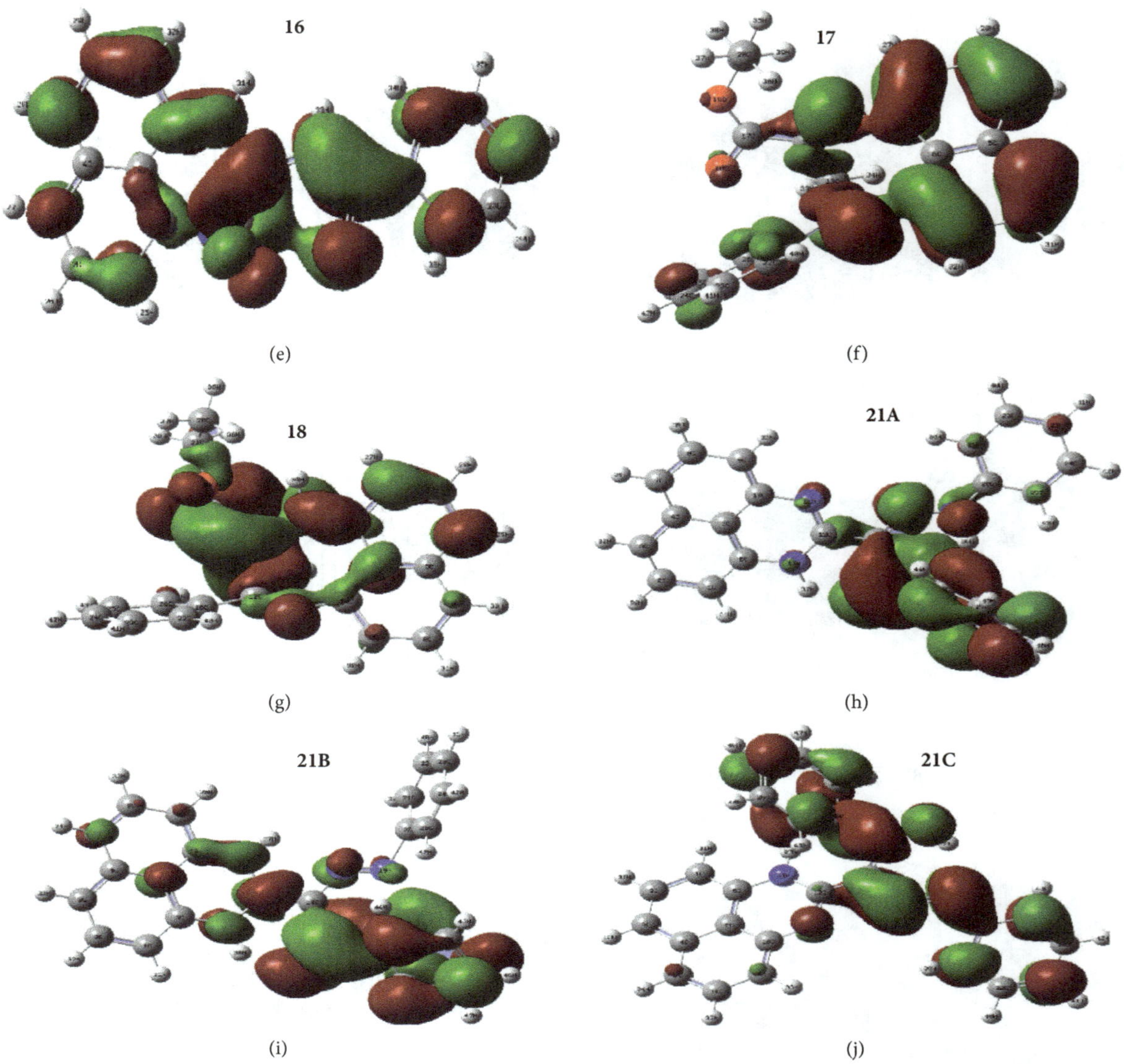

Figure 3: (a–j) Frontier orbitals regarding the LUMO level for optimized structures.

Table 3: The activity of the tested compounds **12a–e** and **21a–e** against two fungi species.

Compound number	*Aspergillus niger*	*Geotrichum candidum*	Compound number	*Aspergillus niger*	*Geotrichum candidum*
12a	19.4 ± 0.63 (83.3%)	18.1 ± 0.58 (71.8%)	**21a**	17.3 ± 1.2 (74.2%)	19.3 ± 1.2 (76.6%)
12b	23.7 ± 0.72 (101.7%)	24.4 ± 1.2 (96.8%)	**21b**	15.5 ± 2.1 (66.5%)	14.2 ± 2.1 (56.3%)
12c	19.6 ± 0.63 (84.4%)	21.6 ± 0.63 (85.71%)	**21c**	24.1 ± 1.2 (103.4%)	23.6 ± 1.2 (93.7%)
12d	22.8 ± 1.2 (97.9%)	20.7 ± 0.72 (82.1%)	**21d**	19.2 ± 0.72 (82.4%)	17.3 ± 0.72 (68.7%)
12e	19.4 ± 0.63 (83.3%)	25.6 ± 0.63 (101.6%)	**21e**	12.9 ± 0.72 (55.4%)	18.5 ± 0.72 (73.4%)
Amphotericin B	23.3 ± 0.58	25.2 ± 0.72	**Amphotericin B**	23.3 ± 0.58	25.2 ± 0.72

Data are expressed in the form of mean ± SD.

The plates were then incubated at 37°C for 24 h and observed for antimicrobial activity. The diameters of the inhibition zone were measured and compared with the respective reference drug. The observed inhibition zones were measured in millimeter beyond well diameter. Each treatment was replicated three times, and the average value is recorded. Also, the percentage value of inhibition zones compared to reference drugs was recorded (Tables 3 and 4).

Table 4: The activity of the tested compounds **12a–e** and **21a–e** against four G^+ bacteria and four G^- bacteria fungi species.

Compound number	Gram-positive bacteria				Gram-negative bacteria			
	Staphylococcus aureus	*Staphylococcus epidermidis*	*Bacillus subtilis*	*Streptococcus pyogenes*	*Pseudomonas aeruginosa*	*Escherichia coli*	*Klebsiella pneumoniae*	*Salmonella typhimurium*
12a	19.3 ± 0.58 (81.4%)	17.4 ± 0.58 (77.7%)	26.7 ± 1.2 (82.4%)	NA	NA	24.3 ± 0.63 (95.7%)	21.9 ± 0.58 (96.9%)	20.4 ± 0.72 (87.6%)
12b	22.8 ± 0.58 (96.2%)	21.6 ± 0.72 (96.4)	29.3 ± 0.58 (90.4%)	NA	NA	24.2 ± 0.72 (95.3%)	20.3 ± 0.63 (89.8%)	21.6 ± 0.63 (92.7%)
12c	22.6 ± 0.85 (95.4%)	19.6 ± 1.2 (87.5%)	28.3 ± 0.63 (87.3%)	NA	NA	25.2 ± 0.58 (99.2%)	18.6 ± 1.2 (82.3%)	19.9 ± 0.63 (85.4%)
12d	21.2 ± 0.63 (89.5%)	19.6 ± 0.58 (87.5%)	27.3 ± 0.53 (84.3%)	NA	NA	24.2 ± 1.2 (95.3%)	22.1 ± 0.44 (97.8%)	19.9 ± 0.72 (85.4%)
12e	23.1 ± 0.63 (97.5%)	22.7 ± 0.72 (101.3%)	28.1 ± 0.53 (86.7%)	NA	NA	18.5 ± 1.2 (72.8%)	19.2 ± 1.2 (85.0%)	20.4 ± 0.58 (87.6%)
21a	16.7 ± 0.63 (70.5%)	14.4 ± 0.72 (64.3%)	19.2 ± 0.63 (59.3%)	NA	NA	17.8 ± 1.2 (70.1%)	17.3 ± 1.2 (76.5%)	19.6 ± 0.58 (84.1%)
21b	15.2 ± 1.2 (64.1%)	19.8 ± 0.67 (88.4%)	24.3 ± 0.58 (75.0%)	NA	NA	17.9 ± 0.58 (70.5%)	13.3 ± 0.58 (58.8%)	18.4 ± 1.2 (79.0%)
21c	22.4 ± 0.63 (94.5%)	20.9 ± 0.72 (93.3%)	23.8 ± 0.63 (73.5%)	NA	NA	24.3 ± 1.2 (95.7%)	22.5 ± 1.2 (99.6%)	26.3 ± 0.58 (112.9%)
21d	13.5 ± 1.2 (57.0%)	15.3 ± 0.44 (68.3%)	18.2 ± 0.58 (56.2%)	NA	NA	19.3 ± 0.63 (76.0%)	16.3 ± 0.63 (72.1%)	19.3 ± 0.58 (82.8)
21e	15.3 ± 0.43 (64.6%)	18.6 ± 0.58 (83.0%)	22.4 ± 0.53 (69.1%)	NA	NA	16.4 ± 0.53 (64.6%)	20.4 ± 0.53 (90.3%)	18.7 ± 0.63 (80.3%)
Ampicillin	23.7 ± 0.63	22.4 ± 1.2	32.4 ± 0.72	24.5 ± 0.63	—	—	—	—
Gentamicin	—	—	—	—	22.3 ± 0.58	25.4 ± 1.2	22.6 ± 0.63	23.3 ± 0.58

NA: no activity; data are expressed in the form of mean ± SD.

Table 5: Docking energy values (k cal/mol) for two best tautomers complexed with variable protein receptors.

Ligands	pK_1–pK_2	Receptor	Estimated free energy of binding	Estimated inhibition constant (K_i) (μM)	vdW + bond + desolve energy	Electrostatic energy	Total intercooled energy	Frequency	Interaction surface
Compound **12a**		1ukc	−6.55	15.80	−6.22	−0.07	−6.29	20%	585.749
		1ydo	−5.81	55.13	−5.52	+0.01	−5.50	10%	569.54
		5o22	−4.67	375.04	−4.54	−0.17	−4.71	10%	487.392
		1thg	−4.72	346.39	−4.15	−0.09	−4.24	40%	473.189
		3gdz	−5.35	119.29	−5.07	−0.15	−5.22	10%	581.063
		2w7q	−7.92	1.56	−7.58	−0.03	−7.61	30%	723.71
	10.33–0.97	1af7	−6.89	8.84	−7.97	−0.26	−8.24	30%	796.368
		1bqb	−5.06	196.91	−5.72	+0.08	−5.64	50%	681.073
		3kp3	−5.32	126.03	−5.25	+0.10	−5.15	30%	585.688
		2esr	−6.76	11.05	−6.71	−0.05	−6.76	40%	585.867
Compound **21a**		1ukc	−4.70	359.18	−6.01	−0.07	−6.08	10%	639.464
		1ydo	−5.54	87.07	−6.65	−0.14	−6.79	20%	558.58
		5o22	−4.41	585.48	−5.73	−0.03	−5.76	10%	520.673
		1thg	−4.73	342.60	−6.00	−0.04	−6.04	20%	542.067

$$\text{The \% activity} = \frac{\text{Mean zone of inhibition by test compound (mm)}}{\text{Mean zone of inhibition by standard (mm)}} \times 100. \quad (1)$$

The obtained results of antifungal activity reflected variable activities of the tested compounds. Among the tested perimidine derivatives **12b** and **21c** was the most potent derivatives against *Aspergillus niger* with inhibition zones exceed the *Amphotericin B* (101.7% and 103.4%, respectively). In addition, compound **12e** showed activity with 101.6% inhibition percent in comparison with the reference standard used (Table 3). The other derivatives revealed the activity ranging from excellent to good.

Investigating the Gram-positive bacterial activity results in Table 4, a highly potent activity was recorded for compounds **12e**, **12b**, and **21c** against *S. epidermidis* with

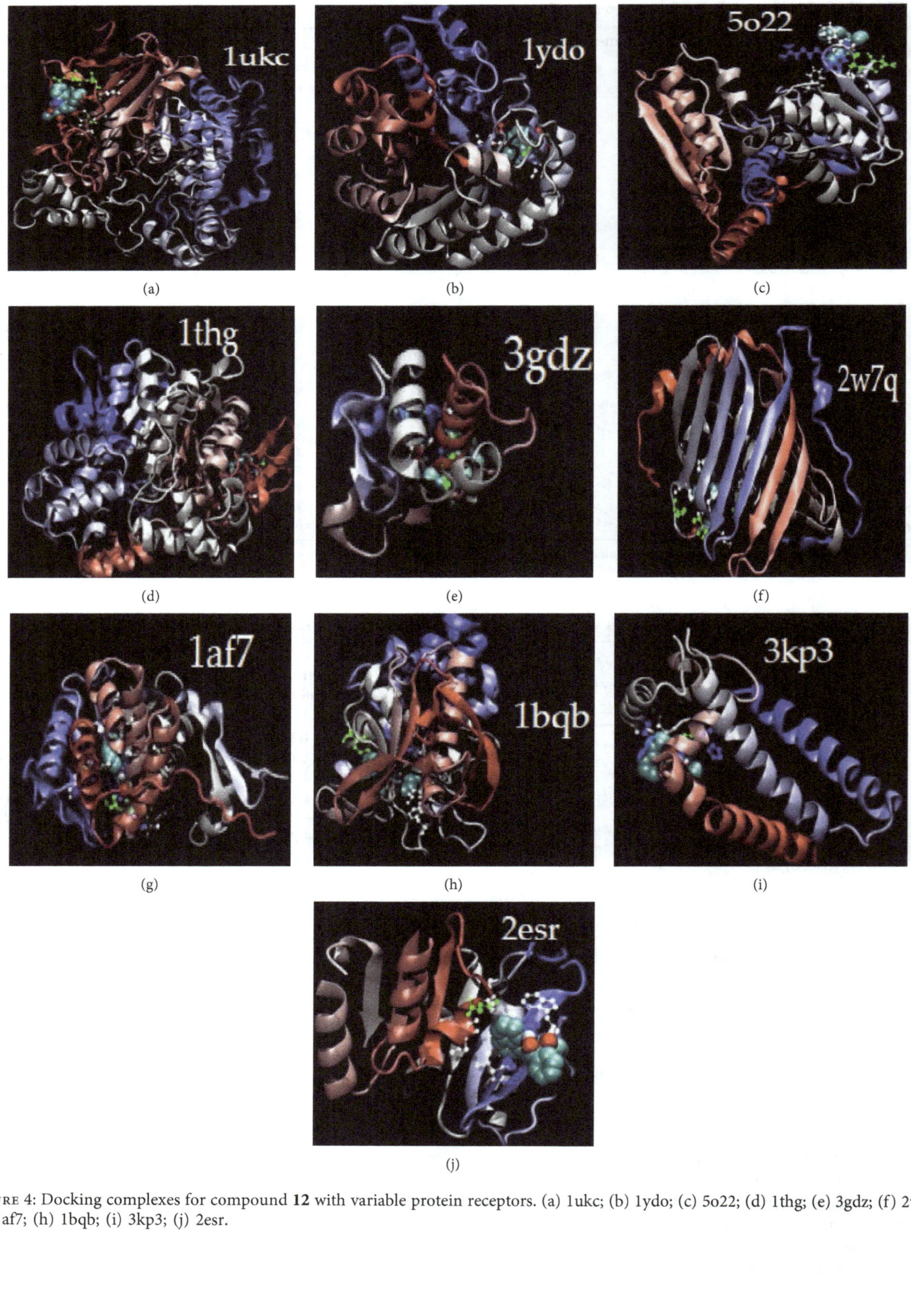

(a) (b) (c) (d) (e) (f) (g) (h) (i) (j)

Figure 4: Docking complexes for compound **12** with variable protein receptors. (a) 1ukc; (b) 1ydo; (c) 5o22; (d) 1thg; (e) 3gdz; (f) 2w7q; (g) 1af7; (h) 1bqb; (i) 3kp3; (j) 2esr.

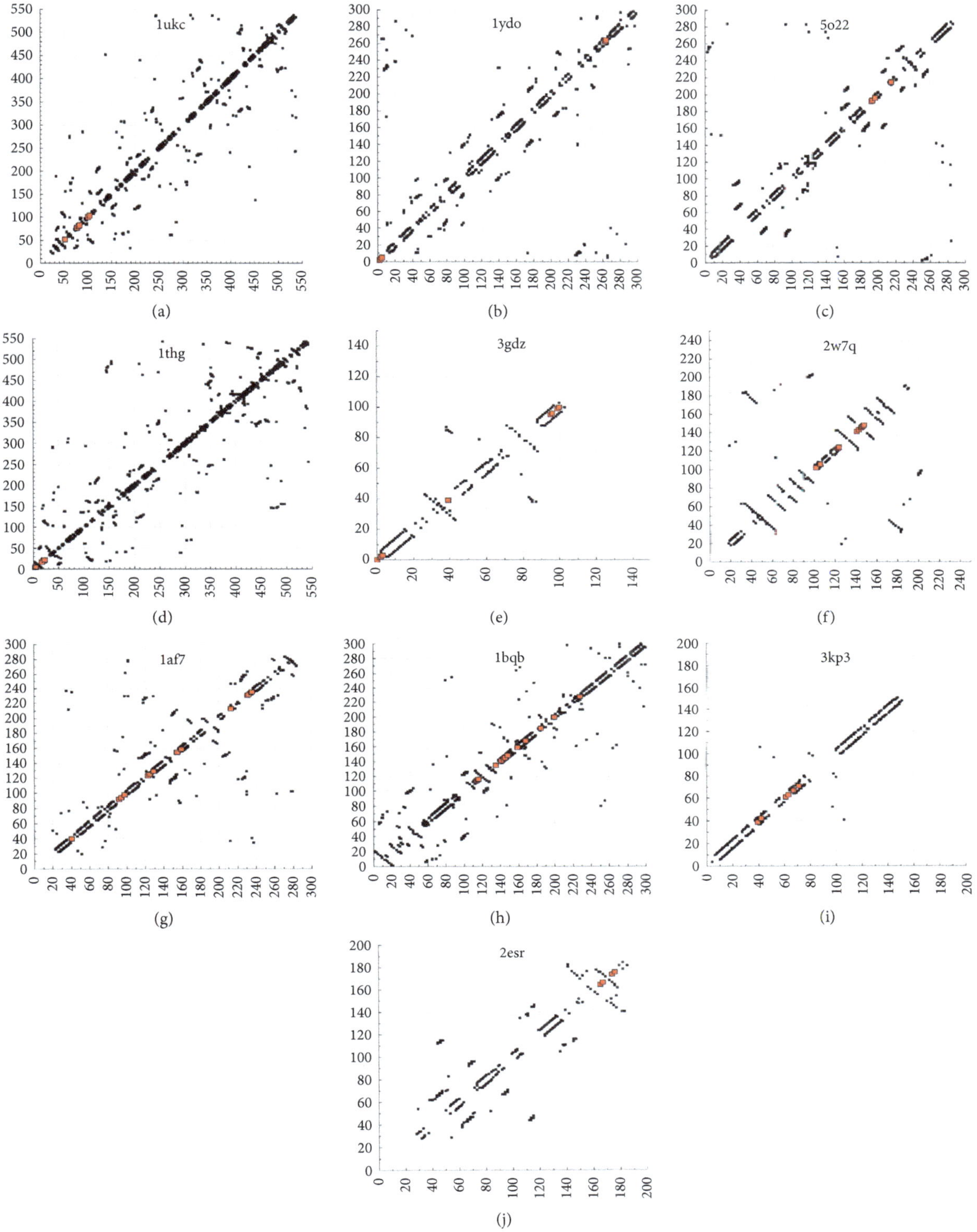

FIGURE 5: Hb plots for the compound **12**-protein complexes. (a) 1ukc; (b) 1ydo; (c) 5o22; (d) 1thg; (e) 3gdz; (f) 2w7q; (g) 1af7; (h) 1bqb; (i) 3kp3; (j) 2esr.

inhibition percentages 101.3%, 96.4%, and 93.3%, respectively. The same derivatives in addition to **12c** showed excellent activity when compared with *ampicillin* (the inhibition percentages ranging from 97.5 to 94.5%).

Focusing on Gram-negative bacteria, all tested perimidine derivatives have no activity against *P. aeruginosa,* and in contrast, compound **21c** is more reactive than *gentamicin* aganist *S. typhimurium* with inhibition percentage

112.9%. The same derivative **21c** also has activity similar to *gentamicin* against the Gram-negative bacteria *Klebsiella pneumonia* and *Escherichia coli* as shown in Table 4. In addition, perimidine derivatives **12a–d** and **21e** was equipotent to *gentamicin* against some species (Table 4).

3.3. Molecular Docking Study. In the recent few decades, advanced development happened in the drug designing industry. Among this study was the computational method used to predicate the inhibition feature of the proposed drug against causative organism proteins. This study was achieved using AutoDock 4.2 tools to establish a comparative insight on the degree of inhibition toward pathogens. The receptors had chosen attributing to the tested bacteria and fungi in the experimental work. 1ukc and 1thg were the protein receptors used which attribute to *Aspergillus niger* and *Geotrichum* candidum fungi. Whenever, 1ydo, 5o22, 3gdz, 2w7q, 1af7, 1bqb, 3kp3, and 2esr were the protein receptors used, they attribute to *Bacillis subtilis, Escherichia coli, Klebsiella pneumoniae, Pseudomonas aeruginosa, Salmonella typhimurium, Staphylococcus aureus, Staphylococcus epidermidis,* and *Streptococcus pyogenes* bacteria. The data extracted upon docking complexes with proven compounds (**12** and **21A**) were displayed in Table 5. The values of free energy of binding and the inhibition constant display a moderate interaction between the two compounds and 1ukc and 1thg receptors. The values of total intercooled energy reflect the stability of docking complexes for such broad interaction surface. The docking complexes (Figure 4; Supplementary Figure 1S) display H-bonding in the corner only with protein helix [28]. This conclusion coincided with a little extent of H-bonding that appeared in the Hb plots (Figure 5; Supplementary Figure 2S) which exhibit interaction over the siding part in broad helix. However, variable features appeared against bacterial pathogen proteins. The best interaction was displayed with 1af7 and 1bqb receptors attributing to *Salmonella typhimurium* and *Staphylococcus aureus* organisms. This is concluded from the reduced energies (Table 5) over broad interacting surfaces. However, moderate inhibition was expected with the two tested compounds (**12** and **21A**) versus rest proteins. The docking complexes (Figure 4; Supplementary Figure 1S) display high occlusion for the two compounds with 1af7 or 1bqb protein helix. This was confirmed upon the Hb plots (Figure 5; Supplementary Figure 2S) which display the best distribution of intra-H-bonding with two reported receptors. But, on the other side, a little extent for H-bonding was recorded upon rest protein helix though broad interaction surfaces. Finally, this simulation process coincided by an excellent way with the experimental screening work which displays a moderate inhibition against most organisms considered.

4. Conclusion

The mechanism of the reaction of 1,8-diaminonaphthalene with ethyl aroylpyrovate was discussed using all possible spectroscopic tools as well as the computational study by applying Guassian09 software. In addition, the antimicrobial activity of all products was screened. The results of such study indicated that three derivatives **12b**, **12e**, and **21c** have impacts on the fungi and bacteria more than the reference drugs used. However, the theoretical study of the antimicrobial activity of two derivatives **12a** and **21a** as examples of the synthesized series using AutoDock 4.2 tools indicated that such derivatives have moderate activity for all fungi and bacteria species.

Conflicts of Interest

The author declares that there are no conflicts of interest.

References

[1] H. Kabiri-fard, S. Balalaie, S. Abbasian, and A. Panjalizadeh, "Reactions of 3-arylhydrazono-2,4-dioxo-4-phenylbutanoates with dinucleophiles," *Journal of Applied Chemistry Research*, vol. 4, no. 14, pp. 33–37, 2010.

[2] C. M. Atkinson, C. W. Brown, and J. C. E. Simpson, "6. Cinnolines and other heterocyclic types in relation to the chemotherapy of trypanosomiasis. Part XI. Some reactions of simple quinoxaline derivatives," *Journal of the Chemical Society (Resumed)*, pp. 26–30, 1956.

[3] B. Kołodziej, M. Morawiak, B. Kamie_nski, and W. Schilf, "The structure investigations of dehydroacetic acid and 1,8-diaminonaphthalene condensation product by NMR, MS, and X-ray measurements," *Journal of Molecular Stracture*, vol. 1112, pp. 81–86, 2016.

[4] W.-Z. Chen, H.-Y. Wie, and D.-Y. Yang, "Visible light-sensitive properties of 1,2-dimethyl-2-(2-nitrophenyl)-2,3-dihydro-1H-perimidine," *Tetrahedron*, vol. 69, no. 13, pp. 2775–2781, 2013.

[5] I. Koca, E. H. Ngren, B. E. Kbrz, and F. Ylmaz, "The synthesis of new pyrrolo[1,2-a]perimidin-10-one dyes via two convenient routes and its characterizations," *Dyes and Pigments*, vol. 95, no. 2, pp. 421–426, 2012.

[6] S. Anga, S. Biswas, R. K. Kottalanka, B. S. Mallik, and T. K. Panda, "Structural and mechanistic insights of substituted perimidine-experimental and computational studies," *Canadian Chemical Transactions*, vol. 2, no. 1, pp. 72–82, 2014.

[7] G. Varsha, V. Arun, P. P. Robinson et al., "Two new fluorescent heterocyclic perimidines: first syntheses, crystal structure, and spectral characterization," *Tetrahedron Letters*, vol. 51, no. 16, pp. 2174–2177, 2010.

[8] C.-K. Wu, T.-J. Liou, P.-S. Tsai, and D.-Y. Yang, "Visible light photoredox catalysis: aerobic oxidation of perimidines to perimidinones," *Tetrahedron*, vol. 70, no. 44, pp. 8219–8225, 2014.

[9] I. A. S. Smellie, A. Fromm, and R. M. Paton, "A new route to 2-substituted perimidines based on nitrile oxide chemistry," *Tetrahedron Letters*, vol. 50, no. 28, pp. 4104–4106, 2009.

[10] A. F. Pozharskii and V. V. Dalnikovskaya, "Perimidines," *Russian Chemical Reviews*, vol. 50, no. 9, pp. 816–835, 1981.

[11] J. M. Herbert, P. D. Woodgate, and W. A. Denny, "Potential antitumor agents 53. Synthesis, DNA binding properties, and biological activity of perimidines designed as minimal DNA-intercalating agents," *Journal of Medicinal Chemistry*, vol. 30,

[12] X. Bu, L. W. Deady, G. J. Finlay, B. C. Baguley, and W. A. Denny, "Synthesis and cytotoxic activity of 7-Oxo-7H-dibenz[f,ij]isoquinoline and 7-Oxo-7H-benzo[e]perimidine derivatives," *Journal of Medicinal Chemistry*, vol. 44, no. 12, pp. 2004–2014, 2001.
[13] V. V. Patil and G. S. Shankarling, "A metal free, eco-friendly protocol for the synthesis of 2,3-dihydro-1H-perimidines using commercially available Amberlyst 15 as a catalys," *Catalysis Communication*, vol. 57, pp. 138–142, 2014.
[14] W. Bauer and M. Akram, "Hair coloring compositions which contain developers and perimidine couplers," vol. 25, U S Patent 5,613,985, 1997.
[15] S. Goswami, D. Sen, and N. K. Das, "A new highly selective, ratiometric and colorimetric fluorescence sensor for Cu2+ with a remarkable red shift in absorption and emission spectra based on internal charge transfer," *Organic Letters*, vol. 12, pp. 856–859, 2010.
[16] P. Herold, A. F. Indolese, M. Studer, H. P. Jalett, U. Siegrist, and H. U. Blaser, "New technical synthesis of ethyl (*R*)-2-Hydroxy-4-phenylbutyrate of high enantiomeric purity," *Tetrahedron*, vol. 56, no. 35, pp. 6497–6499, 2000.
[17] A. Schmidt, T. Habeck, M. K. Kindermann, and M. Nieger, "New pyrazolium-carboxylates as structural analogues of the pseudo-cross-conjugated betainic alkaloid nigellicine," *Journal of Organic Chemistry*, vol. 68, no. 15, pp. 5977–5982, 2003.
[18] S. V. Ryabukhin, A. S. Plaskon, S. S. Bondarenko et al., "Acyl pyruvates as synthons in the Biginelli reaction," *Tetrahedron Letters*, vol. 51, no. 32, pp. 4229–4232, 2010.
[19] M. J. Frisch, *Gaussian 09, Revision D*, Gaussian, Inc., Wallingford, CT, USA, February 2010.
[20] I. Althagafi, M. G. Elghalban, F. Saad et al., "Spectral characterization, CT-DNA binding, DFT/B3LYP, molecular docking and antitumor studies for new nano-sized VO(II)-hydrazonoyl complexes," *Journal of Molecular Liquids*, vol. 242, pp. 662–677, 2017.
[21] J. J. P Stewart, "Optimization of parameters for semiempirical methods V: modification of NDDO approximations and application to 70 elements," *Journal of Molecular Modeling*, vol. 13, no. 12, pp. 1173–1213, 2007.
[22] T. A. Farghaly and H. K. Mahmoud, "Synthesis, tautomeric structures and antitumor activity of new perimidines," *Archive of pharmacy*, vol. 346, no. 5, pp. 392–402, 2013.
[23] T. A. Halgren, "Merck molecular force field. I. Basis, form, scope, parameterization, and performance of MMFF94," *Journal of Computational Chemistry*, vol. 17, no. 5-6, pp. 490–519, 1996.
[24] G. M. Morris, D. S. Goodsell, R. S. Halliday et al., "Automated docking using a lamarckian genetic algorithm and and empirical binding free energy function," *Journal of Computational Chemistry*, vol. 19, no. 14, pp. 1639–1662, 1998.
[25] F. J. Solis and R. J. B. Wets, "Minimization by random search techniques," *Mathematics of Operations Research*, vol. 6, no. 1, pp. 19–30, 1981.
[26] R. R. Phillips, "The Japp-Klingemann reaction," in *Organic Reactions*, pp. 143–178, Wiley, Hoboken, NJ, USA, 2011.
[27] M. Balouiri, M. Sadiki, and S. K. Ibnsouda, "Methods for in vitro evaluating antimicrobial activity: a review," *Journal Pharmaceutical Analysis*, vol. 6, no. 2, pp. 71–79, 2016.
[28] K. Tripathi, R. Muttineni, and S. K. Singh, "Extra precision docking, free energy calculation and molecular dynamics simulation studies of CDK2 inhibitors," *Journal of Theoretical Biology*, vol. 334, pp. 87–100, 2013.

17

Stability of Epidoxorubicin Hydrochloride in Aqueous Solutions: Experimental and Theoretical Studies

Agnieszka Sobczak,[1] Monika A. Lesniewska-Kowiel,[1] Izabela Muszalska,[1] Artur Firlej,[1] Judyta Cielecka-Piontek,[1] Szymon Tomczak,[1] Bolesław Barszcz,[2] Irena Oszczapowicz,[3] and Anna Jelińska[1]

[1]*Department of Pharmaceutical Chemistry, Faculty of Pharmacy, Poznan University of Medical Sciences, Grunwaldzka 6, 60-780 Poznań, Poland*
[2]*Institute of Molecular Physics Polish Academy of Sciences, Smoluchowskiego 17, 60-179 Poznań, Poland*
[3]*Department of Modified Antibiotics, Institute of Biotechnology and Antibiotics, Starościńska 5, 02-515 Warszawa, Poland*

Correspondence should be addressed to Agnieszka Sobczak; asobczak@ump.edu.pl

Academic Editor: Paula G. De Pinho

The first-order degradation kinetics of epidoxorubicin were investigated as a function of pH, temperature, and buffers concentrations. The degradation was followed by HPLC. Buffer catalysis was observed in acetate and phosphate buffers. The pH-rate profiles were obtained at 333, 343, 353, and 363 K. The pH-rate expression was $k_{pH} = k_1 \times a_{H^+} \times f_1 + k_2 \times f_1 + k_3 \times f_2 + (k_4 \times f_2 + k_5 \times f_3) \times a_{OH^-}$, where k_1, k_4, and k_5 are the second-order rate constants ($mol^{-1}\ L\ s^{-1}$) for hydrogen ion activity and for hydroxyl ion activity, respectively, and k_2 and k_3 are the first-order constants (s^{-1}) for spontaneous reaction under the influence of water. Epidoxorubicin demonstrates the greatest stability in the pH range 3–5. The electrostatic molecular potential orbitals HOMO-LUMO were also defined in order to determine the cause of the reactivity of particular epidoxorubicin molecule domains in solutions with various pH values.

1. Introduction

Epidoxorubicin hydrochloride (EP) is the most commonly used anthracycline antitumor antibiotic consisting of a tetracyclic quinolid aglycone connected by a glycoside bond with an amino sugar—L-acosamine. The drug is structurally related to doxorubicin and the difference between them is in the stereochemistry of the C-4′ hydroxyl group of the sugar moiety. Due to its structure, EP demonstrates lower cardiotoxicity and so it can be used in higher doses [1–3]. The lower efficiency and higher toxicity of drugs can be also associated with degradation during their storage or clinical use (e.g., long-term infusions). EP is produced as powder or solutions for injections and administered as long-term infusions. Therefore, it is important to investigate its stability both in the solid phase and in solutions.

The influence of temperature and relative air humidity on the stability of EP as well as of doxorubicin and daunorubicin in the solid state was previously reported [4, 5]. The degradation of EP in the atmosphere of increased relative air humidity was a first-order reaction depending on the substrate concentration, and in dry, hot air (RH 0%; 393 K) a reversible first-order reaction relative to the substrate concentration. The dependences $\ln k = f(1/T)$ and $\ln k = f(\text{RH}\%)$ were described by the equations $\ln k = (35.1 \pm 10.9) - (16250 \pm 3823)(1/T)$ and $\ln k = (3.79 \pm 3.34) \times 10^{-2}(\text{RH}\%) - (12.9 \pm 2.4)$, respectively, and the kinetic and thermodynamic parameters of the above processes were calculated [4].

The photodegradation [6] and hydrolysis of EP under acidic (0.05 M; 363 K) and basic (0.05 M; 303 K) conditions [7] were first-order reactions. The photodegradation rate was inversely proportional to the concentration of the drug and rose as the solvent pH increased. In the case of concentrations which are used in chemotherapy (>500 mcg/mL), protection from light was not necessary [6].

The stability of EP was also assessed in sodium chloride solutions (0.9% NaCl, pH 5.20 and 6.47) and in dextrose (5%; pH 4.36) at 25°C, 4°C, and −20°C, stored in polyvinyl chloride (PCV) minibags, which are no longer in clinical use, and in water for injections, stored at 4°C in polypropylene syringes [8]. The drug was stable (loss below 10%) under the above conditions with the exception of storage in 0.9% NaCl (pH 6.47) at 25°C, when degradation exceeding 10% was observed after 43 days. It was also found that the repeated freezing and defrosting of these solutions stored in minibags did not cause significant degradation and that polypropylene appeared to be the best material for EP storing [8].

During the storage of EP injections at pH 3 and a temperature of 2–8°C for 12 months or at 25°C for 6 months, some degradation products were observed [9]. Three products were isolated and characterized by NMR and LC-MS/MS methods. Product I (dimer of EP) was formed due to the condensation of two EP molecules. Product III (4-(4-amino-5-hydroxy-6-methyl-tetrahydro-pyran-2-yloxy)-2,5,12-trihydroxy-7-methoxy-6,11-dioxo-1,2,3,4,6,11-hexahydro-naphthacene-2-carboxylic acid hydroxymethyl ester) was produced by oxidative rearrangement of EP in the side chain under acidic conditions. After hydrolysis of the ester group of degradation product III, product IV (4-(4-amino-5-hydroxy-6-methyl-tetrahydro-pyran-2-yloxy)-2,5,12-trihydroxy-7-methoxy-6,11-dioxo-1,2,3,4,6,11-hexahydro-naphthacene-2-carboxylic acid) was formed, which can also be directly produced due to the oxidation of the side chain of EP and the elimination of the ester moiety [9].

The literature does not refer to comprehensive kinetic studies of EP in aqueous solutions in a wide range of pH values and temperatures. Thus, the aim of this study was to evaluate the degradation of EP under the influence of pH and temperature and to develop equations permitting the estimation of kinetic and thermodynamic parameters. To separate EP and its degradation products, our previously designed HPLC method was used [4]. The geometric structure of EP molecule, FMOs orbitals (HOMO-LUMO; Highest Occupied Molecular Orbital; Lowest Unoccupied Molecular Orbital), and molecular electrostatic potential were also determined.

2. Experimental

2.1. Materials and Reagents. Epidoxorubicin hydrochloride was synthesized at the Department of Modified Antibiotics, Institute of Biotechnology and Antibiotics, Warsaw, Poland. Sodium laurilsulfate (A.C. reagent, Sigma-Aldrich Logistik GmbH, Germany) and all other chemicals and solvents were obtained from Merck KGaA, Germany, and were of analytical or high-performance liquid chromatographic grade.

2.2. Chromatographic Conditions. Chromatographic separation and quantitative analysis were performed by using the HPLC method [4]. The analytical system consisted of a Shimadzu SPD-20A Prominence UV/VIS detector, a Rheodyne with a 50 μL loop. An LiChrospher RP-18 column (250 mm × 4 mm, 5 μm particle size, Merck, Germany) was used as the stationary phase. The mobile phase consisted of equal volumes of acetonitrile and a solution containing 2.88 g/L of sodium laurilsulfate and 2.25 mL/L of phosphoric acid. The flow rate was 1.0 mL/min and UV detection was performed at 254 nm. As the HPLC method was previously evaluated and validated for a stability study of epidoxorubicin in the solid state [4], selectivity was examined during stability studies in aqueous solutions in this work. Papaverine hydrochloride was used as an internal standard at a concentration of 0.05 mg/mL.

2.3. Kinetic Procedures. The degradation of EP in aqueous solutions was studied at 333 K, 343 K, 353 K, and 363 K in the pH range 0.42–9.95. The pH values of the reaction solutions and those of the buffer standards were measured at reaction temperatures. The pH values of the reaction solutions in HCl were calculated from the equation $\text{pH} = -\log f_{\text{HCl}}\,[\text{HCl}]$. The activity coefficient f_{HCl} was obtained or calculated from the literature data [10]. The ionic strength of all the solutions was adjusted to 0.50 M with a solution of sodium chloride (4 M). Solutions of the desired pH and ionic strength of 0.50 M were heated to the required temperatures and then a sample of EP was added to obtain the initial concentration of 0.2 mg/mL. At selected times, determined by the rate of degradation, samples of the solutions (1.0 mL) were collected and instantly cooled with a mixture of ice and water. Samples with pH above 7.5 were neutralized by using HCl solutions at concentrations ensuring that their pH was approximately 2. To each such sample, 1.0 mL of the internal standard solution (papaverine hydrochloride solution) was added. 50 μL of the solution samples was injected into the column. Due to the fact that anthracyclines are photolabile, all samples to be studied were protected from light.

Microsoft Excel 2000 was used for the calculation of regression parameters.

2.4. Theoretical Studies. In order to interpret the FMOs (frontier molecular orbitals) and EP, quantum-chemical calculations were performed by using the Gaussian 03 package [11]. The GaussView software was utilized to propose the initial geometry of the investigated molecules. The molecular geometries were optimized by means of a density functional theory (DFT) method with the B3LYP hybrid functional and a 6-31G(d,p) basis set.

3. Results and Discussion

3.1. Observed Rate Constants. The observed rate constants of EP degradation in aqueous solutions were determined in a pH range of 0.42–9.95 and described by the equation of a pseudo-first-order reaction:

$$\ln c_t = \ln c_0 - k_{\text{obs}} \times t, \tag{1}$$

where c_t and c_0 are the time-dependent concentration and the initial concentration of EP at times $t > 0$ and $t = 0$, respectively; k_{obs} is the observed rate constant of the pseudo-first-order reaction of EP degradation.

The number of measurements of c_t for each series ranged from 8 to 12.

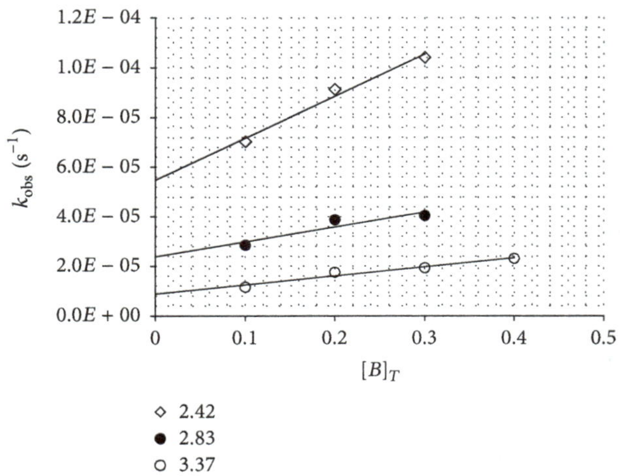

FIGURE 1: The plots of $k_{obs} = f([B]_T)$ for the degradation of EP in phosphate buffer at 353 K.

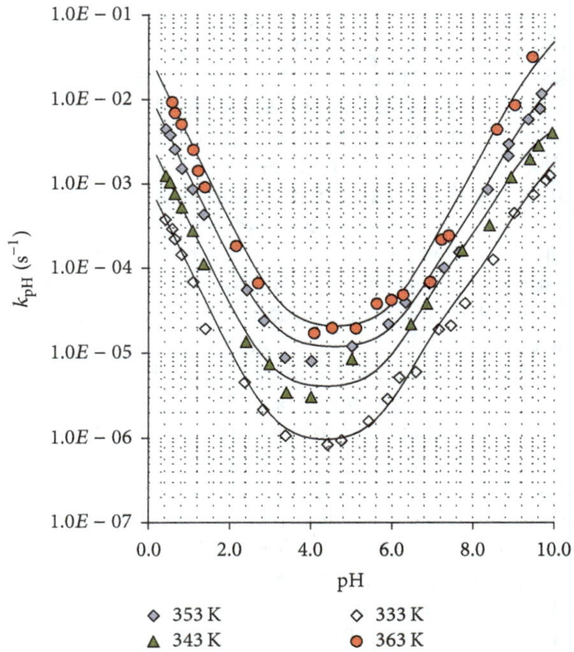

FIGURE 2: The pH-rate profiles for the degradation of EP at 333, 343, 353, and 363 K. The symbols represent experimental points; the lines were calculated from (3).

3.2. Buffer Catalysis. Under the conditions of this study, the rate constants (k_{obs}) depended on the total concentrations of the phosphate and acetate buffers, which indicated that the components of the buffers catalyzed the degradation of EP (general acid-base catalysis). In order to verify that the differences between k_{obs} determined at different buffer concentrations were statistically significant, the parallelism test was used. Under the conditions of general acid-base catalysis, k_{obs} were calculated from the following equation:

$$k_{obs} = k_{pH} + k_B [B]_T, \quad (2)$$

where $[B]_T$ is the total buffer concentration, k_{pH} is the rate constant at zero buffer concentration, and k_B describes the catalytic effect of the buffer components. The catalytic effect of the acetate buffer was investigated at pH 4.02–5.89 while those of phosphate buffers were investigated at pH 2.15–3.40 (Figure 1) and pH 5.91–7.45. Since in the reaction solutions of EP in HCl and borate buffer general acid-base catalysis was not observed, in that pH range, the values of $k_{obs} = k_{pH}$.

3.3. pH-Rate Profiles. The values k_{pH} calculated from (2) under the conditions of general acid-base catalysis (acetate and phosphate buffers) and those obtained for hydrochloric acid as well as borate buffer ($k_{obs} = k_{pH}$) were used to calculate the relationship $\log k_{pH} = f(\text{pH})$ (Figure 2).

The semilogarithmic relationship k_{pH}–pH (Figure 2) indicated that in water solutions at pH 0.42–9.95 the following reactions occurred:

(i) Degradation of the protonated molecules of EP catalyzed by hydrogen ions (k_1)

(ii) Spontaneous hydrolysis of protonated (k_2) and unprotonated (k_3) molecules of EP under the influence of water

(iii) Degradation of unprotonated molecules and monoanions of EP catalyzed by hydroxide ions (k_4 and k_5)

The total reaction rate was equal to the sum of partial reaction rates:

$$k_{pH} = k_1 \times a_{H^+} \times f_1 + k_2 \times f_1 + k_3 \times f_2 + (k_4 \times f_2 + k_5 \times f_3) \times a_{OH^-}, \quad (3)$$

where a_{H^+} is the hydrogen ion activity and f_1–f_3 are the fractions of protonated and unprotonated molecules and monoanions of EP. The f_1–f_3 values were calculated by using the values of pK_a of EP that were about 7.4 and 9.5.

The catalytic rate constants (k_1) were calculated from the plots $k_{pH} = f(a_{H^+})$ by using the values of k_{pH} in the pH range 0.42–3.40. The plots were linear with the positive slope that equaled k_1 (Figure 3).

The catalytic rate constants k_4 and k_5 were calculated from the equation $k'_{pH} = k_{pH} - (k_1 \times a_{H^+} \times f_1) = (k_4 \times f_2 + k_5 \times f_3) \times a_{OH^-}$ using k_{pH} values above pH 8.5 where the concentrations of $f_2 + f_3 \to 1$. The plots $k'_{pH}/(a_{OH^-} \times f_3) = f(f_2/f_3)$ were linear with positive slopes equal to the catalytic rate constant k_4 and the value b corresponded to the catalytic rate constant k_5. The catalytic rate constants k_2 and k_3 were calculated from the following equation: $k'_{pH} = k_{pH} - (k_1 \times a_{H^+} \times f_1 + (k_4 \times f_2 + k_5 \times f_3) \times a_{OH^-}) = k_2 \times f_1 + k_3 \times f_2$, using the k_{pH} values at pH 2.8–7, where the concentrations of $f_1 + f_2 \to 1$.

The plots $k'_{pH}/f_1 = f(f_2/f_1)$ were linear with positive slopes equal to the catalytic rate constant k_3 and the value b corresponded to the catalytic rate constant k_2.

The calculated theoretical profile of $\log k = f(\text{pH})$ and that obtained from the experimental results were nearly identical, indicating that the choice of the equation describing the total rate of EP degradation was correct (Figure 2).

TABLE 1: Catalytic rate constants and thermodynamic parameters for the degradation of EP in aqueous solutions.

Catalytic rate constant	Temperature (K)	$(k \pm \Delta k)$	Parameters of regression $\ln k_i = f(1/T)$	Thermodynamic parameters
k_1 $(\mathrm{mol^{-1}\ L\ s^{-1}})$	333	$(1.02 \pm 0.07) \times 10^{-3}$	$r = -0.9997$	$E_a = 118.8 \pm 9.4$ (kJ mol^{-1})
	343	$(3.44 \pm 0.12) \times 10^{-3}$	$a = -14282 \pm 1130$	$\Delta H^{\neq} = 116.3 \pm 9.4$ (kJ mol^{-1})
	353	$(1.22 \pm 0.11) \times 10^{-2}$	$b = 36.0 \pm 3.3$	$\Delta S^{\neq} = 54.5 \pm 27.0$ (J K^{-1} mol^{-1})
	363	$(3.43 \pm 0.41) \times 10^{-2}$		
k_2 $(\mathrm{s^{-1}})$	333	$(8.82 \pm 8.22) \times 10^{-7}$	$r = -0.9850$	$E_a = 104.1 \pm 55.4$ (kJ mol^{-1})
	343	$(3.79 \pm 2.75) \times 10^{-6}$	$a = -12518 \pm 6666$	$\Delta H^{\neq} = 101.6 \pm 55.4$ (kJ mol^{-1})
	353	$(1.12 \pm 0.85) \times 10^{-5}$	$b = 23.8 \pm 19.2$	$\Delta S^{\neq} = -46.6 \pm 159.5$ (J K^{-1} mol^{-1})
	363	$(1.91 \pm 1.04) \times 10^{-5}$		
k_3 $(\mathrm{s^{-1}})$	333	$(5.02 \pm 0.42) \times 10^{-5}$	$r = -0.9890$	$E_a = 80.7 \pm 36.7$ (kJ mol^{-1})
	343	$(1.57 \pm 0.04) \times 10^{-4}$	$a = -9708 \pm 4416$	$\Delta H^{\neq} = 78.3 \pm 36.7$ (kJ mol^{-1})
	353	$(2.46 \pm 0.11) \times 10^{-4}$	$b = 19.3 \pm 12.7$	$\Delta S^{\neq} = -84.0 \pm 105.7$ (J K^{-1} mol^{-1})
	363	$(6.28 \pm 0.52) \times 10^{-4}$		
k_4 $(\mathrm{mol^{-1}\ L\ s^{-1}})$	333	(4.95 ± 0.65)	$r = -0.9979$	$E_a = 66.8 \pm 13.1$ (kJ mol^{-1})
	343	(9.89 ± 1.21)	$a = -8042 \pm 1579$	$\Delta H^{\neq} = 64.4 \pm 13.1$ (kJ mol^{-1})
	353	(17.6 ± 8.7)	$b = 25.7 \pm 4.5$	$\Delta S^{\neq} = -30.8 \pm 37.8$ (J K^{-1} mol^{-1})
	363	(37.6 ± 19.2)		
k_5 $(\mathrm{mol^{-1}\ L\ s^{-1}})$	333	(0.821 ± 3.110)	$r = -0.9993$	$E_a = 63.8 \pm 7.3$ (kJ mol^{-1})
	343	(1.44 ± 7.17)	$a = -7678 \pm 882$	$\Delta H^{\neq} = 61.4 \pm 7.3$ (kJ mol^{-1})
	353	(3.10 ± 24.78)	$b = 22.8 \pm 2.5$	$\Delta S^{\neq} = -54.8 \pm 21.1$ (J K^{-1} mol^{-1})
	363	(5.39 ± 99.46)		

$\Delta H^{\neq}$ and $\Delta S^{\neq}$ were calculated for 293 K. $E_a = -aR$ (J mol^{-1}); $\Delta H^{\neq} = E_a - RT$ (J mol^{-1}); $\Delta S^{\neq} = R[\ln A - \ln(kT/h)]$ (J K^{-1} mol^{-1}), where k is Boltzmann's constant (1.3807×10^{-23} J K^{-1}), h is Planck's constant (6.626×10^{-34} J s), R is the universal gas constant (8.314 J K^{-1} mol^{-1}), T is temperature in K, a is the vectorial coefficient of the Arrhenius relationship, and A is the frequency coefficient.

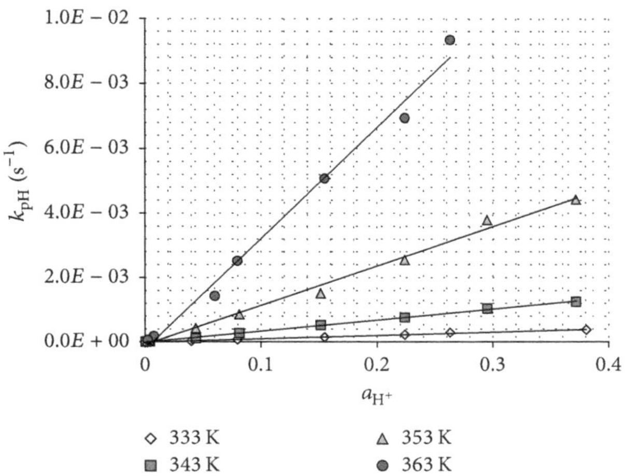

FIGURE 3: The plots of $k_{pH} = f(a_{H^+})$ for the degradation of EP in aqueous solutions at the indicated temperatures.

Based on the Arrhenius relationship $\ln k = \ln A - E_a/RT$, linear plots of $\ln k = f(1/T)$ were used to calculate the energy of activation (E_a), the entropy ($\Delta S^{\neq}$), the enthalpy ($\Delta H^{\neq}$), and the preexponential coefficient (A) for the five partial reactions (Table 1).

The lowest energy of activation was observed in the reactions of unprotonated molecules and monoanions of EP catalyzed by hydroxide ions whereas the highest energy of activation was observed in the reaction of protonated molecules of EP catalyzed by hydrogen ions.

3.4. Theoretical Studies. Based on theoretical calculations, the places in EP molecules that might be connected with the differences in EP reactivity in solutions with various pH values were proposed. The electrons in EP were concentrated in the tetracyclic quinolid aglycone, especially in the C ring. That domain was responsible for the reactivity of EP during base hydrolysis (Figure 4).

The calculations also indicated that the A ring of the tetracyclic quinolid aglycone played a part in EP susceptibility to degradation during acidic hydrolysis. A comparison of electron distribution in the HOMO and LUMO of EP demonstrated that it was similar to electron distribution in daunorubicin and doxorubicin (Figure 5). Previous stability studies of those three anthracycline antibiotics did not show any statistically significant differences between

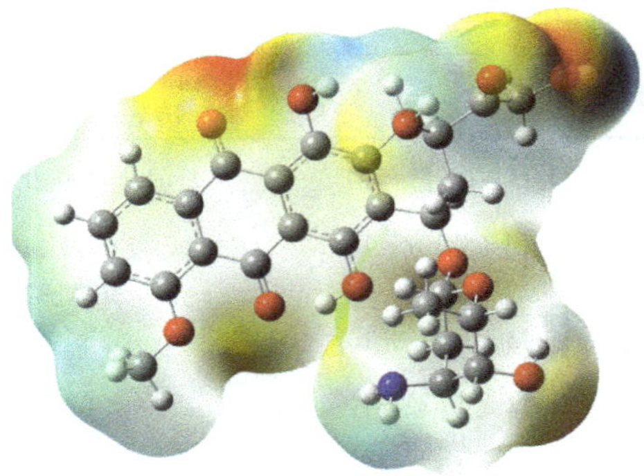

Figure 4: ESP of epidoxorubicin.

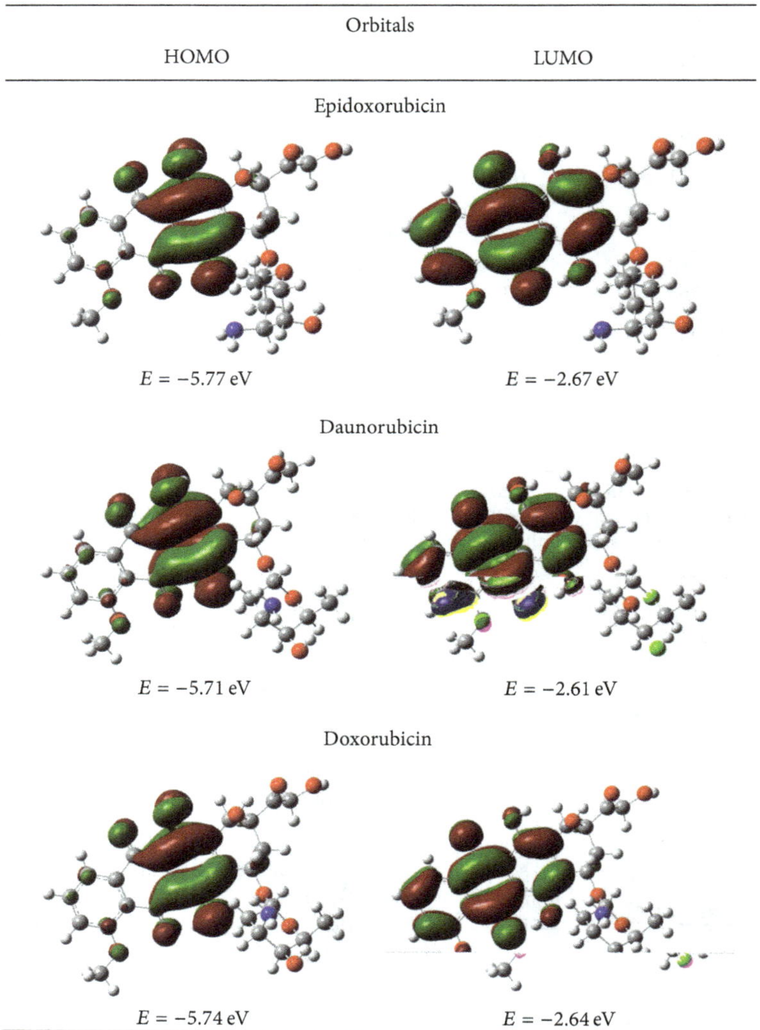

Figure 5: FMOs of EP, DOX, and DAU.

their degradation rate constants in the solid state [4, 5] and revealed a similarity between the relationships log $k_{pH} = f(\mathrm{pH})$, which correlates with theoretical study results [12, 13]. The values of energy gap for the three derivatives were 3.1 eV, proving their similar chemical stability.

Although the stereoisomerism of the C-4′ hydroxyl group of L-acosamine in EP differs from that observed for DOX, the difference does not significantly influence EP reactivity and, consequently, the susceptibility of EP to degradation during acid-base hydrolysis is not affected. That difference in stereoisomerism was crucial for ensuring the greater therapeutic safety of EP.

Conflicts of Interest

The authors declare that there are no conflicts of interest regarding the publication of this paper.

References

[1] G. Minotti, S. Licata, A. Saponiero et al., "Anthracycline metabolism and toxicity in human myocardium: Comparisons between doxorubicin, epirubicin, and a novel disaccharide analogue with a reduced level of formation and [4Fe-4S] reactivity of its secondary alcohol metabolite," *Chemical Research in Toxicology*, vol. 13, no. 12, pp. 1336–1341, 2000.

[2] A. Mordente, E. Meucci, G. E. Martorana, B. Giardina, and G. Minotti, "Human heart cytosolic reductases and anthracycline cardiotoxicity," *IUBMB Life*, vol. 52, no. 1-2, pp. 83–88, 2001.

[3] M. Piekarski and A. Jelińska, "Anthracyclines still prove effective in Anticancer Therapy," *Mini-Reviews in Medicinal Chemistry*, vol. 13, no. 5, pp. 627–634, 2013.

[4] A. Sobczak, A. Jelińska, M. Leśniewska, A. Firlej, and I. Oszczapowicz, "Stability of epidoxorubicin in solid state," *Journal of Pharmaceutical and Biomedical Analysis*, vol. 54, no. 4, pp. 869–872, 2011.

[5] J. Cielecka-Piontek, A. Jelińska, M. Zajac, M. Sobczak, A. Bartold, and I. Oszczapowicz, "A comparison of the stability of doxorubicin and daunorubicin in solid state," *Journal of Pharmaceutical and Biomedical Analysis*, vol. 50, no. 4, pp. 576–579, 2009.

[6] M. J. Wood, W. J. Irwin, and D. K. Scott, "Photodegradation of doxorubicin, daunorubicin and epirubicin measured by high-performance liquid chromatography," *Journal of Clinical Pharmacy and Therapeutics*, vol. 15, no. 4, pp. 291–300, 1990.

[7] A. Jelińska, M. Zając, K. Krzystoń, A. Firlej, and I. Oszczapowicz, "The use of HPLC for determining the stability of epirubicin and the influence of ionic strength on the stability of epirubicin," *Annales Universitatis Mariae Curie-Skłodowska Lublin - Polonia*, vol. 19, no. 2, pp. 45–50, 2006.

[8] M. J. Wood, W. J. Irwin, and D. K. Scott, "Stability of doxorubicin, daunorubicin and epirubicin in plastic syringes and minibags," *Journal of Clinical Pharmacy and Therapeutics*, vol. 15, no. 4, pp. 279–289, 1990.

[9] D. Kumar, R. S. Tomar, S. K. Deolia, R. Srivastava, M. Mitra, and S. Tyagi, "Isolation and characterization of degradation impurities in epirubicin hydrochloride injection," *Journal of Chromatography B*, vol. 869, no. 1-2, pp. 45–53, 2008.

[10] E. Pawełczyk and T. Hermann, *The Fundamentals of Stability of Drugs*, Warsaw, Poland, 1982.

[11] M. J. Frisch, G. W. Trucks, H. B. Schlegel et al., *Gaussian 03, Revision D.01*, Gaussian, Inc., Wallingford, Conn, USA, 2004.

[12] J. H. Beijnen, O. A. G. J. van der Houwen, and W. J. M. Underberg, "Aspects of the degradation kinetics of doxorubicin in aqueous solution," *International Journal of Pharmaceutics*, vol. 32, no. 2-3, pp. 123–131, 1986.

[13] J. H. Beijnen, O. A. G. J. van der Houwen, M. C. H. Voskuilen, and W. J. M. Underberg, "Aspects of the degradation kinetics of daunorubicin in aqueous solution," *International Journal of Pharmaceutics*, vol. 31, no. 1-2, pp. 75–82, 1986.

18

Crystal Structure and Biological Evaluation of Two Novel Organic-Inorganic Hybrid Materials as Antitumor Agents in the Treatment of Liver Cancer

Bao-Hua Song,[1] **Chen Li,**[1] **and Gui-Feng An**[1,2]

[1]*Inner Mongolia University for Nationalities, Tongliao, Inner Mongolia, China*
[2]*The First Affiliated Hospital of Inner Mongolia University for Nationalities, Tongliao, Inner Mongolia, China*

Correspondence should be addressed to Gui-Feng An; anguifeng1224@163.com

Academic Editor: Henryk Kozlowski

Two novel organic-inorganic hybrid materials $\{(Hbiz)_6[As^{III}{}_2As^{V}Mo^{VI}{}_{18}O_{62}]\}\cdot H_2O$ (**1**, biz = benzimidazole) and (dim)$[As^{III}{}_2As^{V}Mo^{VI}{}_{18}O_{62}]$ [**2**, dim = 1,6-bis(imidazol)hexane] have been successfully obtained by using the molybdenum arsenate and different N donor organic compounds and determined through X-ray single-crystal diffraction technique. The *in vitro* cytotoxicity of compounds **1** and **2** was then investigated against three human liver tumor cell lines (SMMC7721, Bel-7402, and MHCC97) by MTT assay. It was found that the two compounds showed potent use as antitumor agents against the aforementioned cell lines.

1. Introduction

Despite remarkable progress in the evolution of biological medicine field and understanding of the origin of the de diseases, the scientists fail to obtain effective methods for treating some diseases such as cancer. Nowadays, it has been widely recognized that cancer is one of the most serious threat to the health of people worldwide which could cause millions of deaths every year [1]. In spite of great efforts currently being devoted to the creation of therapeutic tools, current effective therapeutic is limited to chemotherapy, which is the dominant treatment method. In this case, the creation of effective chemotherapy drugs with low cost, high selectivity, undesirable side effects, and low cytotoxicity is of high importance [2, 3]. The cancer cell resistance to anti-cancer agents and the toxicity of modern chemotherapy make the sciences worldwide search for new drugs and cure methods for such an insidious disease [4].

The design and fabrication of inorganic-organic hybrid materials have drawn much interest in the last twenty years not only for their variable building block but also the excellent properties that inherit from the both organic and inorganic parts, which also leads the function as function materials in many important areas such as fluorescence and guest-dependent sorption [5–7]. Recent literatures have demonstrated that the N donor imidazole ligand is a good coordinating ligand in the synthesis of organic-inorganic hybrid materials, which may adopt many different types of coordinating modes, for instance cocrystal, monodentate chelating, and mutidentate bridging in the *anti-anti*, *syn-syn*, and *syn-anti* arrangement [8, 9]. So far, many metal-imidazole organic-inorganic hybrid materials have been designed and characterized widely. In this work, two novel organic-inorganic hybrid materials $\{(Hbiz)_6[As_2Mo_{18}O_{62}]\}\cdot H_2O$ (**1**, biz = benzimidazole) and (dim)$[As^{III}{}_2As^{V}Mo^{VI}{}_{18}O_{62}]$ (**2**, dim = 1,6-bis(imidazol)hexane) have been successfully obtained by using the molybdenum arsenate and different N donor compounds (Figure 1) under hydrothermal conditions and their corresponding antitumor performance was subsequently studied.

2. Experimental

2.1. Instruments and Chemicals. The Rigaku D/MAX 2500 V diffractometer was used to collect the Powder X-ray diffraction (PXRD) curves. An Oxford Xcalibur E diffractometer was

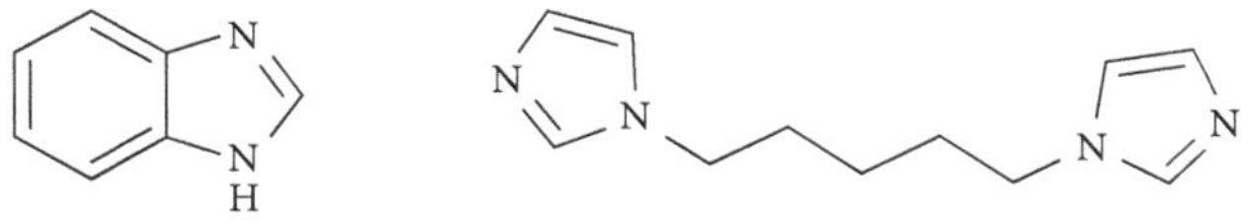

FIGURE 1: Schematic representation of the benzimidazole and 1,6-bis(imidazol)hexane organic compounds used in this research.

employed to collect the single crystal X-ray diffraction data. Perkin-Elmer model 240C instrument was employed to obtain the element analyses (C, H, and N) results. All the chemicals in this study were commercially available and used as received.

SMMC7721, Bel-7402, and MHCC97 human liver cancer cells used in this work were commercially available from the Beijing National Institute for the Control of Biological and Pharmaceutical Products (China).

*2.2. The Preparation Methods of Compounds **1** and **2**.* biz (0.15 mmol), As_2O_3 (0.8 mmol), $Na_2MoO_4{\cdot}2H_2O$ (2 mmol), and H_2O (18 mL) were mixed in a 25 mL of Teflon-lined stainless steel vessel, kept at 140°C for 72 hours. After the reaction time is out, the system is left until the oven return to room temperature. The black block-shape crystals were collected, cleaned with flowing deionized water, and air dried. Yield: 36% (based on biz ligand). Elemental analysis for compound **1** ($C_{42}H_{30}N_{12}As_2Mo_{18}O_{63}$): C, 14.01; H, 0.82; N, 4.61%. Calculated: C, 14.05; H, 0.84; N, 4.69%.

dim (0.2 mmol), As_2O_3 (0.5 mmol), $Na_2MoO_4{\cdot}2H_2O$ (1 mmol), and H_2O (18 mL) were mixed in a 25 mL of Teflon-lined stainless steel vessel, kept at 140°C for 72 hours. After the reaction time is out, the system is left until the oven returns to room temperature. The black block-shape crystals were collected, cleaned with flowing deionized water, and air dried. Yield: 36% (based on dim ligand). Elemental analysis for compound **2** ($C_{11}H_{16}N_{12}As_2Mo_9O_{31}$): C, 9.03; H, 0.84; N, 9.81%. Calculated: C, 8.99; H, 0.93; N, 9.80%.

2.3. Structural Analysis through the X-Ray Diffraction Technique. A graphite monochromatized Mo radiation (0.71073 Å) equipped Oxford Xcalibur E diffractometer with was used to collect the reflection data of **1** and **2** at 298 K. The direct method in the XS program was used to get the initial structural modes and the least-squares method in XL program was used to refine the structural models [10]. All the H atoms were poisoned at their ideal locations and refined using the AFIX commands according to their C atoms attached, and all nonhydrogen atoms in the two structures were treated anistropically. Table 1 shows the crystal data and the structure refinements. Crystal data and structural refinement parameters for compounds **1** and **2** are shown in Table 1.

2.4. Antitumor Activity. The RPMI 1460 medium was employed to grow the three human liver cancer cells (SMMC7721, Bel-7402, and MHCC97). Before the cell growth, the medium was supplemented with 100 μg/mL streptomycin, 100 μg/mL penicillin, and 10% fetal calf serum. A humidified incubator with a 5% CO_2 atmosphere was used to incubate the system at 310 K. Subsequently, the

TABLE 1: Crystal data and structure refinements for compounds **1** and **2**.

	1	2
Formula	$C_{42}H_{30}As_2Mo_{18}N_{12}O_{63}$	$C_{11}H_{16}As_2Mo_9N_4O_{31}$
M_r	3587.54	1713.58
Crystal system	Triclinic	Orthorhombic
Space group	P-1	*Cmcm*
a (Å)	14.192(3)	22.937(2)
b (Å)	14.417(3)	17.9135(16)
c (Å)	23.228(4)	20.0763(17)
α (°)	91.336(2)	90
β (°)	101.346(2)	90
γ (°)	112.992(2)	90
V (Å^3)	4263.2(13)	8249.1(12)
Z	2	8
D_{calc} (g·cm^{-3})	2.795	2.760
Crystal size (mm^3)	0.32 × 0.16 × 0.23	0.26 × 0.26 × 0.26
μ (Mo Kα) (mm^{-1})	3.435	4.331
θ range (°)	2.24 to 28.34	2.274 to 28.272
Reflections collected	39782	38287
Number of unique data (R(int))	21258	5383
Number of data with $I \geq 2\sigma(I)$	16747	4624
R_1	0.0712	0.0658
ωR_2 (all data)	02186	0.2080
CCDC	1810739	1810740

synthesized compounds were dissolved in DMSO at 1 mM immediately before use and diluted just before the addition to the cell culture. After incubation of cells for up to 96 h, the medium was removed from each cell and 15 μL of MTT (0.5 mg/mL) solution was added after the supernatant was removed carefully with the addition of another 100 μL of dimethyl sulfoxide. The 96-well plate was shaken for 5 min, and then the cell viability was detected. The IC_{50} values were determined by plotting the percentage viability versus concentration on a logarithmic graph and reading of the concentration at which 50% of cells viable relative to the control. Each experiment was repeated at least three times to get the mean values.

3. Results and Discussion

3.1. Molecular Structure. An X-ray single crystalline structural investigation demonstrates that compound **1** is built from Wells–Dawson polyoxoanion $[As_2Mo_{18}O_{62}]^{6-}$ (signed as As_2Mo_{18}), six free ligands $Hbiz^+$, and water molecule (Figure 2(a)). Two $[\alpha\text{-}AsMo_9O_{31}]^{3-}$ units are connected with each other through the corner-sharing with elimination of six oxygen atoms to shape the classical Wells–Dawson polyoxoanion, which has only two structurally independent kinds of tungsten (12 equatorial and six polar) with D_{3h} symmetrical point group. The calculation results as revealed by the software bond valence sum (BVS) calculations confirm that all Mo centers show VI oxidation state. The

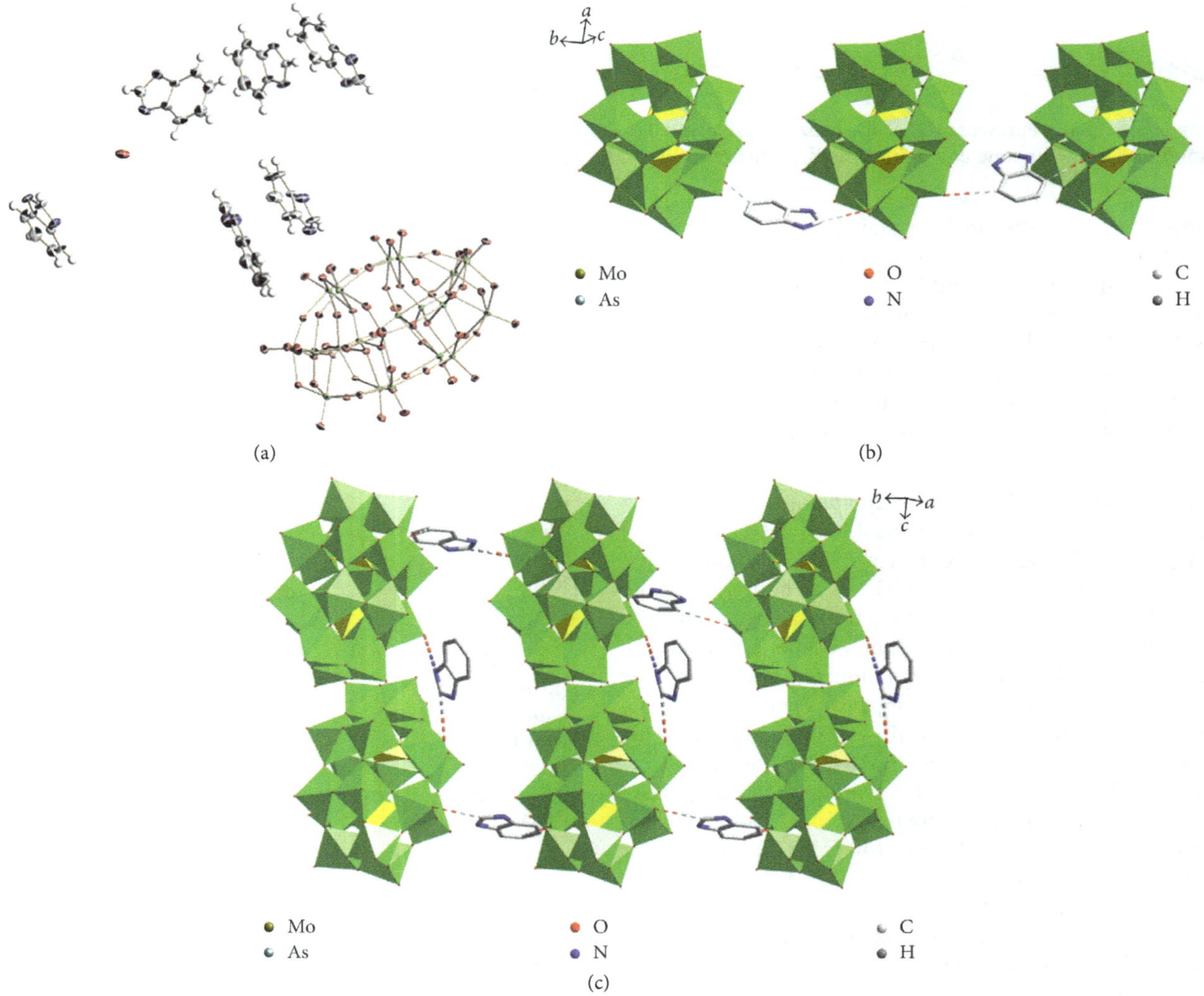

FIGURE 2: (a) An ORTEP drawing for the molecular unit of **1** in 50% thermal ellipsoids; (b) a drawing with polyhedral and ball-stick pattern for the one-dimensional chain in complex **1** by H-bonds; (c) a picture showing the H-bond interaction between the 2D layers.

coordination surroundings of Mo atoms could be regarded as a $\{MoO_6\}$ octahedron with the Mo–O bond lengths ranging from 1.679(1)–1.710(1) Å for edge oxygen donor sites, 1.891(4)–1.917(4) Å for μ_2-associated oxygen donor sites, and 2.369(3)–2.412(2) Å for μ_3-associated oxygen donor sites, which are comparable to the values reported in the literature. In contrast, each As_2Mo_{18} polyoxoanion is coordinating to six free biz ligands by hydrogen bonds along the *a*-axis to lead to the formation of a novel 1-D chain (Figure 2(b)). It is interesting that these ligands further connected with the POMs formed a 2D-layer through hydrogen bonds, and the acceptor-donor distance is 3.262 Å (Figure 2(c)).

X-ray crystal structure determination illuminates that the as-prepared title complex **2** is built from Wells–Dawson polyoxoanion $[As_2Mo_{18}O_{62}]^{6-}$ (abbreviated to As_2Mo_{18}), two As caps and one ligand dim. Two $[\alpha\text{-}AsMo_9O_{31}]^{3-}$ units are connected with each other through the corner-sharing with elimination of six oxygen atoms to shape the classical Wells–Dawson polyoxoanion, which has only two structurally independent kinds of tungsten (12 equatorial and six polar) with D_{3h} symmetrical point group. The calculation results as revealed by the software bond valence sum (BVS) calculations confirm that all Mo centers show VI oxidation state. All the six coordinated Mo atoms reveal a $\{MoO_6\}$ octahedral surrounding with the Mo–O bond lengths ranging from 1.679(1)–1.710(1) Å for the edge O donor sites, 1.891(4)–1.917(4) Å for μ_2-complexing O donor sites, and 2.369(3)–2.412(2) Å for μ_3-complexing O donor sites, which are comparable to the values reported in the literature. In addition, each As_2Mo_{18} polyoxoanion is coordinating to the free dim ligands by hydrogen bonds along the *a*-axis to form an novel 1-D chain (Figure 3(a)). It is interesting that these ligands further connected with the POMs formed a 2D-layer through hydrogen bonds (Figure 3(b)).

3.2. PXRD Patterns. To check the purity of the prepared two compounds, their corresponding power X-ray diffractions were collected at room temperature using their grounded samples. As shown in Figure 4, the peaks in the PXRD curves

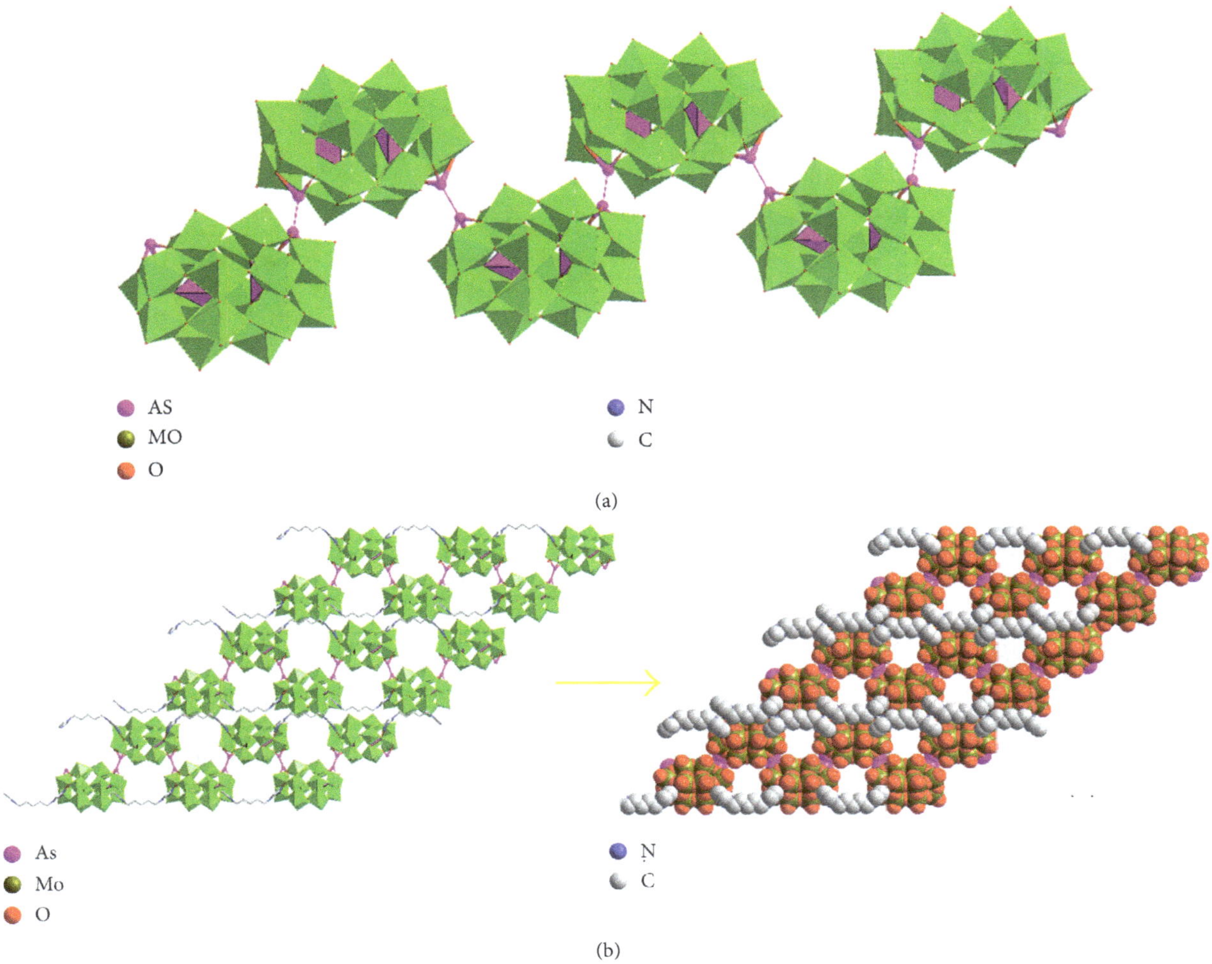

FIGURE 3: (a) The picture with ball-stick and polyhedral fashion shows the 1D chain of **2** by hydrogen bonds; (b) 2D layered framework of **2** formed by hydrogen bonds.

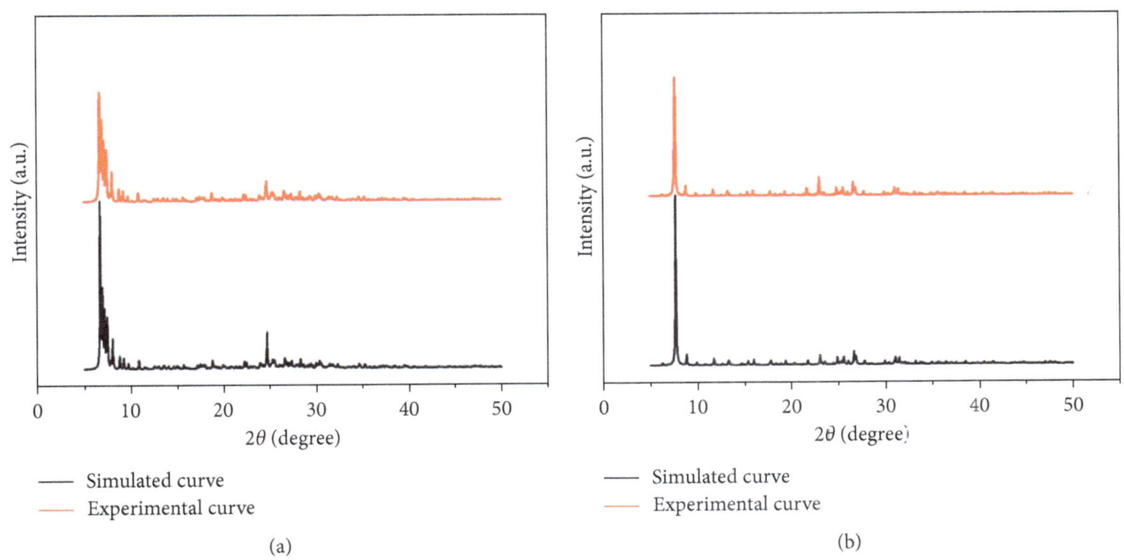

FIGURE 4: The PXRD patterns for **1** (a) and **2** (b).

Table 2: Antiproliferative activity IC_{50} (μM) of compounds **1** and **2**.

Compounds	IC_{50} (μM)[a]		
	SMMC7721	Bel-7402	MHCC97
1	45.6 ± 2.1	49.2 ± 1.9	62.1 ± 1.2
2	50.3 ± 1.8	45.8 ± 2.1	59.5 ± 2.2
Doxorubicin[b]	49.1 ± 2.0	50.1 ± 1.7	60.6 ± 2.1

[a]Drug concentration required to reduce cancer cell growth by 50%; [b]the anticancer drug doxorubicin was chosen as the positive control.

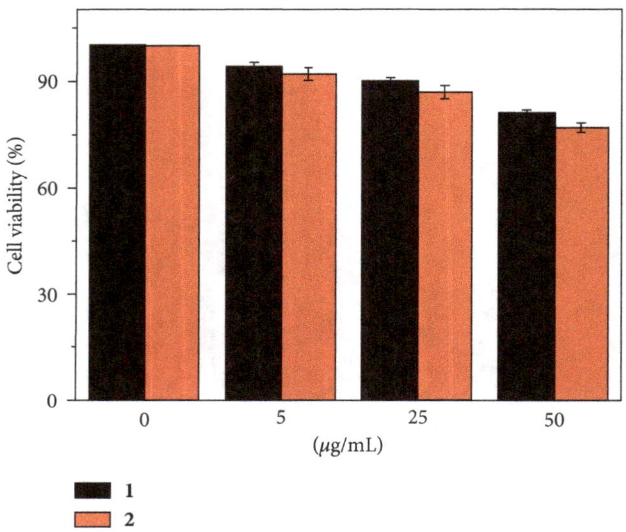

Figure 5: Cell viabilities of the oral epidermal cells after incubation with **1** and **2** at different concentrations.

are in good match with the simulated ones from the crystal data, revealing their phase purity to some extent. The slight difference of some peak intensities between the calculated one and the simulated one might be due to the preferred orientation of the crystal faces.

3.3. In Vitro Cytotoxicity Study against the Cancer Cells. Three human liver cancer cells (MHCC97, Bel-7402, and SMMC7721) were used to access the cancer cell growth inhibition activity for the two new metal-organic coordination complexes. The exposure time was set to 48 h, and the drug doxorubicin (DOX) was employed as the reference control with its antiproliferative effects compared to those of the two complexes. DMSO was used to dissolve the as-prepared two complexes at the concentration level of 1 mM, which need to be diluted just before the addition to the cell culture.

The data, which show the values for three times of separated experiments carried out in duplicate, are represented in the form of means ± SEM and are shown in Table 2. A descending order could be observed for the antiproliferative activity of the test complexes toward each type of the tumor cell lines used, which might be ascribed to the measured concentration required to reduce cancer cell proliferation by 50% (IC_{50} μM). From the results above, we can conclude that compounds **1** and **2** demonstrate comparable growth inhibition activity toward the three tumor cell lines ($IC_{50} = 45–62\,\mu$M) with the reference drug DOX, which illuminate their potential use as anticancer drugs.

In order to explore the cytotoxicity of the two compounds toward the human normal cell *in vitro*, oral epidermal cells (normal cells) were chosen to incubate with **1** and **2** under different concentrations (0–50 $\mu g{\cdot}mL^{-1}$) for 24 h. As shown in Figure 5, the cell viabilities all remained above 70% even at a concentration of 50 $\mu g{\cdot}mL^{-1}$ of **1** and **2**, so the as-prepared two new metals showed low cytotoxicity in the oral epidermal cells.

4. Conclusion

The results indicate that complex **1** contains six free ligands biz and one $[As_2Mo_{18}O_{62}]^{6-}$ cluster in the molecular lattice, which is further connected with each other to make up a 2D supramolecular structure, while complex **2** consists of one free dim ligand and one $[As_2Mo_{18}O_{62}]^{6-}$ cluster in the molecular lattice and demonstrates a 2D supramolecular structure formed by hydrogen bonds. The *in vitro* cytotoxicity experiments demonstrate that as-prepared compounds **1** and **2** revealed potent cytotoxicity toward the three human liver tumor cell lines (SMMC7721, Bel-7402, and MHCC97) as similar to that of the reference drug, which also indicates their potential use as antitumor drugs.

Conflicts of Interest

The authors declare that they have no conflicts of interest.

Acknowledgments

Grant from the Natural Science Foundation of Inner Mongolia (2017MS(LH)0819) is greatly acknowledged.

References

[1] X. Du, Z. Shi, Z. Peng et al., "Acetoacetate induces hepatocytes apoptosis by the ROS-mediated MAPKs pathway in ketotic cows," *Journal of Cellular Physiology*, vol. 232, no. 12, pp. 3296–3308, 2017.

[2] Y. Song, N. Li, J. Gu et al., "β-Hydroxybutyrate induces bovine hepatocyte apoptosis via an ROS-p38 signaling pathway," *Journal of Dairy Science*, vol. 99, no. 11, pp. 9184–9198, 2016.

[3] X. Sun, X. Yuan, L. Chen et al., "Histamine induces bovine rumen epithelial cell inflammatory response *via* NF-κB pathway," *Cell Physiology Biochemistry*, vol. 42, no. 3, pp. 1109–1119, 2017.

[4] X. Du, Y. Zhu, Z. Peng et al., "High concentrations of fatty acids and β-hydroxybutyrate impair the growth hormone-mediated hepatic JAK_2-$STAT_5$ pathway in clinically ketotic cows," *Journal of Dairy Science*, vol. 101, no. 4, pp. 1–12, 2018.

[5] J. Li, X. H. Ji, and J. T. Li, "Two new inorganic anions directed Zn(II)-tetrazole frameworks: syntheses, structures and pho-

toluminescent properties," *Journal of Molecular Structure*, vol. 1147, pp. 22–25, 2017.

[6] J. Li and J. T. Li, "A luminescent porous metal–organic framework with Lewis basic pyridyl sites as a fluorescent chemosensor for TNP detection," *Inorganic Chemistry Communications*, vol. 89, pp. 51–54, 2018.

[7] W. Gao, F. Liu, X. M. Zhang, J. P. Liu, and Q. Y. Gao, "Four 3D coordination polymers based on layers with single syn-anti carboxylate bridges: synthesis, structures, and magnetic properties," *RSC Advance*, vol. 8, no. 25, pp. 14101–14108, 2018.

[8] D. Chen, Y. Zhong, C. Zhang, D. Xu, and Z. Lin, "Dual-ligand approach for the solvent-free synthesis of indium-based coordination polymers," *Inorganic Chemistry Communications*, vol. 92, pp. 74–77, 2018.

[9] P. Köse Yaman and O. Z. Yeşilel, "Hydrothermal synthesis and characterization of cobalt(II), nickel(II) and zinc(II) coordination polymers with 2,2′-dimethylglutarate and 1,2-bis(4-pyridyl)ethane," *Polyhedron*, vol. 148, pp. 189–194, 2018.

[10] G. M. Sheldrick, "Crystal structure refinement with SHELXL," *Acta Crystallographica Section C-Structural Chemistry*, vol. 71, no. 1, pp. 3–8, 2015.

19

On the Interactions of Fused Pyrazole Derivative with Selected Amino Acids: DFT Calculations

Kornelia Czaja,[1] Jacek Kujawski,[1] Elżbieta Jodłowska-Siewert,[1] Paulina Szulc,[1] Tomasz Ratajczak,[2] Dominika Krygier,[3] Marcin K. Chmielewski,[3] and Marek K. Bernard[1]

[1]*Department of Organic Chemistry, Faculty of Pharmacy, Poznan University of Medical Sciences, Grunwaldzka 6 Street, 60-780 Poznań, Poland*
[2]*Centre of New Technologies, University of Warsaw, Banacha 2C, 02-097 Warsaw, Poland*
[3]*Institute of Bioorganic Chemistry, Polish Academy of Sciences, Z. Noskowskiego Str. 12/14, 61-704 Poznań, Poland*

Correspondence should be addressed to Jacek Kujawski; jacekkuj@ump.edu.pl

Academic Editor: James W. Gauld

Due to the increasing prevalence of neoplasms, there is a permanent need for new selective cytostatic compounds. Anticancer drugs can act in different ways, affecting protein expression and synthesis, including disruption of signaling pathways within cells. Continuing our previous research aiming at elucidating the mechanism of pyrazole's anticancer activity, we carried out in silico studies on the interactions of fused pyrazole derivative with alanine, lysine, glutamic acid, and methionine. The objective of the study is to improve our understanding of the possible interactions of pyrazole derivatives with the above-mentioned amino acids. For this purpose, we apply the DFT formalism (optimization using the B3LYP, CAM-B3LYP, PBE0, and M06L functionals) and interaction energy calculations (counterpoise corrected method based on the basis set superposition error, BSSE) together with QTAIM approach and estimation of the ^{1}H NMR chemical shifts of analyzed pyrazole derivative using different basis sets and DFT functionals in CPCM solvation model (and water used as a solvent).

1. Introduction

Currently, there is a growing demand for new cytostatic compounds that act selectively. Drugs with anticancer properties exert them through different modes of action [1]. These include disruption of DNA replication and thus inhibition of cell proliferation and inhibition of translation of target proteins, such as kinases that regulate signaling pathways within cells. Eventually, these compounds induce apoptosis of neoplastic cells. So far, numerous oncogenes have been described and many of them are dysregulated protein kinases. Decreasing the activity of upregulated kinases results in a stunted tumor growth [2–5]. This is why kinase inhibitors have attracted significant attention for targeted therapy of many cancers.

Heterocycles containing pyrazole are a part of a group of anticancer drugs on which we focused our attention. This type of anticancer compounds includes those which interact with active sites of kinases, for example, through binding with amino acids. Alanine (Ala), lysine (Lys), glutamic acid (Glu), and methionine (Met) existing in these cavities are especially important as they can interact with pyrazole derivatives [2–5].

According to the results obtained by our group previously, certain condensed pyrazole derivatives exert a cytostatic activity and induce apoptosis in the HT29 human colorectal neoplastic cells [6, 7]. In our subsequent papers, we hypothesized that they could disrupt the intracellular Mg^{2+} ions balance through the action of pyridinic nitrogen within the pyrazole moiety [8, 9]. An NMR experiment elucidating the interaction of chosen amino acids and the condensed pyrazole derivative showed that this interaction primarily happens between the pyrrolic nitrogen atom (NH) and carbon fragment B (Scheme 1) [7–9].

Continuing our studies that aimed at elucidating the potential mode of pyrazole cytostatic activity, we conducted further theoretical studies investigating simulated interactions of fused derivative **7** (synthesis is given in Scheme 1,

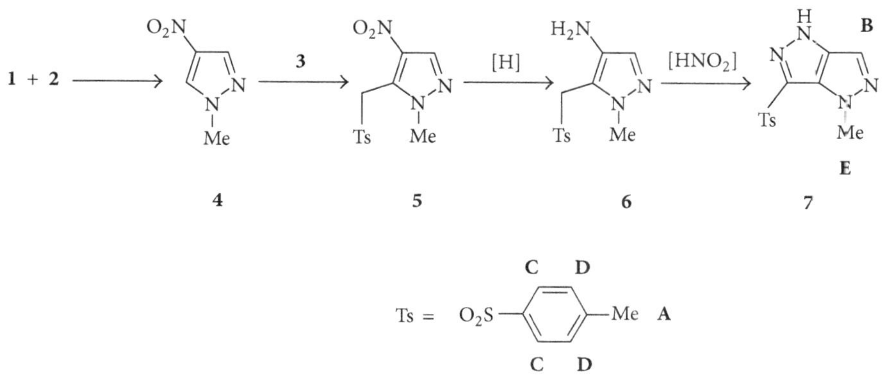

SCHEME 1

1: 4-nitro-1*H*-pyrazole; 2: $(CH_3)_2SO_4$ [8]; 3: Ts-CH_2-Cl; [H]: reduction of the nitro group; [HNO_2]: diazotization and cyclization; A-E: protons of the final product considered for NMR investigations) with certain amino acids (Ala, Lys, Glu, and Met). The objective of the study is to improve our understanding of the tendency regarding mechanism of interactions between pyrazole derivatives and selected amino acids. We focused our attention on the electrostatic contacts, especially hydrogen bond formation between the azole and amino acids. To achieve that, we decided to choose the DFT formalism and interaction energy calculations with QTAIM approach and calculation of the ^{1}H NMR chemical shifts of compound **7** with several basis sets and DFT functionals in conductor-like polarizable continuum model (CPCM) of solvation, so that the influence of the solvent environment in cell is also accounted for. What is more, we aimed to determine the methods and basis sets that gave the closest results to the experimental shifts in the ^{1}H NMR spectrum and the most stable and preferable conformation of compound **7**.

2. Materials and Methods

We carried out density functional calculations and optimized the compound geometries at the DFT [10] level of theory using the *Gaussian 09 D.01* program [11], B3LYP/6-31G(d,p) [12], (b) CAM-B3LYP/6-31G(d,p) [13], (c) B3LYP/6-311+G(d,p) [14, 15], (d) PBE0/6-31G(d,p) [16, 17], and (e) M06L/6-31G(d,p) [18] approaches in the CPCM of solvation (water molecules as solvent). The ideal gas, rigid rotor, and harmonic oscillator approximations were applied to calculate the vibrational frequencies and thermodynamic properties. We obtained no imaginary frequencies within the vibrational spectra and therefore confirmed the energy minimum for all conformers. We created the conformers through the rotations of C1-S1-C2 bond of azole and bonds connecting N-H moiety of azole with carboxyl or amine group of the particular amino acid in dihedral angle increments of 20°.

AIMAll Professional package for topological analysis was used for QTAIM calculations [19]. For this purpose, the "Proaim" basin integration approach was used using "superfine" interatomic surface mesh and "very high" outer angular quadrature. Thanks to the atomic (integrated) Laplacian values, we could monitor the accuracy of basin integrations. What is more, the Poincaré-Hopf rule was satisfied for all carried out calculations. The ρ and $\nabla^2\rho$ parameters at the bond critical points (BCPs) concerning dipole polarizability tensors were calculated based on the optimized structures (taking into consideration the already calculated interaction energy) using the AIM approach and applying B3LYP/6-31G(d,p) (adducts **7AlaB**, **7LysB**, **7GluB**, and **7MetB**), B3LYP/6-311+G(d,p) (adducts **7GluB6** and **7MetB6**), PBE0/6-31G(d,p) (adducts **7LysP**), and M06L/6-31G(d,p) (adduct **7AlaM**) levels of theory, respectively. The components of molecular dipole polarizability tensors were calculated as the set of first derivatives of the dipole moment components with respect to the components of an applied electric field. We calculated the value of NMR shielding for proton (H^{ref}) in TMS at the B3LYP/6-31G(d,p) level of theory (CPCM of solvation and water as solvent). The above-mentioned method was also used for calculation of TMS (reference compound) and the investigated **7**–amino acid adducts. Then the values obtained for TMS allowed calculating the chemical shifts of **7** and **7**–amino acid adducts, in accordance with the following equation: $\delta_i = \sigma_{ref} - \sigma_i$, with δ_i being chemical shift of i-nuclei of **7**–amino acid adducts and σ_{ref} and σ_i being the calculated isotropic magnetic shielding tensors of TMS and **7**–amino acid adducts, respectively [8, 20–22]. An average of the estimated chemical shifts was calculated for the methyl groups protons A or E and also these of C or D (Scheme 1). We calculated interaction energy of all optimized adducts applying counterpoise corrected method based on the basis set superposition error (BSSE) at B3LYP/6-31G(d,p) and B3LYP/6-311++G(2d,3p) levels of theory, respectively [23, 24].

The *ChemCraft 1.7* software was utilized for the visualization of all optimized conformers and adducts [25]. The calculations were carried out using resources provided by the Polish Grid Infrastructure (PL-Grid) and Wrocław Center for Networking and Supercomputing (*Bem* cluster, WCSS Grant number 327/2014).

3. Results and Discussion

The analysis of azole **7** coordination with magnesium ions proved that the pyrrolic nitrogen NH with a labile

Table 1: Interaction energies between compound 7 andalanine (**7Ala**), lysine (**7Lys**), glutamic acid (**7Glu**), and methionine (**7Met**). Suffixes added to the names of adducts mean that the structure was optimized at functional: **B**, **B-C**, **B-N**, **B-αN**: B3LYP/6-31G(d,p); **B6**: B3LYP/6-311+G(d,p); **C**: CAM-B3LYP/6-31G(d,p); **P**, **P-N**, **P-αN**: PBE0/6-31G(d,p); **M**: M06L/6-31G(d,p). Cartesian coordinates of all adducts are gathered in the Supplementary Material.

Adduct	Interaction energy (kcal mol^{-1})	
	B3LYP/6-31G(d,p)	B3LYP/6-311++G(2d,3p)
7AlaB	**−9.36**	**−8.77**
7AlaB6	−9.24	−8.67
7AlaC	−9.28	−8.68
7AlaP	−9.33	−8.74
7AlaM	−8.56	−8.07
7AlaB-N	−10.85	−8.73
7LysB	−5.11	−4.73
7LysB6	−14.88	−14.73
7LysC	−15.70	−14.81
7LysP	**−15.96**	**−15.06**
7LysM	−15.46	−14.57
7LysP-N	1.05	1.41
7LysP-αN	−8.69	−7.20
7GluB	**−15.88**	**−14.85**
7GluB6	−10.90	−9.97
7GluC	−15.53	−14.46
7GluP	−15.88	−14.84
7GluM	−14.30	−13.34
7GluB-αN	−9.70	−7.71
7GluB-C	−9.94	−9.14
7MetB	**−13.69**	**−11.67**
7MetB6	−8.63	−8.39
7MetC	−13.19	−11.11
7MetP	−13.19	−11.11
7MetM	−3.08	−2.37
7MetB-αN	−8.42	−7.79

exchangeable proton is the principal position in this interaction [8]. Moreover, the pH alterations are unimportant for small molecules that, like compound 7, contain practically one protonation site. Additionally, we prove that such compounds do not tautomerize (tautomerism involving protonation of the nitrogen atoms can be neglected) [7–9]. Based on this information, we decided to examine the interaction of the pyrrolic nitrogen with the following amino acids: alanine, lysine, glutamic acid, and methionine. The geometry of the azole–amino acids adducts was optimized using four different functionals, including hybrid, GGA, and meta-GGA functionals, that is, B3LYP, CAM-B3LYP, PBE0, and M06L. The data concerning the lowest (in bold) and highest interaction energies (CPCM of solvation and water as solvent, counterpoise corrected method including the basis set superposition BSSE error, *Gaussian G09 D.01* software [11]) for the azole–amino acids adducts are shown in Table 1.

Interaction energy between two atoms or molecules A and B can be generally calculated to be a result of subtraction of the energies of components A and B from the energy of the fused AB structure:

$$\Delta E = E\left(\mathrm{AB}, r_c\right) - E\left(\mathrm{A}, r_e\right) - E\left(\mathrm{B}, r_e\right). \quad (1)$$

The r_c index reflects the geometry of the fused AB structure, whereas the r_e index reflects that of separate A and B. However, this equation often results in incorrect estimates, particularly when hydrogen bonding or dispersion interactions are present within the fused structures. The error associated with dispersion interactions can be lessened by using smaller basis sets, which make the AB structure more stable than A and B separately using the basis set superposition error (BSSE). This result is achieved thanks to the expansion of the wavefunction over fewer basis functions for A and B molecules compared to the AB structure [23]. Using very big basis sets with the BSSE is impossible for the majority of investigated compounds. Therefore, counterpoise method (CP) can be applied, which can estimate the BSSE size. Then the AB structure is described in the same way, but A and B have basis sets which are of equal size, the same as that of the basis set for the AB structure. Thus, interaction energy is calculated using CP in the following equation:

$$\Delta E = E\left(\mathrm{AB}, r_c\right)^{\mathrm{AB}} - E\left(\mathrm{A}, r_e\right)^{\mathrm{AB}} - E\left(\mathrm{B}, r_e\right)^{\mathrm{AB}}. \quad (2)$$

The AB indices in (2) show that absolute basis sets for the calculations concerning all the components are the same and are equal to those of the AB structure [23]. Hence, it can be concluded that such methodology significantly better describes the interaction phenomenon within the analyzed herein adducts.

The energy of interactions within the azole 7–alanine adduct **7AlaB**, which was optimized at the B3LYP/6-31G(d,p) level of theory, was −9.36 when calculated at the B3LYP/6-31G(d,p) or −8.77 kcal mol^{-1} when calculated at the B3LYP/6-311++G(2d,3p) level of theory. It resulted mainly from one strong hydrogen contact O-H···N ($l_{\text{O-H}}$ = 0.989 Å, $d_{\text{H-N}}$ = 1.892 Å, and θ = 160.1°; Figure 1), although the distance between pyrrolic proton (N2-H3 in Figure 1) of azole and hydroxyl group of alanine was 3.433 Å. Moreover, a potential weak hydrogen bond binding the carbonyl oxygen and phenylic proton (d = 2.236 Å, Figure 1) could contribute to some extent to the overall interaction energy.

A similar geometry was found for the same optimized adduct involving alanine and compound **7** but at the M06L/6-31G(d,p) level of theory (adduct **7AlaM**, Figure S2, in Supplementary Material available online at https://doi.org/10.1155/2017/8124323). The calculated energy equaled −8.56 or −8.75 kcal·mol^{-1} depending on the B3LYP/6-31G(d,p) or B3LYP/6-311++G(2d,3p) approach, respectively. The distance d between the carboxylic OH group and pyridinic nitrogen was 1.906 Å ($l_{\text{O-H}}$ = 0.986 Å and θ = 173.0°). As above, a weak contact involving carbonyl oxygen and phenylic *ortho* hydrogen was also observed (d = 2.146 Å). In this case, the distance between pyrrolic proton of azole and hydroxyl group of alanine was 3.796 Å.

Although the interactions between amino acid carboxylic group and heterocyclic nitrogen are significantly stronger

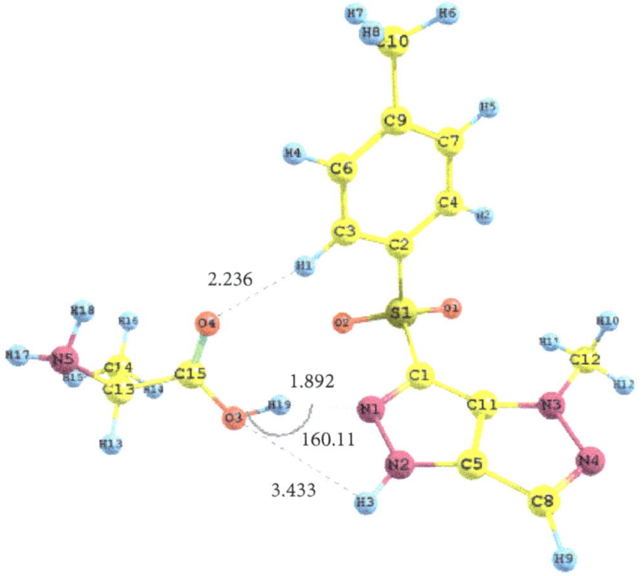

FIGURE 1: The optimized 7–alanine adduct **7AlaB** (B3LYP/6-31G(d,p)).

than the interactions involving amino group of amino acid and heterocyclic nitrogen [20], the adduct 7–Ala was additionally optimized regarding the interaction of the alanine amino group and pyrazole nitrogen atoms (adduct **7AlaB-N**, B3LYP/6-31G(d,p) level of theory; Figure S3, Supplementary Material).

The estimated interaction energy [B3LYP/6-31G(d,p), counterpoise corrected method] was −10.85 kcal/mol. However, this apparently lower energy increased to −8.73 kcal·mol^{-1} after calculation at the B3LYP/6-311++G(2d,3p) level of theory (counterpoise corrected method). The weaker interaction between the alanine amino group and pyrazole **7** is due to one detected hydrogen bond N-H$\cdots$NH$_2$ ($l_{N\text{-}H}$ = 1.037 Å, $d_{H\text{-}N}$ = 1.870 Å, and θ = 161.2°).

The more negative energy (strongest interaction), that is, −15.96 kcal·mol^{-1} [interaction energy calculated for adduct optimized at the B3LYP/6-31G(d,p) level of theory] or −15.06 kcal·mol^{-1} [interaction energy calculated for adduct optimized at the B3LYP/6-311++G(2d,3p) level of theory], respectively, concerning the interaction between azole **7** and lysine (**7–Lys**, Table 1), was observed for adduct **7LysP** (Figure 2, geometry optimized at the PBE0/6-31G(d,p) level of theory). This energy originated from the formation of two strong hydrogen bonds: N-H$\cdots$O=C ($l_{N\text{-}H}$ = 1.027 Å, $d_{H\text{-}O}$ = 1.816 Å, and θ = 147.6°) and O-H$\cdots$N ($l_{O\text{-}H}$ = 1.027 Å, $d_{H\text{-}N}$ = 1.726 Å, and θ = 169.4°).

Instead, the interaction energy for adduct **7LysB** (Figure S4, Supplementary Material; optimized at the B3LYP/6-31G(d,p) level of theory) was merely −5.11 kcal·mol^{-1} and resulted from the only one strong hydrogen bond being formed: N-H$\cdots$O=C ($l_{N\text{-}H}$ = 1.026 Å, $d_{H\text{-}O}$ = 1.851 Å, and θ = 147.7°).

The interaction energies between the lysine amine group and azole **7** for optimized adduct **7LysP-N** at the PBE0/6-31G(d,p) level of theory (Figure S5, Supplementary Material) were similar and amounted to a low value of ca 1 kcal·mol^{-1}. This effect was connected with the presence of one relatively weak hydrogen bond: N-H$\cdots$NH$_2$ ($l_{N\text{-}H}$ = 1.011 Å, $d_{H\text{-}N}$ = 2.367 Å, and θ = 163.6°).

The interaction between the amine group of lysine and azole **7** pyrrolic nitrogen (adduct **7LysP-αN**, Figure S6, Supplementary Material) optimized at the PBE0/6-31G(d,p) level of theory resulted in a strong hydrogen contact formation, N-H$\cdots$NH$_2$ ($l_{N\text{-}H}$ = 1.036 Å, $d_{H\text{-}N}$ = 1.846 Å, and θ = 167.7°), with interaction energy of −8.69 [obtained at the B3LYP/6-31G(d,p) level of theory] or −7.20 kcal·mol^{-1} [obtained at the B3LYP/6-311++G(2d,3p) level of theory].

The above discussion led to a conclusion that adduct **7LysP** (optimized at the PBE0/6-31G(d,p) level of theory) involving azole **7** and lysine has the strongest linking due to the interaction between the amino acid carboxylic group and pyrrolic nitrogen of compound **7**.

The application of B3LYP/6-31G(d,p) (adduct **7GluB**, Figure 3) or PBE0/6-31G(d,p) (adduct **7GluP**) approach for optimization of adduct **7–Glu** led to rotamers of almost identical energies of −15.88 or −14.85 kcal·mol^{-1} using B3LYP/6-31G(d,p) or B3LYP/6-311++G(2d,3p) approaches (Table 1). The most significant contribution to the overall interaction energy of adducts **7GluB** and **7GluP** came from two strong hydrogen bonds, namely, N-H$\cdots$O=C ($l_{N\text{-}H}$ = 1.023 Å, $d_{H\text{-}O}$ = 1.906 Å, and θ = 144.5°) and O-H$\cdots$N ($l_{O\text{-}H}$ = 1.006 Å, $d_{H\text{-}N}$ = 1.729 Å, and θ = 169.7°). Conversely, the interaction energy for adduct **7GluB6** (Figure S7, Supplementary Material; geometry optimized at the B3LYP/6-311+G(d,p) level of theory) was notably lower, that is, −10.9 kcal·mol^{-1} (energy determined at the B3LYP/6-31G(d,p) level of theory, Table 1). Here, the most important contribution originated from one strong hydrogen bond: N-H$\cdots$O=C ($l_{N\text{-}H}$ = 1.017 Å, $d_{H\text{-}O}$ = 1.961 Å, and θ = 148.5°). Adduct **7GluB-C** (optimized at the B3LYP/6-31G(d,p) level of theory, Figure S9, Supplementary Material) regarding interaction of azole with peripheral carboxyl group of Glu had similar interaction energy values between −9 and −10 kcal·mol^{-1} (Table 1) which derived from one strong hydrogen bond: N-H$\cdots$O=C ($l_{N\text{-}H}$ = 1.022 Å, $d_{H\text{-}O}$ = 1.863 Å, and θ = 175.3°).

Analogously to the above-described complex of compound **7** with lysine, the interaction energy involving the glutamic acid α amino group and pyrazole nitrogen [adduct **7GluB-αN**, B3LYP/6-31G(d,p) level of theory, Figure S8, Supplementary Material] was significantly lower, namely, −9.7 [B3LYP/6-31G(d,p) level of theory] or −7.71 [B3LYP/6-311++G(2d,3p) approach] kcal mol^{-1}. Its relatively low value came from the presence of only one hydrogen contact: N-H$\cdots$NH$_2$ ($l_{N\text{-}H}$ = 1.032 Å, $d_{H\text{-}N}$ = 1.886 Å, and θ = 163.8°).

The highest (considering absolute value) interaction energy (Table 1) for the complex of azole **7** with methionine (**7–Met**) was obtained for adduct **7MetB** optimized at the B3LYP/6-31G(d,p) level of theory and amounted to −13.69 [B3LYP/6-31G(d,p) level of theory] or −11.67 [B3LYP/6-311++G(2d,3p) level of theory] kcal mol^{-1}. This structure was stabilized by two strong hydrogen bonds involving the

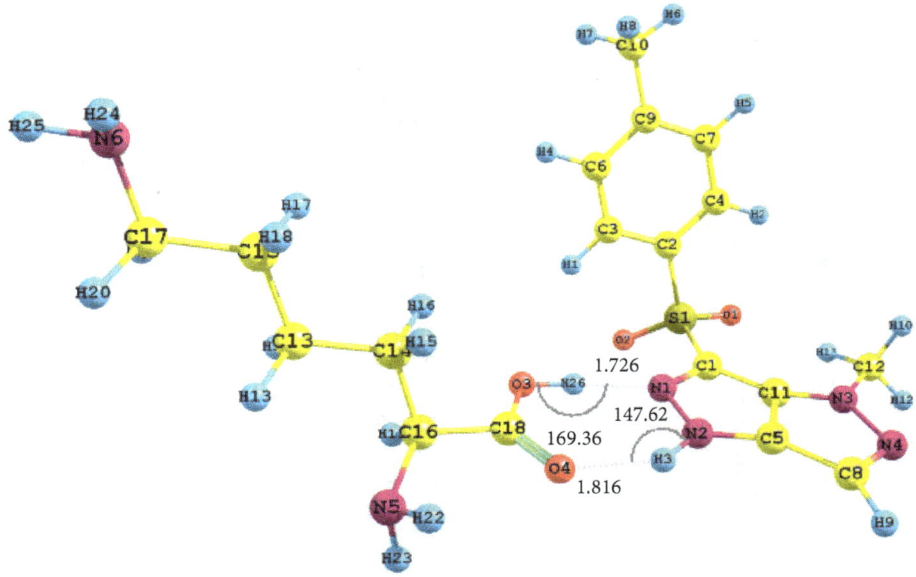

FIGURE 2: The optimized **7**–lysine adduct **7LysP** (PBE0/6-31G(d,p)).

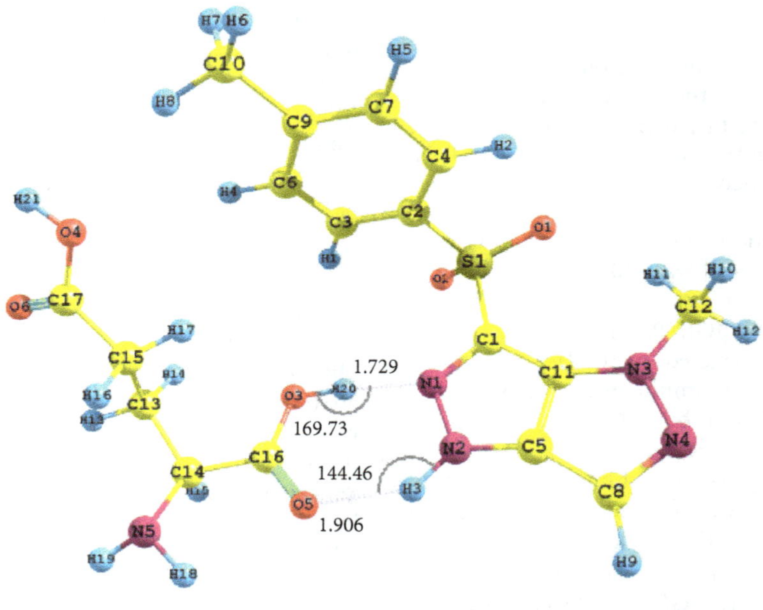

FIGURE 3: The optimized **7**–glutamic acid adduct **7GluB** (B3LYP/6-31G(d,p)).

pyrrolic nitrogen atom (N2-H3 in Figure 4) of azole **7** and amine group (N5 in Figure 4) of amino acid or the hydroxyl function of methionine and sulfonyl group of condensed azole **7**, namely, N-H$\cdots$$NH_2$ ($l_{N\text{-}H}$ = 1.033 Å, $d_{H\text{-}O}$ = 1.976 Å, and θ = 148.6°) or O-H$\cdots$$O_2S$ ($l_{O\text{-}H}$ = 0.981 Å, $d_{H\text{-}N}$ = 1.990 Å, and θ = 144.6°).

Another structure, **7MetB6**, of the adduct **7Met** was obtained using B3LYP/6-311+G(d,p) level of theory (Figure S10, Supplementary Material). The interaction energy calculated at B3LYP/6-31G(d,p) or B3LYP/6-311++G(2d,3p) level of theory was lower than that described in the preceding paragraph, that is, −8.63 or −8.39 kcal mol^{-1}, respectively, and resulted mainly from one hydrogen contact between the methionine carboxylic group and sulfone functionality: O-H$\cdots$$O_2S$ ($l_{O\text{-}H}$ = 0.986 Å, $d_{H\text{-}N}$ = 1.760 Å, and θ = 172.3°).

The optimization of the adduct **7MetB-αN** at the B3LYP/6-31G(d,p) level of theory (Figure S11, Supplementary Material) resulted in a relatively low interaction energy, that is, −8.42 [interaction energy calculated at the B3LYP/6-31G(d,p) level of theory] or −7.79 kcal mol^{-1} [interaction energy calculated at the B3LYP/6-311++G(2d,3p) level of theory]. The interaction involved the methionine α amine group and the condensed pyrazole system led to two relatively

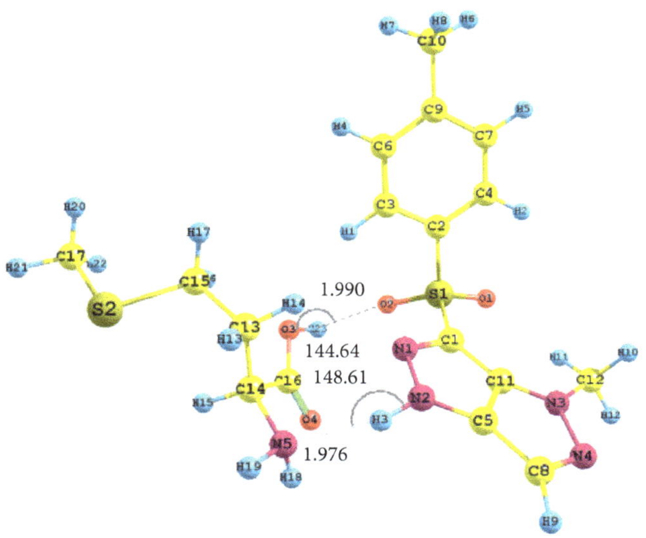

FIGURE 4: The optimized 7-methionine adduct **7MetB** (B3LYP/6-31G(d,p)).

weak hydrogen bonds: N-H$\cdots$NH$_2$ (l_{N-H} = 1.021 Å, d_{H-O} = 3.123 Å, and θ = 106.9°) or NH$_2\cdots$O$_2$S (l_{N-H} = 1.019 Å, d_{H-N} = 3.687 Å, and θ = 142.2°).

3.1. AIM Analysis. The application of AIM method includes determination of type of bond (ionic or covalent) and its properties [26] also when it comes to hydrogen bonding. In our AIM calculations, we focused on the interactions of the amino acids carboxylic group with condensed azole 7. Particular attention was devoted to adducts with the highest and lowest interaction energies.

As a principle, there is a relation between the ρ parameter at the bond critical point (BCP) and the strength of the hydrogen bond. The fact whether the Laplacian $\nabla^2\rho$ at the BCP is positive or negative determines whether the hydrogen bond dominates the covalent ($\nabla^2\rho < 0$) or electrostatic ($\nabla^2\rho > 0$) interactions. The values for the electron density topological parameters ρ and $\nabla^2\rho$ (related with a physical basis via local statement of the virial theorem for bonding interactions [26]) for the analysed adducts are given in Tables 2 and S1–S3 (Supplementary Material). It is noteworthy that the AIM parameters in the literature were correlated with the strength of analyzed bond [27]; however, in cited paper only one bond within the structure of the analyte was considered, whereas in our models we have detected even two hydrogen contacts within the structure of the particular azole-amino acid rotamer being analyzed.

In case of all analyzed adducts, the absolute value of bond ellipticity (ε) was ca 0.01–0.04, which indicates cylindrical, directional character of investigated contacts.

The $\nabla^2\rho$ values were positive but relatively small; thus they indicate that the adducts were connected through strong hydrogen bonds of electrostatic character. Moreover, the hydrogen bonds of adducts azole 7-amino acid with lower (more negative) interaction energies have higher values of

TABLE 2: Electron densities ρ and Laplacian fields $\nabla^2\rho$ [eÅ] determined for the **7Lys** adducts with a hydrogen bond; nd: no data.

Parameter	Adduct	Hydrogen bond	
		N-H$\cdots$O=C	O-H$\cdots$N
ρ	**7LysB**	0.032433	0.048718
$\nabla^2\rho$	**7LysB**	0.096056	0.108884
ρ	**7LysP**	0.035598	nd
$\nabla^2\rho$	**7LysP**	0.103572	nd

electron densities (Table 1). This demonstrates that adducts **7AlaB**, **7LysB6**, **LysM**, and **7MetB** are connected by stronger hydrogen bonds in comparison with adducts **7AlaM**, **7LysP**, **7GluB6**, and **7MetB6**, respectively.

3.2. NMR Analysis. The mechanism of interactions of metal ions and organic compounds, such as synthesized drugs and proteins, is of utmost significance from the biological point of view. Various analytical methods can be applied to investigate them, including ^{1}H NMR spectroscopy. Noteworthy is the fact that NMR spectroscopy has found its use for the determination of interactions of some ligands with proteins and allows assessing quantitatively these interactions if the equilibrium constants and changes in chemical shifts are taken into account [8, 20, 21, 26, 28–31]. As a result, we decided to focus only on ^{1}H NMR spectroscopy in our study. Albeit compound 7 has four nitrogen atoms within its structure, the low abundance of the ^{15}N isotope in comparison with the ^{1}H isotope as well as the significant signals broadening due to the large quadrupole moment of ^{14}N renders the nitrogen NMR spectroscopy impractical [8, 30, 31]. Considering the low utility of ^{13}C NMR spectroscopy for the investigations of interactions between metal ions and small molecule organic compounds [8, 9, 29] as well as the low natural abundance of ^{13}C, we decided to use ^{1}H NMR for the studies depicted in our paper. The theoretical ^{1}H NMR spectrum was generated for all azole-amino acid adducts using the gauge-including atomic orbital (GIAO) method [22] as implemented in *Gaussian G09 D.01.* In order to maintain clarity, we considered four adducts with the lowest (more negative) energy, previously optimized in CPCM of solvation (water as solvent) at the B3LYP/6-31G(d,p) (Tables S4, S9, S13, and S18; Supplementary Material), B3LYP/6-311+G(d,p) (Tables S5, S10, S14, and S19; Supplementary Material), CAM-B3LYP/6-31G(d,p) (Tables S6, S11, S15, and S20; Supplementary Material), PBE0/6-31G(d,p) (Table 1 and Tables S7, S16, and S21; Supplementary Material), and M06L/6-31G(d,p) (Tables S8, S12, S17, and S22; Supplementary Material) level of theory.

The theoretical values of chemical shifts correlate well with the experimental ones for pyrazole derivative 7, which is shown by low mean absolute deviation (MAD) [8]. These low MAD values for the theoretical ^{1}H NMR spectra of the adducts azole 7-amino acids (Tables 3 as well as S4, S16, and S18; Supplementary Material) correspond to the method and basis used for optimization of the adducts **7AlaB** (B3LYP/6-31G(d,p) approach), **7LysP** (PBE0/6-31G(d,p) approach),

Table 3: Experimental (δ_{exp}) and calculated chemical shifts (rotamers **7LysP**–**7LysP$_c$**) for compound **7** within the **7Lys** adduct; absolute errors (δ_{7LysP}–δ_{7LysPc}), average absolute error (δ), and relative percentage errors ($\Delta\delta$); calculated NMR shielding for proton H_{ref} = 31.6651 ppm for TMS (PBE0/6-31G(d,p)/GIAO); MAD = 0.27.

Locant	δ_{exp}	7LysP	7LysP$_a$	7LysP$_b$	7LysP$_c$	δ_{7LysP}	δ_{7LysPa}	δ_{7LysPb}	δ_{7LysPc}	δ	$\Delta\delta$
A	2.37	2.30	2.29	2.32	2.32	0.08	0.09	0.06	0.06	0.07	3
B	7.65	7.31	7.32	7.30	7.31	0.34	0.33	0.35	0.34	0.34	4
NH	13.67	13.16	13.16	13.17	13.11	0.51	0.51	0.50	0.56	0.52	4
C	7.85	8.26	8.24	8.26	8.28	0.41	0.39	0.41	0.42	0.41	5
D	7.43	7.49	7.48	7.51	7.51	0.05	0.05	0.08	0.08	0.07	1
E	4.08	4.32	4.33	4.31	4.31	0.24	0.26	0.23	0.24	0.24	6

7GluB (B3LYP/6-31G(d,p) approach), and **7MetB** (B3LYP/6-31G(d,p) approach) with the highest (more negative) interaction energies calculated at the B3LYP/6-311++G(2d,3p) level of theory (Table 1) and emphasize the importance of hydrogen bonds within these adducts.

From theoretical point of view, fascinating conclusions can be drawn from the calculations involving spin-spin coupling constants between the nuclei of hydrogen bond donor and acceptor as calculated at the B3LYP/6-311++G(2d,3p) level of theory. The conclusions are related with the adducts with the highest (more negative) and lowest (more positive) interaction energies (Figures S12–S19 and Table S23, Supplementary Material). For the interactions of azole 7 and alanine, the coupling constants involving atoms connected by hydrogen bonds O-H⋯N have values of ca −2 Hz. The increase in length of this bond in adducts **7AlaB** and **7AlaM** leads to an increase in both the torsion angle θ of the hydrogen bond donor and negative value of the dihedral angle τ. Conversely, the increase in length of C-H⋯O=C bond (adducts **7AlaM** and **7AlaB**) resulted in an insignificant decrease of the angle θ and more negative value of the dihedral angle τ as well as a decrease in coupling constant from 2 to 1 Hz. It should be added that the hydrogen contacts in adduct **7AlaB** are characterized by greater values of electron density ρ calculated by the QTAIM method (Table S1, Supplementary Material).

The N-H⋯O=C hydrogen contact in azole 7–lysine adduct (adducts **7LysB** and **7LysP**) can be characterized by comparable values of the bond length, torsion angle, and coupling constant for hydrogen bond acceptor–donor (ca 5 Hz), as well as electron density ρ (Table 2). However, the dihedral angle τ has a negative value for adduct **7LysB**.

The higher interaction energy of adduct **7LysP** in comparison with complex **7LysB** is probably due to the already noted smaller distance of the hydroxyl group (carboxylic function) of lysine from the pyrazole N2 nitrogen atom, as well as the second hydrogen contact O-H⋯N.

Comparison of the interaction energy of **7Met** adducts **7MetB** and **7MetB6** reveals that the higher energy of adduct **7MetB** is connected with the presence of a second hydrogen bond O-H⋯O_2S, despite the fact that this contact has lower electron density ρ and is longer compared to adduct **7MetB6** (Tables S3 and S23, Supplementary Material). However, this is associated with a smaller value of torsion angle θ and dihedral angle τ as well as coupling constant that decreases to ca 4 Hz (adduct **7MetB**) in comparison with 7 Hz (adduct **7MetB6**).

The interaction of azole 7 with glutamic acid leads to adducts **7GluB** and **7GluB6** that are linked through a N-H⋯O=C hydrogen bond. This hydrogen contact has similar values of the bond length, torsion angle θ, and coupling constant (ca 4 Hz, Table S23, Supplementary Material), although the dihedral angle τ (O-H⋯N) in adduct **7GluB** is shorter and has somewhat higher value of torsion angle θ and lower coupling constant, that is, −1.18 Hz versus −9.38 Hz, in comparison with the same bond parameters in adduct **7GluB6**. It should be added that the electron densities ρ in the paths of both hydrogen bonds between the nuclear attractors of adduct **7GluB** attain higher values.

4. Conclusions

The paper has focused attention on the interactions of the cytotoxic condensed derivative of pyrazole 7 with selected amino acids present in the receptor cavities of human protein kinases. The most significant interactions involve the amino acids carboxylic groups with azole 7 pyrrolic nitrogen. Results of interaction energy calculations among the 7–amino acid adducts regarding method and basis set used for initial optimization protocol for each of the analyzed adducts are gathered in Table 1. From this point of view, we observed that for **7Ala**, **7Glu**, and **7Met** adducts the optimization at the B3LYP/6-31G(d,p) level of theory was undoubtedly a reasonable choice. However, for the **7Lys** adduct, the PB0 functional resulted with optimized structure, within which more significant interactions were detected.

We have shown that the low MAD values for the theoretical ^{1}H NMR spectra of the adducts azole 7–amino acids correspond to the method and basis used for optimization of the adducts **7AlaB**, **7LysP**, **7GluB**, and **7MetB6** with the highest interaction energies.

The results presented in the previous sections show that, for adducts with similar geometric parameters and coupling constants, the electron densities ρ of the paths of bonds between nuclear attractors (adducts **7Lys**) have a decisive influence on the interaction strength. For the remaining adducts, that is, **7Ala**, **7Glu**, and **7Met**, a second hydrogen bond significantly affects the interaction. Furthermore, an increase in the hydrogen bond length in these adducts is accompanied by a decrease in the value of donor torsion angle θ and coupling constant.

Conflicts of Interest

The authors declare that they have no conflicts of interest.

Acknowledgments

The calculations were carried out using resources provided by the Polish Grid Infrastructure (PL-Grid) and Wrocław Center for Networking and Supercomputing (*Bem* cluster, WCSS Grant no. 327/2014). The authors are grateful to Dr. Jarosław J. Panek (Faculty of Chemistry, University of Wrocław, Poland) and Dr. Dariusz Kędziera (Faculty of Chemistry, Nicolaus Copernicus University in Toruń, Poland) for help in the NMR and DFT calculations and for manuscript correction.

References

[1] C. Avendaño and J. C. Menendéz, *Medicinal Chemistry of Anticancer Drugs*, Elsevier, Amsterdam, The Netherlands, 2nd edition, 2015.

[2] S. S. Abd El-Karim, M. M. Anwar, N. A. Mohamed, T. Nasr, and S. A. Elseginy, "Design, synthesis, biological evaluation and molecular docking studies of novel benzofuran-pyrazole derivatives as anticancer agents," *Bioorganic Chemistry*, vol. 63, pp. 1–12, 2015.

[3] W. Aman, J. Lee, M. Kim, S. Yang, H. Jung, and J.-M. Hah, "Discovery of highly selective CRAF inhibitors, 3-carboxamido-2H-indazole-6-arylamide: in silico FBLD design, synthesis and evaluation," *Bioorganic & Medicinal Chemistry Letters*, vol. 26, pp. 1188–1192, 2016.

[4] L. Ye, X. Ou, Y. Tian et al., "Indazoles as potential c-met inhibitors: design, synthesis and molecular docking studies," *European Journal of Medicinal Chemistry*, vol. 65, pp. 112–118, 2013.

[5] H.-P. Buchstaller, H.-M. Eggenweiler, C. Sirrenberg et al., "Fragment-based discovery of hydroxy-indazole-carboxamides as novel small molecule inhibitors of Hsp90," *Bioorganic and Medicinal Chemistry Letters*, vol. 22, no. 13, pp. 4396–4403, 2012.

[6] E. Totoń, E. Ignatowicz, M. K. Bernard, J. Kujawski, and M. Rybczyńska, "Evaluation of apoptotic activity of new condensed pyrazole derivatives," *Journal of Physiology and Pharmacology*, vol. 64, no. 1, pp. 115–123, 2013.

[7] J. Kujawski, K. Czaja, M. Doskocz et al., "On the interactions of indazole derivative with nucleosides - toward modeling the cytotoxic activity mechanism," *Computational and Theoretical Chemistry*, vol. 1059, pp. 45–50, 2015.

[8] K. Czaja, J. Kujawski, U. Girreser, J. J. Panek, M. Doskocz, and M. K. Bernarda, "Possible interactions between fused pyrazole derivative and magnesium ions - NMR experiments and theoretical calculations," *Arkivoc*, vol. 4, pp. 22–43, 2016.

[9] J. Kujawski, M. Doskocz, H. Popielarska et al., "Interactions between indazole derivative and magnesium cations-NMR investigations and theoretical calculations," *Journal of Molecular Structure*, vol. 1047, pp. 292–301, 2013.

[10] P. Hohenberg and W. Kohn, "Inhomogeneous electron gas," *Physical Review*, vol. 136, pp. B864-B871, 1964.

[11] M. J. Frisch, G. W. Trucks, H. B. Schlegel et al., *Gaussian 09 D.01. Revision A.1*, Gaussian, Inc, Wallingford, Conn, USA, 2009.

[12] A. D. Becke, "Density-functional thermochemistry. III. The role of exact exchange," *The Journal of Chemical Physics*, vol. 98, no. 7, pp. 5648–5652, 1993.

[13] T. Yanai, D. Tew, and N. A. Handy, "A new hybrid exchange-correlation functional using the Coulomb-attenuating method (CAM-B3LYP)," *Chemical Physics Letters*, vol. 393, pp. 51–57, 2004.

[14] C. Lee, W. Yang, and R. G. Parr, "Development of the colle-salvetti correlation-energy formula into a functional of the electron density," *Physical Review*, vol. 37, no. 2, pp. 785–789, 1988.

[15] F. Eckert and A. Klamt, "Fast Solvent Screening via Quantum Chemistry: COSMO-RS Approach," *AIChE Journal*, vol. 48, no. 2, pp. 369–385, 2002.

[16] J. P. Perdew, K. Burke, and M. Ernzerhof, "Generalized gradient approximation made simple," *Physical Review Letters*, vol. 77, no. 18, pp. 3865–3868, 1996.

[17] J. P. Perdew, K. Burke, and M. Ernzerhof, "Generalized gradient approximation made simple [Phys. Rev. Lett. 77, 3865 (1996)]," *Physical Review Letters*, vol. 78, no. 7, p. 1396, 1997.

[18] Y. Zhao and D. G. Truhlar, *The Journal of Chemical Physics*, vol. 125, no. 194101, pp. 1–18, 2006.

[19] T. A. Keith, *AIMAll, Version 14.11.23*, TK Gristmill Software, Overland Park, Kan, USA, 2014.

[20] J. Kujawski, M. K. Bernard, E. Jodłowska, K. Czaja, and B. Drabińska, "On the interactions of leflunomide and teriflunomide within receptor cavity — NMR studies and energy calculations," *Journal of Molecular Modeling*, vol. 21, p. 105, 2015.

[21] J. Kujawski, K. Czaja, E. Jodłowska et al., "Structural and spectroscopic properties of econazole and sulconazole - Experimental and theoretical studies," *Journal of Molecular Structure*, vol. 1119, pp. 250–258, 2016.

[22] K. Wolinski, J. F. Hinton, and P. Pulay, "Efficient implementation of the gauge-independent atomic orbital method for NMR chemical shift calculations," *Journal of the American Chemical Society*, vol. 112, no. 23, pp. 8251–8260, 1990.

[23] S. F. Boys and F. Bernardi, "The calculation of small molecular interactions by the differences of separate total energies. Some procedures with reduced errors," *Molecular Physics*, vol. 19, article 553, 1970.

[24] A. Famulari, R. Specchio, M. Sironi, and M. Raimondi, "New basis set superposition error free ab initio MO-VB interaction potential: molecular-dynamics simulation of water at critical and supercritical conditions," *The Journal of Chemical Physics*, vol. 108, article 3296, 3303 pages, 1998.

[25] http://www.chemcraftprog.com.

[26] R. F. W. Bader, *Atoms in Molecules: A Quantum Theory*, Clarendon, Oxford, UK, 1994.

[27] S. J. Grabowski, "An estimation of strength of intramolecular hydrogen bonds — ab initio and AIM studies," *Journal of Molecular Structure*, vol. 562, pp. 137–143, 2001.

[28] J. Kujawski, K. Czaja, T. Ratajczak, E. Jodłowska, and M. K. Chmielewski, "Investigations on synperiplanar and antiperiplanar isomers of losartan: Theoretical and experimental NMR studies," *Molecules*, vol. 20, no. 7, pp. 11875–11890, 2015.

[29] L. Fielding, "NMR methods for the determination of protein–ligand dissociation constants," *Progress in Nuclear Magnetic Resonance Spectroscopy*, vol. 51, article 219, 2007.

[30] F. Wada, T. Kadonaga, and S. Matsuo, "15N abundance in nitrogen of naturally occuring substances and global assessment of denitrification from isotropic viewpoint," *Geochemical Journal*, vol. 9, article 139, 1975.

[31] E. Anders and N. Grevesse, "Abundances of the elements: meteoritic and solar," *Geochimica et Cosmochimica Acta*, vol. 53, article 197, 1998.

Biogenetic Conversion of Wyerone and Dihydrowyerone into Wyerone Epoxide in *Vicia faba* Cotyledons and Screening of Antibacterial Activity

Nedhal A. Al-Douri,[1] Randa N. Haddadin,[2] Ashok K. Shakya,[1] and Ghaleb A. Oriquat[1]

[1]*Pharmacy Department, Faculty of Pharmacy and Medical Sciences, Al-Ahliyya Amman University, Al Salt Road, Amman 19328, Jordan*
[2]*Pharmaceutics and Pharmaceutical Technology Department, School of Pharmacy, The University of Jordan, Amman, Jordan*

Correspondence should be addressed to Nedhal A. Al-Douri; n.douri@ammanu.edu.jo

Academic Editor: Philippe Jeandet

Background. Vicia faba is a plant that belongs to the family Fabaceae. Cotyledons of this plant produce compounds called phytoalexins as a result of fungal and bacterial infection or stress factors. The phytoalexins are furanoacetylenic compounds. They include wyerone, wyerol, wyerone acid, their dihydro derivatives, and wyerone epoxide. These compounds have antimicrobial activities mainly against fungi. *Objective.* The purpose of this study was to elucidate the biogenetic conversion of wyerone and dihydrowyerone to wyerone epoxide in *V. faba* and to investigate the antibacterial activities of some of these phytoalexins. *Materials and Methods.* Seeds of *Vicia faba* were used. Labelled wyerone and dihydrowyerone were obtained by treating $CuCl_2$-induced cotyledons with sodium (2-^{14}C) acetate and separated by TLC and HPLC. Labelled wyerone and dihydrowyerone were then applied to induced bean cotyledons to establish any possible interconversion to wyerone epoxide. Antibacterial activity of wyerone, wyerone acid, and wyerone epoxide was investigated by disc diffusion test against a panel of microorganisms. Zones of inhibition were reported. *Results.* The radiolabeling studies showed that 12.4% of ^{14}C-wyerone, and 6.01% of ^{14}C-dihydrowyerone were incorporated into wyerone epoxide. This indicates that wyerone epoxide was most probably derived from wyerone and dihydrowyerone. In addition, a new compound, 11-hydroxywyerone was isolated for the first time. Additionally, this study showed that wyerone, wyerone epoxide, and wyerone acid had no antibacterial activity against Gram-negative bacteria but were active against Gram-positive bacteria.

1. Introduction

Plants were found to have physical and biochemical responses towards pathogen infections. They evolve coordinate events by relying on their immune system to recognize and actively respond to these infections. One example of the defense mechanisms is the accumulation of antimicrobial secondary metabolites, which are called the phytoalexins [1, 2]. These phytoalexins are synthesized by plants as a result of infections (fungal, bacterial, or viral) [3, 4] or due to other forms of stress, which could be chemical elicitors (e.g., mercury salts, copper) or physical events like UV light [5]. Fabaceae plants usually synthesize certain isoflavonoid phytoalexins, but certain species accumulate nonisoflavanoid compounds instead of or in addition to isoflavanoid [6, 7].

Species of *Vicia* synthesize seven phytoalexins which are furanoacetylenes in structure as a result of fungal or bacterial infection [8, 9]. The major component is the furanoacetylene ester wyerone (I) accompanied by structurally similar derivatives. These include wyerone acid (II); the alcohol, wyerol (III); the 11, 12-dihydroanalogues, e.g., dihydrowyerone (IV), dihydrowyerol (V), and dihydrowyerone acid (VI); and the wyerone epoxide (VII)

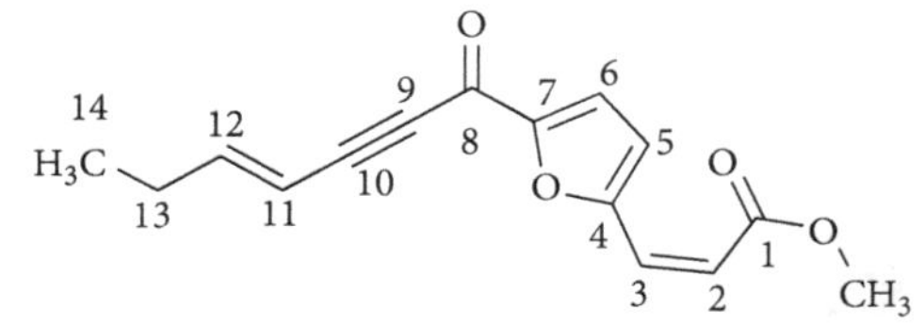	Wyerone (I)
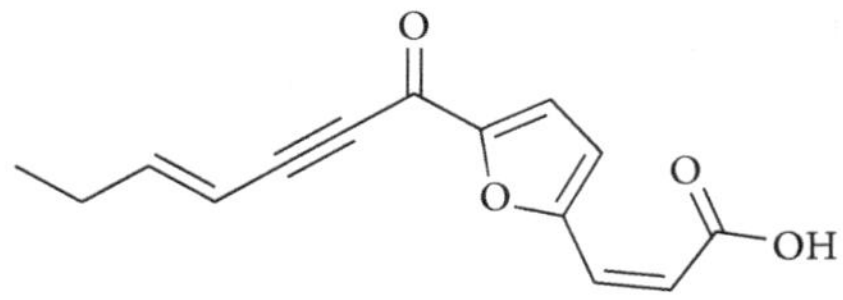	Wyerone acid (II)
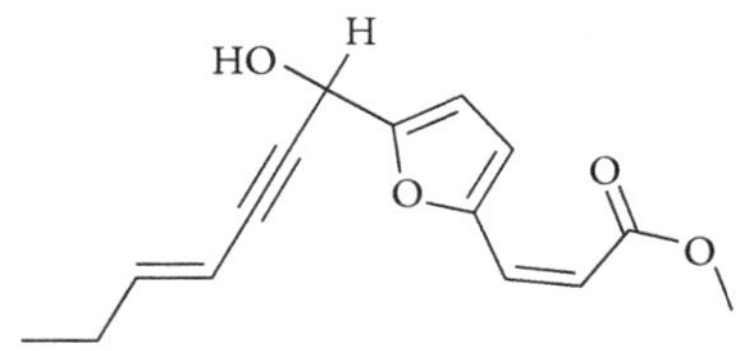	Wyerol (III)
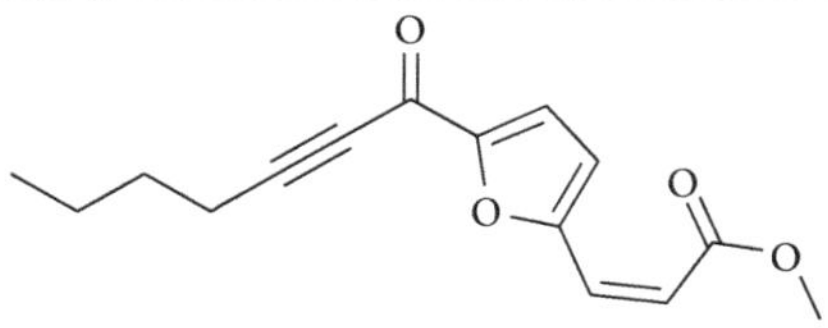	Dihydrowyerone (IV)
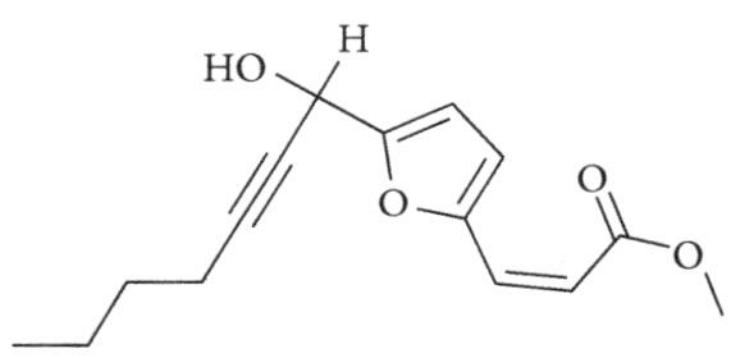	Dihydrowyerol (V)
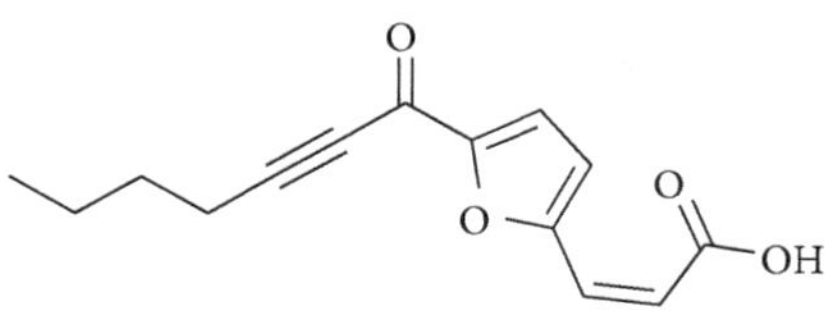	Dihydrowyerone acid (VI)
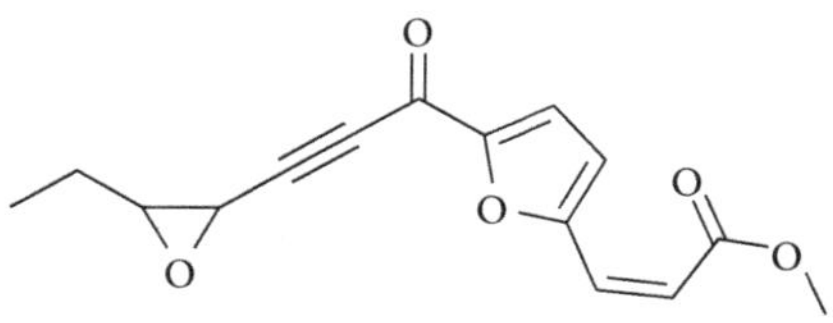	Wyerone epoxide (VII)
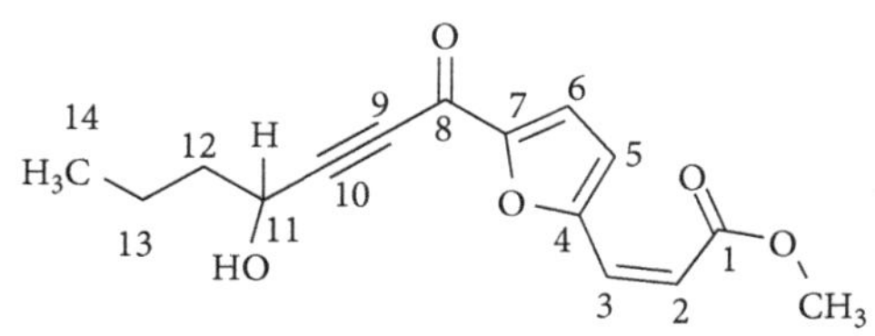	11-Hydroxywyerone (new metabolite)

Figure 1: Structures of phytoalexins of *Vicia faba* (I, II, III, IV, V, VI, and VII) and the new metabolite 11-hydroxywyerone.

(Figure 1) [10–13]. The accumulation of wyerone and its derivatives in cotyledon, pod, and leaves of *Vicia faba*, infected by Botrytis spp. fungi, were evaluated previously. Wyerone acid was observed to be the main phytoalexin in leaves and pods, while wyerone was predominant in cotyledons. The accumulation of wyerone and wyerone epoxide in pod endocarp has led to the suggestion that wyerone may act as precursor for wyerone epoxide. This was supported by the observed conversion of wyerone to wyerone epoxide during fractionation of pod tissue [14–16]. The biosynthesis of wyerone in *Vicia faba* was elucidated previously and found to be derived from acetate malonate pathway via fatty acid derivatives [17], and the conversion of wyerone to wyerone acid had been elucidated earlier [18].

The aim of this study is to explore possible biogenetic conversions in *V. faba* cotyledons from wyerone and dihydrowyerone to wyerone epoxide. During the separation of phytoalexins, a new compound, 11-hydroxywyerone was isolated. This study excluded any fungal conversions, since such fungal transformation of wyerone derivatives has been studied extensively [14, 19, 20].

The antibacterial activity of phytoalexins has largely been confined to isoflavonoids from Fabaceae, with one report of this activity of wyerone in comparison with other isoflavonoid phytoalexins [21] was published. Moreover, most of the antimicrobial activity studies focused on antifungal rather than antibacterial activity [3, 4]. In the present study, we tried to explore more on this issue regarding three of these phytoalexins, wyerone, wyerone acid, and wyerone epoxide against five Gram-positive and two Gram-negative bacteria. The effect of wyerone acid and wyerone epoxide was not evaluated previously on these bacteria. These bacteria studied were human and plant pathogens and a saprophyte.

2. Materials and Methods

2.1. Materials. Seeds of *Vicia faba* (broad bean, cv. Aquadulce) were procured from Sinclair McGill. 2-^{14}C-sodium acetate, 58 mCi/mmol, was procured from New England Nuclear (NEN, Boston, MA), ^{14}C-Toluene (Internal liquid scintillation counting (LSC) standard) was procured from Perkin-Elmer, and Silica gel (0.5 mm, GF254 TLC plates) was procured from Merck. All other chemicals were of analytical grade and procured from Sigma-Aldrich (St. Louis, MO, USA).

2.2. Bacterial Strains and Media. Human pathogens included *Bacillus cereus* ATCC 14579, *Micrococcus luteus* ATCC 49732, and *Klebsiella pneumoniae* ATCC 13883. Plant pathogens included *Rhodococcus fascians* ATCC 12974, *Curtobacterium flaccumfaciens pathovar oortii* ATCC 25283, and *Pseudomonas syringae pathovar phaseolicola* ATCC 19304, and the saprophyte used was *Bacillus megaterium* ATCC 14581. These microorganisms were procured from the American Type Culture Collection (ATCC). Media used were procured from Oxoid (UK). Amoxicillin was a kind gift from Hikma pharmaceuticals (Amman, Jordan).

2.3. Plant Material, Feeding Techniques, and Isolation of Phytoalexins. The phytoalexins were synthesized using the procedure described earlier by Al-Douri and Dewick [17, 18]. Seeds of *V. faba* were immersed in ethanol for 10 minutes, washed with water, soaked in dilute detergent (Quadralene, UK) for 5 minutes, sterilized by immersion with sodium hypochlorite (2%) for 5 minutes, and washed again. Damaged seeds distinguished by eye during sterilization were discarded. The healthy seeds were imbibed in distilled water for 24 hours. The seed coat was removed, and 100 g of the obtained cotyledons were separated and placed flat side up on a moist tissue paper in glass trays (Figure 2). The chemical elicitor, aqueous $CuCl_2$ (3 mM), was spotted onto the surface of these cotyledons using a syringe. Sufficient amount of elicitor solution was added to keep the whole surface moistened. The glass trays containing the cotyledons were covered by a foil paper and kept at 25°C for several days in a dark cabinet. Extra $CuCl_2$ was added when necessary to keep the surface moist. In our laboratory, a five-day time period was found the best to give the highest production of phytoalexins. When the five days elapsed, the inducer solution was removed by a syringe and replaced by drops of sodium (2-^{14}C) acetate solution (1 mCi) as a precursor for synthesis of the phytoalexins to be studied. The cotyledons were left for another two days adding drops of distilled water to keep the cotyledons moist. After this period, the cotyledons were homogenized in a blender with 200 ml dichloromethane. This was repeated ten times after filtering the extract. Dichloromethane extract was concentrated using a rotary evaporator.

2.4. Purification of the Different Compounds. The extract was applied on TLC chromatography (silica gel GF254; hexane-acetone, 2 : 1) to separate the zones representing the phytoalexins [17, 18]. ^{14}C-phytoalexin mixture was detected as a deep blue fluorescent band under UV254. The band was scratched and eluted by acetone. The resulting mixture was further purified by second TLC using chloroform-light petroleum (2 : 1) and then by chloroform: methanol (25 : 1). The resulting band, the purified phytoalexins, was scratched, eluted with acetone, and dried. A sample from this phytoalexins mixture was subjected to Proton NMR analysis (^{1}HNMR) to identify the structure of the separated phytoalexins. High-performance liquid chromatograph (HPLC) connected to a variable wavelength Perkin-Elmer LC55 UV detector was used to separate the labelled phytoalexins. Samples were dissolved in MeOH prior to injection. Reversed-phase columns (Spherisorb S5, ODS-2 hypersil), 25 cm × 8 mm (semipreparative), were employed with solvent methanol : acetonitrile : water (30 : 20 : 50), at flow rates of 1 ml/min or 2 ml/min [22]. Radioactivity of the compounds isolated was measured by SL 4000 liquid scintillation counter with dioxin-based liquid scintillator (BDH). ^{14}C-toluene internal standard was used for measurement of efficiencies.

2.5. Feeding Experiment of the Labelled (Wyerone and Dihydrowyerone) into the Induced Cotyledons to Produce Wyerone Epoxide. Approximately 1 mg samples of ^{14}C-wyerone and ^{14}C-dihydrowyerone obtained from the previous experiment (Section 2.4) were added as dispersions in 1-2 mL of water with the aid of few drops of Tween 20. Each was placed on 20 g of induced *V.* faba cotyledons. Extraction of the cotyledons was carried out as described above to

Figure 2: *V. faba* cotyledons treated with aqueous $CuCl_2$ (3 mM) and placed flat side up. Scale 2 cm.

isolate and purify the ^{14}C-wyerone epoxide. However, ^{14}C-Wyerone epoxide was detected as a small blue fluorescent band under UV254 below the band of ^{14}C-wyerone/^{14}C-dihydrowyerone mixture. The band was scratched and eluted with acetone. A sample was analysed by ^{1}HNMR to identify the compound. In this TLC, a new band was observed under UV254. It was scratched, eluted with acetone, and purified again according to the procedure mentioned in Section 2.4. The compound was identified by ^{1}HNMR and ^{13}CNMR. Radioactivity of the compounds isolated was measured by liquid scintillation counter as described above.

2.6. Screening for Antibacterial Activity. Gram-positive and Gram-negative bacteria were selected to evaluate the antibacterial activity of the separated phytoalexins. These bacteria were chosen to represent plant pathogens, human pathogens, and a saprophyte. Bacterial suspensions of 0.5 McFarland (*ca* 1×10^8 CFU/ml) were prepared and diluted to get 10^6 CFU/ml. These suspensions were each spread separately on Mueller–Hinton agar plates (9 cm diameter). The separated phytoalexins from cotyledons of *V. faba* were dissolved in methanol. Blank paper discs (9 mm) were loaded gradually with phytoalexin methanolic solutions to reach a final amount of 60 μg/disc. The discs were left to dry by air. The loaded discs were placed on the inoculated Mueller–Hinton agar plates. Discs loaded with the broad-spectrum antibiotic, amoxicillin (60 μg), were used as a control, and discs loaded with methanol alone were used as negative control. The plates were incubated at 30°C for 24 hour for all strains except *Pseudomonas syringae* (26°C for 24–36 hr). Zones of inhibition were measured and reported. The test was performed in duplicate.

3. Results and Discussion

3.1. Biosynthetic Conversion, Isolation, and Purification of Compounds. In the first feeding experiment using ^{14}C-sodium acetate, the phytoalexin mixture separated by TLC chromatography was detected as deep blue fluorescent band under UV254, with R_f value 0.31. ^{1}HNMR analysis revealed the presence of approximately 8% of ^{14}C-dihydrowyerone in the mixture of ^{14}C-wyerone and ^{14}C-dihydrowyerone. This is in agreement with previous results reported in our laboratory [17, 18]. However, the mixture purified by TLC was reported previously to contain about 13% of the dihydro derivative [20]. The resulting phytoalexin mixture was separated by semi-preperative HPLC. This system was previously used for quantitative analysis of seven wyerone derivatives isolated from *V. faba* cotyledons as reported in the literature [22]. The purified ^{14}C-wyerone obtained was quantified by UV spectroscopy (Perkin-Elmer 554 Spectrophotometer). It has UV absorbance λ_{max} in ethanol at 224, 290, and 350 nm compared with those reported in the literature at 226, 291, and 351 nm (€ = 27,000) [22]. The purified ^{14}C-dihydrowyerone was quantified by UV spectroscopy and produced a spectrum (λ_{max} ethanol, 340, 235 nm) entirely consistent with that reported [18, 22].

The quantitation of the isolated ^{14}C-wyerone and ^{14}C-dihydrowyerone was determined using the reported extinction coefficient (€ = 27,000) [22]. Accordingly, the yields of the mixture of the labelled wyerone/dihydrowyerone were 20–28 mg/100 g fresh tissue of cotyledons. The biosynthesis of wyerone was proven previously to be from acetate units via acetate pathway. This was done by a series of experiments applying radiolabeled ^{14}C- and ^{13}C-isotopes techniques [17]. In the present study, ^{14}C–sodium acetate was incorporated efficiently into wyerone (Table 1). This is in agreement with earlier reports [17, 18]. The radiolabel analysis of the isolated ^{14}C-wyerone/^{14}C-dihydrowyerone is shown in Table 2.

3.2. Separation of ^{14}C-Wyerone Epoxide, ^{14}C-Wyerone Acid, and a New Compound. After the feeding experiment of ^{14}C-Wyerone and ^{14}C-dihydrowyerone into the induced cotyledons (Section 2.5), ^{14}C-wyerone epoxide was detected as a small blue fluorescent band under UV254 below the labelled wyerone/dihydrowyerone mixture, with R_f value of 0.26. The purified ^{14}C-Wyerone epoxide was quantified by UV λ_{max} 350 using an extension coefficient (€ = 27,000) [15]. Radiolabel measurements of this compound are shown in Table 3. The results of this experiment indicated that wyerone was an effective precursor of wyerone epoxide.

^{14}C-Wyerone acid was separated using the same solvent with R_f value of 0.12. The compounds were isolated and identified using HPLC and ^{1}HNMR. The results of isolation and identification are consistent with previous reports [11, 13]. Nevertheless, Wyerone acid was isolated in order to test its antibacterial activity.

The new band which was observed under UV254 below the band of ^{14}C-wyerone and ^{14}C-wyerone epoxide on TLC had an R_f value of 0.21. This band was collected and purified using the same solvents used above. The resulting compound was identified as a new moiety, 11-hydroxywyerone, which is separated for the first time (Figure 1). ^{1}HNMR and ^{13}CNMR for this compound are the following: (250 MHz, $CDCl_3$): δ1.02 (3H, t, J = 7.1 Hz), 1.38–1.57 (2H,m), 1.58–1.70, (3H, q/br), 3.79 (3H, s), 4.16 (1H, t, J = 7.0 Hz), 6.69 (1H, d, J = 16.5 Hz), 7.34 (1H, d, J = 3.5 Hz), 7.45 (1H, d, J = 3.5 Hz), 7.69 (1H, d, J = 16.5 Hz).

^{13}CNMR (62.5 MHz, $CDCl_3$) δ (ppm): 13.90, 18.49, 31.03, 51.57, 62.01, 83.30, 85.97, 115.84, 119.16, 121.38, 129.80, 153.62, 154.83, 164.69, and 166.45.

The radioactivity of the new compound was measured and found to be 6.97×10^7 Sp.act. dpm/mM.

TABLE 1: Mean radioactivity measurement of ^{14}C-labelled wyerone and ^{14}C-dihydrowyerone mixture from feeding experiments with sodium (2-^{14}C) acetate. The results are average of two trials.

Compound	Yield (mg)	Sp. act dpm/mM	Dilution	Incorporation %
Mixture of ^{14}C-wyerone + ^{14}C-dihydrowyerone	16.91	6.66×10^6	1920	0.182

TABLE 2: Mean radioactivity measurement of separated ^{14}C-wyerone and ^{14}C-dihydrowyerone from first feeding experiment obtained after separation by HPLC. The results are average of two trials.

Compound	Yield (mg)	Sp. act dpm/mM	Dilution	Incorporation %
^{14}C-wyerone	10.3	7.99×10^7	1600	0.129
^{14}C-dihydrowyerone	1.014	9.56×10^7	1340	0.015

TABLE 3: Incorporation data of ^{14}C-labelled wyerone and ^{14}C-dihydrowyerone into ^{14}C-wyerone epoxide (according to Section 2.5).

Compound fed	Compound isolated	μg/g fr.wt.	Sp.act. dpm/mM	Dilution	Incorporation %
^{14}C-wyerone	^{14}C-wyerone epoxide	75	6.52×10^6	16	12.4
^{14}C-dihydrowyerone	^{14}C-wyerone epoxide	13.9	2.87×10^6	3.32	6.01

Incorporation data showed the highest conversion of ^{14}C-wyerone (12.4%) into ^{14}C-wyerone epoxide. The results in this study indicated that it is possible that a metabolic grid existed for these compounds in *V. faba*, in this case, hydroxylation of the ester followed by oxidation of the secondary alcohol. These might be controlled by relatively specific enzymes. Supporting this hypothesis would require further feeding experiment with this new labelled compound and isolation of wyerone epoxide. The incorporation value should be higher than that in the present study. Another aspect of this hypothesis is the involvement of the enzyme which catalyzes this first reaction in the biosynthetic pathway from 11-hydroxywyerone into the epoxide, which will need further elaboration as in other studies [23]. Experiments are still in progress.

The results also indicated a high incorporation (6.01%) of ^{14}C-dihydrowyerone into ^{14}C-wyerone epoxide. In each experiment, recovery of the precursor fed was considerably less than that of wyerone indicating further metabolism or degradation of the applied compound. In this case, dihydrowyerone may be transformed into wyerone epoxide quite efficiently by dehydrogenation to produce the 11, 12-alkene first, followed by the same pathway as for wyerone itself.

3.3. Antimicrobial Activity. The antimicrobial activity of the three phytoalexins against five Gram-positive and two Gram-negative bacteria were evaluated. These bacteria represent human pathogens, plant pathogens, and a saprophyte (Table 4). The 60 μg amount of phytoalexin per disc was chosen to be tested based on an amount of compound versus zone of inhibition experiments performed to determine the minimum amount of phytoalexins that produced the largest zone of inhibition (data not shown). For most bacteria-phytoalexins combination, 50–60 μg/disc was the lowest amount to produce the largest zone of inhibition. Above this amount, a plateau was reached. Therefore, 60 μg/disc was selected to be tested for the three compounds against the bacteria. Amoxicillin showed the highest zone of inhibition against all the tested microorganisms. The negative control (methanol on disc) showed no inhibition zone. The three phytoalexins showed no antimicrobial activity against the Gram-negative bacteria but had noticeable activity against the Gram-positive bacteria. Assuming that the three phytoalexins have similar diffusion coefficients, then, these phytoalexins have comparable activity against Gram-positive bacteria, since at the same dose (60 μg/disc), the zones of inhibition against plant pathogens and human pathogens are very close (range between 14 and 16 mm). The saprophyte *Bacillus megaterium* showed the least susceptibility to the three phytoalexins with zones of inhibition (12-13 mm). These results are in line with the previous report. Gnanamanickam and Mansfield [21] had shown that wyerone and some flavonoid phytoalexins were more toxic against Gram-positive than Gram-negative bacteria. Several other researches were conducted to investigate the antifungal activity of phytoalexins and only one studied their antibacterial effect [4, 5, 14]. To the best of our knowledge, this is the first study to report antibacterial activity of wyerone acid and wyerone epoxide against the microorganisms selected in this research. These results could open the door for developing new atypical antimicrobial agents.

4. Conclusion

The current study tried to elucidate the biogenetic conversion of some of the furanoacetyene phytoalexins of *V. faba*, particularly the biogenetic conversion among wyerone, dihydrowyerone, and wyerone epoxide. The conversions were studied through feeding experiments with labelled precursors to yield the phytoalexins. The results from the incorporation and dilution data showed that wyerone epoxide is most probably derived from wyerone. Moreover, dihydrowyerone could be a precursor to wyerone epoxide through more sequence of events,

Table 4: Mean zones of inhibition in (mm), $n = 2$, for the three phytoalexins and amoxicillin (positive control) tested against different bacterial genera.

	Gram stain	Pathogenicity	Wyerone	Wyerone acid	Wyerone epoxide	Amoxicillin
Bacillus cereus ATCC 14579	Positive	Human	16	16	15	25
Bacillus megaterium ATCC 14581	Positive	Saprophyte	13	12	12	25
Curtobacterium flaccumfaciens pathovar oortii (Corynebacterium oorti) ATCC 25283	Positive	Plant	16	15	16	34
Klebsiella pneumoniae ATCC 13883	Negative	Human	0	0	0	23
Micrococcus luteus ATCC 49732	Positive	Human	15	14	15	33
Pseudomonas syringae pathovar phaseolicola (Pseudomonas phaseolicola) ATCC 19304	Negative	Plant	0	0	0	20
Rhodococcus fascians ATCC 12974	Positive	Plant	16	15	15	33

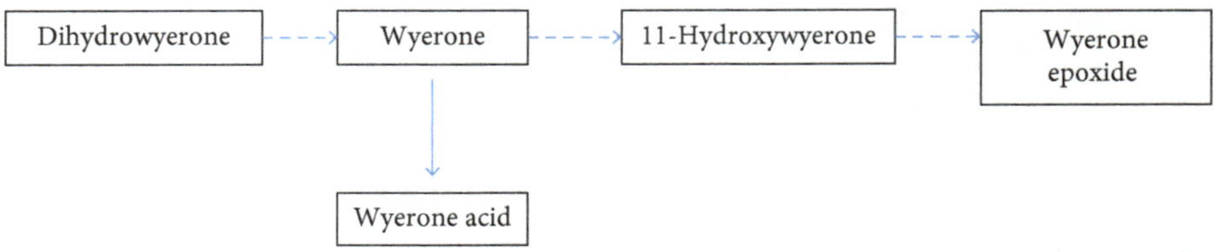

Figure 3: Isolation of a new compound, 11-hydroxywyerone, during the biosynthetic conversion of wyerone into wyerone epoxide. Dotted lines: the proposed pathway in the present study. Solid line: the previously reported conversion of wyerone into wyerone acid.

which is explained by less incorporation value. It is not possible at this stage to surmise about the natural sequence leading to wyerone epoxide. Nevertheless, we may suggest that the epoxidation is a late process of the biogenesis starting from wyerone and dihydrowyerone, which could be passing through the new compound 11-hydroxywyerone (Figure 3). However, further work is required to elicit these steps by feeding of this new compound to the cotyledons and measuring the incorporation data to wyerone epoxide. On the contrary, the three phytoalexins showed noticeable activity against the Gram-positive bacteria with no activity against the Gram-negative bacteria chosen. This activity could be promising to explore new sources for antimicrobial agents.

Conflicts of Interest

The authors declare that there are no conflicts of interest regarding the publication of this paper.

Acknowledgments

The authors wish to thank Prof. Dr. Paul M. Dewick (Retired), University of Nottingham, UK, Dean of Faculty of Pharmacy and Medical Sciences and the Dean of Research and Higher Education, Al-Ahliyya Amman University, Amman, Jordan, for providing necessary facilities and support.

References

[1] P. Jeandet, C. Clement, E. Courot, and S. Cordelier, "Modulation of phytoalexins biosynthesis in engineered plant for disease resistance," *International Journal of Molecular Sciences*, vol. 14, no. 7, pp. 14136–14170, 2013.

[2] R. L. Arruda, A. T. Santana Paz, M. T. Freitas Bara, M. V. de Carvalho Barros Côrtes, M. C. Corsi de Filippi, and E. C. Cardoso da Conceição, "An approach on phytoalexins: function, characterization and biosynthesis in plants of the family Poaceae," *Ciência Rural*, vol. 46, no. 7, pp. 1206–1216, 2016.

[3] P. Jeandet, "Structure, chemical analysis, biosynthesis, metabolism, molecular engineering, and biological functions of phytoalexins," *Molecules*, vol. 23, no. 1, p. 61, 2018.

[4] I. Ahuja, R. Kissen, and A. M. Bones, "Phytoalexins in defense against pathogens," *Trends in Plant Science*, vol. 17, no. 2, pp. 73–90, 2012.

[5] J. Kuc and W. Currier, "Phytoalexins, plants, and human health," *Advances in Chemistry*, Chapter 16, vol. 149, pp. 356–368, American Chemical Society, Washington, DC, USA, 1976.

[6] M. Wink, "Evolution of secondary metabolites in legumes (Fabaceae)," *South African Journal of Botany*, vol. 89, pp. 164–175, 2013.

[7] S. Soylu and M. Kemal, "*Induction of phytoalexin accumulation in Broad Bean* (*Vicia faba* L.) cotyledons following treatments with biotic and abiotic elicitors," *Turkish Journal of Agriculture and Forestry*, vol. 26, pp. 343–348, 2002.

[8] D. J. Robenson, "Furanoacetylene and isoflavonoid phytoalexins in Lens cullinaris," *Phytochemistry*, vol. 17, no. 4, pp. 807-808, 1978.

[9] D. J. Robenson and J. B. Harborne, "A chemical dichotomy in phytoalexins induction within the tribe Vicieae of the Leguminosea," *Phytochemistry*, vol. 19, no. 11, pp. 2359–2365, 1980.

[10] A. M. Tarrad, Y. Y. El-Hyatemy, and S. A. Omer, "Wyerone derivatives and activities of peroxidase and polyphenooxidase in faba bean leaves as induced by chocolate spot disease," *Plant Science*, vol. 89, no. 2, pp. 161–165, 1993.
[11] R. M. Letcher, D. A. Widdoswson, B. J. Deverall, and J. W. Mansfield, "Identification and activity of wyerone acid as a phytoalexins in broad bean after infection by botrytis," *Phytochemistry*, vol. 9, no. 2, pp. 249–252, 1970.
[12] G. P. Chapman "Systems for cytogenetic analysis in Vicia *faba* L," in *Proceedings of a Seminar in the EEC Programme of Coordination of Research on Plant Productivity*, G. P. Chapman, and S. Tarawali, Eds., Ashford, UK, April 1984.
[13] J. W. Mannsfield and D. A. Widdowson, "The metabolism of wyerone acid (a phytoalexins from Vicia faba L.) by Botrytis faba and *B. cinerea*," *Physiological Plant Pathology*, vol. 3, no. 3, pp. 393–404, 1973.
[14] S. Rossall, J. W. Mansfield, and R. A. Hutson, "Death of Botrytis cinerea and B. faba following exposure to wyerone derivatives in vitro and during infection development in broad bean leaves," *Physiological Plant Pathology*, vol. 16, no. 1, pp. 135–138, 1980.
[15] J. A. Hargreaves, J. W. Mansfield, D. T. Coxon, and K. R. Price, "Wyerone epoxide as a phytoalexins in Vicia faba and its metabolism by Botrytis cinerea and B. faba in vitro," *Phytochemistry*, vol. 15, no. 7, pp. 1119–1121, 1976.
[16] J. A. Hargreaves, J. W. Mansfield, and S. Rossal, "Changes in phytoalexin concentration in tissues of the broad bean plant (Vicia faba) following inoculation with species of Botrytis," *Physiological Plant Pathology*, vol. 11, no. 3, pp. 227–242, 1977.
[17] N. A. Al-Douri and P. M. Dewick, "Biosynthesis of the furanoacetylene phytoalexin wyerone in Vicia faba," *Zeitschrift für Naturforschung C*, vol. 41, no. 1-2, pp. 34–38, 1986.
[18] N. A. Al-Douri, "Biosynthetic conversion of antifungal phytoalexins from broad beans cotyledons," *Bothalia*, vol. 44, no. 5, pp. 161–166, 2014.
[19] S. Rossall and J. W. Mansfield, "Relationships between germ-tube growth by Botrytis cinerea and B. faba and their metabolism of the phytoalexins wyerone acid and wyerone epoxide," *Transactions of the British Mycological Society*, vol. 82, no. 3, pp. 571–576, 1984.
[20] J. W. Mansfield, A. E. A. Porter, and R. V. Smallman, "Dihydrowyerone derivatives as component of furanoacetylenic phytoalexinic phytoalexin response of tissues of Vicia faba," *Phytochemistry*, vol. 19, no. 6, pp. 1057–1061, 1980.
[21] S. Gnanamanickam and J. Mansfield, "Selective toxicity of wyerone and other phytoalexins to Gram positive bacteria," *Phytochemistry*, vol. 20, no. 5, pp. 997–1000, 1981.
[22] A. E. A. Porter, R. V. Smallman, and J. W. Mansfield, "Analysis of furanoacetylenic phytoalexins from the broad bean plant by high performance liquid chromatography," *Journal of Chromatography A*, vol. 172, no. 1, pp. 498–504, 1979.
[23] T. Hashimoto, A. Hayashi, Y. Amano et al., "Hyoscyamine 6 beta-hydroxylase, an enzyme involved in tropane alkaloid biosynthesis, is localized at the pericycle of the root," *Journal of Biological Chemistry*, vol. 5, no. 7, pp. 4648–4653, 1991.

21

Investigation into Polyurethane at Varying Dose Rates of Ionizing Radiation for Clinical Application

Shelley L. Cooke [1] and Abby R. Whittington [1,2]

[1]*Department of Materials Science and Engineering, Virginia Polytechnic Institute and State University, Blacksburg, VA 24061, USA*
[2]*Department of Chemical Engineering, Virginia Polytechnic Institute and State University, Blacksburg, VA 24061, USA*

Correspondence should be addressed to Abby R. Whittington; awhit@vt.edu

Academic Editor: Yves Grohens

Polyurethanes (PUs) are commonly used materials for medical devices. These devices are exposed repeatedly to radiation when patients undergo radiotherapy treatments. It has been found that peripherally inserted central catheters (PICCs) and central venous catheters (CVCs) fail at an increased rate (14.7% and 8.8%, respectively) when radiated. Currently, little research is available on increased failure seen in conjunction with radiation, but complex *in vivo* environments within a human patient make it difficult to isolate effects of individual variables. This research investigated effects of radiation in an aqueous environment to determine whether radiation combined with a mimicked *in vivo* environment is sufficient to change PU devices. The following dose rates were used in this study: 3.2 Gy·min^{-1}, 4.5 Gy·min^{-1}, 44 Gy·min^{-1}, and 833 Gy·min^{-1}. Samples were characterized in four main ways: cellular response, physical changes, chemical changes, and mechanical changes. Results reveal normal cellular response at all dose rates, indicating dose rate does not alter cellular adhesion or proliferation, and biocompatibility of the material is not being altered. Results from physical, chemical, and mechanical effects confirm that varying dose rates alone do not initiate material changes, which negates the hypothesis that varying dose rates of radiation contribute to the complications in PICC and CVCs.

1. Introduction

Polyurethane biomaterials are often studied for use in fabricating 3D tissue scaffolds, catheters, blood contact materials (heart valves and artificial veins), hospital bedding, injection molded implants, and other short-term implants [1–3]. Polyurethane catheters, specifically PICCs and CVCs, have been shown to fail at higher rates when exposed to radiotherapy in cancer patient [4, 5], 14.7% and 8.8%, respectively [4]. Postinsertion complication rate of non-irradiated PICC lines is 30.4% [4–7]. Previous investigations in our lab (data not shown) have indicated that varying total therapeutic radiation doses at a constant dose rate does not alter PU enough to cause instability in an aqueous environment. However, altering radiotherapy dose rate is commonly used in treating different types of cancer or targeting specific tumor locations on the body [8] and may be a contributing factor to material stability *in vivo* [6]. Some researchers and clinicians have suggested that radiation should be applied prior to PICC line insertion to reduce complications believed to occur through radiation exposure (thrombosis, infection, mechanical occlusion/puncture, catheter movement, and extravasation), and this study aims to investigate whether radiation may be the cause of any complications [6,9–11].

The leading causes of PU catheter complications are infection and thrombosis (blood clotting). Radiation may play a role leading up to these biological responses and will be investigated using cell proliferation and adhesion studies on PU films [7]. Occlusion and puncture are mainly affected by the placement of the catheter in the body and may arise due to improper placement causing pinching of the catheter from arm movement or pinching between rib and clavicle, rupture due to pressure from liquid moving through catheter and the body's natural immune reaction to foreign materials causing catheter-associated thrombosis [9–12]. This study investigated these complications using biocompatibility tests combined with mechanical or chemical to provide indication of strength or chemical structure changes in the material which can aid in the movement of catheters contributing to complications [12]. Our current investigation does not focus on how material properties of catheters change only on whether or not they need to be removed.

The International Organization for Standardization (ISO 10555 and ISO 10993) and the United States Food and Drug Administration (FDA) regulations emphasize three major areas for catheter approval: toxicity/biocompatibility, mechanical properties, and thermal properties. All three were evaluated in this study in combination with chemical changes to determine if radiation causes significant changes leading to catheter failure [13, 14].

Currently, there is debate among clinicians on whether varying dose rate is actually effective in radiotherapy treatment. In order to clearly see changes between dose rates, a high sterilization dose was used to evaluate material changes as therapeutic radiation itself has been shown to not cause significant changes in PU materials under aqueous conditions chosen to mimic *in vitro* conditions [13, 15]. Polyurethane irradiated at sterilization doses or higher (World Health Organization standard for biomedical sterilization is 25 kGy before material implantation) has shown significant alteration and should aid in determining whether varying dose rate also produces changes in polyurethane [16–20]. However, it is important to point out that this current study is focused on the changes occurring between dose rates.

The aqueous environment is a key factor in this research because it is an initial step in preclinical approval, mimicking an *in vivo* environment [13, 15]. FDA and ISO guidelines for catheter preclinical studies require that catheter performance testing be simulated in *in vivo* conditions, such as at body temperature, and/or in water or saline environments [15, 21].

To apply different dose rates during radiotherapy, multiple LINAC instruments can be used [22]. Instrument energy level settings range between 6 and 22 MeV (approximately 6–22 $Gy{\cdot}min^{-1}$), where low-range instruments use single 4 or 6 MeV sources while high-range instruments may use multiple sources at once [22, 23]. The low ranges are represented by both 3.2 $Gy{\cdot}min^{-1}$ and 4.5 $Gy{\cdot}min^{-1}$ dose rates that were evaluated in this study while the higher dose rates, 44 $Gy{\cdot}min^{-1}$ and 833 $Gy{\cdot}min^{-1}$, are the representative of multiple sources combined and high-dose rates [24]. The literature states that polyurethanes show alteration at high total doses; however, changes between varying dose rates have not been evaluated and may have implications in varying dose rates at low therapeutic doses as well [16, 19, 25].

In this study, investigation was based on ISO standards to determine whether radiation changes in polyurethane samples were significant enough to cause samples to fall outside regulations set for device approval. This study addresses a concern voiced by clinicians that radiation may cause increased complications [4–7], and the principal goal was to isolate effects of radiation dose rates on polyurethane in a controlled aqueous environment.

2. Materials and Methods

2.1. Polyurethane Sample Preparation. Polyurethane pellets (polyether urethane; Texin 950) were gifted from Covestro™ AG (further information on this material is proprietary). Pellets were cryoground into powder using a three-step cool-grind-cool method repeated three times. Once sufficient polymer was ground, 2 g powder was pressed into films using a hot press at 365°F with 4,500 lbft in −2 pressure for 2 min. After 2 min, a slight increase in pressure was used to remove any air bubbles, and samples were removed. Samples were then cut into 6.35 mm diameter circles with 0.03 mm thickness for testing (Figure 1). All samples were fabricated from the same batch of polyurethane pellets and pressed in the same day to eliminate batch-to-batch differences.

2.2. Material Radiation. Polyurethane films were irradiated in water at varying dose rates from different aged Cs-137 sources (22, 10, 0.1, and 1 years; Table 1). Each dose rate was applied to a constant total dose of 25 kGy. Samples were irradiated at Oak Ridge National Laboratory using the High Flux Isotope Reactor, Gamma Irradiation Facility. Characterization methods were based on prior studies and ASTM procedure F561-13, which describes retrieval and recommended analysis of medical devices, associated tissues, and fluids. Once irradiated, films were cut into pieces for testing by various characterization methods described below. Sets of films stored in water at room temperature for the duration of the radiation experiment were compared to dry as-fabricated. During control studies, it was found that the aqueous environment did not alter the polyurethane material resulting in all methods, specifically characterizing radiation effects. Films submersed in water showed no significant change from as-fabricated samples, and it can be concluded that any changes observed are the result of varying radiation dose. Preliminary results also confirm that overall sterilization dose significantly alters the polyurethane films (data not shown).

2.3. Cell Cytotoxicity and Proliferation. Polyurethane films were sterilized via 70% ethanol and then patted dry with Kimwipes™. Samples were then immersed in alpha Minimum Essential Medium (α-MEM, Life Technologies) with 10 % fetal bovine serum (FBS) and 1% penicillin-streptomycin for 12 hr. Soaked samples were placed in 48-well plates. Each sample was seeded with 5,000-mouse fibroblast cells (3T3, source ATCC) in α-MEM. This cell line was chosen because the fibroblast cells synthesize an extracellular matrix and collagen, common tissues found near polyurethane medical devices [26]. The plate was then incubated for 1 hr at 37°C with 5% CO_2 to allow attachment. Once attached, cells were grown for 1, 4, and 7 days to determine fibroblast proliferation and cell cytotoxicity

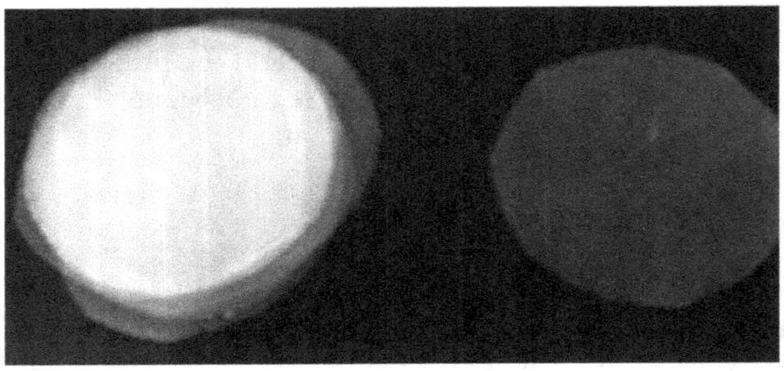

FIGURE 1: Fabricated polyurethane test films.

TABLE 1: Radiation sources and corresponding dose rates.

Age of source (# of years)	Dose rate (Gy·min^{-1})
22	3.2
10	4.5
1	44
0.1	833

within wells. A sample size of three was used for each time point and factor.

CellTiter 96® AQueous Non-Radioactive Cell Proliferation Assay (MTS, Promega, USA) was performed to determine cell proliferation, as previously described [27]. For adhesion studies, 1, 4, and 7 day polyurethane film samples were removed from wells and placed in new 48-well plates with 200 μL of fresh media and 20 μL MTS. All wells were incubated for 2 hr at 37°C and 5% CO_2 to allow colorimetric reactions to occur with cells. All media were then transferred to a 96-well plate and read using a UV-VIS spectrophotometer at 490 nm (BioTek, Multi-Mode Plate Reader, Synergy Mx, USA). An empty polystyrene well was used as a positive control for all studies. A calibration curve was created using a standard cell count (102–106) and was used to determine the number of viable cells from each sample.

2.4. Size Exclusion Chromatography (SEC). Light scattering size exclusion chromatography (SEC; Agilent Technologies, USA) was used to determine polyurethane molecular weight. Samples ($n = 3$) were run in tetrahydrofuran (THF) at 1 mL·min^{-1} at 30°C on two Agilent PLgel 10 μm MIXED-B columns connected in series with a Wyatt Dawn Helios 2 light scattering detector and a Wyatt Optilab Rex refractive index detector. An MW 21,000 polystyrene reference standard was used for calibration.

2.5. Attenuated Total Reflectance Fourier Transform Infrared Spectroscopy (ATR-FTIR). A Nicolet model 8700 Fourier-transform infrared (FTIR) instrument was employed to investigate polyurethane bond changes. FTIR spectra were recorded over a range of 500 to 4,500 nm with 126 scans per run and a resolution of 4 cm^{-1} sample size of 10. A background scan was performed for each sample to remove baseline noise. Peak deconvolution was performed to quantify peak area under each curve. Percent change was calculated using Origin® 8.0 graphing software. Seven major peaks were isolated for PU; all peak heights were measured and then averaged. A sample size of 10 films was used for each factor.

2.6. Dynamic Mechanical Analysis (DMA). Dynamic mechanical analysis (TA Q800) was used to investigate thermal degradation at different sample conditions. PU samples were heated from −50°C to 140°C at a rate of 2°C min^{-1}. Storage modulus, loss modulus, and Tan Delta were calculated with a sample size of three.

2.7. Field-Emission Scanning Electron Microscopy (FESEM). A LEO (Zeiss) 1550, field-emission environmental scanning electron microscope (FESEM), was used to determine microscopic surface differences between irradiated films. Films were sputter coated with gold using a high-vacuum Leica sputter coater. A sample size of two films was used, and representative images were taken at multiple locations on each film.

2.8. Statistical Analysis. Statistical analysis was performed for all tests, and corresponding graphs were created using Origin® 8.0. A p value <0.05 was used in all t-tests, and ANOVA was computed to determine significant statistical differences.

3. Results

Polyurethane results are discussed in separate sections: cellular responses (adhesion and proliferation/migration), physical changes (FESEM and visual), chemical changes (FTIR-ATR and SEC), and thermal/mechanical responses (DMA). Results compare dose rates to determine if certain dose rates alter material properties more than others.

3.1. Cellular Responses. Mammalian cell response was investigated by looking at the cell number adhered on the films each day and then normalized to a tissue culture treated polystyrene control, and variability was calculated at 1, 4, and 7 days. Polyurethane samples displayed a decreased adherence after day 1 but had significantly lower growth than a polystyrene positive control (Figure 2). In day 1 (Figure 2), it was observed that polyurethane samples were 30% or less as adherent as polystyrene, while by day 7, the polyurethane only had ~10% adherence compared to the polystyrene-positive control. While both days 4 and 7 for the polyurethane had a significantly lower percentage (18% and 17% less, respectively) of cells than day 1, no statistical differences were seen between radiation doses within individual days, indicating that dose rate has no effect on cell adhesion but overall cells do not adhere to polyurethane films (Figure 2). This decrease in cellular attachment is not an indicator of cytotoxicity, as cells do proliferate in proximity to the films during this time. These results reveal a positive effect for this investigation since PICC and CVC lines are temporary implants, and adhesion is not desired.

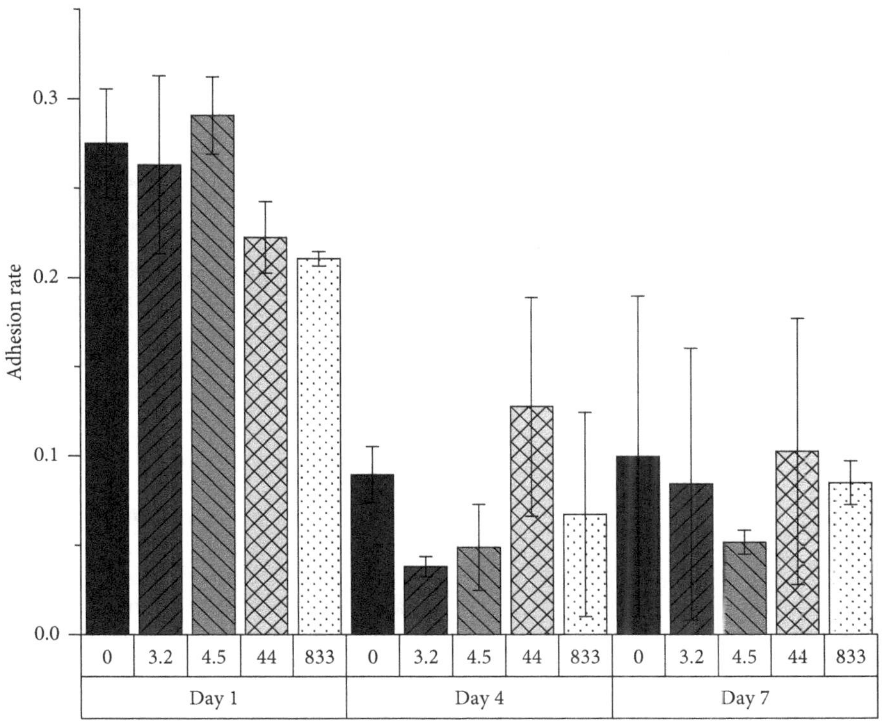

FIGURE 2: Polyurethane adhesion study of normalized cell attachment on samples ($n = 3$, $p < 0.05$, error bars depict standard deviations).

3.1.1. Physical Changes. Physical changes were detected using FESEM to determine if surface changes were occurring. Representative FESEM images (Figure 3) further elucidated microscopic differences. In all FESEM images, lines and minor surface particulates due to film pressing were seen, but little visual difference was found between dose rates.

ISO 10555 standards state that "with a minimum of ×2.5 magnification the external surface of the effective length of the catheter shall appear free of extraneous matter" [13]. In this study, images showed that at ×2 magnification, little to no differences were seen between dose rates. Surface changes indicate that dose rates do not affect polyurethane, and ISO guidelines for catheters were still met even with changes in dose rates.

3.2. Chemical Changes from Irradiation

3.2.1. Size Exclusion Chromatography (SEC). SEC was used to measure the number average molecular weight, polydispersity, and refractive index. SEC results revealed no significant difference in number average molecular weight or polydispersity (Figure 4; Table S1). However, significant changes were seen in refractive index (Figure 4). Refractive index results were significantly different between all dose rates except 44 and 833 Gy·min^{-1}. Refractive index indicates a change in the speed in which light moved through the polyurethane and can indicate changes in composition (chemical structure/crystallinity) or purity (remnants of polymer, precipitates, modifiers, stabilizers, etc.) [28, 29]. In the case of polyurethane samples for this study, the slight changes seen in refractive index are not due to molecular weight changes or crystallinity changes (see mechanical results) and are likely due to small impurities within the material. These impurities did not affect mechanical properties and are not likely to contribute to catheter movement or complication [10, 12]. Methods used to view catheter movement (X-ray spectroscopy or electrocardiography) would also not be affected by changes in refractive index, and it can be concluded that although changes are seen, they would not cause failure in a clinical setting [10, 30, 31].

3.2.2. Fourier-Transform Infrared Spectroscopy. FTIR confirmed the results seen in SEC and indicated no major change in chemical structure. Seven peaks were isolated in the polyurethane spectra (data not shown). No new peaks were observed forming or disappearing, indicating that the polyurethane films do not undergo polymer backbone changes. The height of each peak can indicate a change in the number of bonds within the material and indicates that the material is remaining the same throughout all dose rate exposures (Table 2). Although slight variation is seen, FTIR results can be skewed due to the overlapping of peaks, and it is accepted that these slight changes are due to overlapping of peaks rather than significant changes in the material itself [32]. Control samples ($n = 20$) analyzed using FTIR were shown to have equal to or higher variations than each sample peak. This suggesting that variation within samples does not

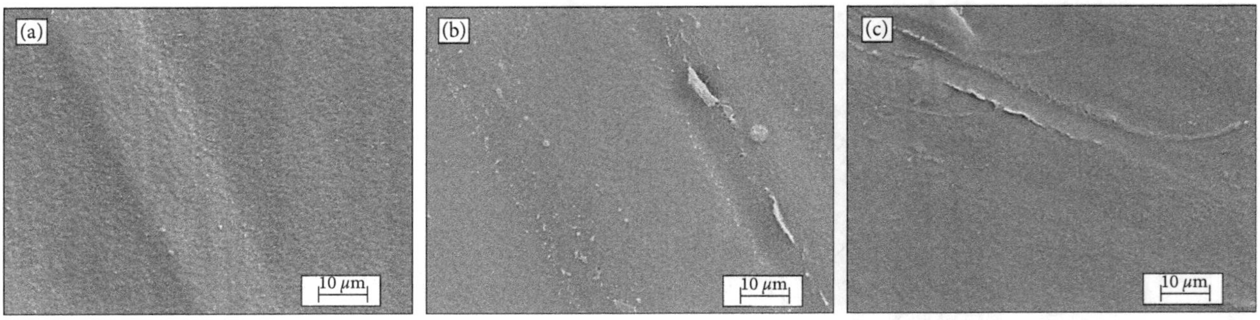

FIGURE 3: Field-emission scanning electron microscopy of polyurethane films sputter coated with gold at (a) $0\,Gy{\cdot}min^{-1}$, control (b) $3.2\,Gy{\cdot}min^{-1}$, and (c) $833\,Gy{\cdot}min^{-1}$ dose rates. Scale bars represent $10\,\mu m$.

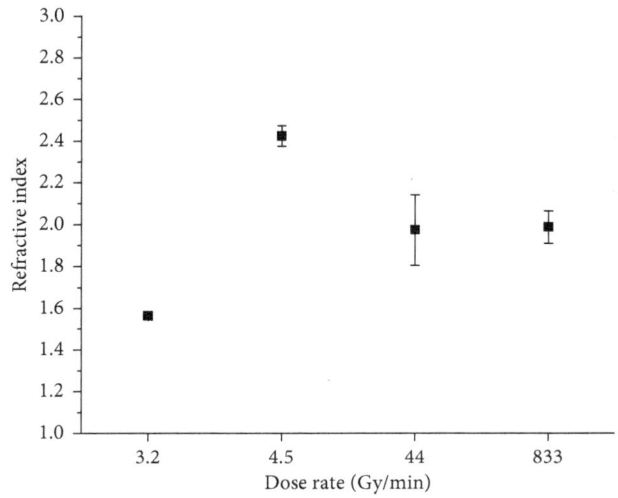

FIGURE 4: Refractive index changes show significant differences between all samples except between 44 and $833\,Gy{\cdot}min^{-1}$ ($n = 3$, $p < 0.05$, error displayed in standard deviations).

indicate significant changes between varying dose rates but rather overlapping peaks and/or variation between sample films.

3.3. *Mechanical and Thermal Response*

3.3.1. Dynamic Mechanical Analysis (DMA). DMA was used to investigate mechanical and thermal changes in polyurethane samples. Two main types of information were collected from these tests: Tan Delta and storage modulus. Tan Delta is used to determine glass transition temperature and can indicate a shift in soft segments to hard segments [33]. Storage modulus is the elastic modulus indicating changes in energy storage and material elasticity [33]. Storage modulus was analyzed at room temperature (25°C) and body temperature (37°C) (Table 3). No statistical differences were found between any of the dose rates evaluated. This was expected based on the chemical results previously discussed.

High variation was seen in the storage modulus data as well as within preliminary studies using nonirradiated, control samples. Power analysis was run using preliminary data to determine the number of samples for this study. This variation can be contributed largely to the variation in the material pellets themselves, combined with material fabrication (hand-pressed films), and would require further samples testing in the future. However, no trend was observed in the storage modulus data; it would have been expected for storage modulus to show an increase with increasing dose [34, 35]. Due to the high variation within the groups, there was no true significance determined between the dose rates.

4. Discussion

This study found that varying dose rates alone in an aqueous environment does not alter cellular, physical, chemical, or mechanical properties. The belief that varying radiation dose rates causes changes *in vivo* is not supported by this study. These data are in agreement with Zhang et al., who investigated physical properties of catheters (fracture, leakage, and rupture) [6], as well as with previous research in our laboratory on therapeutic levels of radiation [36]. Radiation by itself or in an aqueous environment does not appear to cause the increased failure documented with use of PICC and CVC lines. It is important to note that polyether urethane used in this study is highly resistant to hydrolytic degradation [37]; control samples were not significantly different than the as-prepared polyurethane films (confirmed through FTIR, DMA, and SEC; data not shown), indicating that the aqueous environment employed in this study did not change properties of test films. The conditions in the study were selected to meet basic FDA regulations for testing in an *in vitro* environment.

Biocompatibility (ISO 10933) states that catheters shall be free of biological hazards. The first test listed is cytotoxicity, which states that cell culture techniques shall be used to determine cell death, inhibition of growth, colony formation, and other measured effects [14]. This study confirmed that cell death was not occurring using 24 hr live-dead testing (data not shown) and also confirmed that cell growth/proliferation was not inhibited. Cells grew as quickly as with the positive controls used. However, cellular adhesion shows a decrease over time, which is a positive outcome for catheter use since catheters during radiotherapy

TABLE 2: FTIR Peak area data showing slight differences in bonds ($n = 20$).

	C-O-C, sym	N-H, C-N	N-H, C-N	C=O stretch	Sym. CH_2	Asym. CH_2	N-H bond
$Gy{\cdot}min^{-1}$	1073	1216	1524	1700	2853	2933	3301
3.2	0.44 ± 0.01	0.64 ± 0.1	0.72 ± 0.01	0.60 ± 0.01	0.49 ± 0.01	0.40 ± 0.01	0.25 ± 0.0
4.5	0.54 ± 0.04	0.66 ± 0.04	0.72 ± 0.03	0.57 ± 0.01	0.38 ± 0.01	0.38 ± 0.02	0.23 ± 0.01
44	0.51 ± 0.04	0.65 ± 0.04	0.70 ± 0.02	0.57 ± 0.02	0.46 ± 0.04	0.36 ± 0.11	0.24 ± 0.01
833	0.51 ± 0.01	0.59 ± 0.01	0.71 ± 0.01	0.59 ± 0.01	0.45 ± 0.02	0.39 ± 0.01	0.25 ± 0.0

TABLE 3: DMA results indicating significant changes between factors ($n = 3$, $p < 0.05$, error displayed in standard deviations).

$Gy{\cdot}min^{-1}$	Area (Tan Delta)	37°C (storage, MPa)	25°C (storage, MPa)
3.2	2.23 ± 0.2	51.15 ± 17.8	82.59 ± 24.6
4.5	1.91 ± 0.02	70.20 ± 33.9	115.47 ± 48.6
44	2.22 ± 0.1	56.22 ± 29.2	101.42 ± 56.0
833	2.21 ± 0.2	44.05 ± 19.5	69.21 ± 29.9

are temporary implants, and therefore, decreased cell adhesion is preferred.

Cellular response depends on a multitude of properties in an *in vitro* environment [38–40]. Some of these include biological factors (proteins/biological fluids), surface topography, surface chemistry, mechanical properties, and even absorption properties of the material itself [37, 39]. Adhesion studies in this research found the majority of the cells only adhered for the first 24 hours. Polyurethane films were initially soaked in culture media before seeding the cells, so the initial contact between cell and material was between cells, and the media absorbed onto polyurethane implants. Cells do not begin to lay down a cellular matrix until several days later [38, 40]. The culture media used in this study were used to promote cell growth. It was expected that cells would be initially interested in the material surface. Once the cells have thoroughly absorbed proteins on the surface of polyurethane samples, they came into contact with the polyurethane surface, and ultimately did not adhere to polyurethane films. Other research has discussed adhesion and suggested it is likely a combination of properties (chemistry, topography, and mechanics) that promote cell adhesion and makes it difficult to isolate effects [39].

Overall, cellular responses indicated positive results for temporary catheter implantation, and no changes were observed between any dose rates tested, signifying that varying dose rates did not alter polyurethane properties sufficiently to alter cellular response.

Microscopy results also confirmed the polymer stability seen in cellular, chemical, and mechanical tests. Visually, surfaces appeared similar, meeting the ISO standards of "free from extraneous matter". In relation to cellular response, surface topography plays a significant role in long-term cell adhesion while short-term cell adhesion (24 hr) primarily responds to surface chemistry [38]. Surface topography did not change with varying dose rates, which supports the cellular adhesion results.

Along with surface topography, surface chemistry can affect both adhesion and proliferation. It was expected that chemical changes would reflect findings that dose rates do not cause significant alteration to polyurethane films. Size-exclusion chromatography (SEC) was used to detect changes in molecular structure of the polyurethane polymer backbone. No changes in molecular weight or polydispersity indicate that dose rate is not causing backbone scission, and distribution of molecular mass remains the same. However, changes were seen in refractive index, which can be dependent on characteristics such as soft segments versus hard segments, chain extenders, and isocyanate used (methyl diphenyl diisocyanate, MDI or toluene diisocyanate, and TDI) [41–43]. Since the polyurethane used in this study is from the same batch of material, chain extenders and isocyanate should be constant, indicating that slight changes seen in refractive index are due to changes in soft versus hard segments. This indicates difference in amorphous and crystalline regions [42, 43]. However, the Tan Delta shown in the DMA results indicated no significant shifts between soft and hard segments at any dose rate. It is possible that the small sample weight in SEC resulted in a poor representation of material composition due to increased variability. This is also seen in the 44 $Gy{\cdot}min^{-1}$ sample. However, refractive index results are within range of normal polyurethane materials, and it can be concluded that polyurethane films at all dose rates maintained chemical structure stability.

DMA results support clinical use of polyurethane films exposed at varying dose rates. Changes seen in DMA are regularly based on the ratio of hard segments to soft segments in the material [44]. Hard segments are isocyanate or diisocyanate linked to soft segments (ether-polyol) through a urethane linkage [45]. DMA shows no change in Tg or storage modulus at 25°C or 37°C, indicating that ratios of hard segments to soft segments did not change supporting the absence of any chemical backbone changes seen in SEC data. Dose rates did not change material strength at room or body temperature, thus maintaining stability of catheters under clinical *in vitro* conditions.

5. Conclusions

Results from this study found that varying dose rates does not alter polyurethane material properties. Polyurethanes are common in clinical settings and have a high probability of being irradiated at various dose rates and different total doses *in vivo*. This study answered the question currently being posed by clinicians and researchers as to whether radiation is the main factor contributing to early failure in PICC and CVC lines; radiation alone does not cause material failure in polyurethane catheter applications. To the authors' knowledge, this research provides the first investigation into varying dose rates in an aqueous setting for clinical applications and has laid the groundwork to further investigate different medical devices and variables that may be exposed to radiation *in vivo*.

Conflicts of Interest

The authors declare that there are no conflicts of interest regarding the publication of this article.

Acknowledgments

The authors would like to thank Geoffrey Deichert and Caleb Robison at Oak Ridge National Lab for conducting radiation exposures. The authors would also like to thank Jeffrey Weaver, Scott Radzinski, and Dr. John Matson for running size exclusion chromatography on polyurethane samples.

Supplementary Materials

Table S1: comparison of SEC results ($n = 3$, error displayed in standard deviations). It includes numerical data collected from size exclusion chromatography. Results indicate no significant change in polydispersity or molecular weight ($p < 0.05$, error in standard deviation). Figure S1: representative FTIR spectra of 0, 3.2, and 833 Gy/min polyurethane samples. It compares resulting spectra for 0, 3.2, and 833 Gy/min collected from Fourier-transform infrared spectroscopy. The spectra reveal little visual change between 3.2 and 833 Gy/min but expected change from 0 Gy/min due to overall sterilization dose. Major peak area and wavelength for all dose rates were further analyzed using peak deconvolution and discussed in Section 4.2 of the article. (*Supplementary Materials*)

References

[1] S. Lyu and D. Untereker, "Degradability of polymers for implantable biomedical devices," *International Journal of Molecular Sciences*, vol. 10, no. 9, pp. 4033–4065, 2009.

[2] R. J. Zdrahala and I. J. Zdrahala, "Biomedical applications of polyurethanes: a review of past promises, present realities, and a vibrant future," *Journal of Biomaterials Applications*, vol. 14, no. 1, pp. 67–90, 1999.

[3] A. Burke, *Biomaterials: From Molecules to Engineered Tissues*, Kluwer Academic/Plenum, New York, NY, USA, 2004.

[4] J. Bratton, P. Johnstone, and K. McMullen, "Outpatient management of vascular access devices in children receiving radiotherapy: complications and morbidity," *Pediatric Blood and Cancer*, vol. 61, no. 3, pp. 499–501, 2014.

[5] S. A. Jain, S. N. Shukla, S. S. Talati, S. K. Parikh, S. J. Bhatt, and V. Maka, "A retrospective study of central venous catheters GCRI experience," *Indian Journal of Medical and Paediatric Oncology*, vol. 34, no. 4, pp. 238–241, 2013.

[6] J. Zhang, S. Zhang, L. Li et al., "Effects of ionizing radiation on physical properties of peripherally inserted central catheter," *PLoS One*, vol. 11, no. 9, Article ID e0162837, 2016.

[7] T. Seckold, S. Walker, and T. Dwyer, "A comparison of silicone and polyurethane picc lines and postinsertion complication rates: a systematic review," *Journal of Vascular Access*, vol. 16, no. 3, pp. 167–177, 2015.

[8] H. B. Stone, C. N. Coleman, M. S. Anscher, and W. H. McBride, "Effects of radiation on normal tissue: consequences and mechanisms," *The Lancet Oncology*, vol. 4, no. 9, pp. 529–536, 2003.

[9] K. H. Polderman and A. R. Girbes, "Central venous catheter use," *Intensive Care Medicine*, vol. 28, no. 1, pp. 1–17, 2002.

[10] R. Gnannt, B. L. Connolly, D. A. Parra, J. Amaral, R. Moineddin, and A. S. Thakor, "Variables decreasing tip movement of peripherally inserted central catheters in pediatric patients," *Pediatric Radiology*, vol. 46, no. 11, pp. 1532–1538, 2016.

[11] V. Vidal, C. Muller, A. Jacquier et al., "Évaluation prospective des complications des PICCs," *Journal de Radiologie*, vol. 89, no. 4, pp. 495–498, 2008.

[12] A. Earhart, "Central lines: recognizing, preventing, and troubleshooting complications," *Journal of the American Association*, vol. 8, no. 11, 2013.

[13] International Organization for Standardization, *10555: Intravascular Catheters—Sterile and Single Use Catheters*, International Organization for Standardization, Geneva, Switzerland, 2017.

[14] International Organization for Standardization, *ISO 10993: Biological evaluation of Medial Devices*, International Organization for Standardization, Geneva, Switzerland, 2010.

[15] Center for Devices and Radiological Health, Class II Special Controls Guidance Document for Certain Percutaneous Transluminal Coronary Angioplasty (PTCA) Catheters," in *Medical Devices*, United States Food and Drug Administration Silver Spring, MA, USA, 2010, https://www.fda.gov/MedicalDevices/ucm225145.htm.

[16] N. Hirata, K. I. Matsumoto, T. Inishita et al., "Gamma-ray irradiation, autoclave and ethylene oxide sterilization to thermosetting polyurethane: sterilization to polyurethane," *Radiation Physics and Chemistry*, vol. 46, no. 3, pp. 377–381, 1995.

[17] H. Shintani and A. Nakamura, "Degradation and crosslinking of polyurethane irradiated by gamma-rays," *Polymer Degradation and Stability*, vol. 32, no. 2, pp. 191–208, 1991.

[18] H. Shintani and A. Nakamura, "Mechanism of degradation and crosslinking of polyurethane when irradiated by gamma-rays," *Journal of Applied Polymer Science*, vol. 42, no. 7, pp. 1979–1987, 1991.

[19] K. Gorna and S. Gogolewski, "The effect of gamma radiation on molecular stability and mechanical properties of biodegradable polyurethanes for medical applications," *Polymer Degradation and Stability*, vol. 79, no. 3, pp. 465–474, 2003.

[20] J. Haugen, M. Brunner, F. Pellkofer, J. Aigner, J. Will, and E. Wintermantel, "Effect of different γ-irradiation doses on cytotoxicity and material properties of porous polyether-urethane polymer," *Journal of Biomedical Materials Research Part B: Applied Biomaterials*, vol. 80B, no. 2, pp. 415–423, 2007.

[21] J. P. Santerre, K. Woodhouse, G. Laroche, and R. S. Labow, "Understanding the biodegradation of polyurethanes: from classical implants to tissue engineering materials," *Biomaterials*, vol. 26, no. 35, pp. 7457–7470, 2005.

[22] E. B. Podgoršak, *Radiation Oncology Physics*, International Atomic Energy Agency, Vienna, Austria, 2005.
[23] M. J. Gazda and L. R. Coia, "Cancer management: a multidisciplinary approach: medical, surgical, and radiation oncology," in *Principles of Radiation Therapy*, Chapter 2, SCP Oncology Group, New York, NY, USA, 2005.
[24] E. J. Hall and D. J. Brenner, "The dose-rate effect revisited: radiobiological considerations of importance in radiotherapy," *International Journal of Radiation Oncology Biology Physics*, vol. 21, no. 6, pp. 1403–1414, 1991.
[25] S. Chandra, S. Sharma, H. Ali et al., "Effect of γ irradiation on the impact response of rigid polyurethane foam," *Journal of Polymer Engineering*, vol. 36, no. 8, pp. 829–835, 2016.
[26] B. Alberts, A. Johnson, J. Lewi et al., "Molecular biology of the cell," in *Fibroblasts and Their Transformations: The Connective-Tissue Cell Family*, Garland Science, New York, NY, USA, 4th edition, 2002.
[27] Promega, *96® AQueous One Solution Cell Proliferation Assay*, Technical Bulletin, London, UK, 2005.
[28] H. Mark, "Purity and identity of polymers," *Analytical Chemistry*, vol. 20, no. 2, pp. 104–110, 1948.
[29] ASTM D542-14, *Standard Test Method for Index of Refraction of Transparent Organic Plastics*, ASTM International, West Conshohocken, PA, USA, 2014, http://www.astm.org.
[30] L. Kelly, "A practical guide to safe PICC placement," *British Journal of Nursing*, vol. 22, no. 5, pp. S13–S19, 2013.
[31] J. Jin, C. Chen, R. Zhao, A. Li, Y. Shentu, and N. Jiang, "Repositioning Techniques of malpositioned peripherally inserted central catheters," *Journal of Clinical Nursing*, vol. 22, no. 13-14, pp. 1791–1804, 2013.
[32] J. B. Huang and M. W. Urban, "Effect of overlapping bands on intensities in ATR FT-IR spectra," *Applied Spectroscopy*, vol. 46, no. 6, pp. 1014–1019, 1992.
[33] K. P. Menard and N. R. Menard, "Dynamic mechanical analysis in the analysis of polymers and rubbers," in *Encyclopedia of Polymer Science and Technology*, John Wiley & Sons, Inc., Hoboken, NJ, USA, 2015.
[34] M. M. Ghobashy and Z. I. Abdeen, "Radiation crosslinking of polyurethanes: characterization by FTIR, TGA, SEM, XRD, and raman spectroscopy," *Journal of Polymers*, vol. 2016, Article ID 9802514, 9 pages, 2016.
[35] H. Shintani, H. Kikuchi, and A. Nakamura, "Effects of gamma-ray irradiation on the change of characteristics of polyurethane," *Polymer Degradation and Stability*, vol. 32, no. 1, pp. 17–30, 1991.
[36] S. L. Cooke and A. R. Whittington, "Influence of therapeutic radiation on polycaprolactone and polyurethane biomaterials," *Materials Science and Engineering: C*, vol. 60, pp. 78–83, 2016.
[37] E. M. Christenson, J. M. Anderson, and A. Hiltner, "Biodegradation mechanisms of polyurethane elastomers," *Corrosion Engineering, Science and Technology*, vol. 42, no. 4, pp. 312–323, 2007.
[38] K. Anselme, L. Ploux, and A. Ponche, "Cell/material interfaces: influence of surface chemistry and surface topography on cell adhesion," *Journal of Adhesion Science and Technology*, vol. 24, no. 5, pp. 831–852, 2010.
[39] X. Yao, R. Peng, and J. Ding, "Cell–material interactions revealed via material techniques of surface patterning," *Advanced Materials*, vol. 25, no. 37, pp. 5257–5286, 2013.
[40] C. J. Wilson, R. E. Clegg, D. I. Leavesley, and M. J. Pearcy, "Mediation of biomaterial-cell interactions by adsorbed proteins: a review," *Tissue Engineering*, vol. 11, no. 1-2, pp. 1–18, 2005.
[41] P. C. Painter and M. M. Coleman, *Essentials of Polymer Science and Engineering*, DEStech Publications, Inc., Lancaster, PA, USA, 2009.
[42] A. Puszka and A. Kultys, "The influence of soft segments on some properties of new transparent segmented polyurethanes," *Polymers for Advanced Technologies*, vol. 28, no. 12, pp. 1937–1944, 2017.
[43] S. Ioan, D. Macocinschi, D. Filip, and A. Taranu, "Specific refractive index increments of segmented poly(ester urethane) s," *Polymer Testing*, vol. 21, no. 7, pp. 757–762, 2002.
[44] C. Prisacariu, "Thermal behaviour of polyurethane elastomers," in *Polyurethane Elastomers: From Morphology to Mechanical Aspects*, pp. 61–101, Springer, Vienna, Austria, 2011.
[45] C. Prisacariu, "Chemistry of polyurethane elastomers," in *Polyurethane Elastomers: From Morphology to Mechanical Aspects*, pp. 1–22, Springer, Vienna, Austria, 2011.

22

Synthesis, Characterization, and Antibacterial and Anti-Inflammatory Activities of New Pyrimidine and Thiophene Derivatives

Siham Lahsasni,[1] Dunya A. M. Al-Hemyari,[1] Hazem A. Ghabbour,[2,3] Yahia Nasser Mabkhoot,[1] Fadilah S. Aleanizy,[1] Asma A. Alothman,[1] and Zainab M. Almarhoon[1]

[1]*Department of Chemistry, King Saud University, P.O. Box 2455, Riyadh 11451, Saudi Arabia*
[2]*Department of Pharmaceutical Chemistry, College of Pharmacy, King Saud University, Riyadh 11451, Saudi Arabia*
[3]*Department of Medicinal Chemistry, Faculty of Pharmacy, University of Mansoura, Mansoura 35516, Egypt*

Correspondence should be addressed to Siham Lahsasni; slahsasni@ksu.edu.sa

Academic Editor: Gabriel Navarrete-Vazquez

Substituted[4,5]thieno[2,3-*d*]thiazolo[3,2-*a*]pyrimidin-5-one (**3a–b**) and pyrimidin-5(6*H*)-imine (**3c–e**) were synthesized via reaction of the starting compounds, ethyl 2-amino-substituted[*b*]thiophene-3-carboxylate (**2a–c**) and 2-amino-substituted [*b*]thiophene-3-carbonitrile (**2d–f**), respectively, with 2-bromothiazole. Synthesis of (bromo-substituted[*b*]thiophen-2-yl) alkanamide derivatives (**4a–e**) and thieno[2,3-*d*][1,3]oxazin-4-imine derivative (**5**) was accomplished via reaction of the starting compounds with bromoalkyl chloride through nucleophilic substitution; however, for the synthesis of compound **5**, nucleophilic substitution was followed by nucleophilic addition to a nitrile group to form the oxazinimine ring. 1-(3-cyano-substituted[*b*]thiophen-2-yl)-3-(4-(trifluoromethyl)phenyl)thiourea derivatives (**6a–c**) were obtained via reaction of the starting compounds (**2d–f**) and 4-(trifluoromethyl phenyl)isothiocyanate. The lead compounds (**2d–f**) rapidly reacted with 4-(trifluoromethyl)benzaldehyde or 4-(2-pyridyl)benzaldehyde in acidic medium to yield compounds (**7a–f**) in large quantities. X-ray crystallography of compounds **4c** and **7e** confirmed their structures. Antimicrobial studies revealed that compound **6a** was equally potent to ampicillin against *Bacillus* strains. Moreover, compounds **3e**, **4d**, and **6a** possessed greater anti-inflammatory potency than that of the standard drug.

1. Introduction

Thiophene, thiazole, and pyrimidine derivatives have been used as therapeutic drugs over years. Figure 1 shows some potent drugs containing thiazole or thiophene ring.

Dasatinib is a dual Src/Abl [1] and *pan*-Src [2] kinase inhibitor, zopolrestat [3] and lidorestat [4] are aldose reductase inhibitors used for the treatment of diabetic complications, and raloxifene is the first clinically available selective estrogen receptor modulator (SERM) used to prevent both osteoporosis and breast cancer [5–8]. In addition, clopidogrel is an antiplatelet agent used to inhibit blood clots in coronary artery disease and to prevent heart attacks and strokes in patients with heart or circulatory diseases [9–11].

The most efficient protocol for the synthesis of these thiophene derivatives is intramolecular cyclization via nucleophilic displacement [12–15], Gewald method [16, 17], thio-Claisen rearrangement [18], and dehydrophotocyclization [19, 20].

Within the scope of these diverse synthetic methods and utility of thiophene-based systems and in continuation to our interest in the design of bioactive heterocycles [21–23], we aimed at developing novel heterocyclic compounds containing thiophene, thiazole, and pyrimidine rings, identifying their structures using infrared (IR), ^{1}H nuclear

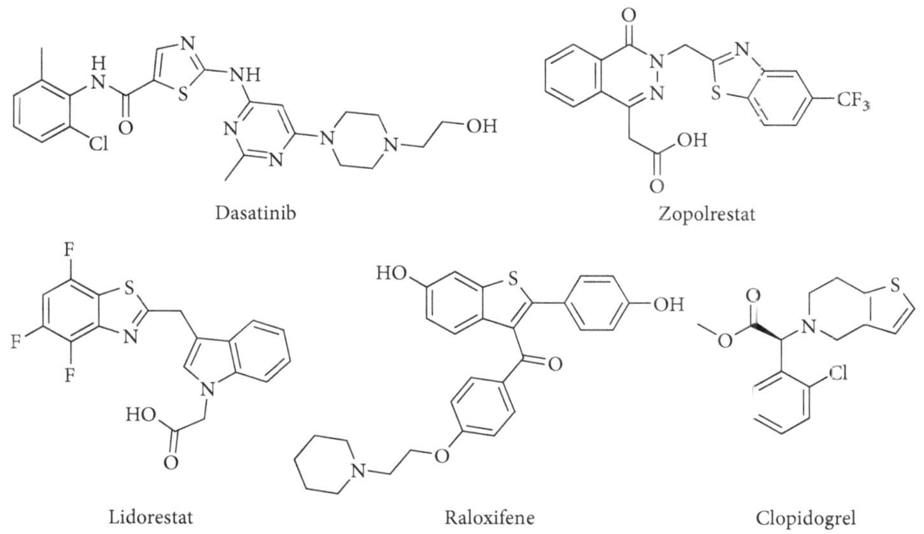

FIGURE 1: Some drugs containing thiazole, pyrimidine, or thiophene ring.

magnetic resonance (^{1}H NMR), ^{13}C NMR spectroscopy, and X-ray, and evaluating their biological effects, in particular, their antimicrobial and anti-inflammatory activities.

2. Materials and Methods

2.1. General Information. Melting points (mp) were determined using an electrothermal digital melting point apparatus and were uncorrected. IR spectra (KBr discs) were recorded using FTIR plus 460 or Pye Unicam SP-1000 spectrophotometer. ^{1}H NMR spectra were recorded using Varian Gemini-200 (200 MHz) and Jeol AS (500 MHz) instruments. Dimethyl sulfoxide- (DMSO-) d6 was used as a solvent, and tetramethylsilane (TMS) was used as an internal standard. Chemical shifts were expressed as δ ppm. Mass spectra were recorded using Hewlett Packard 5988 A GC/MS system and GCMS-QP 1000 Ex Shimadzu instrument. Vario EL III Elemental CHNS analyzer at the Microanalytical Data Unit of Cairo University was used to obtain the analytical data.

2.2. Chemistry

*2.2.1. Synthetic Procedure for Substituted [4,5]Thieno[2,3-d]thiazolo[3,2-a]pyrimidine Derivatives (**3a–e**).* A mixture of the appropriate lead compound (**2b–f**) (5 mmol), an excess amount of 2-bromothiazole (approximately 1 mL), and few drops of hydrochloric acid in absolute ethanol (3 mL) was refluxed for 3–5 h. The product started to precipitate shortly once the reaction started (within approximately 15 min). The precipitate was filtered, washed with absolute ethanol, dried, and recrystallized from absolute ethanol.

*6,7,8,9-Tetrahydro-5H-benzo[4,5]thieno[2,3-d]thiazolo[3,2-a]pyrimidin-5-one (**3a**).* White solid; yield: 30%; mp: 221–222°C; IR (KBr, υ_{max}, cm^{-1}): 3079.1 (CH aromatic), 2930.47 (CH aliphatic), 1675.45 (C=O), and 1563.55, 1516.11 (C=C aromatic); ^{1}H NMR (δ, ppm, $CDCl_3$): 1.878 (*m*, 4H, *J* = 6 Hz, H7-8), 2.77 (*t*, 2H, *J* = 6 Hz, H9), 3.049 (*t*, 2H, *J* = 6 Hz, H6), 6.87 (*d*, 1H, *J* = 4.8 Hz, H2), and 7.98 (*d*, 1H, *J* = 4.8 Hz, H3); ^{13}C NMR (δ, ppm, $CDCl_3$): 22.253, 22.95, 25.158, 25.618 (C6-9), 109.571 (C2), 116.155 (C5a), 121.652 (C5a′), 130.973 (C3), 131.479 (C9a), 154.951 (C10a), 157.419 (C11a), and 163.835 (C5); and DART-TOF-MS (*m/z*): 263.03 $[M + H]^+$; anal. calcd. for $C_{12}H_{10}N_2OS_2$ (262.02): C, 54.94; H, 3.84; N, 10.68; O, 6.10; and S, 24.44; found: C, 55.04; H, 3.74; N, 10.60; O, 6.17; and S, 24.53.

*7,8,9,10-Tetrahydrocyclohepta[4,5]thieno[2,3-d]thiazolo[3,2-a]pyrimidin-5(6H)-one (**3b**).* White solid; yield: 38%; mp: 145–147°C; IR (KBr, υ_{max}, cm^{-1}): 3130.08 (CH aromatic), 2917.76 (CH aliphatic), 1674.41 (C=O), and 1563.5, 1509.46 (C=C aromatic); ^{1}H NMR (δ, ppm, $CDCl_3$): 1.72 (*m*, 4H, H7, H9), 1.897 (*m*, 2H, H8), 2.838 (*t*, 2H, *J* = 5.4 Hz, H10), 3.365 (*t*, 2H, *J* = 4.8 Hz, H6), 6.855 (*d*, 1H, *J* = 4.8 Hz, H2), and 7.984 (*d*, 1H, *J* = 4.8 Hz, H3); ^{13}C NMR (δ, ppm, $CDCl_3$): 27.212, 27.672, 27.925, 29.895, 32.486 (C6-10), 109.494 (C2), 116.853 (C5a), 121.667 (C5a′), 135.634 (C3), 136.438 (C10a), 155.403 (C11a), 156.982 (C12a), and 162.317 (C5); and DART-TOF-MS (*m/z*): 277.05 $[M + H]^+$; anal. calcd. for $C_{13}H_{12}N_2OS_2$ (276.04): C, 56.50; H, 4.38; N, 10.14; O, 5.79; and S, 23.20; found: C, 56.43; H, 4.40; N, 10.18; O, 5.73; and S, 23.29.

*7,8-Dihydrocyclopenta[4,5]thieno[2,3-d]thiazolo[3,2-a]pyrimidin-5(6H)-imine (**3c**).* Dark green solid; yield: 22%; mp < 300°C; IR (KBr υ_{max}, cm^{-1}): 3241.38 (NH), 3073.4 (CH aromatic), 1643.81 (C=N), and 1590.95, 1509.8 (C=C aromatic); ^{1}H NMR (δ, ppm, DMSO-d_6): 2.469 (*m*, 2H, H7), 3.005 (*t*, 2H, H8), 3.15 (*t*, 2H, H6), 8.05 (*d*, 1H, *J* = 4.8 Hz, H2), 8.653 (*d*, 1H, *J* = 4.8 Hz, H3), and 9.172 (*s*, 1H, NH); and ^{13}C NMR (δ, ppm, DMSO-d_6): 27.664, 29.496, 29.971 (C6-8), 108.682 (C2), 118.210 (C5a), 122.395 (C5a′), 136.416 (C8a), 141.138 (C9a), 148.359 (C3), 157.121 (C5), and 168.972 (10a); and DART-TOF-MS (*m/z*): 248.03 $[M + H]^+$; anal. calcd. for $C_{11}H_9N_3S_2$ (247.02): C, 53.42; H, 3.67; N, 16.99; and S, 25.92; found: C, 53.47; H, 3.64; N, 17.05; and S, 25.89.

*6,7,8,9-Tetrahydro-5H-benzo[4,5]thieno[2,3-d]thiazolo[3,2-a]pyrimidin-5-imine (**3d**).* Yellow solid; yield: 88%;

mp < 300°C; IR (KBr, υ_{max}, cm^{-1}): 3281.42 (NH), 3080.16 (CH aromatic), 2929.63 (CH aliphatic), 1643.47 (C=N), and 1564.15, 1504.9 (C=C aromatic); ^{1}H NMR (δ, ppm, DMSO-d_6): 1.835 (*s*, 4H, H7, H8), 2.824 (*s*, 2H, H6), 2.993 (*s*, 2H, H9), 8.002 (*d*, 1H, *J* = 4.8 Hz, H2), 8.619 (*d*, 1H, *J* = 4.8 Hz, H3), and 9.002 (*s*, 1H, NH); ^{13}C NMR (δ, ppm, DMSO-d_6): 21.945, 22.144, 25.471, 25.931 (C6-9), 111.372 (C2), 118.003 (C5a), 122.319 (C5a′), 128.091 (C9a), 135.664 (C10a), 148.596 (C3), 157.573 (C5), and 164.303 (C11a); and DART-TOF-MS (*m/z*): 262.05 $[M+H]^+$; anal. calcd. for $C_{12}H_{11}N_3S_2$ (261.04): C, 55.15; H, 4.24; N, 16.08; and S, 24.53; found: C, 55.09; H, 4.31; N, 16.13; and S, 24.57.

*7,8,9,10-Tetrahydrocyclohepta[4,5]thieno[2,3-d]thiazolo[3,2-a]pyrimidin-5(6H)-imine (**3e**)*. Light brown solid; yield: 70%; mp < 300°C; IR (KBr, υ_{max}, cm^{-1}): 3270.42 (NH), 3053.25 (CH aromatic), 2906.04 (CH aliphatic), 1639.62 (C=N), and 1556.71, 1501.3 (C=C aromatic); ^{1}H NMR (δ, ppm, DMSO-d_6): 1.714, 1.866, 2.953, 3.12 (*m*, 10H, H6-10), 8.02 (*d*, 1H, H2), 8.63 (*d*, 1H, H3), and 9.186 (*s*, 1H, NH); ^{13}C NMR (δ, ppm, DMSO-d_6): 26.485, 26.914, 28.838, 28.93, 30.67 (C6-10), 112.192 (C2), 117.865 (C5a), 122.434 (C5a′), 133.319 (C10a), 139.765 (C11a), 148.941 (C3), 157.204 (C5), and 163.107 (C12a); and DART-TOF-MS (*m/z*): 276.07 $[M+H]^+$; anal. calcd. for $C_{13}H_{13}N_3S_2$ (275.06): C, 56.70; H, 4.76; N, 15.26; and S, 23.28; found: C, 56.74; H, 4.72; N, 15.34; and S, 23.33.

*2.2.2. Synthetic Procedure for Compounds (**4a–e**, **5**).* Compounds **2b–f** (1.7 mmol) were refluxed in bromoalkanoyl chloride for 2–3 h. Then, the mixture was left at room temperature until the product precipitated. The precipitate was filtered, washed with absolute ethanol, dried, and recrystallized from a mixture of absolute ethanol and methanol.

*Ethyl-2-(5-bromopentanamido)-5,6-dihydro-4H-cyclopenta[b]thiophene-3-carboxylate (**4a**)*. Greenish white solid; yield: 25%; mp: 57–59°C; IR (KBr, υ_{max}, cm^{-1}): 3271.77 (NH), 2944.81 (CH aliphatic), 1669.36 (C=O), and 1558.46, 1520.71 (C=C aromatic); ^{1}H NMR (δ, ppm, $CDCl_3$): 1.353 (*t*, 3H, *J* = 6.6 Hz, $-CH_2\underline{CH_3}$), 1.916 (*m*, 4H, H3′-4′), 2.356 (*m*, 2H, *J* = 6.6 Hz, H5), 2.491 (*t*, 2H, *J* = 6.6 Hz, H2′), 2.82 (*t*, 2H, H6), 2.867 (*t*, 2H, H4), 3.419 (*t*, 2H, *J* = 5.4 Hz, H5′), 4.29 (*q*, 2H, *J* = 6.6 Hz, $-\underline{CH_2}CH_3$), and 11.004 (*s*, 1H, NH); ^{13}C NMR (δ, ppm, $CDCl_3$): 14.271 ($-CH_2CH_3$), 22.481, 23.731, 27.908, 28.828, 30.262, 31.940, 32.929, 35.589 (C4-6, C2′-5′), 60.441 ($-\underline{C}H_2CH_3$), 108.083 (C3), 132.169 (C6a), 141.214 (C3a), 151.341 (C=O ester), 166.143 (C=O amide), and 169.11 (C2); and DART-TOF-MS (*m/z*): 374.04 $[M+H]^+$; anal. calcd. for $C_{15}H_{20}BrNO_3S$ (373.03): C, 48.13; H, 5.39; Br, 21.35; N, 3.74; O, 12.82; and S, 8.57; found: C, 48.18; H, 5.24; Br, 21.41; N, 3.69; O, 12.77; and S, 8.52.

*5-Bromo-N-(3-cyano-4,5,6,7-tetrahydrobenzo[b]thiophen-2-yl)pentanamide (**4b**)*. White solid; yield: 74%; mp: 149–150°C; IR (KBr, υ_{max}, cm^{-1}): 3468.26 (NH), 2935.12 (CH aliphatic), 2214.8 (CN), 1699.82 (C=O), and 1548.87, 1455.74 (C=C aromatic); ^{1}H NMR (δ, ppm, $CDCl_3$): 1.818–1.967 (*m*, 8H, H5-6, H3′-4′) 2.569, 2.614 (*m*, 4H, H4, H7), 2.533 (*t*, 2H, H2′), 3.433 (*t*, 2H, H5′), and 9.277 (*s*, 1H, NH); ^{13}C NMR (δ, ppm, $CDCl_3$): 22.115, 23.065, 23.74, 23.885, 23.954, 31.942, 32.946, 34.794 (C4-7, C2′-5′), 92.316 (C3), 114.814 (CN), 128.083 (C7a), 130.689 (C3a), 147.507 (C2), and 169.829 (C=O amide); and DART-TOF-MS (*m/z*): 341.03 $[M+H]^+$; anal. calcd. for $C_{14}H_{17}BrN_2OS$ (340.02): C, 49.27; H, 5.02; Br, 23.41; N, 8.21; O, 4.69; and S, 9.39; found: C, 49.32; H, 4.98; Br, 23.49; N, 8.17; O, 4.73; and S, 9.31.

*5-Bromo-N-(3-cyano-5,6,7,8-tetrahydro-4H-cyclohepta[b]thiophen-2-yl)pentanamide (**4c**)*. Beige solid; yield: 82%; mp: 127–129°C; IR (KBr, υ_{max}, cm^{-1}): 3486.48 (NH), 2924.92 (CH aliphatic), 2213.55 (CN), 1696.13 (C=O), and 1544.89, 1441.07 (C=C aromatic); ^{1}H NMR (δ, ppm, $CDCl_3$): 1.656 (*m*, 4H, H5, H7), 1.85–1.951 (*m*, H2′-4′), 2.507 (*m*, 2H, H6), 2.697 (*s*, 4H, H4, H8), 3.431, 3.567 (*m*, 2H, H5′), and 8.575 (*s*, 1H, NH); ^{13}C NMR (δ, ppm, $CDCl_3$): 22.483, 23.725, 27.32, 27.994, 29.044, 31.758, 31.888, 32.003, 34.947 (C4-8, C2′-5′), 95.106 (C3), 115.159 (CN), 131.77 (C8a), 135.465 (C3a), 145.031 (C2), and 169.623 (C=O amide); and DART-TOF-MS (*m/z*): 355.05 $[M+H]^+$; anal. calcd. for $C_{15}H_{19}BrN_2OS$ (354.04): C, 50.71; H, 5.39; Br, 22.49; N, 7.88; O, 4.50; and S, 9.02; found: C, 50.78; H, 5.34; Br, 22.41; N, 7.93; O, 4.47; and S, 8.96.

*Ethyl-2-(6-bromohexanamido)-5,6,7,8-tetrahydro-4H-cyclohepta[b]thiophene-3-carboxylate (**4d**)*. White solid; yield: 43%; mp: 66–67°C; IR (KBr, υ_{max}, cm^{-1}): 3360.95 (NH), 2915.27 (CH aliphatic), 1675.80 (C=O), and 1529.22 (C=C aromatic); ^{1}H NMR (δ, ppm, $CDCl_3$): 1.379 (*t*, 3H, *J* = 7.2 Hz, $-CH_2\underline{CH_3}$), 1.508 (*m*, 2H, H4′), 1.594, 1.638 (4H, H5, H7), 1.756 (*m*, 2H, H3′), 1.826 (*s*, 2H, H6), 1.888 (*m*, 2H, H5′), 2.456 (*t*, 2H, *J* = 7.2 Hz, H2′), 2.699 (*m*, 2H, H8), 3.015 (*m*, 2H, H4), 3.396 (*t*, 2H, *J* = 6.6 Hz, H6′), 4.33 (*q*, 2H, *J* = 7.2 Hz, $-\underline{CH_2}CH_3$), and 11.2 (*s*, 1H, NH); ^{13}C NMR (δ, ppm, $CDCl_3$): 14.25 ($-CH_2\underline{C}H_3$), 24.442, 26.936, 27.688, 27.795, 28.247, 28.561, 32.21, 32.417, 33.444, 36.564 (C4-8, C2′-6′), 60.649 ($-\underline{C}H_2CH_3$), 112.668 (C3), 130.843 (C8a), 136.270 (C3a), 145.484 (C=O ester), 166.748 (C=O amide), and 169.554 (C2); and DART-TOF-MS (*m/z*): 416.09 $[M+H]^+$; anal. calcd. for $C_{18}H_{26}BrNO_3S$ (415.08): C, 51.92; H, 6.29; Br, 19.19; N, 3.36; O, 11.53; and S, 7.70; found: C, 51.88; H, 6.34; Br, 19.25; N, 3.31; O, 11.47; and S, 7.79.

*6-Bromo-N-(3-cyano-5,6,7,8-tetrahydro-4H-cyclohepta[b]thiophen-2-yl)hexan-amide(**4e**)*. Brown solid; yield: 69%; mp: 104–106°C; IR (KBr, υ_{max}, cm^{-1}): 3420.51 (NH), 2925.32 (CH aliphatic), 2217.32 (CN), 1688.82 (C=O), and 1552.66, 1439.5 (C=C aromatic); ^{1}H NMR (δ, ppm, $CDCl_3$): 1.529 (*m*, 2H, H4′), 1.66 (*m*, 4H, H5, H7), 1.765 (*m*, 2H, H3′), 1.85 (*m*, 2H, H6), 1.899 (*m*, 2H, H5′), 2.496 (*t*, 2H, *J* = 7.2 Hz, H2′), 2.694 (*m*, 4H, H4, H8), 3.409 (*t*, 2H, *J* = 6.6 Hz, H6′), and 8.91 (*s*, 1H, NH); ^{13}C NMR (δ, ppm, $CDCl_3$): 24.322, 27.327, 27.665, 27.994, 29.044, 29.067, 31.996, 32.364, 33.406, 35.706 (C4-8, C2′-6′), 95.075 (C3), 115.151 (CN), 131.716 (C8a), 135.411 (C3a), 145.07 (C2), and 169.921 (C=O amide); and DART-TOF-MS (*m/z*): 369.07 $[M+H]^+$; anal. calcd. for $C_{16}H_{21}BrN_2OS$ (368.06): C, 52.03; H, 5.73; Br, 21.64; N, 7.59; O, 4.33; and S, 8.68; found: C, 52.08f H, 5.64; Br, 21.69; N, 7.51; O, 4.37; and S, 8.62.

*2-(5-Bromopentyl)-5,6,7,8-tetrahydro-4H-benzo[4,5]thieno[2,3-d][1,3]oxazin-4-imine (**5**)*. White solid; yield: 40%; mp: 174–175°C; IR (KBr, υ_{max}, cm^{-1}): 2936.82 (CH aliphatic), 1664.9 (C=N), and 1588.12 (C=C); ^{1}H NMR (δ, ppm, DMSO-d_6): 1.376 (*m*, 2H, H3′), 1.641–1.816 (*m*, 8H, H2′, H4′, H7-8), 2.553 (*t*, 2H, J = 6.6 Hz, H1′), 2.686 (*s*, 2H, H6), 2.821 (*s*, 2H, H6), 3.508 (*t*, 2H, J = 6.6 Hz, H5), and 12.177 (*s*, 1H, NH); ^{13}C NMR (δ, ppm, DMSO-d_6): 22.246, 22.966, 24.821, 25.726, 26.4, 27.343, 32.264, 34.058, 35.469 (C6-9, C1′-5′), 120.778 (C4a), 130.965 (C4a′), 131.395 (C9a), 157.787 (C10a), 159.013 (C4), and 163.559 (C2); and DART-TOF-MS (*m/z*): 355.05 [M + H]$^+$; anal. calcd. for $C_{15}H_{19}BrN_2OS$ (354.04): C, 50.71; H, 5.39; Br, 22.49; N, 7.88; O, 4.50; and S, 9.02; found: C, 50.76; H, 5.32; Br, 22.52; N, 7.85; O, 4.53; and S, 8.96.

*2.2.3. General Synthetic Procedure for 1-(3-Cyano-substituted [b]thiophen-2-yl)-3-(4-(trifluoromethyl)phenyl)thiourea (**6a–c**).* A mixture of the appropriate compound **2d–f** (1.3 mmol) with 1 equivalent of 4-(trifluoromethyl)phenyl isothiocyanate in absolute ethanol was stirred at room temperature for 2–7 h. The precipitate was filtered, washed with absolute ethanol, and dried.

*1-(3-Cyano-5,6-dihydro-4H-cyclopenta[b]thiophen-2-yl)-3-(4-(trifluoromethyl)phenyl)thiourea (**6a**).* Light brown solid; yield: 31%; mp: 172–174°C; IR (KBr, υ_{max}, cm^{-1}): 3327.76, 3234.11 (NH), 2212.49 (CN), and 1589.8, 1560.47 (C=C); ^{1}H NMR (δ, ppm, DMSO-d_6): 2.323 (*s*, 2H, H5), 2.72 (*s*, 2H, H6), 2.814 (*s*, 2H, H4), 7.701 (*s*, 2H, H2′, H6′), 7.818 (*s*, 2H, H3′, H5′), 10.71 (*s*, 1H, NH), and 10.95 (*s*, 1H, NH); ^{13}C NMR (δ, ppm, DMSO-d_6): 27.786, 28.039, 29.649 (C4-5), 115.036 (CN), 123.139 (C3′, C5′), 126.236 (C2′, C6′), and 130.613 (C3a); and DART-TOF-MS (*m/z*): 368.05 [M + H]$^+$; anal. calcd. for $C_{16}H_{12}F_3N_3S_2$ (367.04): C, 52.31; H, 3.29; F, 15.51; N, 11.44; and S, 17.45; found: C, 52.27; H, 3.35; F, 15.55; N, 11.38; and S, 17.40.

*1-(3-Cyano-4,5,6,7-tetrahydrobenzo[b]thiophen-2-yl)-3-(4-(trifluoromethyl)phenyl)thiourea (**6b**).* Light beige solid; yield: 49%; mp: 187–189°C; IR (KBr, υ_{max}, cm^{-1}): 3328.78, 3235.86 (NH), 2210.19 (CN), and 1590.92, 1560.13 (C=C); ^{1}H NMR (δ, ppm, DMSO-d_6): 1.74 (*s*, 4H, H5-6), 2.498 (*m*, 2H, H7), 2.585 (*s*, 2H, H4), 7.715 (*d*, 2H, J = 7.8 Hz, H2′, H6′), 7.838 (*d*, 2H, J = 7.8 Hz, H3′, H5′), 10.765 (*s*, 1H, NH), and 11.016 (*s*, 1H, NH); ^{13}C NMR (δ, ppm, DMSO-d_6): 22.146, 23.05, 23.809, 23.848 (C4-7), 96.041 (C3), 114.791 (CN), 123.07 (C3′, C5′), 123.775 (CF_3), 125.584 (C4′), 126.297 (C2′, C6′), 128.581 (C3a), 130.988 (C7a), 142.862 (C1′), 148.895 (C2), and 176.728 (C=S); and DART-TOF-MS (*m/z*): 382.07 [M + H]$^+$ anal. calcd. for $C_{17}H_{14}F_3N_3S_2$ (381.06): C, 53.53; H, 3.70; F, 14.94; N, 11.02; and S, 16.81; found: C, 53.57; H, 3.64; F, 15.0; N, 11.06; and S, 16.75.

*1-(3-Cyano-5,6,7,8-tetrahydro-4H-cyclohepta[b]thiophen-2-yl)-3-(4-(trifluoro-methyl)phenyl)thiourea (**6c**).* Dark beige solid; yield: 32%; mp: 169–171°C; IR (KBr, υ_{max}, cm^{-1}): 3326.54 (NH), 2211.97 (CN), and 1582.92, 1554.87 (C=C); ^{1}H NMR (δ, ppm, DMSO-d_6): 1.572 (*s*, 4H, H5, H7), 1.781 (*s*, 2H, H6), 2.633–2.674 (*m*, 4H, H4, H8), 7.705 (*d*, 2H, J = 8.4 Hz, H2′, H6′), 7.828 (*d*, 2H, J = 8.4 Hz, H3′, H5′), 10.777 (*s*, 1H, NH), and 10.937 (*s*, 1H, NH);^{13}C NMR (δ, ppm, DMSO-d_6): 27.334, 28.07, 28.783, 31.78 (C4-8), 98.241 (C3), 115.266 (CN), 122.955 (C3′, C5′), 123.77 (CF_3), 125.577 (C4′), 126.274 (C2′, C6′), 131.947(C3a), 135.634 (C8a), 142.855 (C1′), 146.787 (C2), and 176.507 (C=S); DART-TOF-MS (*m/z*): 396.08 [M + H]$^+$; anal. calcd. for $C_{18}H_{16}F_3N_3S_2$ (395.07): C, 54.67; H, 4.08; F, 14.41; N, 10.63; and S, 16.21; found: C, 54.77; H, 3.99; F, 14.46; N, 10.68; and S, 16.16.

*2.2.4. General Synthetic Procedure for Compounds (**7a–f**).* Few drops of sulfuric acid were added to a mixture of the appropriate lead compound **2d–f** (5.6 mmol) and 1 equivalent of 4-(trifluoromethyl)benzaldehyde or 4-(2-pyridyl) benzaldehyde in absolute ethanol. Precipitation occurred once sulfuric acid was added. The precipitate was filtered, washed with absolute ethanol, and dried.

*2-((4-(Trifluoromethyl)benzylidene)amino)-5,6-dihydro-4H-cyclopenta[b]thiophene-3-carbonitrile (**7a**).* Dark brown solid; yield: 41%; mp: 150–152°C; IR (KBr, υ_{max}, cm^{-1}): 2219.12 (CN), 1678.47 (C=N), and 1567.51, 1536.64 (C=C); ^{1}H NMR (δ, ppm, $CDCl_3$): 2.444 (*m*, 2H, H5), 2.873 (*t*, 2H, H6), 2.946 (*t*, 2H, H4), 7.704 (*d*, 2H, J = 7.8 Hz, H3′, H5′), 8.033 (*d*, 2H, J = 7.8 Hz, H2′, H6′), and 8.468 (*s*, 1H, N=CH); ^{13}C NMR (δ, ppm, $CDCl_3$): 27.379, 28.169, 30.239 (C4-6), 103.913 (C3), 114.492 (CN), 124.611 (CF_3), 125.814 (C3′, C5′), 129.44 (C2′, C6′), 133.441 (C4′), 138.087 (C3a), 139.007 (C1′), 145.108 (C6a), 155.649 (C2), and 163.744 (N=CH); and DART-TOF-MS (*m/z*): 321.07 [M + H]$^+$; anal. calcd. for $C_{16}H_{11}F_3N_2S$ (320.06): C, 59.99; H, 3.46; F, 17.79; N, 8.75; and S, 10.01; found: C, 60.07; H, 3.39; F, 17.84; N, 8.80; and S, 9.97.

*2-((4-(Trifluoromethyl)benzylidene)amino)-4,5,6,7-tetrahydrobenzo[b]thiophene-3-carbonitrile (**7b**).* Yellow solid; yield: 78%; mp: 146–147°C; IR (KBr, υ_{max}, cm^{-1}): 2216.15 (CN), 1659.53 (C=N), and 1563.09, 1514.72 (C=C); ^{1}H NMR (δ, ppm, $CDCl_3$): 1.861 (*m*, 4H, H5, H6), 2.664 (*s*, 2H, H7), 2.711 (*s*, 2H, H4), 7.696 (*d*, 2H, J = 7.8 Hz, H3′, H5′), 8.026 (*d*, 2H, J = 7.8 Hz, H2′, H6′), and 8.422 (*s*, 1H, N=CH); ^{13}C NMR (δ, ppm, $CDCl_3$): 21.914, 22.987, 24.282, 25.271 (C4-7), 108.39 (C3), 114.201 (CN), 124.618 (CF_3), 125.799 (C3′, C5′), 129.448 (C2′, C6′), 133.426 (C4′), 133.832 (C3a), 135.572 (C1′), 138.102 (C7a), 156.73 (C2), and 158.569 (N=CH); and DART-TOF-MS (*m/z*): 335.09 [M + H]$^+$; anal. calcd. for $C_{17}H_{13}F_3N_2S$ (334.08): C, 61.07; H, 3.92; F, 17.05; N, 8.38; and S, 9.59; found: C, 61.12; H, 3.86; F, 16.98; N, 8.42; and S, 9.55.

*2-((4-(Trifluoromethyl)benzylidene)amino)-5,6,7,8-tetrahydro-4H-cyclohepta[b]thiophene-3-carbonitrile (**7c**).* Yellow solid; yield: 99%; mp: 113–114°C; IR (KBr, υ_{max}, cm^{-1}): 2221.48 (CN), 1684.98 (C=N), and 1547.17, 1517.48 (C=C); ^{1}H NMR (δ, ppm, $CDCl_3$): 1.7 (*m*, 4H, H5, H7), 1.882 (*s*, 2H, H6), 2.804 (*t*, 4H, H4, H8), 7.7 (*d*, 2H, J = 7.8 Hz, H3′, H5′), 8.026 (*d*, 2H, J = 7.8 Hz, H2′, H6′), and 8.438 (*s*, 1H, N=CH); ^{13}C NMR (δ, ppm, $CDCl_3$): 27.073, 27.755, 29.142, 30.691, 31.956 (C4-8), 110.605 (C3), 114.729 (CN), 122.824 (CF_3), 125.776 (C3′, C5′), 129.356 (C2′, C6′), 133.150 (C4′), 137.558 (C3a), 138.186 (C1′), 140.624 (C8a), 156.538 (C2), and 156.576 (N=CH); and DART-TOF-MS (*m/z*): 349.1 [M + H]$^+$; anal. calcd. for $C_{18}H_{15}F_3N_2S$ (348.09): C,

62.06; H, 4.34; F, 16.36; N, 8.04; and S, 9.20; found: C, 62.12; H, 4.30; F, 16.45; N, 8.0; and S, 9.25.

2-((4-(Pyridin-2-yl)benzylidene)amino)-5,6-dihydro-4H-cyclopenta[b]thiophene-3-carbonitrile (***7d***). Yellow solid; yield: 29%; mp: 191–193°C; IR (KBr, υ_{max}, cm^{-1}): 2219.27 (CN), and 1581.62, 1548.56 (C=C); ^{1}H NMR (δ, ppm, $CDCl_3$): 2.433 (*m*, 2H, *J* = 7.2 Hz, H5), 2.869 (*t*, 2H, *J* = 7.2 Hz, H6), 2.934 (*t*, 2H, *J* = 7.2 Hz, H4), 7.273 (*t*, 1H, H4″), 7.794 (*m*, 2H, *J* = 7.2 Hz, H5″, H6″), 8.032 (*d*, 2H, *J* = 7.8 Hz, H2′, H6′), 8.106 (*d*, 2H, *J* = 7.8 Hz, H3′, H5′), 8.488 (*s*, 1H, N=CH), and 8.719 (*d*, 1H, H3″); ^{13}C NMR (δ, ppm, $CDCl_3$): 27.356, 28.215, 30.223 (C4-6), 102.779 (C3), 114.814 (CN), 120.946 (C6″), 122.794 (C4″), 127.294, 129.869 (C2′-3′, C5′-6′), 135.404 (C3a), 136.906 (C1′), 137.887 (C5″), 142.809 (C6a), 144.787 (C4′), 149.884 (C3″), 156.185 (C1″), 157.236 (C2), and 164.986 (N=CH); and DART-TOF-MS (*m/z*): 330.11 $[M+H]^+$; anal. calcd. for $C_{20}H_{15}N_3S$ (329.10): C, 72.92; H, 4.59; N, 12.76; and S, 9.73; found: C, 72.88; H, 4.55; N, 12.81; and S, 9.69.

2-((4-(Pyridin-2-yl)benzylidene)amino)-4,5,6,7-tetrahydro benzo[b]thiophene-3-carbonitrile (***7e***). Yellow solid; yield: 83%; mp: 191–193°C; IR (KBr, υ_{max}, cm^{-1}): 2214.28 (CN), and 1582.55, 1552.3 (C=C); ^{1}H NMR (δ, ppm, $CDCl_3$): 1.853 (*m*, 4H, H5-6), 2.68 (*d*, 4H, H4, H7), 7.272 (*m*, 1H, H4″), 7.788 (*s*, 2H, H5″, H6″), 8.019 (*d*, 2H, *J* = 7.8 Hz, H2′, H6′), 8.094 (*d*, 2H, *J* = 7.8 Hz, H3′, H5′), 8.433 (*s*, 1H, N=CH), and 8.719 (*s*, 1H, H3″); ^{13}C NMR (δ, ppm, $CDCl_3$): 21.984, 23.05, 24.299, 25.25 (C4-5), 107.256 (C3), 114.538 (CN), 120.969 (C6″), 122.817 (C4″), 127.278 (C3′, C5′), 129.869 (C2′, C6′), 132.797 (C3a), 135.235 (C1′), 135.434 (C5″), 136.96 (C7a), 142.739 (C4′), 149.853 (C3″), 156.147 (C1″), 158.27 (C2), and 159.749 (N=CH); and DART-TOF-MS (*m/z*): 344.12 $[M+H]^+$; anal. calcd. for $C_{21}H_{17}N_3S$ (343.11): C, 73.44; H, 4.99; N, 12.24; and S, 9.33; found: C, 73.48; H, 4.94; N, 12.31; and S, 9.27.

2-((4-(Pyridin-2-yl)benzylidene)amino)-5,6,7,8-tetrahydro-4H-cyclohepta [b]thiophene-3-carbonitrile (***7f***). Yellow solid; yield: 86%; mp: 199–200°C; IR (KBr, υ_{max}, cm^{-1}): 2218.75 (CN), and 1582.43, 1546.89 (C=C); ^{1}H NMR (δ, ppm, $CDCl_3$): 1.703 (*m*, 4H, H5, H7), 1.88 (*s*, 2H, H6), 2.801 (*m*, 4H, H4, H8), 7.275 (*s*, 1H, H4″), 7.795 (*m*, 2H, H5″, H6″), 8.026 (*d*, 2H, *J* = 7.8 Hz, H2′, H6′), 8.102 (*d*, 2H, *J* = 7.8 Hz, H3′, H5′), 8.463 (*s*, 1H, N=CH), and 8.722 (*s*, 1H, H3″); ^{13}C NMR (δ, ppm, $CDCl_3$): 27.126, 27.809, 29.15, 30.653, 31.994 (C4-8), 109.532 (C3), 115.036 (CN), 120.946 (C6″), 122.771 (C4″), 127.286 (C3′, C5′), 129.777 (C2′, C6′), 135.519 (C3a), 136.462 (C1′), 136.922 (C5″), 140.279 (C8a), 142.701 (C4′), 149.861 (C3″), 156.224 (C1″), 157.726 (C2), and 158.155 (N=CH); and DART-TOF-MS (*m/z*): 358.14 $[M+H]^+$; anal. calcd. for $C_{22}H_{19}N_3S$ (357.13): C, 73.92; H, 5.36; N, 11.75; and S, 8.97; found: C, 73.97; H, 5.29; N, 11.81; and S, 9.02.

2.3. X-Ray Crystallographic Analysis. Compounds **4c** and **7e** were obtained as single crystals by slow evaporation of the ethanol solution of the pure compounds at room temperature. Crystallographic data were collected using a Bruker APEX-II D8 Venture area diffractometer, equipped with graphite monochromatic Mo Kα radiation, $\lambda = 0.71073$, and Cu Kα radiation, $\lambda = 1.54178$ Å at 293 (2) K. Cell refinement and data reduction were carried out using Bruker SAINT. SheLXT was used to determine the structure [24, 25]. The final refinement was carried out by the full-matrix least-squares technique with anisotropic thermal data for -non-hydrogen atoms on F-CCDC 1823351 and 1534088 that contain the supplementary crystallographic data for these compounds obtained free of charge from the Cambridge Crystallographic Data Centre (http://www.ccdc.cam.ac.uk/data_request/cif).

2.4. Biology

2.4.1. Evaluation of Antimicrobial Activity. All selected organisms (0.5 McFarland standards) were thoroughly mixed with sterilized Mueller–Hinton agar (MHA). This suspension (25 mL) was placed into Petri dishes (90 mm diameter) and left to cool and solidify by placing the Petri dishes on a cool horizontal surface. A 10 mm diameter well was holed on both sides of the agar plate by using a sterilized hollow cylinder as a template. The formulations and control antibiotics (1 mg/mL) were placed into each well (50 μL) to permit diffusion. All plates were incubated at 37 ± 0.5°C for 24 h in aerobic conditions; the test was performed in triplicate. The antimicrobial activities of the selected formulations and control antibiotics against the tested microorganisms were compared. The diameter of the inhibition zone was measured with a gauge and expressed in mm (mean ± standard deviation (SD)).

2.4.2. Evaluation of Anti-Inflammatory Activity. Fresh whole human blood was collected and mixed with equal volumes of sterilized Alsever's solution (2% dextrose, 0.8% sodium citrate, 0.05% citric acid, 0.42% sodium chloride, and 100 mL of distilled water). This blood solution was centrifuged at 3,000 rpm for 10 min and then washed three times with an equal volume of normal saline. The volume of blood was measured, and it was reconstituted with normal saline to prepare 10% *v/v* suspension. The reaction mixture consisted of 1 mL of the test sample in normal saline at different concentrations: 0.5 mL of 10% human red blood cell (HRBC) suspension, 1mL of 0.2 M phosphate buffer, and 1 mL of hypotonic saline. They were incubated at 37°C for 30 min and centrifuged at 3,000 rpm for 30 min. Hemoglobin content in the supernatant was determined spectrophotometrically at 560 nm. Each experiment was performed in triplicate. Diclofenac sodium was used as a standard, and distilled water was used as a control. The blood control represented 100% lysis or zero percent stability.

3. Results and Discussion

3.1. Chemistry. Using the Gewald method for thiophene synthesis, we synthesized the lead compounds **(2a–f)** (Scheme 1) [26, 27].

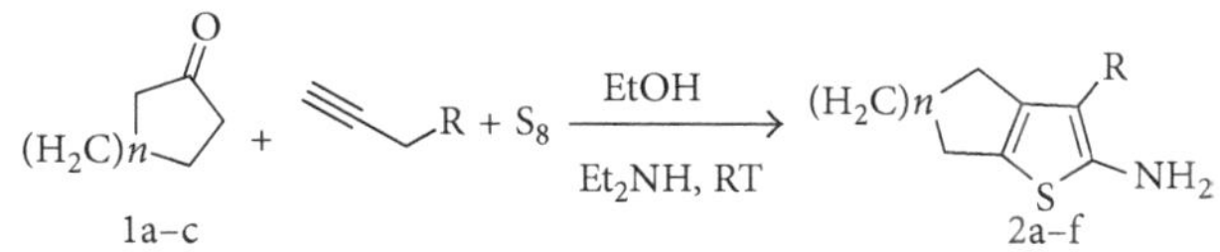

1a–c

a $n = 1$
b $n = 2$
c $n = 3$

2a–f

a $n = 1$, R = –COOEt
b $n = 2$, R = –COOEt
c $n = 3$, R = –COOEt
d $n = 1$, R = –CN
e $n = 2$, R = –CN
f $n = 3$, R = –CN

Scheme 1: Gewald thiophene synthesis.

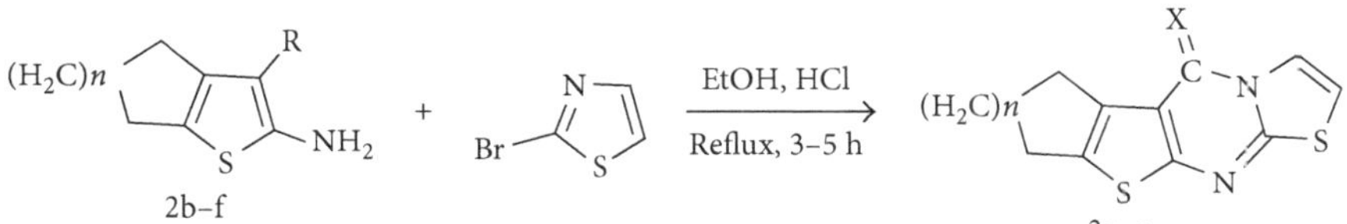

2b–f

b $n = 2$, R = –COOEt
c $n = 3$, R = –COOEt
d $n = 1$, R = CN
e $n = 2$, R = CN
f $n = 3$, R = CN

3a–e

a $n = 2$, X = O
b $n = 3$, X = O
c $n = 1$, X = NH
d $n = 2$, X = NH
e $n = 3$, X = NH

Scheme 2: Synthesis of thieno[2,3-(d)]thiazolo[3,2-(a)]pyrimidine derivatives (**3a–e**).

A mixture of the appropriate compound (**2b–f**), 2-bromothiazole, and few drops of hydrochloric acid in ethanol was refluxed for 3–5 h to produce thieno[2,3-*d*]thiazolo[3,2-*a*]pyrimidine derivatives (Scheme 2).

This reaction involves nucleophilic substitution in bromothiazole followed by nucleophilic substitution on the ester group to form the pyrimidinone ring (compounds **3a–b**). However, for compounds **3c–e**, nucleophilic substitution was followed by nucleophilic addition of a cyanide group to the thiazole nitrogen to form the pyrimidinimine ring.

IR, ^{1}H NMR, ^{13}C NMR, and mass spectral data of the polycyclic compound **3a** and new compounds **3b–e** were consistent with the assigned structures. The IR spectrum of compound **3b** showed an absorption band at 1674.41 cm^{-1} for the carbonyl group, whereas there was no amino group absorption band. The ^{1}H NMR spectrum showed the presence of the cycloheptenyl protons and appearance of new signals: two doublet signals at $\delta = 6.855$ and 7.984 ppm (each integrates for one proton corresponding to the two protons of the thiazole ring) with coupling constant (J) = 4.8 Hz, whereas there was no signal for the ethyl group linked to the ester group. In ^{13}C NMR, the three carbons of the thiazole group appeared at $\delta = 109.494$, 121.667, and 156.982 ppm, carbonyl peak appeared at $\delta = 162.317$ ppm, and no peaks for the ethyl group. Finally, the direct analysis in real-time/time-of-flight mass spectrometry (DART-TOF-MS) spectrum showed a molecular ion peak $[M+H]^+$ at $m/z = 277.05$.

The lead compounds (**2b–f**) were refluxed in bromoalkanoyl chloride derivatives (Scheme 3) to produce compounds (**4a–e**) by nucleophilic substitution reaction.

However, the nucleophilic substitution reaction between compound **2e** and 6-bromohexanoyl chloride was followed by nucleophilic addition of a cyanide group to the carbonyl oxygen to form the oxazinimine ring of the novel compound **5** (Scheme 4).

The IR data of compound **4d** showed the appearance of an NH absorption band at 3360.95 cm^{-1} and another absorption band with two heads at 1675.80 cm^{-1}, which represents the two carbonyl groups. In ^{1}H NMR, five new signals appeared: multiplet at $\delta = 1.508$, 1.756, and 1.888 ppm and triplet at $\delta = 2.456$ and 3.396 ppm (each signal integrates for two protons corresponding to the protons of the side chain). In addition, one proton singlet signal for NH appeared at $\delta = 11.2$ ppm. The ^{13}C NMR spectrum showed the appearance of the five carbons of the side chain in the aliphatic range; besides, it showed signals for the cycloheptenyl carbons in the range of $\delta = 24.442$–36.564 ppm and a peak for the amide group at $\delta = 166.748$ ppm. Finally, the DART-TOF-MS spectrum showed a molecular ion peak $[M+H]^+$ at $m/z = 416.09$.

For compound **5**, the disappearance of the cyanide absorption band in the IR spectrum, appearance of one proton signal for NH at $\delta = 12.177$ ppm in the ^{1}H NMR spectrum, and appearance of carbon peaks at $\delta = 159.013$ and 163.559 ppm for C4 and C2, respectively, in the ^{13}C NMR spectrum proved the formation of the oxazinimine ring. The six protons of the side chain appeared in the ^{1}H NMR spectrum as multiplet signals at $\delta = 1.376$ and 1.641–1.816 ppm for H2–H4′ and two triplet signals at $\delta = 2.533$ and 3.508 ppm for H1′ and H5′, respectively (each signal integrates for two protons). Moreover, the ^{13}C NMR spectrum showed the five carbons of the side

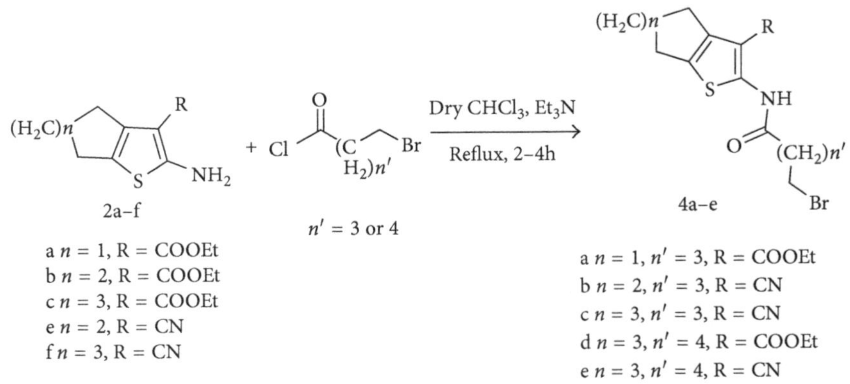

Scheme 3: Synthesis of compounds **4a–e**.

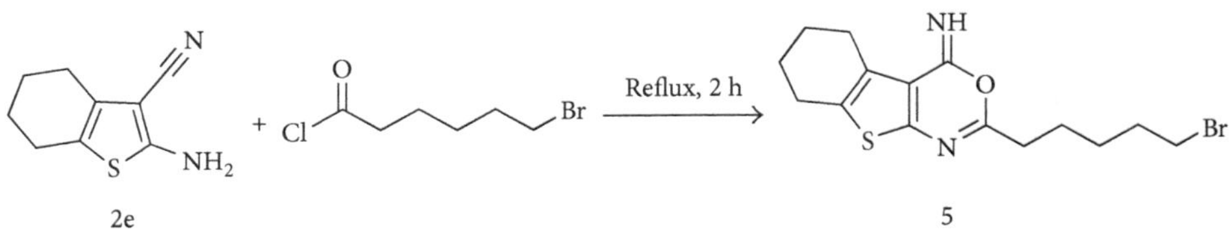

Scheme 4: Synthesis of thieno[2,3-(d)][1,3]oxazin-4-imine derivative (**5**).

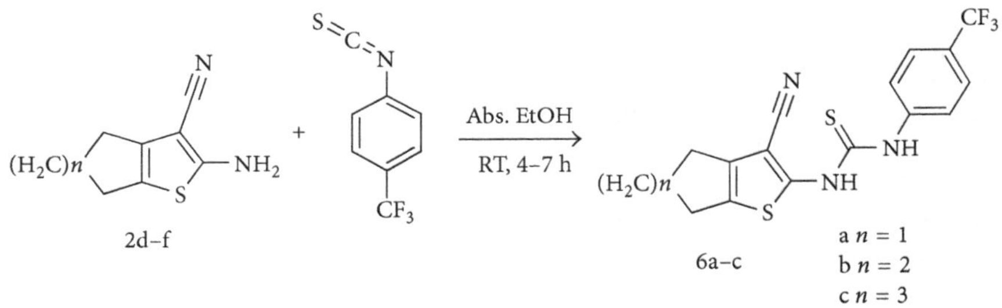

Scheme 5: Synthesis of thiophene-thiourea derivatives (**6a–c**).

chain in the aliphatic range. Finally, the molecular weight was confirmed by the appearance of a molecular ion peak $[M+H]^+$ at $m/z = 355.05$ in the DART-TOF-MS spectrum.

A mixture of the appropriate lead compound (**2d–f**) and 4-(trifluoromethyl phenyl) isothiocyanate in absolute ethanol was stirred at 25°C for 2–7 h. The new thiophene-thiourea derivatives (**6a–c**) were obtained by nucleophilic addition (Scheme 5).

Structures of compounds (**6a–c**) were confirmed by the presence of a cyanide group and appearance of (4-(trifluoromethyl)phenyl)thiourea bands and peaks in the IR, ^{1}H NMR, and ^{13}C NMR spectra. Herein, for compound **6b**, the IR spectrum showed two absorption bands for the two NH groups at 3328.78 and 3235.86 cm^{-1} and an absorption band for cyanide at 2210.19 cm^{-1}. The ^{1}H NMR spectrum showed a doublet signal (2 H) with $J = 7.8$ Hz at $\delta = 7.715$ ppm for H2′ and H6′, doublet signal (2 H) with $J = 7.8$ Hz at $\delta = 7.838$ ppm for H3′ and H5′, and two singlet signals for NH at $\delta = 10.765$ and 11.016 ppm (each integrates for one proton). The ^{13}C NMR spectrum showed a peak for cyanide at $\delta = 114.791$ ppm; additionally, the phenyl ring carbons appeared as follows: C3′ and C5′ at $\delta = 123.07$ ppm, C2′ and C6′ at $\delta = 126.297$ ppm, C4′ at $\delta = 130.988$ ppm, and C1′ at $\delta = 142.862$ ppm. CF_3 appeared at $\delta = 123.77$ ppm, whereas the thionyl group appeared at $\delta = 176.728$ ppm. Finally, the molecular weight was confirmed by the appearance of a molecular ion peak $[M+H]^+$ at $m/z = 382.07$ in the DART-TOF-MS spectrum.

Addition of few drops of sulfuric acid to a mixture of the appropriate lead compound (**2d–f**) and 4-(trifluoromethyl) benzaldehyde or 4-(2-pyridyl) benzaldehyde in absolute ethanol resulted in instant precipitation of the product (Scheme 6).

All structures of the novel synthesized compounds (**7a–f**) were confirmed by the existence of a cyanide group, disappearance of NH_2, and appearance of 4-(trifluoromethyl) benzylidene or 4-(pyridin-2-yl)benzylidene bands and peaks in the IR, ^{1}H NMR, and ^{13}C NMR spectra. For instance, compound **7d** showed a cyanide absorption band at

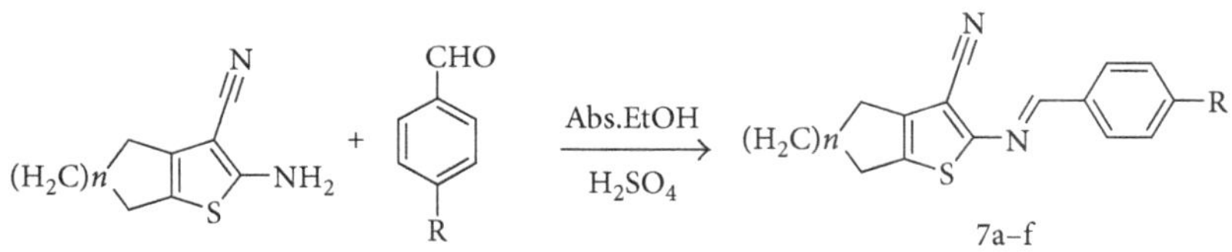

2d–f

R = $-CF_3$ or 2-pyridyl

7a–f

a n = 1, R = –CF3
b n = 2, R = –CF3
c n = 3, R = –CF3
d n = 1, R = 2-pyridyl
e n = 2, R = 2-pyridyl
f n = 3, R = 2-pyridyl

SCHEME 6: Synthesis of thiophene-3-carbonitrile derivatives (**7a–f**).

TABLE 1: Experimental detail of the crystalline structures of **4c** and **7e**.

Crystal data	**4c**	**7e**
Chemical formula	$C_{15}H_{19}BrN_2OS$	$C_{21}H_{17}N_3S$
Molecular weight	355.29	343.44
Crystal system, space group	Triclinic, $P-1$	Triclinic, $P-1$
Temperature (K)	293	293
a, b, c (Å)	5.7305 (7), 9.0388 (13), 15.825 (2)	6.6424 (6), 11.5386 (10), 11.5868 (10)
α, β, γ (°)	97.214 (5), 97.557 (5), 90.726 (5)	98.088 (3), 93.063 (3), 98.051 (3)
V (Å^3)	805.76 (18)	868.13 (13)
Z	2	2
Radiation type	Mo $K\alpha$	Mo $K\alpha$
μ (mm^{-1})	2.68	0.19
Crystal size (mm)	$0.55\times0.42\times0.04$	$0.4\times0.29\times0.1$
Data collection		
Diffractometer	Bruker APEX-II D8 venture diffractometer	Bruker APEX-II D8 venture diffractometer
Absorption correction	Multiscan SADABS Bruker 2014	Multiscan SADABS Bruker 2014
T_{min}, T_{max}	0.893, 0.921	0.927, 0.982
Number of measured, independent, and observed $[I>2\sigma(I)]$ reflections	16032, 3700, and 1871	33436, 3412, and 2377
R_{int}	0.152	0.103
Refinement		
$R[F^2>2\sigma(F^2)]$, wR(F^2), S	0.114, 0.341, 1.66	0.066, 0.201, 1.08
Number of reflections	3700	3412
Number of parameters	176	226
Number of restraints	0	1
$\Delta\rho_{max}, \Delta\rho_{min}$ (e Å^{-3})	1.37, −1.74	0.34, −0.33
CCDC number	1823351	1534088

2219.27 cm^{-1} and disappearance of the NH_2 bands in the IR spectrum. The ^{1}H NMR spectrum showed signals of the pyridyl ring protons as follows: a triplet signal (1 H) at $\delta=7.273$ ppm for H4″, multiplet signal (2 H) at $\delta=7.794$ ppm for H5″ and H6″, and doublet signal at $\delta=8.719$ ppm for H3″. In addition, the four protons of the benzene ring appeared as two doublet signals (each integrates for two protons) with $J=7.2$ Hz at $\delta=8.032$ and 8.106 ppm. Moreover, an N=CH singlet signal (1 H) appeared at $\delta=8.488$ ppm. ^{13}C NMR showed a cyanide peak ($\delta=$ 114.814 ppm), pyridyl ring peaks (C6″ at $\delta=120.946$ ppm, C4″ peak at $\delta=122.794$ ppm, C5″ peak at $\delta=137.887$ ppm, C3″ peak at $\delta=149.884$ ppm, and C1″ peak at $\delta=$ 156.185 ppm), benzene ring peaks (C2′ and C3′ at $\delta=127.294$, C5′ and C6′ at 129.869 ppm, C1′ at $\delta=136.906$ ppm, and C4′ at $\delta=144.787$ ppm), and N=CH peak at $\delta=164.986$ ppm. DART-TOF-MS confirmed the molecular weight of the expected structure, as evidenced by the appearance of a molecular ion peak $[M+H]^+$ at $m/z=330.11$.

3.2. X-Ray Crystallography. The structures of three of the synthesized compounds (**4c**, **7e**) were examined by X-ray crystallography. The crystallographic data and refinement information are summarized in Table 1. As shown in Figure 2, the asymmetric units contained one independent molecule.

3.3. Biology

3.3.1. Evaluation of Antimicrobial Activity. An *in vitro* antimicrobial study was performed using the agar diffusion method to evaluate the ability of the synthesized compounds to inhibit microbial growth, as previously described by

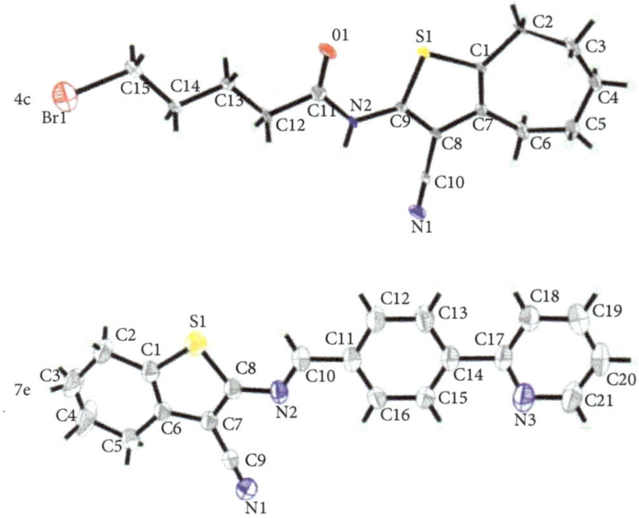

FIGURE 2: ORTEP diagrams of the titled compounds **4c** and **7e**. Displacement ellipsoids are plotted at the 40% probability level for non-H atoms.

TABLE 2: The zones of inhibition (mm) of the synthesized compounds and standard antimicrobial drugs.

Compound number	Zone of inhibition (mm)						
	Staphylococcus aureus	*Bacillus*	*Escherichia coli*	*Klebsiella*	*Salmonella*	*Pseudomonas*	*Candida*
3a	11 ± 0.186	—	—	—	—	—	—
3b	—	—	—	—	—	—	—
3c	—	—	—	—	—	—	—
3e	—	—	—	—	—	—	—
5b	—	—	—	—	—	—	—
5	16 ± 0.136	13 ± 0.501	—	12 ± 0.42	12 ± 0.075	—	—
4b	—	—	—	—	—	14 ± 0.056	—
4c	—	—	—	—	—	10 ± 0.138	—
6a	15 ± 0.322	15 ± 0.894	—	—	—	—	—
6b	10 ± 0.27	—	—	—	—	—	—
6c	13 ± 0.273	12 ± 0.071	—	—	—	—	—
Ampicillin	30 ± 0.057	15 ± 0.069	25 ± 0.07	13 ± 0.062	34 ± 0.059	18 ± 0.054	—
Fluconazole	—	—	—	—	—	—	38 ± 0.037

Bonev et al. [28]. The antimicrobial activities of the synthesized compounds (**3a–c**, **3e**, **4e**, **4b-c**, **5**, and **6a–c**) were tested against some selected microorganisms, including *Staphylococcus aureus, Bacillus* species, *Escherichia coli, Klebsiella* species, *Salmonella* species, *Pseudomonas* species, and *Candida albicans* (Table 2).

As shown in Table 2, results of the antimicrobial activity studies revealed that the test compounds displayed variable inhibitory effects against the growth of the tested bacteria. Interestingly, compound **6a** was equally potent to ampicillin against *Bacillus* species. The antimicrobial potencies of other derivatives, particularly compound **5** (against *Klebsiella* and *Bacillus*), were potentially comparable to those of ampicillin. The antifungal activity study revealed that the tested compounds exhibited no activity against *Candida albicans.*

3.3.2. Evaluation of Anti-Inflammatory Activity. Compounds **3a–c**, **3e**, **4a–e**, **5**, **6a–c**, and **7a–f** were screened *in vitro* for anti-inflammatory activity, using a method previously described by Mahajan et al. [29]. The anti-inflammatory activity of the thiazole and thiophene derivatives (Table 3) showed that compounds **3e**, **4b**, and **6a** possessed more potent anti-inflammatory activity than that of diclofenac sodium. Compounds **4d** and **7b** showed good anti-inflammatory activity, whereas compounds **5**, **7d**, and **7f** showed moderate activity.

4. Conclusions

In this study, we successfully synthesized some thiazole, pyrimidine, and thiophene derivatives. Among them, fifteen new target compounds were prepared. In addition, seven compounds (**4c**, **7e**) were successfully obtained as pure crystals. Antimicrobial activity studies revealed that the test compounds displayed broad antibacterial spectrum and good potency. Interestingly, compound **6a** was equally potent to ampicillin against *Bacillus* strains. According to the

TABLE 3: *In vitro* anti-inflammatory activity of the synthesized compounds and standard drug (diclofenac sodium).

Compound number	Spectrophotometer + RBC	Spectrophotometer + RBC (% hemolysis)
3a	0.011 ± 0.0068	101.13 ± 0.0031
3b	0.013 ± 0.000516	102.1542 ± 0.021
3c	0.009 ± 0.000473	102.6048 ± 0.00402
3e	0.179 ± 0.01775	101.6216 ± 0.0005
4d	0.011 ± 0.0068	101.13 ± 0.0031
4e	0.098 ± 0.001227	102.686 ± 0.00211
5	0.101 ± 0.01862	101.9084 ± 0.0002
4a	0.017 ± 0.0075	101.908 ± 0.002404
4d	0.219 ± 0.0225	102.727 ± 0.0126
4c	0.032 ± 0.002	102.441 ± 0.01308
6a	0.146 ± 0.001366	102.523 ± 0.0187
6b	0.01 ± 0.00493	102.154 ± 0.00312
6c	0.046 ± 0.001789	102.236 ± 0.0107
7a	0.023 ± 0.000894	102.0723 ± 0.02135
7b	0.113 ± 0.001033	102.318 ± 0.0042
7c	0.029 ± 0.003225	102.113 ± 0.028
7d	0.066 ± 0.00324	102.2 ± 0.00301
7e	0.052 ± 0.0041	102.154 ± 0.0014
7f	0.06 ± 0.00319	101.908 ± 0.01405
Diclofenac sodium	0.132 ± 0.002251	102.851 ± 0.00076
Control	0.261 ± 0.002683	—
Blank	0.428 ± 0.002442	—

anti-inflammatory activity test, compounds **3e**, **4b**, and **6a** possessed greater anti-inflammatory potency than that of the standard drug. Therefore, compound **6a** was shown to exhibit both antibacterial and anti-inflammatory activities, which could be beneficial in the treatment of various diseases.

Conflicts of Interest

The authors declare that they have no conflicts of interest.

Acknowledgments

The authors thank the Deanship of Scientific Research and RSSU at King Saud University for their technical support.

References

[1] L. J. Lombardo, F. Y. Lee, P. Chen et al., "Discovery of *N*-(2-Chloro-6-methyl- phenyl)-2-(6-(4-(2-hydroxyethyl)- piperazin-1-yl)-2-methylpyrimidin-4- ylamino)thiazole-5-carboxamide (BMS-354825), a dual Src/Abl kinase inhibitor with potent antitumor activity in preclinical assays," *Journal of Medicinal Chemistry*, vol. 47, no. 27, pp. 6658–6661, 2004.

[2] J. Das, P. Chen, D. Norris et al., "2-Aminothiazole as a novel kinase inhibitor template. Structure–activity relationship studies toward the discovery of *N*-(2-Chloro-6-methylphenyl)-2-[[6-[4-(2-hydroxyethyl)-1-piperazinyl)]-2-methyl-4-pyrimidinyl]amino)]-1,3-thiazole-5-carboxamide (dasatinib, BMS-354825) as a potent pan-Src kinase inhibitor," *Journal of Medicinal Chemistry*, vol. 49, no. 23, pp. 6819–6832, 2006.

[3] B. L. Mylari, E. R. Larson, T. A. Beyer et al., "Novel, potent aldose reductase inhibitors: 3,4-dihydro-4-oxo-3-[[5-(trifluoromethyl)-2-benzothiazolyl]methyl]-1-phthalazineacetic acid (zopolrestat) and congeners," *Journal of Medicinal Chemistry*, vol. 34, no. 1, pp. 108–122, 1991.

[4] M. C. Van Zandt, M. L. Jones, D. E. Gunn et al., "Discovery of 3-[(4,5,7-trifluorobenzothiazol-2-yl)methyl]indole-*N*-acetic acid (lidorestat) and congeners as highly potent and selective inhibitors of aldose reductase for treatment of chronic diabetic complications," *Journal of Medicinal Chemistry*, vol. 48, no. 9, pp. 3141–3152, 2005.

[5] B. Ettinger, D. M. Black, B. H. Mitlak et al., "Reduction of vertebral fracture risk in postmenopausal women with osteoporosis treated with raloxifene: results from a 3-year randomized clinical trial. multiple outcomes of raloxifene evaluation (MORE) investigators," *JAMA*, vol. 282, no. 7, pp. 637–645, 1999.

[6] S. R. Cummings, S. Eckert, K. A. Krueger et al., "The effect of raloxifene on risk of breast cancer in postmenopausal women: results from the MORE randomized trial. Multiple outcomes of raloxifene evaluation," *JAMA*, vol. 281, no. 23, pp. 2189–2197, 1999.

[7] V. G. Vogel, J. P. Costantino, D. L. Wickerham et al., "Effects of tamoxifen vs raloxifene on the risk of developing invasive breast cancer and other disease outcomes: the NSABP study of tamoxifen and raloxifene (STAR) P-2 trial," *JAMA*, vol. 295, no. 23, pp. 2727–2741, 2006.

[8] V. G. Vogel, J. P. Costantino, D. L. Wickerham et al., "Update of the National Surgical Adjuvant Breast and Bowel Project Study of Tamoxifen and Raloxifene (STAR) P-2 Trial: preventing breast cancer," *Cancer Prevention Research*, vol. 3, no. 6, pp. 696–706, 2010.

[9] J. M. Herbert, D. Frehel, E. Vallee et al., "A novel antithrombotic agent," *Cardiovascular Drug Reviews*, vol. 11, no. 2, pp. 180–198, 1993.

[10] J. M. Herbert, A. Tissinier, G. Defreyn, and J. P. Maffrand, "Inhibitory effect of clopidogrel on platelet adhesion and intimal proliferation after arterial injury in rabbits," *Arteriosclerosis, Thrombosis, and Vascular Biology*, vol. 13, no. 8, pp. 1171–1179, 1993.

[11] M. Gent, "A randomised, blinded, trial of clopidogrel versus aspirin in patients at risk of ischaemic events (CAPRIE). CAPRIE Steering Committee," *The Lancet*, vol. 348, no. 9038, pp. 1329–1339, 1996.

[12] K. Starčević, G. Karminski-Zamola, I. Piantanida, M. Zinić, L. Sǔman, and M. Kralj, "Photo-induced switch of a DNA/RNA inactive molecule into a classical intercalator," *Journal of the American Chemical Society*, vol. 127, no. 4, pp. 1074-1075, 2005.

[13] R. Romagnoli, P. G. Baraldi, M. D. Carrion et al., "Design, synthesis and biological evaluation of thiophene analogues of chalcones," *Bioorganic and Medicinal Chemistry*, vol. 16, no. 10, pp. 5367–5376, 2008.

[14] I. C. F. R. Ferreira, M. R. P. Queiroz, M. Vilas-Boas, L. M. Estevinho, A. Begouin, and G. Kirsch, "Evaluation of the antioxidant properties of diarylamines in benzo[*b*]thiophene series by free radical scavenging activity and reducing power,"

Bioorganic and Medicinal Chemistry Letters, vol. 16, no. 5, pp. 1384–1387, 2006.

[15] I. L. Pinto, R. L. Jarvest, and H. T. Serafinowska, "The synthesis of 5-alkoxy and 5-amino substituted thiophenes," *Tetrahedron Letters*, vol. 41, pp. 1597–1600, 2000.

[16] P. G. Baraldi, M. G. Pavani, J. C. Shryock et al., "Synthesis of 2-amino-3-heteroaroylthiophenes and evaluation of their activity as potential allosteric enhancers at the human A1 receptor," *European Journal of Medicinal Chemistry*, vol. 39, no. 10, pp. 855–865, 2004.

[17] A. D. Pillai, P. D. Rathod, F. P. Xavier, K. K. Vasu, H. Padh, and V. Sudarsanam, "Design, synthesis, and pharmacological evaluation of some 2-[4-morpholino]-3-aryl-5-substituted thiophenes as novel anti-inflammatory agents: generation of a novel anti-inflammatory pharmacophore," *Bioorganic and Medicinal Chemistry*, vol. 12, no. 17, pp. 4667–4671, 2004.

[18] F. M. Moghaddam and H. Zali-Boinee, "A versatile one-pot synthesis of 2,3,5-tri-substituted thiophenes from thiomorpholides," *Tetrahedron Letters*, vol. 44, no. 33, pp. 6253–6255, 2003.

[19] V. H. Rawal, R. J. Jones, and M. P. Cava, "Photocyclization strategy for the synthesis of antitumor agent CC-1065: synthesis of dideoxy PDE-I and PDE-II. Synthesis of thiophene and furan analogs of dideoxy PDE-I and PDE-II," *Journal of Organic Chemistry*, vol. 52, no. 1, pp. 19–28, 1987.

[20] K. Starčević, M. Kralj, I. Piantanida, L. Šuman, K. Pavelić, and G. Karminski-Zamola, "Synthesis, photochemical synthesis, DNA binding and antitumor evaluation of novel cyano- and amidinosubstituted derivatives of naphtho-furans, naphthothiophenes, thieno-benzofurans, benzodithiophenes and their acyclic precursors," *European Journal of Medicinal Chemistry*, vol. 41, no. 8, pp. 925–939, 2006.

[21] H. Z. Shams, Y. M. Elkholy, R. A. Azzam, and R. M. Mohareb, "Synthetic potentialities of thiophene systems in heterocyclic synthesis: a novel synthesis of thieno[2,3-*b*]pyridine derivatives," *Phosphorus, Sulfur, and Silicon and the Related Elements*, vol. 155, no. 1, pp. 215–233, 1999.

[22] H. Z. Shams, R. M. Mohareb, M. H. Helal, and A. E. Mahmoud, "Synthesis, structure elucidation, and biological evaluation of some fused and/or pendant thiophene, pyrazole, imidazole, thiazole, triazole, triazine, and coumarin systems based on cyanoacetic 2-[(benzoylamino)thioxomethyl]-hydrazide," *Phosphorus, Sulfur, and Silicon and the Related Elements*, vol. 182, no. 2, pp. 237–263, 2007.

[23] S. M. Sherif, R. M. Mohareb, H. Z. Shams, and H. M. Gaber, "A convenient synthesis of polyfunctionally substituted benzo[*b*]thiophen-2-yl-pyrimidine, -pyrazole, -isoxazole and pyridazine derivatives," *Journal of Chemical Research*, no. 11, pp. 434-435, 1995.

[24] G. M. Sheldrick, "A short history of SHELX," *Acta Crystallographica Section A Foundations of Crystallography*, vol. 64, no. 1, pp. 112–122, 2008.

[25] G. M. Sheldrick, *SHELXTL-PC (Version 5.1)*, Siemens Analytical Instruments, Inc., Madison, WI, USA, 1997.

[26] K. Gewald, "Zur reaktion von α-oxo-mercaptanen mit nitrilen," *Angewandte Chemie*, vol. 73, no. 3, pp. 114–114, 1961.

[27] K. Gewald, E. Schinke, and H. Böttcher, "Heterocycles from CH-acidic nitriles VII 2-aminothiophene, carbonyl compounds, and sulfur," *Chemische Berichte*, vol. 99, no. 1, pp. 94–100, 1966.

[28] B. Bonev, J. Hooper, and J. Parisot, "Principles of assessing bacterial susceptibility to antibiotics using the agar diffusion method," *Journal of Antimicrobial Chemotherapy*, vol. 61, no. 6, pp. 1295–1301, 2008.

[29] P. Mahajan, M. Nikam, A. Chate et al., "Synthesis, and antitubercular activity of new benzo[*b*]thiophenes," *Phosphorus, Sulfur, and Silicon and the Related Elements*, vol. 190, no. 11, pp. 1803–1813, 2015.

23

Rapid Separation and Determination of Metronidazole Benzoate and Other Antiprotozoal Drugs by Pressurized Capillary Electrochromatography

Sulan Liao[1] **and Qilin Weng**[2]

[1]*College of Tea and Food Science, Wuyi University, Wuyishan 354300, China*
[2]*Information Technology & Laboratory Management Center, Wuyi University, Wuyishan 354300, China*

Correspondence should be addressed to Sulan Liao; slliao@wuyiu.edu.cn

Academic Editor: Manuela Curcio

A novel method for the rapid separation and determination of five polar 5-nitroimidazoles in water from different sources by pressurized capillary electrochromatography has been developed. Compared with the gradient elution mode and the packed nonpolar columns which were usually utilized for the separation of 5-nitroimidazoles, a simple isocratic elution mode and low-cost homemade polar molecularly imprinted polymer monolith were used in the experiment. Electrochromatographic conditions such as pH of buffer, organic modifier, concentration of buffer, and separation voltage were optimized. At 320 nm UV wavelengths, the five 5-nitroimidazoles could be baseline-separated rapidly in less than 11 min with the separation voltage of +20 kV in 10 mmol/L sodium dihydrogen phosphate-disodium hydrogen phosphate buffer solution (pH 4.82) containing 30% acetonitrile. Under the optimum conditions, the linear ranges of the metronidazole, secnidazole, tinidazole, ornidazole, and metronidazole benzoate were 0.50–100.00, 0.50–100.00, 0.80–500.00, 0.80–100.00, and 5.00–500.00 μg/mL, respectively, and the detection limits of these analytes were 0.11–0.73 μg/mL. Column efficiencies of 43 000, 36 000, 34 000, 14 000, and 29 000 plates/m were obtained for metronidazole, secnidazole, tinidazole, ornidazole, and metronidazole benzoate, respectively. The recoveries of different water samples were about 85.0–95.8%. Additionally, the proposed method has been successfully applied to the rapid separation of 5-nitroimidazoles in the locally available pure milk sample by simple pretreatment.

1. Introduction

Structurally related antiprotozoal drugs, namely, nitroimidazole derivatives including metronidazole (MNZ), secnidazole (SNZ), tinidazole (TNZ), ornidazole (ONZ), and metronidazole benzoate (MB), etc., are a class of veterinary drugs and are used widely as antibacterial and antiprotozoan drugs in poultry and swine dysentery [1]. In some countries, they are also incorporated into feed products as feed additives to prevent diseases and promote growth in livestock and aquaculture. However, because of their potential mutagenic and carcinogenic properties, the abuse of them is very harmful to human health [2, 3]. In the EU, nitroimidazole application to food-producing animals and other products has been prohibited [4, 5]. So, a rapid, simple, and reliable analytical method is needed to prevent and control the illegal use of them. Moreover, the rapid analysis method for MB and other polar nitroimidazole derivatives still represents a challenge and holds a significant interest for separation scientists.

In addition to spectrophotometry [6–9], electrochemistry, voltammetry [10], and thin-layer chromatography [11], the methods commonly used for the separation and determination of MB or MB-mixed other compounds in various matrices, such as blood plasma, urine, drug, and aqueous solution samples, were high-performance liquid chromatography (HPLC) [12–17] and ultra performance liquid chromatography (UPLC) [18] using highly or totally aqueous mobile phases combined with nonwetted nonpolar packed columns (such as C_8 and C_{18}) and gradient elution mode, which can lead to a number of issues such as bubble formation. They are also approach-complicated and time-consuming. Upon the literature survey, it was observed that there was no validated method for separation and determination of MB and other 5-nitroimidazoles in milk or other samples.

Capillary electrochromatography (CEC) fuses the high separation efficiency of CE with the various retention mechanisms of liquid chromatography (LC). As a high efficiency, prominent selectivity, and novel-promising miniaturized chromatography technique, CEC gains much attention in recent years [19, 20]. However, because of too high salt concentration or intrinsic limitation of the packed column, some problems such as "dry out" and bubble formation were frequently encountered in "pure" CEC. Pressurized capillary electrochromatography (pCEC) is a powerful separation system in which a mobile phase is driven by both a pressurized flow and an electroosmotic flow (EOF), which can improve stability and reproducibility of the experimental system by avoiding the above problems [21].

Molecular-imprinted polymers (MIPs) have seen increased use in contaminant or trace analysis as suitable materials for applications where analyte selectivity is essential [22]. MIP monoliths represent a novel method and a promising tool to be used in separation processes; however, an example that uses the MIP monolithic column in CEC for the separation of a group of structurally related compounds was relatively rare in the literature [23, 24]. In recent years, they all confirmed that these MIP polymers with mixed functional monomers yielded better selectivity than that of using single-functional monomers [25–27]. In our laboratory, a homemade MIP monolithic column using (S)-ornidazole as template situ polymerization of 2-hydroxyethyl methacrylate (HEMA), dimethylaminoethyl methacrylate (DMAEMA), and ethylene dimethacrylate (EDMA) was prepared [23], which offers unique pore structure with high permeability and good selectivity, and applied for rapid separation of MB and other 5-nitroimidazole derivatives in isocratic elution high pCEC for the first time. The effects of separation conditions were studied in detail, respectively. The mechanism of separation was simply discussed, and the potential applicant of this method in water and milk sample was also investigated.

2. Materials and Methods

2.1. Materials. MNZ, TNZ, ONZ, and MB were supplied by the National Institute for the Control of Pharmaceutical and Biological Products of China (Beijing, China). (S)-ONZ was purchased from Xin An Medicine Co., Ltd (Xi'an, China). SNZ was purchased from the Xi'an Ziguang Institute of Biochemistry (Xi'an, China). Functional monomers, crosslinker, and 3-trimethoxysilyl propyl methacrylate were purchased from Aldrich (Milwaukee, WI, USA). AIBN was purchased from the Forth Chemical Reagent Plant (Shanghai, China). Toluene and dodecanol were purchased from the Tianjin Chemical Reagent Plant (Tianjin, China). Phosphate, disodium hydrogen phosphate, sodium dihydrogen phosphate, HPLC-grade methanol, and ACN were purchased from the Chemical Reagent Corporation (Shanghai, China). Each of the standard stock solutions (1.0 mg mL^{-1}) was made in methanol and stored refrigerated in dark bottles at −20°C. Water was purified with a Millipore Milli-Q purification system (Milford, MA, USA). The fused-silica capillaries with a dimension of 100 μm (ID) and 375 μm (OD) were purchased from the Yongnian Optic Fiber Plant (Hebei, China).

2.2. Preparation of MIP Monoliths. In order to improve the stability of the MIP monoliths, the inner wall of the capillary was treated with γ-MAPs, according to the procedure reported previously [28]. The mole ratio of template ((S)-ONZ) to functional monomers (HEMA and DMAEMA) was 1 : 1.5. The mole ratio of HEMA to DMAEMA was 1 : 1. The mole ratio of functional monomers (HEMA and DMAEMA) to crosslinker (EDMA) was 1 : 4. The concentration of toluene in a binary porogenic solvent mixture (toluene and dodecanol) was 12% (wt.%). According to the abovementioned ratios, (S)-ONZ, functional monomers, crosslinker, and AIBN (1 wt.% with respect to the monomers) were dissolved in porogenic solvents to form homogenous solution, which was sonicated for 15 min and purged with nitrogen for 10 min to remove oxygen. The polymerization mixture was injected into the pretreated capillary. The ends of the capillary were sealed with rubber stoppers and submerged in a 56°C water bath for 24 h. The obtained MIP monolith was washed with methanol/acetic acid (9/1, v/v) using an HPLC pump to remove the template and the residual reagents. The detection window was made by burning off 1-2 mm of the coated polymer outside after the polymer bed.

2.3. Apparatus and Procedures. The Trisep™ 2010GV pCEC system (Unimicro Technologies, Inc., Pleasanton, CA, USA) consisted of a UV/Vis detector (190–800 nm), two microvolume pumps, a high-voltage power supply (+30 and −30 kV), a microfluid manipulation module (including a six-port injector), and a data acquisition module [29]. The samples injected were delivered to the injection valve and introduced in the internal 0.8 μL sample loop and then carried to the four-port split valve by the mobile phase flow. After splitting in the four-port valve, the flow entered the capillary column under constant supplementary pressure controlled by a back-pressure regulator [26]. A positive voltage was applied to the outlet of the column, and the inlet of the column was connected to the split valve and grounded. The isocratic elution system was used in all experiments. Scanning electron micrography (SEM) of the monolith was carried out on an XL30E scanning electron microscope (Philips, The Netherlands). All the buffer solutions were prepared using double-distilled water and filtered through 0.22 μm membrane. Before use, mobile phase solution was degassed in an ultrasonic bath for 20 min. Applied voltage was firstly ramped from 0 to + 10 kV (or + 20 kV) and then operated at +10 kV (or +20 kV). UV detection was performed at 320 nm. Before pCEC experiments, the monolith was conditioned with the mobile phase for 1 h and equilibrated for about 30 min after every mobile phase changed.

2.4. Sample Preparation. Take 5 mL of the milk sample in a 50 mL centrifuge tube and then add 20 mL of aqueous trichloroacetic acid (2%, v/v). After 1 min of vortex shaking, place in an ultrasonic bath for 20 min, shake for 10 min, and then centrifuge for 10 min at 4000 rpm. Filter the supernatant with a 0.22 μm membrane for pCEC analysis.

3. Results and Discussion

3.1. Characteristics and Selectivity of the MIP Monolith. Based upon our experiments, a low-polar porogen mixture consisting of toluene-dodecanol was found to yield a suitable, rigid porous MIP monolith. The SEM images show aggregates of nanometer-sized globular particles which are surrounded by interconnected large through-pores that permit bulk flow through the capillary (Figure 1). The small size of the globular units offers good diffusion properties, and thus the rapid access of the solute to the imprinted sites in globules gives a shorter separation time. Furthermore, the covalent attachment points of the polymer to the capillary wall can also be observed, which assured definite stability of the stationary phase.

A nonimprinted polymer (NIP) monolith (in the absence of template) was prepared and treated in an identical manner with MIP monolith. In our experiment, the NIP monolith was found to have less flow resistance than the MIP monolith. The selectivity of the MIP monolith was assessed by parallel experiments using MIP and NIP monoliths (Figure 2). Shorter retention time and poorer resolution for the five 5-nitroimidazole drugs in NIP monolith were observed (Figure 2). The experiment results also indicated that the selectivity depended on not only chromatographic retention but also imprinting cavities on the MIP monolith. These results confirmed that the template molecule has a great influence on the morphology and performance of the result columns. Deng et al. have also observed the same phenomena [30].

3.2. Optimization of the Chromatographic Separation

3.2.1. Effect of the pH. The effect of the pH value of buffer solution on the molecular recognition abilities was studied by varying the pH value of sodium dihydrogen phosphate-disodium hydrogen phosphate solutions from 4.50 to 5.80 in aqueous solution before mixing with 30% ACN (Figure 3). The pKa values of the studied 5-nitroimidazole drugs (MNZ pKa 2.52, SNZ pKa 1.16, TNZ pKa 1.82, and ONZ pKa 2.3) were lower than 3.0. The pKa value of DMAEMA is about 8.4, and the HEMA is a neutral compound. When the pH value of buffer solution was in the range from 4.50 to 5.80, the studied 5-nitroimidazole drugs remained unprotonated and had a migration nearly identical to the neutral solute. The DMAEMA in the polymer matrix is a main EOF promoter. With the increase of the pH value of the buffer solution from 4.50 to 5.80, the protonated degree of DMAEMA slightly decreased, thus causing the decrease of the EOF and the longer retention times. The increase of the pH is beneficial to the hydrogen-bonding formation between the analyte and functional groups in the imprinted cavities thus the ability of molecular recognition increased, and then the resolution between MNZ and SNZ increased. So, pH 4.82 was chosen because it provided the appropriate resolution and analysis time.

3.2.2. Effect of ACN Concentration. In pCEC, the organic modifier in the mobile phase is an important parameter. ACN is the most widely used organic modifier in pCEC. The effect of the mobile phase of ACN concentration on separation of the five 5-nitroimidazole drugs was studied, and different concentrations of ACN, in the range of 30–70% (v/v), were added to a 10 mM sodium dihydrogen phosphate-disodium hydrogen phosphate solution with pH 4.82 (Figure 4). It is well known that the hydrogen-bonding interactions in the process of molecular recognition are stronger with the increase of ACN concentration in the hydro-organic mobile phase. However, with the increase of ACN concentration, the decrease of the viscosity of the solution and the increase of zeta potential caused the increase of EOF; so the retention times was shorter, resulting in a poorer recognition. In addition, it could be seen from Figure 4 that the separation degree and retention of 5-nitroimidazole drugs decreased with the increase of ACN concentration. From Figure 4, with the decrease of ACN concentration, it can be seen that MB has longer retention time than ONZ, and the difference of retention time between MB and ONZ also increases. From Figure 2 (MIP monolith and NIP monolith), it can be seen that MB has always longer retention time than ONZ. It can be deduced that the retention of 5-nitroimidazole drugs was due to hydrophobic interaction, hydrogen-bonding and electrostatic interactions with the imprint cavities in the polymer. So, 30% ACN was the optimized organic content.

3.2.3. Effect of Applied Voltage. The effect of applied voltage on the separation performance of the column was investigated by applying voltages +5 kV, +10 kV, +15 kV, and +20 kV, using the mobile phase containing 30% ACN in 10 mM sodium dihydrogen phosphate-disodium hydrogen phosphate solution with pH 4.82 under constant pressure (Figure 5). Methanol was used as the marker of the EOF mobility. The EOF linearly increased with the increase of the voltage from +5 kV to +20 kV with a correlation coefficient of 0.9428, which suggested the Joule heating was negligible in the system under the experimental conditions. When the EOF velocity was faster, the mobility times of analytes would be shorter and rapid separation could be realized on the MIP monolith (Figure 5). However, too high voltage would cause large current and lead to a greater Joule heating. So, +20 kV was the optimized voltage. Good flow characteristics of the MIP monolith and a supplement pressure of 1000 psi avoid air bubble formation. Under the optimum conditions, the column efficiency (plates/m) was 43 000 for MNZ, 36 000 for SNZ, 34 000 for TNZ, 14 000 for ONZ, and 29 000 for MB.

FIGURE 1: SEM of the end of the MIP monolithic stationary phase in a fused-silica capillary column.

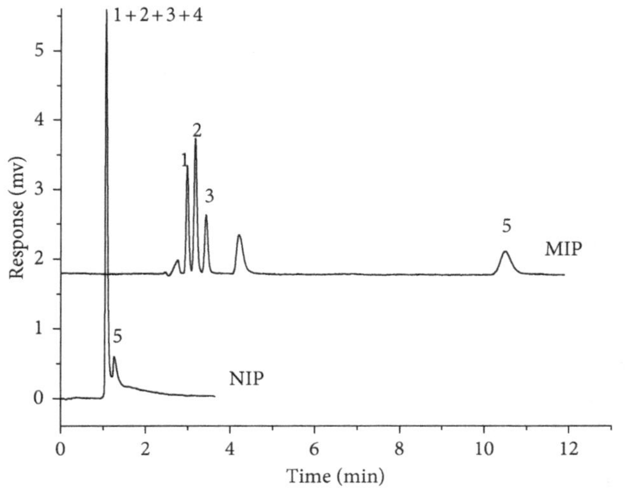

FIGURE 2: Comparison between MIP monolith and NIP monolith. Capillary column: 51 cm column length; 30 cm effective length. Electrochromatographic conditions: mobile phase, ACN/10 mM sodium dihydrogen phosphate-disodium hydrogen phosphate buffer (30/70, v/v, pH 4.82); voltage: +20 kV; flow rate: 0.1 ml/min; supplement pressure: 1000 psi. 1 MNZ, 2 SNZ, 3 TNZ, 4 ONZ, and 5 MB.

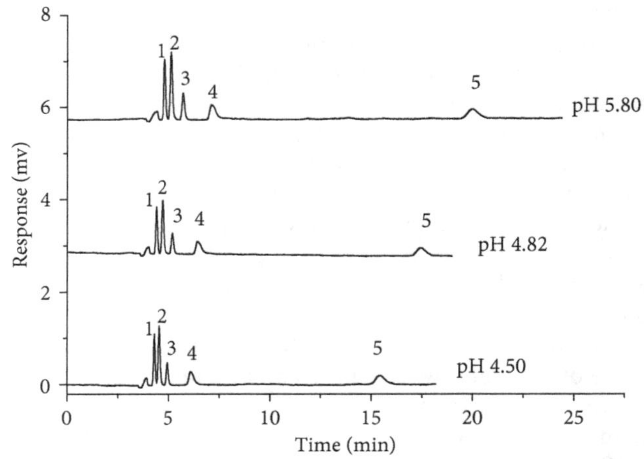

FIGURE 3: Effect of pH of buffer solution on separation of 5-nitroimidazole compounds. MIP monolith: 51 cm column length; 30 cm effective length. Electrochromatographic conditions: mobile phase, ACN/10 mM sodium dihydrogen phosphate-disodium hydrogen phosphate buffer (30/70, v/v); voltage: +10 kV; flow rate: 0.1 ml/min; supplement pressure: 1000 psi. 1 MNZ, 2 SNZ, 3 TNZ, 4 ONZ, and 5 MB.

3.3. *Linear Response Range, Detection Limit, and Precision.*

Under the optimized experimental conditions, the linear regression analysis of five 5-nitroimidazoles was constructed by plotting the peak area against analyte concentration. The linear ranges, correlation coefficients, relative standard deviations, and the detection limits are summarized in Table 1. The reliability and reproducibility of the chromatographic performance of the MIP monolith was examined using 0.05 mg/mL of five 5-nitroimidazoles for four repeated injections, and the percent of relative standard deviation (RSD) values illustrated that the results were highly reproducible.

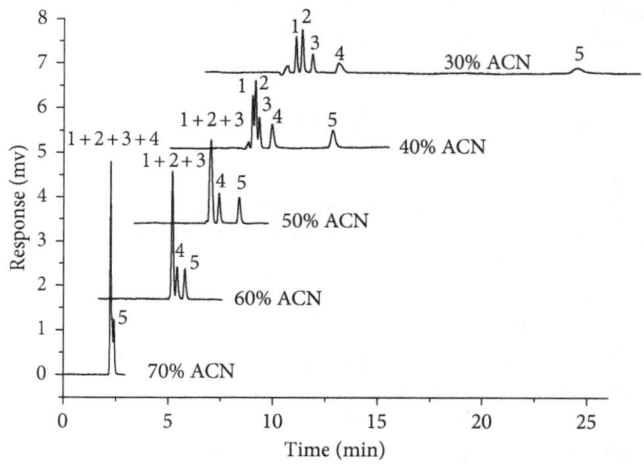

FIGURE 4: Effect of content of ACN on separation of 5-nitroimidazole compounds. Electrochromatographic conditions: mobile phase: ACN/10 mM sodium dihydrogen phosphate-disodium hydrogen phosphate buffer (pH 4.82). Other conditions and the order of peak can be seen in Figure 3.

3.4. *Application for Separation of 5-Nitroimidazoles in Real Water Sample.*

In order to assess the specificity of the method, a water blank sample was analysed. No interference peaks were observed around the retention time of the 5-nitroimidazoles. These results demonstrate that real sample

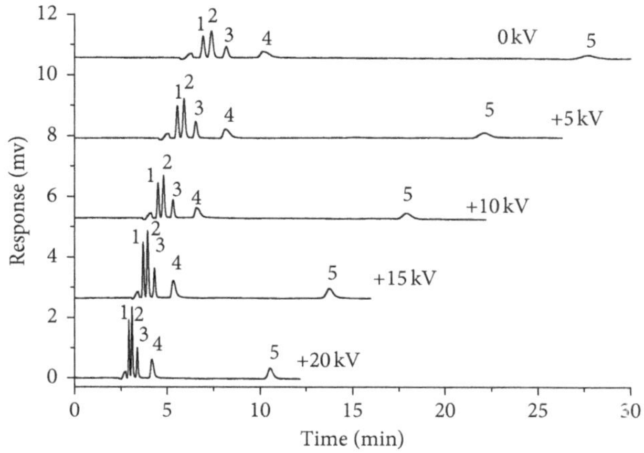

FIGURE 5: Effect of the applied voltage on separation of 5-nitroimidazole compounds. Electrochromatographic conditions: pH 4.82 and applied voltage 0 kV, +5 kV, +10 kV, +15 kV, and +20 kV. Other conditions and the order of peak can be seen in Figure 3.

TABLE 1: Linear range, detection limit, and precisions.

Analyte	Linear range (μg/mL)	Correlation coefficient (R^2)	Detection limits[a] (μg/mL)	% RSD time	% RSD area
MNZ	0.50–100	0.9952	0.11	1.0	1.69
SNZ	0.50–100	0.9952	0.15	1.21	2.11
TNZ	0.80–500	0.9990	0.22	1.47	1.48
ONZ	0.80–100	0.9984	0.13	1.72	2.37
MB	5–500	0.9955	0.73	2.57	1.84

[a]Calculated as three times the signal-to-noise ratio.

matrices had no effect on the performance of the proposed method. In order to verify the applicability of the MIP-pCEC method, the developed method was applied to the analysis of 0.05 mg/mL 5-nitroimidazoles mixture in the river (Minjiang River, Fuzhou, China) and pond water samples (Pond near Fuzhou University, Fuzhou, China), respectively. Each water sample spiked with five kinds of 5-nitroimidazoles was filtered through a 0.45 μm membrane and then analysed using the optimized chromatographic conditions. Recoveries were estimated by comparing the concentrations from spiked water samples determined by the calibration equations and the corresponding added concentrations. For the river water sample, the recoveries were found to be 93.2% for MNZ, 95.8% for SNZ, 85.0% for TNZ, 93.6% for ONZ, and 90.6% for MB, respectively. For the pond water sample, the recoveries were found to be 95.0% for MNZ, 95.8% for SNZ, 86.5% for TNZ, 93.7% for ONZ, and 94.6% for MB, respectively. The RSDs for four repeated injections were less than 3.8%.

3.5. Application for Separation of 5-Nitroimidazoles in Milk Sample. In order to enlarge the applications of the obtained MIP monolith, the monolith was applied to the rapid separation of five 5-nitroimidazole drugs in the locally available pure milk sample (AoNiu). The sample was simple pretreated according to the procedure stated in Section 2.4, and the analysis was performed using the optimized chromatographic conditions. The chromatogram is shown in Figure 6. The added standards were successfully separated from the milk sample. The result confirmed that the development method for the fast separation of the real sample is feasible (Figure 6).

4. Conclusion

In the present work, a homemade MIP monolith which contained polar tertiary amine groups and alcohol groups in the imprinted cavities was firstly and successfully applied to the rapid separation and determination of five 5-nitroimidazoles in water samples and milk samples. The effects of electrochromatographic conditions, such as voltage, the composition of the mobile phase on the separation of 5-nitroimidazoles have been systemically investigated and also suggested that the separation mechanism of analytes on the MIP monolith was mainly based on the interplays of hydrophobic, hydrogen-bonding, and electrostatic interactions and imprinting cavities on the MIP monolith. Under the optimized conditions, five 5-nitroimidazoles could be baseline-separated by a simple isocratic elution and high pCEC mode within 11 min on the MIP monolith. The whole procedures demonstrated a good recovery, and the MIP monolith is robust enough after more than one month and 300 electrochromatographic runs. Compared with the present method for separation of nitroimidazole drugs, the proposed method shows its advantages in the simple of column preparation, environmental-friendly, low cost, fast analysis, and high efficiency. However, due to the low concentrations of 5-nitroimidazoles in some real analytical samples with environmental and biological origin, our

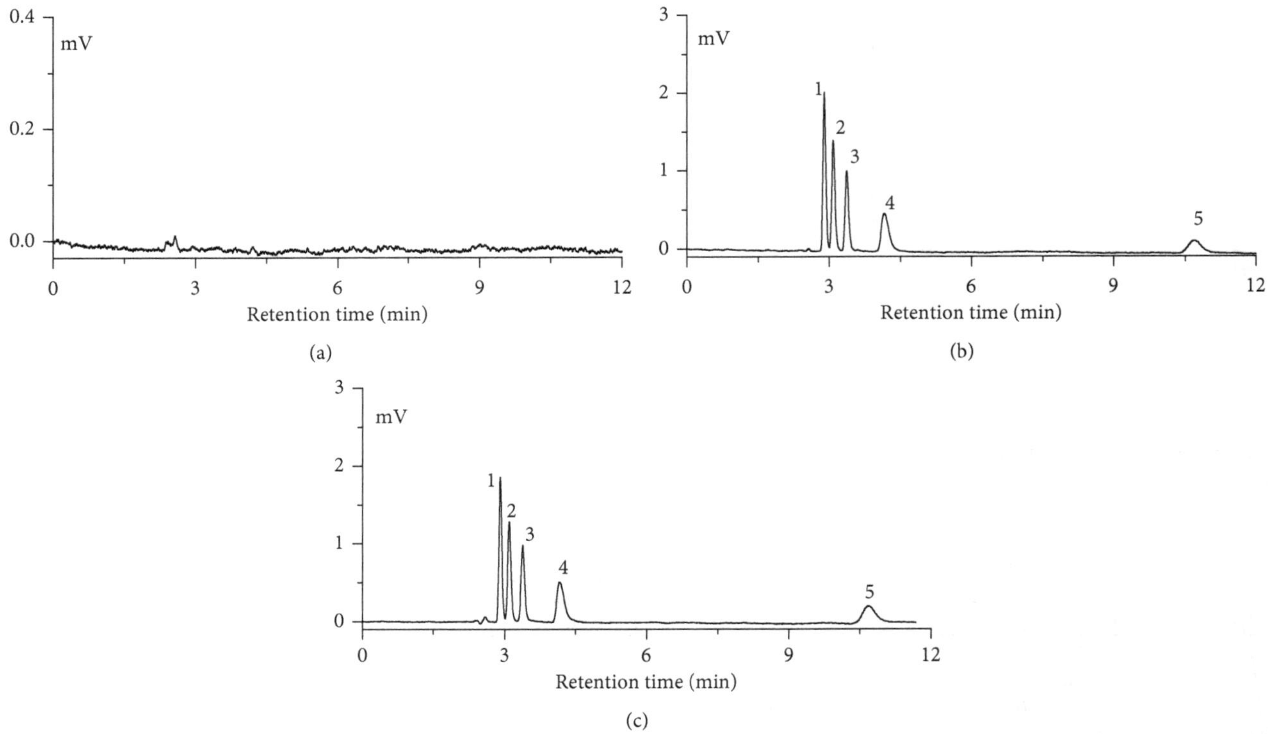

FIGURE 6: Separation of five 5-nitroimidazole drugs in the milk sample. Electrochromatographic conditions: pH 4.82; voltage: +20 kV. Other conditions and the order of peak can be seen in Figure 3. (a) Blank milk sample. (b) Spiked milk sample. (c) Standard sample.

future work will explore the combination of the present method with other sensitive detections or preconcentration.

Conflicts of Interest

The authors declare that there are no conflicts of interest.

Acknowledgments

This work was supported by the National Natural Science Foundation (21407118, 20907009, 40976071, and 20905079), Program for New Century Excellent Talents in University of Fujian Province (HX2016-23), Key Science & Technology Project of Fujian Province (2015N0035), and Key Project of Education and Scientiifc Research for Middle-Aged and Young of Fujian Province (JA13314). The authors are also grateful to Professor Zenghong Xie and Xucong Lin of the Institute of Food Safety and Environmental Monitoring of Fuzhou University (P.R. China) for their technical assistance.

References

[1] M. Hoffer and E. Grunberg, "Synthesis and antiprotozoal activity of 1-(3-chloro-2-hydroxypropyl)-substituted nitroimidazoles," *Journal of Medicinal Chemistry*, vol. 17, no. 9, pp. 1019-1020, 1974.

[2] C. E. Voogd, "On the mutagenicity of nitroimidazoles," *Mutation Research/Reviews in Genetic Toxicology*, vol. 86, no. 3, pp. 243–277, 1981.

[3] W. Raether and H. Hänel, "Nitroheterocyclic drugs with broad spectrum activity," *Parasitology Research*, vol. 90, no. 1, pp. S19–S39, 2003.

[4] Anon., Council Regulation (EEC) No 2377/90 of 26 June 1990, "Laying down a community procedure for the establishment of maximum residue limits of veterinary medicinal products in foodstuffs of animal origin," *Official Journal of the European Communities*, no. L224, p. 136, 1990.

[5] Anon, Commission Regulation (EU) No 37/2010 of 22 December 2009, "On pharmacologically active substances and their classification regarding maximum residue limits in foodstuffs of animal origin," *Official Journal of the European Communities*, no. L15, pp. 1–72, 2010.

[6] K. Farhadi and S. Bahar, "Kinetic-spectrophotometric determination of metronidazole benzoate in surfactant medium," *Journal of the Chinese Chemical Society*, vol. 54, no. 6, pp. 1521–1528, 2007.

[7] E. A. Abdelaleem and N. S. Abdelwahab, "Simultaneous determination of some antiprotozoal drugs in different combined dosage forms by mean centering of ratio spectra and multivariate calibration with model updating methods," *Chemistry Central Journal*, vol. 6, no. 27, pp. 1–8, 2012.

[8] A. A. Salem and H. A. Mossa, "Method validation and determinations of levofloxacin, metronidazole and sulfamethoxazole in an aqueous pharmaceutical, urine and blood plasma samples using quantitative nuclear magnetic

resonance spectrometry," *Talanta*, vol. 88, no. 17, pp. 104–114, 2012.

[9] A. K. Mishra, A. Kumar, A. Mishra, and H. V. Mishra, "Development of ultraviolet spectroscopic method for the estimation of metronidazole benzoate from pharmaceutical formulation," *Journal of Natural Science, Biology, and Medicine*, vol. 5, no. 2, pp. 261–264, 2014.

[10] R. Joseph and K. Girish Kumar, "Electrochemical reduction and voltammetric determination of metronidazole benzoate at modified carbon paste electrode," *Analytical Letters*, vol. 42, no. 14, pp. 2309–2321, 2009.

[11] E. A. Abdelaleem and N.. S. Abdelwahab, "Simultaneous determination of some antiprotozoal drugs in their binary and ternary mixtures with mebeverine hydrochloride in different dosage forms," *Journal of Liquid Chromatography & Related Technologies*, vol. 36, no. 11, pp. 1528–1539, 2013.

[12] P. P. Pashankov and L. L. Kostova, "Reversed-phase high-performance liquid-chromatography of metronidazole benzoate in suspension dosage form," *Journal of Chromatography*, vol. 394, no. 2, pp. 382–387, 1987.

[13] M. S. Ali, R. S. Chaudhary, and M. A. Takieddin, "Simultaneous determination of metronidazole benzoate, methylparaben, and propylparaben by high-performance liquid chromatography," *Drug Development and Industrial Pharmacy*, vol. 25, no. 10, pp. 1143–1147, 1999.

[14] D. K. Bempong, R. T. Manning, and L. Bhattacharyya, "A stability-indicating HPLC assay for metronidazole benzoate," *Journal of Pharmaceutical and Biomedical Analysis*, vol. 38, no. 4, pp. 776–780, 2005.

[15] A. Mishal and D. Sober, "Stability indicating reversed-phase liquid chromatographic determination of metronidazole benzoate and diloxanide furoate as bulk drug and in suspension dosage form," *Journal of Pharmaceutical and Biomedical Analysis*, vol. 39, no. 3-4, pp. 819–823, 2005.

[16] K. S. Syed, R. Maheswari, P. Shanmugasundaram, and M. Vijeyaanandhi, "Simultaneous estimation of chlorhexidine gluconate, metronidazole, lignocaine hydrochloride and triamcinolone acetonide in combined dosage form by RP-HPLC," *Asian Journal of Chemistry*, vol. 20, no. 2, pp. 1130–1136, 2008.

[17] S. H. SaeedulHassan and M. T. Ansari, "Quantitation of metronidazole in pharmaceutical suspension using high performance liquid chromatographic method," *Pakistan Journal of Zoology*, vol. 43, no. 5, pp. 909–914, 2011.

[18] S. A. Ahmad, M. Ahmed, M. A. Qadi et al., "Simultaneous determination of diloxanide furoate, metronidazole benzoate, methyl paraben and propyl paraben by uplc-dad in pharmaceutical suspension," *Latin American Journal of Pharmacy*, vol. 35, no. 7, pp. 1626–1633, 2016.

[19] N. S. Ye, X. Wang, and Q. Y. Liu, "Covalent bonding of schiff base network-1 as a stationary phase for capillary electrochromatography," *Analytica Chimica Acta*, vol. 1028, pp. 113–120, 2018.

[20] D. Polikarpova, D. Makeeva, and L. Kartsova, "Nano-sized anion-exchangers as a stationary phase in capillary electrochromatography for separation and on-line concentration of carboxylic acids," *Talanta*, vol. 188, pp. 744–749, 2018.

[21] J. Wang, C. Y. Ding, J. H. Xiao et al., "Quinine-modified polymer monolithic column with reversed-phase/strong anion-exchange mixed-mode for pressurized capillary electrochromatography," *Electrophoresis*, vol. 39, no. 12, pp. 1504–1511, 2018.

[22] B. S. Batlokwa, J. Mokgadi, R. Majors, C. Turner, and N. Torto, "A novel molecularly imprinted polymer for the selective removal of chlorophyll from heavily pigmented green plant extracts prior to instrumental analysis," *Journal of Chemistry*, vol. 2013, Article ID 540240, 4 pages, 2013.

[23] S. L. Liao, X. C. Wang, X. C. Lin, and Z. H. Xie, "Preparation and characterization of a molecularly imprinted monolithic column for pressure-assisted CEC separation of nitroimidazole drugs," *Electrophoresis*, vol. 31, no. 16, pp. 2822–2830, 2010.

[24] Q. Q. Wang, X. Zhang, and J. Li, "Synthesis and evaluation of organic-inorganic hybrid molecularly imprinted monolith column for selective recognition of acephate and phosphamidon in vegetables," *Advances in Polymer Technology*, vol. 36, no. 4, pp. 401–408, 2017.

[25] S. L. Liao, S. Y. Chen, Q. L. Liu, L. B. Chen, and X. M. Li, "Preparation and application of ornidazole magnetic imprinted polymers with dual functional monomers," *Chinese Journal of Analytical Chemistry*, vol. 46, no. 1, pp. 100–106, 2018.

[26] S. L. Liao, X. C. Wang, X. C. Lin, X. P. Wu, and Z. H. Xie, "A molecularly imprinted monolith for the fast chiral separation of antiparasitic drugs by pressurized CEC," *Journal of Separation Science*, vol. 33, no. 14, pp. 2123–2130, 2010.

[27] K. K. Zhi, A. J. Dong, and X. Yang, "Preparation and adsorption properties study of glucose magnetic molecularly imprinted polymers with dual functional monomers," *Acta Chimica Sinica*, vol. 74, no. 2, pp. 199–207, 2016.

[28] X. P. Wu, L. Wang, Z. H. Xie et al., "Rapid separation and determination of carbamate insecticides using isocratic elution pressurized capillary electrochromatography," *Electrophoresis*, vol. 27, no. 4, pp. 768–777, 2006.

[29] B. H. Xiong, L. H. Zhang, Y. K. Zhang, H. F. Zou, and J. D. Wang, "Capillary electrochromatography with monolithic poly(styrene-co-divinylbenzene-co-methacrylic acid) as the stationary phase," *Journal of High Resolution Chromatography*, vol. 23, no. 1, pp. 67–72, 2000.

[30] Q. L. Deng, Z. H. Lun, H. Shao, C. Yan, and R. Y. Gao, "Molecularly imprinted polymers as tools for the screening of felodipine from dihydropyridine calcium antagonists by pressurized capillary electrochromatography," *Analytical and Bioanalytical Chemistry*, vol. 382, no. 1, pp. 51–58, 2005.

Synthesis, Characterization, and Antimicrobial Studies of Novel Series of 2,4-*Bis*(hydrazino)-6-substituted-1,3,5-triazine and Their Schiff Base Derivatives

Hessa H. Al-Rasheed,[1] Essam N. Sholkamy,[2] Monirah Al Alshaikh,[1] Mohammed R. H. Siddiqui,[1] Ahmed S. Al-Obaidi,[2] and Ayman El-Faham[1,3]

[1]*Department of Chemistry, College of Science, King Saud University, P.O. Box 2455, Riyadh 11451, Saudi Arabia*
[2]*Department of Botany and Microbiology, College of Science, King Saud University, P.O. Box 2455, Riyadh 11451, Saudi Arabia*
[3]*Chemistry Department, Faculty of Science, Alexandria University, P.O. Box 426, Ibrahimia, Alexandria 12321, Egypt*

Correspondence should be addressed to Ayman El-Faham; aymanel_faham@hotmail.com

Academic Editor: José M. G. Martinho

The present work represents the synthesis, characterization, and antimicrobial studies of novel series of 2,4-*bis*(hydrazino)-6-substituted-1,3,5-triazine and their Schiff base derivatives. IR, NMR (^{1}H and ^{13}C), elemental analysis, and LC-MS characterized the prepared compounds. The biological activity of the target products was evaluated as well. Twenty-two of the prepared compounds were selected according to their solubility in aqueous DMSO. Only eight compounds showed good activity against the selected pathogenic bacteria and did not show antagonistic effect against fungus *Candida albicans*. Two compounds **4k** and **5g** have wide-range effect presently in Gram-positive and Gram-negative bacteria while other compounds (**4f**, **4i**, **4m**, **5d**, **6i**, and **6h**) showed specific effect against the Gram-negative or Gram-positive bacteria. The minimum inhibitory concentration (MIC, μg/mL) of **4f**, **4i**, **4k**, and **6h** compounds against *Streptococcus mutans* was 62.5 μg/mL, 100 μg/mL, 31.25 μg/mL, and 31.25 μg/mL, respectively. The MIC of **4m**, **4k**, **5d**, **5g**, and **6h** compounds against *Staphylococcus aureus* was 62.5 μg/mL, 31.25 μg/mL, 31.25 μg/mL, 100 μg/mL, and 62.5 μg/mL, respectively. The MIC of **4k**, **5g**, and **6i** compounds against *Salmonella typhimurium* was 31.25 μg/mL, 100 μg/mL, and 62.5 μg/mL, respectively. The MIC of **6i** compound against *Escherichia coli* was 62.5 μg/mL.

1. Introduction

Hydrazones are considered as extraordinary class of compounds in the Schiff bases family, characterized by the presence of -C=N-N-. The two connected nitrogen atoms are unlike nature and the C-N double bond that is conjugated with a lone pair of the terminal nitrogen atom is mainly responsible for their physical and chemical properties [1–7]. This class of compounds is usually used as polydentate chelating agents that can form a variety of complexes with a range of transition and inner transition metals which have attracted the attention of many researchers [8, 9]. Huge derivatives of hydrazones and their metal complexes have been studied and used in different applications, such as metal ions extraction and microdetermination of metal ion [10, 11]. In addition, they have different biological activity, for example, antimicrobial activity [12–14], antifungal activity [15, 16], antitumor activity [17, 18], and insecticides [19].

On the other hand, the growing interest in s-triazine derivatives specially its chlorinated derivative (cyanuric chloride) and its efficient reaction with hydrazine to render mono-, di-, or tri- substituted hydrazino derivatives has increased [20–22]. Several prepared and well explored hydrazino-s-triazine derivatives were reported and considered with special interest in coordination chemistry and supramolecular chemistry [23, 24] as well as their enormous potential in the field of material chemistry [25, 26], complexation with large metal ions [27–36], and pharmaceutical industry [37–43].

In view of this, the present work represents the synthesis and characterization of novel series of Schiff base derived from 2,4-bis(hydrazino)-6-substituted-s-triazine and their antibacterial and antifungal activity.

2. Experimental Section

2.1. Chemistry

2.1.1. Materials. The solvents used were of analytical reagent grade. NMR (^{1}H and ^{13}C) spectra were recorded on a JEOL 400, 600 MHz spectrometer at room temperature. The chemical shifts were measured using internal standard δ = 0 ppm. Elemental analyses were performed on Perkin-Elmer 2400 elemental analyzer, and the values found were within ±0.3% of the theoretical values. Melting points were recorded on a Mel-Temp apparatus in an open capillary and are uncorrected. Fourier transform infrared spectroscopy (FTIR) spectra were recorded on Nicolet 560 spectrometer from KBr discs. The reaction was followed up and the purity was checked using TLC (silica gel-protected aluminum sheets type 60 GF254, Merck). 2-(4-Iodophenyl)-3-(4-nitrophenyl)-5-phenyltetrazolium chloride (INT) (2 mg/mL) was purchased from Sigma and used for the bacterial growth test.

*2.1.2. General Method for the Synthesis of 2,4-Dihydrazino-6-substituted-1,3,5-triazine Derivatives **3a–e**.* Compounds **3a–e** were prepared according to the reported procedure with some modifications [21, 44, 45]. Hydrazine hydrate (80%) in 20 mL acetonitrile was added to a solution of **2a–e** (20 mmol) in acetonitrile (50 mL) at room temperature. The reaction mixture refluxed for 3 h, and then the excess of hydrazine and solvent was removed under vacuum and the crude white precipitate was filtered off, washed with acetonitrile and ether, and then dried to afford the desired products with high yields and purities. The products were used directly in the next step.

*2.1.3. General Procedure for Synthesis of 2,4-Bis(2-benzylidene-hydrazinyl)-6-substituted-1,3,5-triazine Derivatives **4–8**.* 2,4-Dihydrazino-6-substituted-1,3,5-triazine derivatives **3a–e** (2 mmoles) were added portion-wise to a hot solution of different aldehyde or ketone (4 mmoles) in 30 mL ethanol in the presence of 2-3 drops of acetic acid. The reaction mixture was refluxed for 3-4 hours and the progress of the reaction was followed by TLC using ethyl acetate-hexane 2 : 1. The reaction was left to cool down to room temperature and then the product was filtered and dried.

*(1) N-Benzyl-4,6-bis(2-benzylidenehydrazinyl)-1,3,5-triazine-2-amine (**4a**, Supporting Information, Figure S1 in Supplementary Data).* Beige solid in yield 95%; mp 150–152°C; IR (KBr, cm^{-1}): 3233 (NH), 1679 (C=N), 1587, 1517 (C=C); ^{1}H NMR (DMSO-d$_6$): δ = 4.52 (d, 2H, J = 6.0 Hz, CH$_2$NH), 7.18 (t, 1H, J = 7.2 Hz, Ar), 7.27 (t, 2H, J = 8.0 Hz, Ar), 7.30–7.42 (m, 8H, Ar), 7.63 (d, 4H, J = 8.0 Hz, Ar), 8.14 (s, 1H, CH), 11.12 (brs, 1H, NH) ppm; ^{13}C NMR (DMSO-d$_6$): δ = 43.8, 126.9, 127.6, 128.6, 130.1, 136.8, 140.7, 143.0, 157.9, 164.6 ppm; Anal. Calc. for $C_{24}H_{22}N_8$ (422.50): C, 68.23; H, 5.25; N, 26.52. Found: C, 68.51; H, 5.12; N, 26.37.

*(2) 3,3′-((2,2′-(6-(Benzylamino)-1,3,5-triazine-2,4-diyl)bis(hydrazin-2-yl-1-ylidene))bis(methanylylidene))diphenol (**4b**, Supporting Information, Figure S2 in Supplementary Data).* White solid in yield 80%; mp 273–275°C; IR (KBr, cm^{-1}): 3371 (OH), 3304 (NH), 1579 (C=N), 1520, 1388 (C=C); ^{1}H NMR (DMSO-d6): δ = 4.54 (d, 2H, J = 6.4 Hz, CH$_2$NH), 6.77 (dd, 2H, J^4 = 2.4, J^3 = 8.2 Hz, Ar), 7.04 (d, 2H, J = 8.0 Hz, Ar), 7.08 (s, 2H, Ar), 7.22 (t, 3H, J = 8.0 Hz, Ar), 7.32 (t, 2H, J = 7.6 Hz, Ar), 7.38 (s, 2H, Ar), 8.06 (s, 1H, CH), 9.54 (s, 1H, OH), 10.82 (brs, 1H, NH) ppm; ^{13}C NMR (DMSO-d$_6$): δ = 43.8, 112.9, 116.8, 118.3, 126.9, 127.6, 128.6, 130.1, 136.8, 140.7, 143.0, 157.9, 164.6, 166.6 ppm; Anal. Calc. for $C_{24}H_{22}N_8O_2$ (454.48): C, 63.43; H, 4.88; N, 24.66. Found: C, 63.21; H, 5.01; N, 24.81.

*(3) N-Benzyl-4,6-bis(2-(3-methoxybenzylidene)hydrazinyl)-1,3,5-triazine-2-amine (**4c**, Supporting Information, Figure S3 in Supplementary Data).* White solid in yield 97%; mp 173–175°C; IR (KBr, cm^{-1}): 3253 (NH), 1584 (C=N), 1515, 1382 (C=C); ^{1}H NMR (DMSO-d$_6$): δ = 3.80 (d, 6H, J = 3.8 Hz, 2OCH$_3$), 4.53 (d, 2H, J = 6.0 Hz, CH$_2$NH), 6.90 (dd, 2H, J^4 = 2.4, J^3 = 8.4 Hz, Ar), 7.17–7.24 (m, 3H, Ar), 7.28–7.36 (m, 6H, Ar), 7.38 (s, 2H, Ar), 8.10 (s, 1H, CH), 10.90 (brs, 1H, NH) ppm; ^{13}C NMR (DMSO-d$_6$): δ = 43.8, 55.5, 111.4, 115.4, 119.7, 127.0, 127.6, 128.6, 130.2, 136.9, 140.7, 142.7, 159.9, 164.7, 166.5 (2-HN-C=N-) ppm; Anal. Calc. for $C_{26}H_{26}N_8O_2$ (482.54): C, 64.72; H, 5.43; N, 23.22. Found: C, 64.99; H, 5.68; N, 23.00.

*(4) N-Benzyl-4,6-bis(2-(4-methylbenzylidene)hydrazinyl)-1,3,5-triazine-2-amine (**4d**, Supporting Information, Figure S4 in Supplementary Data).* Beige solid in yield 97%; mp 153–155°C; IR (KBr, cm^{-1}): 3345 (NH), 1681 (C=N), 1604, 1534 (C=C); ^{1}H NMR (DMSO-d$_6$): δ = 2.33 (s, 6H, 2CH$_3$), 4.55 (d, 2H, J = 5.6 Hz, CH$_2$NH), 7.22 (d, 4H, J = 9.2 Hz, Ar), 7.31 (t, 3H, J = 8.0 Hz, Ar), 7.38 (s, 2H, Ar), 7.56 (d, 4H, J = 8.0 Hz, Ar), 8.13 (s, 1H, CH), 11.05 (brs, 1H, NH) ppm; ^{13}C NMR (DMSO-d$_6$): δ = 21.4, 43.8, 126.9, 127.1, 127.7, 128.6, 129.8, 132.4, 139.0, 139.5, 142.6, 159.9, 164.7, 166.5 ppm; Anal. Calc. for $C_{26}H_{26}N_8$ (450.54): C, 69.31; H, 5.82; N, 24.87. Found: C, 69.01; H, 5.89; N, 25.00.

*(5) N-Benzyl-4,6-bis(2-(4-methoxybenzylidene)hydrazinyl)-1,3,5-triazine-2-amine (**4e**, Supporting Information, Figure S5 in Supplementary Data).* White solid in yield 99%; mp 152–154°C; IR (KBr, cm^{-1}): 3346 (NH), 1677 (C=N), 1607, 1509 (C=C); ^{1}H NMR (DMSO-d$_6$): δ = 3.78 (s, 6H, 2OCH$_3$), 4.51 (d, 2H, J = 6.8 Hz, CH$_2$NH), 6.97 (d, 4H, J = 8.0 Hz, Ar), 7.31 (t, 1H, J = 8.0 Hz, Ar), 7.37 (s, 2H, Ar), 7.57 (d, 4H, J = 8.0 Hz, Ar), 8.08 (s, 1H, CH), 10.70 (brs, 1H, NH) ppm; ^{13}C NMR (DMSO-d$_6$): δ = 43.8, 55.6, 112.9, 114.6, 126.9, 128.0, 128.3, 128.5, 140.8, 142.7, 159.9, 164.7, 166.5, 165.0, 172.6 ppm; Anal. Calc. for $C_{26}H_{26}N_8O_2$ (482.54): C, 64.72; H, 5.43; N, 23.22. Found: C, 64.98; H, 5.65; N, 23.47.

*(6) 3,3′-((2,2′-(6-(Benzylamino)-1,3,5-triazine-2,4-diyl)bis(hydrazin-2-yl-1-ylidene))bis(methanylylidene))diphenol (**4f**, Supporting Information, Figure S6 in Supplementary Data).* Beige solid in yield 85%; mp 240–242°C; IR (KBr, cm^{-1}): 3433 (OH), 3251 (NH), 1679 (C=N), 1605, 1503 (C=C); ^{1}H NMR (DMSO-d$_6$): δ = 4.55 (d, 2H, J = 5.6 Hz, CH$_2$NH), 6.82 (d, 4H,

J = 8.0 Hz, Ar), 7.23 (t, 1H, J = 8.0 Hz, Ar), 7.32 (t, 2H, J = 7.6 Hz, Ar), 7.36 (s, 2H, Ar), 7.52 (d, 4H, J = 8.0 Hz, Ar), 8.08 (s, 1H, CH), 9.91 (s, 1H, OH), 11.0 (brs, 1H, NH) ppm; ^{13}C NMR (DMSO-d_6): δ = 47.5, 112.3, 114.8, 116.1, 126.9, 128.7, 129.1, 136.9, 142.8, 160.5, 166.5 ppm; Anal. Calc. for $C_{24}H_{22}N_8O_2$ (454.48): C, 63.43; H, 4.88; N, 24.66. Found: C, 63.68; H, 4.77; N, 24.91.

*(7) N-Benzyl-4,6-bis(2-(4-bromobenzylidene)hydrazinyl)-1,3,5-triazine-2-amine (**4g**, Supporting Information, Figure S7 in Supplementary Data).* Beige solid in yield 97%; mp 167–169°C; IR (KBr, cm^{-1}): 3348 (NH), 1680 (C=N), 1608, 1540 (C=C); ^{1}H NMR (DMSO-d_6): δ = 4.55 (d, 2H, J = 6.0 Hz, CH_2NH), 6.76 (d, 4H, J = 7.6 Hz, Ar), 7.12 (s, 4H, Ar), 7.23 (t, 1H, J = 7.2 Hz, Ar), 7.32 (t, 2H, J = 8.0 Hz, Ar), 7.36 (s, 2H, Ar), 8.0 (s, 1H, CH), 11.10 (brs, 1H, NH) ppm; ^{13}C NMR (DMSO-d_6): δ = 43.9, 122.8, 127.1, 127.6, 128.6, 128.7, 132.1, 133.4, 134.5, 140.3, 142.3, 161.2, 164.2 ppm; Anal. Calc. for $C_{24}H_{20}Br_2N_8$ (454.48): C, 49.68; H, 3.47; N, 19.31. Found: C, 49.83; H, 3.61; N, 19.52.

*(8) N-Benzyl-4,6-bis(2-(4-chlorobenzylidene)hydrazinyl)-1,3,5-triazine-2-amine (**4h**, Supporting Information, Figure S8 in Supplementary Data).* White solid in yield 90%; mp 155–157°C; IR (KBr, cm^{-1}): 3350 (NH), 1676 (C=N), 1610, 1542 (C=C); ^{1}H NMR (DMSO-d_6): δ = 4.53 (d, 2H, J = 6.0 Hz, CH_2NH), 7.21 (t, 1H, J = 7.6 Hz, Ar), 7.31 (t, 2H, J = 7.6 Hz, Ar), 7.36 (s, 2H, Ar), 7.49 (d, 4H, J = 8.0 Hz, Ar), 7.65 (d, 4H, J = 8.0 Hz, Ar), 8.12 (s, 1H, CH), 11.02 (brs, 1H, NH) ppm;^{13}C NMR (DMSO-d_6): δ = 45.2, 122.6, 127.0, 128.4, 129.3, 132.3, 134.5, 141.0, 142.7, 161.0, 164.3. Anal. Calc. for $C_{24}H_{20}Cl_2N_8$ (490.12): C, 58.66; H, 4.10; N, 22.80. Found: C, 58.43; H, 4.32; N, 23.03.

*(9) N-Benzyl-4,6-bis(2-(4-fluorobenzylidene)hydrazinyl)-1,3,5-triazine-2-amine (**4i**, Supporting Information, Figure S9 in Supplementary Data).* Beige solid in yield 94%; mp 172–174°C; IR (KBr, cm^{-1}): 3356 (NH), 1690 (C=N), 1606, 1504 (C=C); ^{1}H NMR (DMSO-d_6): δ = 4.55 (d, 2H, J = 6.0 Hz, CH_2NH), 7.21–7.35 (m, 7H, Ar), 7.37 (s, 2H, Ar), 7.71 (s, 4H, Ar), 8.15 (s, 1H, CH), 11.10 (brs, 1H, NH) ppm;^{13}C NMR (DMSO-d_6): δ = 43.9, 116.2, 127.1, 127.6, 128.6, 128.9, 131.1, 140.2, 142.5, 160.9, 163.5, 163.9 ppm; Anal. Calc. for $C_{24}H_{20}F_2N_8$ (458.47): C, 62.87; H, 4.40; N, 24.44. Found: C, 62.64; H, 4.55, N, 24.69.

*(10) 4,4′-((2,2′-(6-(Benzylamino)-1,3,5-triazine-2,4-diyl)bis(hydrazin-2-yl-1-ylidene))bis(methanylylidene))bis(N,N-dimethylaniline) (**4j**, Supporting Information, Figure S10 in Supplementary Data).* Yellow solid in yield 95%; mp 190-191°C; IR (KBr, cm^{-1}): 3334 (NH), 1681 (C=N), 1605, 1527 (C=C); ^{1}H NMR (DMSO-d_6): δ = 2.96 (d, 12H, J = 9.6 Hz, 2N$(CH_3)_2$), 4.55 (d, 2H, J = 6.4 Hz, CH_2NH), 6.73 (d, 4H, J = 7.2 Hz, Ar), 7.22 (t, 1H, J = 7.2 Hz, Ar), 7.31 (t, 2H, J = 7.2 Hz, Ar), 7.37 (s, 2H, Ar), 7.49 (d, 4H, J = 8.0 Hz, Ar), 8.02 (s, 1H, CH), 10.85 (brs, 1H, NH) ppm; ^{13}C NMR (DMSO-d_6): δ = 43.8, 112.2, 122.5, 127.2, 127.5, 128.7, 129.9, 138.2, 151.6, 152.4, 160.5 ppm; Anal. Calc. for $C_{28}H_{32}N_{10}$ (508.62): C, 66.12; H, 6.34; N, 27.54. Found: C, 66.37; H, 6.55; N, 27.67.

*(11) 4,4′-((2,2′-(6-(Benzylamino)-1,3,5-triazine-2,4-diyl)bis(hydrazin-2-yl-1-ylidene))bis(methanylylidene))bis(benzene-1,2-diol) (**4k**, Supporting Information, Figure S11 in Supplementary Data).* Beige solid in yield 99%; mp 200–202°C; IR (KBr, cm^{-1}): 3385 (OH), 3341 (NH), 1681 (C=N), 1615, 1513 (C=C); ^{1}H NMR (DMSO-d_6): δ = 4.53 (d, 2H, J = 5.6 Hz, CH_2NH), 7.21 (t, 1H, J = 8.0 Hz, Ar), 7.31 (t, 2H, J = 7.6 Hz, Ar), 7.37 (s, 2H, Ar), 7.55–7.66 (m, 6H, Ar), 8.11 (s, 1H, CH), 9.50 (s, 1H, OH), 10.85 (brs, 1H, NH) ppm;^{13}C NMR (DMSO-d_6): δ = 43.9, 114.3, 116.1, 120.1, 122.3, 127.2, 127.7, 128.7, 140.0, 146.0, 147.8, 149.3, 160.9, 172.5 ppm; Anal. Calc for $C_{24}H_{22}N_8O_4$ (486.48): C, 59.25; H, 4.56; N, 23.03. Found: C, 59.02; H, 4.66; N, 23.25.

*(12) N-Benzyl-4,6-bis(2-(1-(p-tolyl)ethylidene)hydrazinyl)-1,3,5-triazine-2-amine (**4l**, Supporting Information, Figure S12 in Supplementary Data).* White solid in yield 87%; mp 103–105°C; IR (KBr, cm^{-1}): 3335 (NH), 1691 (C=N), 1587, 1499 (C=C); ^{1}H NMR (DMSO-d_6): δ = 2.26 (s, 6H, 2CH_3), 2.32 (s, 6H, 2CH_3), 4.56 (d, 2H, J = 6.0 Hz, CH_2NH), 7.19–7.23 (m, 5H, Ar), 7.30 (t, 2H, J = 7.2 Hz, Ar), 7.40 (d, 2H, J =7.6 Hz, Ar), 7.71 (s, 4H, Ar), 9.58 (brs, 1H, NH) ppm; ^{13}C NMR (DMSO-d_6): δ = 14.0, 21.2, 43.8, 126.4, 126.9, 127.8, 128.5, 129.2, 136.4, 138.4, 140.9, 148.3, 165.5, 166.7 ppm; Anal. Calc. for $C_{28}H_{30}N_8$ (478.59): C, 70.27; H, 6.32; N, 23.41. Found: C, 70.50; H, 6.44; N, 23.67.

*(13) 4,4′-((2,2′-(6-(Benzylamino)-1,3,5-triazine-2,4-diyl)bis(hydrazin-2-yl-1-ylidene))bis(ethan-1-yl-1-ylidene))diphenol (**4m**, Supporting Information, Figure S13 in Supplementary Data).* Beige solid in yield 91%; mp. 260–262°C; IR (KBr, cm^{-1}): 3422 (OH), 3317 (NH), 1673 (C=N), 1608, 1502 (C=C); ^{1}H NMR (DMSO-d_6): δ = 2.23 (s, 6H, 2CH_3), 4.56 (d, 2H, J = 5.6 Hz, CH_2NH), 6.78 (d, 4H, J = 8.8 Hz, Ar), 7.21 (t, 1H, J = 7.2, Ar), 7.30 (t, 2H, J = 7.2, Ar), 7.38 (d, 2H, J = 7.2, Ar), 7.66 (s, 4H, Ar), 9.70 (s, 1H, OH) ppm; ^{13}C NMR (DMSO-d_6): δ = 14.1 (2CH_3), 43.9 (CH_2NH), 115.4, 127.1, 127.8, 128.1, 128.6, 129.9, 140.2, 149.3, 158.7, 163.9 ppm; Anal. Calc. for: $C_{26}H_{26}N_8O_2$ (482.54): C, 64.72; H, 5.43; N, 23.22. Found: C, 64.55; H, 5.31; N, 23.00.

*(14) N-Benzyl-4,6-bis(2-benzylidenehydrazinyl)-1,3,5-triazine-2-amine (**5a**, Supporting Information, Figure S14 in Supplementary Data).* White solid in yield 69%; mp 138–140°C; IR (KBr, cm^{-1}): 3225 (NH), 1675 (C=N), 1588, 1516 (C=C); ^{1}H NMR (DMSO-d_6): δ = 3.06 (s, 3H, CH_3N), 4.83 (s, 2H, CH_2N), 7.22–7.46 (m, 11H, Ar), 7.62 (brs, 4H, Ar), 8.01 (s, 1H, CH), 11.20 (brs, 1H, NH) ppm;^{13}C NMR (DMSO-d_6): δ = 34.2, 51.3, 126.9, 127.5, 127.9, 128.9, 129.2, 129.7, 135.4, 138.6, 142.8, 164.6, 165.9 ppm; Anal. Calc. for $C_{25}H_{24}N_8$ (436.51): C, 68.79; H, 5.54; N, 25.67. Found: C, 69.10; H, 5.33; N, 25.98.

*(15) N-Benzyl-N-methyl-4-(2-(4-methylbenzylidene)hydrazinyl)-6-(-2-(4-methylbenzylidene)hydrazinyl)-1,3,5-triazine-2-amine (**5b**, Supporting Information, Figure S15 in Supplementary Data).* White solid in yield 86%; mp 253–255°C; IR (KBr, cm^{-1}): 3229 (NH), 1608 (C=N), 1553, 1527 (C=C); ^{1}H NMR (DMSO-d_6): δ = 2.33 (s, 6H, 2CH_3), 3.10 (s, 3H, CH_3N), 4.87 (s, 2H, CH_2N), 7.21–7.26 (m, 4H, Ar), 7.31–7.38

(m, 5H, Ar), 7.53 (d, 4H, J = 8.0 Hz, Ar) 8.11 (s, 1H, CH), 11.90 (brs, 1H, NH) ppm; ^{13}C NMR (DMSO-d$_6$): δ = 21.4, 34.1, 51.1, 126.8, 127.4, 128.1, 128.8, 129.8, 132.8, 138.7, 139.1, 142.8, 164.7, 165.9 ppm; Anal. Calc. for $C_{27}H_{28}N_8$ (464.56): C, 69.80; H, 6.08; N, 24.12. Found: C, 70.03; H, 6.33; N, 24.42.

*(16) N-Benzyl-4-(2-(4-methoxybenzylidene)hydrazinyl)-6-(2-(4-methoxy benzylidene)hydrazinyl)-N-methyl-1,3,5-triazine-2-amine (**5c**, Supporting Information, Figure S16 in Supplementary Data).* Beige solid in yield 58%; mp 245–247°C; IR (KBr, cm^{-1}): 3354 (NH), 1678 (C=N), 1605, 1511 (C=C); ^{1}H NMR (DMSO-d$_5$): δ = 3.13 (s, 3H, CH$_3$N), 3.80 (s, 3H, OCH$_3$), 3.81 (s, 3H, OCH$_3$), 4.90 (s, 2H, CH$_2$N), 6.98 (d, 4H, J = 8.0 Hz, Ar), 7.21–7.27 (m, 1H, Ar), 7.30–7.39 (m, 4H, J = 8.0 Hz, Ar), 7.57 (d, 4H, J = 8.0 Hz, Ar), 8.12 (s, 1H, CH), 10.80 (brs, 1H, NH) ppm; ^{13}C NMR (DMSO-d$_6$): δ = 34.2, 55.8, 114.7, 127.4, 128.9, 129.6, 132.8, 139.5, 142.8, 164.7, 172.6. Anal. Calc. for $C_{27}H_{28}N_8O_2$ (496.58): C, 65.31; H, 5.68; N, 22.57. Found: C, 65.55; H, 5.81; N, 22.79.

*(17) 4,4′-(2,2′-(6-(Benzyl(methyl)amino)-1,3,5-triazine-2,4-diyl)bis(hydrazin-2-yl-1-ylidene))bis(methanylylidene)diphenol (**5d**, Supporting Information, Figure S17 in Supplementary Data).* Beige solid in yield 63%; mp 216–218°C; IR (KBr, cm^{-1}): 3407 (OH), 3345 (NH), 1680 (C=N), 1599, 1512 (C=C); ^{1}H NMR (DMSO-d$_6$): δ = 3.05 (s, 3H, CH$_3$N), 4.82 (s, 2H, CH$_2$N), 6.77 (brs, 4H, Ar), 7.23–7.36 (m, 5H, Ar), 7.44 (s, 4H, Ar), 8.01 (brs, 1H, CH), 9.78 (s, 1H, NH), 10.70 (s, 1H, OH) ppm; ^{13}C NMR (DMSO-d$_6$): δ = 34.2, 51.3, 116.0, 126.5, 127.5, 127.8, 128.5, 128.9, 138.8, 143.2, 150.3, 159.0, 164.5, 172.5 ppm; Anal. Calc. for $C_{25}H_{24}N_8O_2$ (468.51): C, 64.09; H, 5.16; N, 23.92. Found: C, 64.36; H, 5.00; N, 24.10.

*(18) N-Benzyl-4-(2-(4-bromobenzylidene)hydrazinyl)-6-(2-(4-bromobenzylidene)hydrazinyl)-N-methyl-1,3,5-triazine-2-amine (**5e**, Supporting Information, Figure S18 in Supplementary Data).* White solid in yield 99%; mp 227–229°C; IR (KBr, cm^{-1}): 3227 (NH), 1548 (C=N), 1524, 1404 (C=C); ^{1}H NMR (DMSO-d$_6$): δ = 3.10 (s, 3H, CH$_3$N), 4.86 (s, 2H, CH$_2$N), 7.21–7.26 (m, 1H, Ar), 7.31–7.39 (m, 4H, Ar), 7.59–7.66 (m, 8H, Ar), 8.11 (s, 1H, CH), 11.10 (brs, 1H, NH) ppm; ^{13}C NMR (DMSO-d$_6$): δ = 34.2, 51.2, 122.6, 127.5, 128.0, 128.7, 128.8, 132.2, 134.8, 138.6, 141.5, 164.7, 165.9 ppm; Anal. Calc. for $C_{25}H_{22}Br_2N_8$ (594.30): C, 50.52; H, 3.73; N, 18.85. Found: C, 50.19; H, 3.99; N, 18.61.

*(19) N-Benzyl-4-(2-(4-chlorobenzylidene)hydrazinyl)-6-(2-(4-chlorobenzylidene)hydrazinyl)-N-methyl-1,3,5-triazine-2-amine (**5f**, Supporting Information, Figure S19 in Supplementary Data).* White solid in yield 82%; mp 225–227°C; IR (KBr, cm^{-1}): 3332 (NH), 1584 (C=N), 1514, 1370 (C=C); ^{1}H NMR (DMSO-d$_6$): δ = 3.10 (s, 3H, CH$_3$N), 4.87 (s, 2H, CH$_2$N), 7.21–7.27 (m, 1H, Ar), 7.30–7.38 (m, 4H, Ar), 7.47 (d, 4H, J = 8.0 Hz, Ar), 7.65 (d, 4H, J = 8.0 Hz, Ar), 8.13 (s, 1H, CH), 11.10 (brs, 1H, NH) ppm; ^{13}C NMR (DMSO-d$_6$): δ = 34.2, 51.2, 127.5, 128.0, 128.4, 128.9, 129.3, 133.9, 134.4, 138.6, 141.4, 164.7, 165.9 ppm. Anal. Calc. for $C_{25}H_{22}Cl_2N_8$ (505.41): C, 59.41; H, 4.39; N, 22.17. Found: C, 59.66; H, 4.49; N, 22.00.

*(20) N-Benzyl-4-(2-(4-fluorobenzylidene)hydrazinyl)-6-(2-(4-fluorobenzylidene)hydrazinyl)-N-methyl-1,3,5-triazine-2-amine (**5g**, Supporting Information, Figure S20 in Supplementary Data).* White solid in yield 69%; mp 210–212°C; IR (KBr, cm^{-1}): 3332 (NH), 1602 (C=N), 1500, 1372 (C=C); ^{1}H NMR (DMSO-d$_6$): δ = 3.10 (s, 3H, CH$_3$N), 4.87 (s, 2H, CH$_2$N), 7.21–7.29 (m, 5H, Ar), 7.31–7.38 (m, 4H, Ar), 7.69 (s, 4H, Ar), 8.14 (s, 1H, CH), 11.0 (brs, 1H, NH) ppm; ^{13}C NMR (DMSO-d$_6$): δ = 34.2, 51.3, 116.2, 127.4, 128.1, 128.8, 128.9, 130.6, 138.7, 141.6, 162.1, 163.8, 164.7, 165.0 ppm. Anal. Calc. for $C_{25}H_{22}F_2N_8$ (472.50): C, 63.55; H, 4.69; N, 23.72. Found: C, 63.87; H, 4.89; N, 23.99.

*(21) N-Benzyl-N-methyl-4-(2-(1-(p-tolyl)ethylidene)hydrazinyl)-6-(2-(1-(p-tolyl)ethylidene)hydrazinyl)-1,3,5-triazine-2-amine (**5h**, Supporting Information, Figure S21 in Supplementary Data).* White solid in yield 46%; mp 150–152°C; IR (KBr, cm^{-1}): 3353 (NH), 1711 (C=N), 1586, 1494 (C=C); ^{1}H NMR (DMSO-d$_6$): δ = 2.25 (s, 6H, 2CH$_3$), 2.32 (s, 6H, 2CH$_3$), 3.11 (s, 3H, CH$_3$N), 4.90 (s, 2H, CH$_2$N), 7.19–7.27 (m, 5H, Ar), 7.31–7.36 (m, 2H, Ar), 7.57–7.75 (m, 8H, Ar), 11.0 (brs, 1H, NH) ppm; ^{13}C NMR (DMSO-d$_6$): δ = 13.9, 21.2, 34.1, 51.1, 126.3, 127.4, 128.1, 128.8, 129.2, 136.4, 138.4, 138.8, 147.9, 165.5, 166.0 ppm; Anal. Calc. for: $C_{29}H_{32}N_8$ (492.62): C, 70.71; H, 6.55; N, 22.75. Found: C, 70.99; H, 6.76; N, 22.98.

*(22) 4,4′-(2,2′-(6-(Benzyl(methyl)amino)-1,3,5-triazine-2,4-diyl)bis(hydrazin-2-yl-1-ylidene))bis(ethan-1-yl-1-ylidene)diphenol (**5i**, Supporting Information, Figure S22 in Supplementary Data).* Beige solid in yield 89%; mp 205–207°C; IR (KBr, cm^{-1}): 3445 (OH), 3347 (NH), 1674 (C=N), 1597, 1503 (C=C); ^{1}H NMR (DMSO-d$_6$): δ = 2.25 (s, 6H, 2CH$_3$), 3.11 (s, 3H, CH$_3$N), 4.90 (s, 2H, CH$_2$N), 6.78 (d, 4H, J = 8.0 Hz, Ar), 7.25–7.27 (m, 1H, Ar), 7.30–7.38 (m, 4H, Ar), 7.68 (d, 4H, J = 8.0 Hz, Ar), 9.70 (s, 1H, OH), 11.0 (brs, 1H, NH) ppm; ^{13}C NMR (DMSO-d$_6$): δ = 21.5, 34.3, 51.3, 115.4, 127.5, 127.8, 128.0, 128.9, 131.1, 140.5, 145.4, 158.7, 166.2, 172.5 ppm; Anal. Calc. for $C_{27}H_{28}N_8O_2$ (496.56): C, 65.31; H, 5.68; N, 22.57. Found: C, 65.63; H, 5.33; N, 22.89.

*(23) 4-(4,6-bis(2-Benzylidenehydrazinyl)-1,3,5-triazine-2-yl)morpholine (**6a**, Supporting Information, Figure S23 in Supplementary Data).* White solid in yield 88%; mp 273–275°C; IR (KBr, cm^{-1}): 3228 (NH), 1596 (C=N), 1559, 1510 (C=C); ^{1}H NMR (DMSO-d$_6$): δ = 3.66 (d, 4H, J = 4.4 Hz, 2 OCH$_2$-), 3.78 (s, 4H, 2 NCH$_2$-), 7.36 (t, 2H, J = 7.2 Hz, Ar), 7.42 (t, 4H, J = 7.2 Hz, Ar), 7.65 (d, 4H, J = 7.6 Hz, Ar), 8.14 (s, 1H, CH), 11.0 (brs, 1H, NH) ppm; ^{13}C NMR (DMSO-d$_6$): δ = 43.8, 66.5, 126.9, 129.2, 129.6, 135.5, 142.8, 164.8, 165.3 ppm; Anal. Calc. for $C_{21}H_{22}N_8O$ (402.45): C, 62.67; H, 5.51; N, 27.84. Found: C, 62.88; H, 5.71; N, 28.00. *(m/z)* Calcd: 402.45; LC-MS [M + H]; Found: 403 *(Supporting Information, Figure S24 in Supplementary Data).*

*(24) 4-(4,6-Bis(2-(4-methylbenzylidene)hydrazinyl)-1,3,5-triazine-2-yl)morpholine (**6b**, Supporting Information, Figure S25 in Supplementary Data).* White solid in yield 92%; mp 298–300°C; IR (KBr, cm^{-1}): 3225 (NH), 1601 (C=N), 1559, 1503 (C=C); ^{1}H NMR (DMSO-d$_6$): δ = 2.32 (s, 6H, 2CH$_3$),

3.64 (brs, 4H, 2 OCH_2-), 3.77 (brs, 4H, 2 NCH_2-), 7.23 (d, 4H, J = 8.4 Hz, Ar), 7.55 (d, 4H, J = 7.2 Hz, Ar), 8.10 (s, 1H, CH), 10.90 (brs, 1H, NH) ppm; ^{13}C NMR (DMSO-d_6): δ = 20.4, 43.4, 66.1, 126.6, 129.4, 132.2, 138.9, 143.3, 164.0 ppm; Anal. Calc. for: $C_{23}H_{26}N_8O$ (430.51): C, 64.17; H, 6.09; N, 26.03. Found: C, 64.00; H, 6.21; N, 26.29.

*(25) 4-(4,6-Bis(2-(4-methoxybenzylidene)hydrazinyl)-1,3,5-triazine-2-yl)morpholine (**6c**, Supporting Information, Figure S26 in Supplementary Data).* White solid in yield 89%; mp. 253–255°C; IR (KBr, cm^{-1}): 3226 (NH), 1605 (C=N), 1565, 1514 (C=C); ^1H NMR (DMSO-d_6): δ = 3.65 (brs, 4H, 2 OCH_2-), 3.78 (brs, 4H, 2 NCH_2-), 3.81 (s, 6H, 2OCH_3), 6.98 (d, 4H, J = 8.0 Hz, Ar), 7.58 (d, 4H, J = 8.8 Hz, Ar), 8.10 (s, 1H, CH), 10.80 (brs, 1H, NH) ppm; ^{13}C NMR (DMSO-d_6): δ = 43.3, 55.2, 66.1, 114.2, 127.6, 127.9, 142.3, 160.1, 164.2, 165.0 ppm; Anal. Calc. for $C_{23}H_{26}N_8O_3$ (462.51): C, 59.73; H, 5.67; N, 24.23. Found: C, 59.92; H, 5.80; N, 24.47.

*(26) 4,4′-(2,2′-(6-Morpholino-1,3,5-triazine-2,4-diyl)bis(hydrazin-2-yl-1-ylidene)bis(methanylylidene))diphenol (**6d**, Supporting Information, Figure S27 in Supplementary Data).* White solid in yield 94%; mp 275–277°C; IR (KBr, cm^{-1}): 3435 (OH), 3349 (NH), 1677 (C=N), 1582, 1509 (C=C); ^1H NMR (DMSO-d_6): δ = 3.64 (brs, 4H, 2 OCH_2-), 3.76 (s, 4H, 2 NCH_2-), 6.82 (d, 4H, J = 5.2 Hz, Ar), 7.49 (s, 4H, Ar), 8.05 (s, 1H, CH), 9.85 (brs, 1H, OH), 11.16 (brs, 1H, NH) ppm; ^{13}C NMR (DMSO-d_6): δ = 43.0, 66.0, 115.7, 125.8, 128.3, 145.0, 149.9, 159.0, 163.5 ppm; Anal. Calc. for: $C_{21}H_{22}N_8O_3$ (434.45): 58.06; H, 5.10; N, 25.79. Found: C, 58.33; H, 5.31; N, 25.50. (m/z) Calcd: 434.45; LC-MS (M + H); Found: 435.2.

*(27) 4-(4,6-Bis(2-(4-Bromobenzylidene)hydrazinyl)-1,3,5-triazine-2-yl)morpholine (**6e**, Supporting Information, Figure S28 in Supplementary Data).* White solid in yield 83%; mp. 308–310°C; IR (KBr, cm^{-1}): 3225 (NH), 1593 (C=N), 1561, 1515 (C=C); ^1H NMR (DMSO-d_6): δ = 3.65 (brs, 4H, 2 OCH_2-), 3.77 (brs, 4H, 2 NCH_2-), 7.58 (m, 8H, Ar), 8.10 (s, 1H, CH), 11.10 (brs, 1H, NH) ppm; ^{13}C NMR (DMSO-d_6): δ = 43.5, 66.0, 122.2, 128.3, 131.8, 134.2, 141.3, 164.2, 164.8 ppm; Anal. Calc. for: $C_{21}H_{20}Br_2N_8O$ (560.24): C, 45.02; H, 3.60; N, 20.00. Found: C, 45.31; H, 3.43; N, 20.28. (m/z) Calcd: 560.24; LC-MS [M + H]; Found: 561.1.

*(28) 4-(4,6-Bis(2-(4-Chlorobenzylidene)hydrazinyl)-1,3,5-triazine-2-yl) (**6f**, Supporting Information, Figure S29 in Supplementary Data).* White solid in yield 96%; mp. 305–307°C; IR (KBr, cm^{-1}): 3226 (NH), 1596 (C=N), 1563, 1515 (C=C); ^1H NMR (DMSO-d_6): δ = 3.65 (brs, 4H, 2 OCH_2-), 3.78 (brs, 4H, 2 NCH_2-), 7.48 (d, 4H, J = 8.0 Hz, Ar), 7.66 (d, 4H, J = 8.8 Hz, Ar), 8.12 (s, 1H, CH), 11.10 (brs, 1H, NH) ppm; ^{13}C NMR (DMSO-d6): δ = 43.3, 66.0, 128.0, 128.8, 133.4, 133.9, 141.1, 164.3, 164.8 ppm; Anal. Calc. for: $C_{21}H_{20}Cl_2N_8O$ (471.34): C, 53.51; H, 4.28; N, 23.77. Found: C, 53.77; H, 4.01; N, 23.56. (m/z) Calcd: 471.34; LC-MS [M + H]; Found: 471.2.

*(29) 4-(4,6-Bis(2-(4-Fluorobenzylidene)hydrazinyl)-1,3,5-triazine-2-yl)morpholine (**6g**, Supporting Information, Figure S30 in Supplementary Data).* White solid in yield 91%; mp. 301–303°C; IR (KBr, cm^{-1}): 3229 (NH), 1601 (C=N), 1568, 1515 (C=C); ^1H NMR (DMSO-d_6): δ = 3.65 (brs, 4H, 2 OCH_2-), 3.78 (brs, 4H, 2 NCH_2-), 7.25–7.34 (m, 4H, Ar), 7.69–7.74 (m, 4H, Ar), 8.14 (s, 1H, CH), 11.10 (brs, 1H, NH) ppm; ^{13}C NMR (DMSO-d_6): δ = 43.3, 66.1, 115.8, 128.8, 131.4, 141.4, 161.5, 163.9 ppm; Anal. Calc. for $C_{21}H_{20}F_2N_8O$ (438.43): C, 57.53; H, 4.60; N, 25.56. Found: C, 57.87; H, 4.32; N, 25.23.

*(30) 4-(4,6-Bis(2-(1-(p-Tolyl)ethylidene)hydrazinyl)-1,3,5-triazine-2-yl)morpholine (**6h**, Supporting Information, Figure S31 in Supplementary Data).* The product was obtained as a yellow solid in yield 92%; mp. 235-236°C; IR (KBr, cm^{-1}): 3358 (NH), 1643 (C=N), 1563, 1442 (C=C); ^1H NMR (DMSO-d_6): δ = 2.27 (s, 6H, 2CH_3CN), 2.32 (s, 6H, 2CH_3), δ 3.65 (t, 4H, J = 4.0 Hz, 2 OCH_2-), δ 3.80 (t, 4H, J = 4.4 Hz, 2 NCH_2-), 7.20 (d, 4H, J = 8.0 Hz, Ar), δ 7.72 (d, 4H, J = 7.2 Hz, Ar), δ 11.18 (brs, 1H, NH) ppm; ^{13}C NMR (DMSO-d_6): δ = 13.6, 20.8, 43.3, 66.1, 125.9, 128.8, 135.9, 138.0, 149.9, 165.0 ppm; Anal. Calc. for $C_{25}H_{30}N_8O$ (458.56): C, 65.48; H, 6.59; N, 24.44. Found: C, 65.71; H, 6.33; N, 24.29.

*(31) 4,4′-((2,2′-(6-Morpholino-1,3,5-triazine-2,4-diyl)bis(hydrazin-2-yl-1-ylidene))bis(ethan-1-yl-1-ylidene))diphenol (**6i**, Supporting Information, Figure S32 in Supplementary Data).* White solid in yield 93%; mp. 265–267°C; IR (KBr, cm^{-1}): 3430 (OH), 3360 (NH), 1673 (C=N), 1593, 1503 (C=C); ^1H NMR (DMSO-d_6): δ = 2.28 (s, 6H, 2CH_3CN), 3.66 (brs, 4H, 2 OCH_2-), 3.82 (brs, 4H, 2 NCH_2-), 6.80 (d, 4H, J = 8.0 Hz, Ar), 7.69 (d, 4H, J = 8.4 Hz, Ar), 9.80 (brs, 1H, OH), 11.80 (brs, 1H, NH) ppm; ^{13}C NMR (DMSO-d_6): δ = 14.3, 44.1, 66.4, 115.6, 128.2, 129.0, 152.0, 159.1, 165.5, 172.5 ppm; Anal. Calc. for $C_{23}H_{26}N_8O_3$ (462.50): C, 59.73; H, 5.67; N, 24.23. Found: C, 59.99; H, 5.83; N, 24.51.

*(32) 2,4-Bis(2-benzylidenehydrazinyl)-6-(piperidin-1-yl)-1,3,5-triazine (**7a**, Supporting Information, Figure S33 in Supplementary Data).* White solid in yield 94%; mp. 247–249°C; IR (KBr, cm^{-1}): 3228 (NH), 1677 (C=N), 1550, 1512 (C=C); ^1H NMR (DMSO-d_6): δ = 1.54 (s, 4H, 2CH_2), 1.63 (s, 2H, CH_2), 3.80 (s, 4H, 2CH_2N), 7.43 (brs, 6H, Ar), 7.70 (brs, 4H, Ar), 8.18 (s, 1H, CH), 11.0 (brs, 1H, NH) ppm; ^{13}C NMR (DMSO-d_6): δ = 24.7, 25.9, 44.3, 127.1, 129.2, 129.7, 134.9, 142.5, 165.1 ppm; Anal. Calc. for $C_{22}H_{24}N_8$ (400.49): C, 65.98; H, 6.04; N, 27.98. Found: C, 65.79; H, 6.19; N, 28.14. (m/z) Calcd: 400.49; LC-MS [M + H]; Found: 401 *(Supporting Information, Figure S50 in Supplementary Data).*

*(33) 2,4-Bis(2-(4-methylbenzylidene)hydrazinyl)-6-(piperidin-1-yl)-1,3,5-triazine (**7b**, Supporting Information, Figure S34 in Supplementary Data).* White solid in yield 89%; mp. 282–284°C; IR (KBr, cm^{-1}): 3235 (NH), 1673 (C=N), 1567, 1509 (C=C); ^1H NMR (DMSO-d_6): δ = 1.58 (s, 4H, 2CH_2), 1.62 (d, 2H, J = 4.4 Hz, CH_2), 2.32 (s, 6H, 2CH_3), 3.78 (s, 4H, 2CH_2N), 7.24 (d, 4H, J = 7.2 Hz, Ar), 7.54 (d, 4H, J = 7.6 Hz, Ar), 8.10 (s, 1H, CH), 10.90 (brs, 1H, NH) ppm; ^{13}C NMR (DMSO-d_6): δ = 20.9, 24.3, 25.5, 43.6, 126.4, 129.3, 132.2, 138.7, 142.4, 164.0 ppm; Anal. Calcd for $C_{24}H_{28}N_8$ (428.53): C, 67.27; H, 6.59; N, 26.15. Found: C, 67.43; H, 6.71; N, 26.38.

*(34) 2,4-Bis(2-(4-methoxybenzylidene)hydrazinyl)-6-(piperidin-1-yl)-1,3,5-triazine (**7c**, Supporting Information, Figure S35 in Supplementary Data).* White solid in yield 94%; mp. 257–259°C; IR (KBr, cm^{-1}): 3341 (NH), 1665 (C=N), 1570, 1512 (C=C); ^1H NMR (DMSO-d_6): δ = 1.52 (s, 4H, $2CH_2$), 1.62 (s, 2H, CH_2), 3.79 (s, 6H, $2OCH_3$), 3.81 (s, 4H, $2CH_2N$), 7.0 (d, 4H, J = 8.8 Hz, Ar), 7.60 (d, 4H, J = 8.0 Hz, Ar), 8.09 (s, 1H, CH), 10.90 (brs, 1H, NH) ppm; ^{13}C NMR (DMSO-d_6): δ = 24.2, 25.4, 43.8, 55.3, 114.2, 126.5, 128.2, 140.8, 160.4, 161.6 ppm.

*(35) 4,4′-((2,2′-(6-(Piperidin-1-yl)-1,3,5-triazine-2,4-diyl)bis(hydrazin-2-yl-1-ylidene))bis(methanylylidene))diphenol (**7d**, Supporting Information, Figure S36 in Supplementary Data).* Beige solid in yield 99%; mp. 331–333°C; IR (KBr, cm^{-1}): 3435 (OH), 3349 (NH), 1677 (C=N), 1590, 1511 (C=C); ^1H NMR (DMSO-d_6): δ = 1.54 (s, 4H, $2CH_2$), 1.63 (s, 2H, CH_2), 3.78 (s, 4H, $2CH_2N$), 6.86 (s, 4H, Ar), 7.56 (s, 4H, Ar), 8.09 (s, 1H, CH), 9.80 (s, 1H, OH), 10.30 (brs, 1H, NH) ppm; ^{13}C NMR (DMSO-d_6): δ = 24.6, 25.9, 44.5, 116.2, 125.8, 129.3, 142.7, 159.5, 164.0 ppm; Anal. Calcd for: $C_{22}H_{24}N_8O_2$ (432.48): C, 61.10; H, 5.59; N, 25.91. Found: C, 61.33; H, 5.78; N, 25.72. (m/z) Calcd: 432.48; LC-MS [M + H]; Found: 433 *(Supporting Information, Figure S51 in Supplementary Data).*

*(36) 2,4-Bis(2-(4-Bromobenzylidene)hydrazinyl)-6-(piperidin-1-yl)-1,3,5-triazine (**7e**, Supporting Information, Figure S37 in Supplementary Data).* White solid in yield 96%; mp. 288–290°C; IR (KBr, cm^{-1}): 3240 (NH), 1673 (C=N), 1569, 1512 (C=C); ^1H NMR (DMSO-d_6): δ = 1.52 (s, 4H, $2CH_2$), 1.63 (d, 2H, J = 4.4 Hz, CH_2), 3.78 (s, 4H, $2CH_2N$), 7.60 (d, 4H, J = 8.0 Hz, Ar), 7.64 (d, 4H, J = 8.0 Hz, Ar), 8.11 (s, 1H, CH), 11.15 (brs, 1H, NH) ppm; ^{13}C NMR (DMSO-d_6): δ = 24.3, 25.5, 43.5, 122.1, 128.2, 131.7, 134.3, 140.9, 164.3 ppm; Anal. Calcd for $C_{22}H_{22}Br_2N_8$ (558.27): C, 47.33; H, 3.97; N, 20.07. Found: C, 47.61; H, 3.77; N, 20.25.

*(37) 2,4-Bis(2-(4-chlorobenzylidene)hydrazinyl)-6-(piperidin-1-yl)-1,3,5-triazine (**7f**, Supporting Information, Figure S38 in Supplementary Data).* White solid in yield 97%; mp. 293–295°C; IR (KBr, cm^{-1}): 3237 (NH), 1679 (C=N), 1570, 1513 (C=C); ^1H NMR (DMSO-d_6): δ = 1.52 (s, 4H, $2CH_2$), 1.62 (d, 2H, J = 4.4 Hz, CH_2), 3.78 (s, 4H, $2CH_2N$), 7.49 (d, 4H, J = 8.0 Hz, Ar), 7.68 (d, 4H, J = 8.4 Hz, Ar), 8.13 (s, 1H, CH), 11.20 (brs, 1H, NH) ppm; ^{13}C NMR (DMSO-d_6): δ = 24.2, 25.5, 43.7, 128.1, 128.8, 132.6, 135.9, 141.3, 160, 164.0 ppm; Anal. Calcd for $C_{22}H_{22}Cl_2N_8$ (469.37): C, 56.30; H, 4.72; N, 23.87. Found: C, 56.55; H, 4.94; N, 24.03. (m/z) Calcd: 469.37; LC-MS [M]; Found: 469 *(Supporting Information, Figure S52 in Supplementary Data).*

*(38) 2,4-Bis(2-(4-Fluorobenzylidene)hydrazinyl)-6-(piperidin-1-yl)-1,3,5-triazine (**7g**, Supporting Information, Figure S39 in Supplementary Data).* White solid in yield 95%; mp. 275–277°C; IR (KBr, cm^{-1}): 3358 (NH), 1680 (C=N), 1603, 1501 (C=C); ^1H NMR (DMSO-d_6): δ = 1.56 (s, 4H, $2CH_2$), 1.64 (s, 2H, CH_2), 3.82 (s, 4H, $2CH_2N$), 7.32 (brs, 4H, Ar), 7.85 (brs, 4H, Ar), 8.23 (brs, 1H, CH), 12.50 (brs, 1H, NH) ppm; ^{13}C NMR (DMSO-d_6): δ = 24.4, 25.9, 44.7, 116.4, 129.7, 130.8, 144.2, 162.7, 164.2 ppm; Anal. Calcd for $C_{22}H_{22}F_2N_8$ (436.46): C, 60.54; H, 5.08; N, 25.67. Found: C, 60.51; H, 5.23; N, 25.38.

*(39) 2-(Piperidin-1-yl)-4,6-bis(2-(1-(p-tolyl)ethylidene)hydrazinyl)-1,3,5-triazine (**7h**, Supporting Information, Figure S40 in Supplementary Data).* White solid in yield 91%; mp. 222–224°C; IR (KBr, cm^{-1}): 3356 (NH), 1628 (C=N), 1562, 1437 (C=C); ^1H NMR (DMSO-d_6): δ = 1.52 (s, 4H, $2CH_2$), 1.62 (s, 2H, CH_2), 2.28 (s, 6H, $2CH_3CN$), 2.33 (s, 6H, $2CH_3$), 3.81 (s, 4H, $2CH_2N$), 7.19 (d, 4H, J = 8.0 Hz, Ar), 7.71 (d, 4H, J = 8.0 Hz, Ar), 9.70 (brs, 1H, NH) ppm; ^{13}C NMR (DMSO-d_6): δ = 13.9, 21.3, 24.9, 26.0, 43.9, 126.4, 129.3, 136.5, 138.4, 147.9, 164.8, 165.6 ppm; Anal. Calcd for $C_{26}H_{32}N_8$ (456.59): C, 68.39; H, 7.06; N, 24.54. Found: C, 68.65; H, 7.33; N, 24.23.

*(40) 4,4′-((2,2′-(6-(Piperidin-1-yl)-1,3,5-triazine-2,4-diyl)bis(hydrazin-2-yl-1-ylidene))bis(ethan-1-yl-1-ylidene))diphenol (**7i**, Supporting Information, Figure S41 in Supplementary Data).* Beige solid in yield 85%; mp. 288–290°C; IR (KBr, cm^{-1}): 3440 (OH), 3358 (NH), 1672 (C=N), 1597, 1500 (C=C); 1H NMR (DMSO-d_6): δ = 1.55 (s, 4H, $2CH_2$), 1.64 (s, 2H, CH_2), 2.31 (s, 6H, $2CH_3$), 3.84 (s, 4H, $2CH_2N$), 6.82 (brs, 4H, Ar), 7.70 (brs, 4H, Ar), 9.90 (brs, 2H, OH, NH) ppm; ^{13}C NMR (DMSO-d_6): δ = 14.4, 24.6, 26.0, 44.3, 115.6, 128.2, 131.1, 154.2, 159.1, 160.0 ppm; Anal. Calcd for $C_{24}H_{28}N_8O_2$ (460.53): C, 62.59; H, 6.13; N, 24.33. Found: C, 62.82; H, 6.42; N, 24.61. (m/z) Calcd: 460.53; LC-MS [M + 2H]; Found: 463.2 *(Supporting Information, Figure S42 in Supplementary Data).*

*(41) 2,4-Bis(2-benzylidenehydrazinyl)-6-methoxy-1,3,5-triazine (**8a**, Supporting Information, Figure S43 in Supplementary Data).* Beige solid in yield 70%; mp 292–294°C; IR (KBr, cm^{-1}): 3224 (NH), 1620 (C=N), 1586, 1547 (C=C); ^1H NMR (DMSO-d_6): δ = 3.91 (s, 3H, OCH_3), 7.36–7.47 (m, 6H, Ar), 7.67 (d, 4H, J = 7.2 Hz, Ar), 8.19 (s, 1H, CH), 11.35 (brs, 1H, NH), 11.51 (brs, 1H, NH) ppm; ^{13}C NMR (DMSO-d_6): δ = 54.0, 126.8, 128.8, 129.5, 134.7, 144.5, 161.4, 165.5, 171.0 ppm; Anal. Calcd for $C_{18}H_{17}N_7O$ (347.38): C, 62.24; H, 4.93; N, 28.23. Found: C, 62.51; H, 5.04; N, 28.49.

*(42) 2-Methoxy-4,6-bis(2-(4-methylbenzylidene)hydrazinyl)-1,3,5-triazine (**8b**, Supporting Information, Figure S44 in Supplementary Data).* White solid in yield 84%; mp 298–300°C; IR (KBr, cm^{-1}): 3323 (NH), 1616 (C=N), 1586, 1540 (C=C); ^1H NMR (DMSO-d_6): δ = 2.33 (s, 6H, $2CH_3$), 3.89 (s, 3H, OCH_3), 7.25 (d, 4H, J = 7.2 Hz, Ar), 7.56 (d, 4H, J = 7.2 Hz, Ar), 8.14 (s, 1H, CH), 11.28 (brs, 1H, NH), ppm; ^{13}C NMR (DMSO-d_6): δ = 21.5, 54.4, 127.1, 129.8, 132.4, 139.6, 144.3, 165.5, 166.2 ppm; Anal. Calcd for $C_{20}H_{21}N_7O$ (375.44): C, 63.98; H, 5.64; N, 26.12. Found: C, 63.82; H, 5.73; N, 26.33.

*(43) 2-Methoxy-4,6-bis(2-(4-methoxybenzylidene)hydrazinyl)-1,3,5-triazine (**8c**, Supporting Information, Figure S45 in Supplementary Data).* White solid in yield 86%; mp 300–302°C; IR (KBr, cm^{-1}): 3277 (NH), 1612 (C=N), 1589, 1545 (C=C); ^1H NMR (DMSO-d_6): δ = 3.79 (s, 6H, $2OCH_3$), 3.89 (s, 3H, OCH_3), 7.0 (d, 4H, J = 8.0 Hz, Ar), 7.61 (d, 4H, J = 7.2 Hz, Ar),

8.13 (s, 1H, CH), 11.22 (brs, 2H, NH) ppm; ^{13}C NMR (DMSO-d_6): $\delta = 53.9, 55.3, 114.3, 127.2, 128.2, 144.0, 160.4, 164.7$ ppm; Anal. Calcd for $C_{20}H_{21}N_7O_3$ (407.43): C, 58.96; H, 5.20; N, 24.06. Found: C, 58.77; H, 5.02; N, 24.33.

*(44) 4,4′-(2,2′-(6-Methoxy-1,3,5-triazine-2,4-diyl)bis(hydrazin-2-yl-1-ylidene)bis(methanylylidene))diphenol (**8d**, Supporting Information, Figure S46 in Supplementary Data).* Beige solid in yield 83%; mp 294–296°C; IR (KBr, cm^{-1}): 3522 (OH), 3268 (NH), 1618 (C=N), 1545, 1514 (C=C); ^{1}H NMR (DMSO-d_6): $\delta = 3.84$ (s, 3H, OCH_3), 6.77 (d, 4H, $J = 8.0$ Hz, Ar), 7.45 (d, 4H, $J = 7.2$ Hz, Ar), 8.03 (s, 1H, CH), 9.83 (s, 1H, OH), 11.10–11.30 (brs, 2H, NH) ppm; ^{13}C NMR (DMSO-d_6): $\delta = 53.9, 115.7, 125.8, 128.4, 144.0, 158.9, 165.8, 171.0$ ppm; Anal. Calcd for $C_{18}H_{17}N_7O_3$ (379.37): C, 56.99; H, 4.52; N, 25.84. Found: C, 56.76; H, 4.76; N, 26.07.

*(45) 2,4-Bis(2-(4-bromobenzylidene)hydrazinyl)-6-methoxy-1,3,5-triazine (**8e**, Supporting Information, Figure S47 in Supplementary Data).* White solid in yield 87%; mp. 317–319°C; IR (KBr, cm^{-1}): 3318 (NH), 1678 (C=N), 1586, 1535 (C=C); ^{1}H NMR (DMSO-d_6): $\delta = 3.89$ (s, 3H, OCH_3), 7.62 (brs, 8H, Ar), 8.15 (s, 1H, CH), 11.4 (brs, 1H, NH), 11.70 (brs, 1H, NH) ppm; ^{13}C NMR (DMSO-d_6): $\delta = 54.0, 122.6, 128.5, 131.7, 133.9, 142.8, 163.8$ ppm; Anal. Calcd for $C_{18}H_{15}Br_2N_7O$ (505.17): C, 42.80; H, 2.99; N, 19.41. Found: C, 43.01; H, 3.15; N, 19.66.

*(46) 2,4-Bis(2-(4-chlorobenzylidene)hydrazinyl)-6-methoxy-1,3,5-triazine (**8f**, Supporting Information, Figure S48 in Supplementary Data).* Beige solid in yield 86%; mp 308–310°C; IR (KBr, cm^{-1}): 3315 (NH), 1616 (C=N), 1586, 1534 (C=C); ^{1}H NMR (DMSO-d_6): $\delta = 3.91$ (s, 3H, OCH_3), 7.50 (d, 4H, $J = 7.2$ Hz, Ar), 7.69 (d, 4H, $J = 8.0$ Hz, Ar), 8.17 (s, 1H, CH), 11.50 (brs, 2H, NH) ppm; ^{13}C NMR (DMSO-d_6): $\delta = 53.9, 122.5, 128.4, 128.9, 131.7, 133.9, 141.4, 161.0$ ppm; Anal. Calc. for $C_{18}H_{15}Cl_2N_7O$ (416.26): C, 51.94; H, 3.63; N, 23.55. Found: C, 51.86; H, 3.88; N, 23.79.

*(47) 2,4-Bis(2-(4-fluorobenzylidene)hydrazinyl)-6-methoxy-1,3,5-triazine (**8g**, Supporting Information, Figure S49 in Supplementary Data).* Beige solid in yield 69%; mp 323–325°C; IR (KBr, cm^{-1}): 3217 (NH), 1603 (C=N), 1547, 1504 (C=C); ^{1}H NMR (DMSO-d_6): $\delta = 3.90$ (s, 3H, OCH_3), 7.26–7.30 (m, 4H, Ar), 7.71–7.74 (m, 4H, Ar), 8.17 (s, 1H, CH), 11.40 (brs, 2H, NH) ppm; ^{13}C NMR (DMSO-d_6): $\delta = 54.0, 116.0, 128.6, 131.2, 143.2, 161.4, 164.2$ ppm; Anal. Calcd for $C_{18}H_{15}F_2N_7O$ (383.35): C, 56.39; H, 3.94; N, 25.58. Found: C, 56.65; H, 4.05; N, 25.32.

2.2. Antimicrobial Activity. The antibacterial activity of the prepared compounds was assessed against two selected bacteria groups: Gram-positive bacteria, namely, *Streptococcus mutans* (wild strain) and *Staphylococcus aureus* (ATCC 29213), and Gram-negative bacteria, namely, *Escherichia coli* (ATTC 25922) *and Salmonella typhimurium* (ATCC 14028), maintained in Brain Heart Infusion medium at −20°C. One mL of each targeted grown bacterial culture was added to 100 mL of Brain Heart Infusion broth and incubated at 37°C ± 1°C for subculture of all targeted bacterial cultures. Using of sterile physiological solution, concentration of bacterial inoculum was determined to be 10^8 CFU/mL (0.5 McFarland turbidity standards) for antimicrobial tests. In case of fungus *Candida albicans*, the PDA was used as a medium in antagonistic activity against the tested fungi.

2.2.1. Primary Screening of Synthetic Organic Compounds for Bioactivity by Agar Diffusion Disc. The tested compounds were dissolved in DMSO and prepared at concentration 2500 μg/mL. The sterilized Mueller Hinton agar plates seeded with pathogenic bacteria were prepared; 6 mm paper discs with 100 μg of the tested compounds were placed in the plates seeded with tested pathogenic bacteria. The inoculated plates were kept at 37°C for 24 h for bacteria while, in case of fungi, the used medium in antagonistic activity against tested fungi is PDA. Antimicrobial activity was determined by the inhibition zone [46].

2.2.2. Minimum Inhibitory Concentration (MIC) Determination. The bioactivity of the tested compounds was determined by using a microdilution method, using Mueller–Hinton broth. The inoculum was prepared as described previously and the tested compounds were stocked in 0.25% of DMSO (concentration of each compound 1.25 mg/mL). After that, serial dilutions were done by addition of 100 μL of culture broth which was poured in microplate and 25 μL of each concentration of tested compound in the first line of microplate to reach concentrations ranging from 125 to 0.973 μg/mL in presence of the positive and negative control. The microplates were vaccinated with 5 μL of a bacterial inoculum (10^8 CFU/mL) and were kept for 24 h at 37°C in triplicate tests. After incubation period, we added 20 μL INT (0.5 mg/mL) to achieve bacterial growth. The INT containing microplates were incubated at 37°C for 30 min and the collected data were expressed in micrograms per milliliters [47].

3. Results and Discussion

3.1. Chemistry. The first chlorine atom of cyanuric chloride **1** was replaced by benzylamine, *N*-methylbenzylamine, morpholine, piperidine, or methoxy, while two hydrazine groups replaced the second and third chlorine atoms. Accordingly, 2,4-dihydrazino-6-substituted-1,3,5-triazine derivatives **3a–e** were prepared by displacement of the two chlorine atoms by two hydrazine groups.

Cyanuric chloride **1** was first reacted with different amines such as benzylamine, *N*-methylbenzylamine, morpholine, piperidine, or methanol at 0°C for 1-2 h to afford the products **2a–e** (Scheme 1); the spectral data agreed with the reported data [44, 45]. The dichloro derivatives **2a–e** were reacted with hydrazine hydrate (80%) under reflux using acetonitrile as a solvent to render the products of 2,4-dihydrazino-6-substituted-1,3,5-triazine derivatives **3a–e** which were used directly in the next step without further purification (Scheme 1).

The products **4–8** were prepared by condensation of the hydrazine derivatives **3a–e** with different aldehydes or

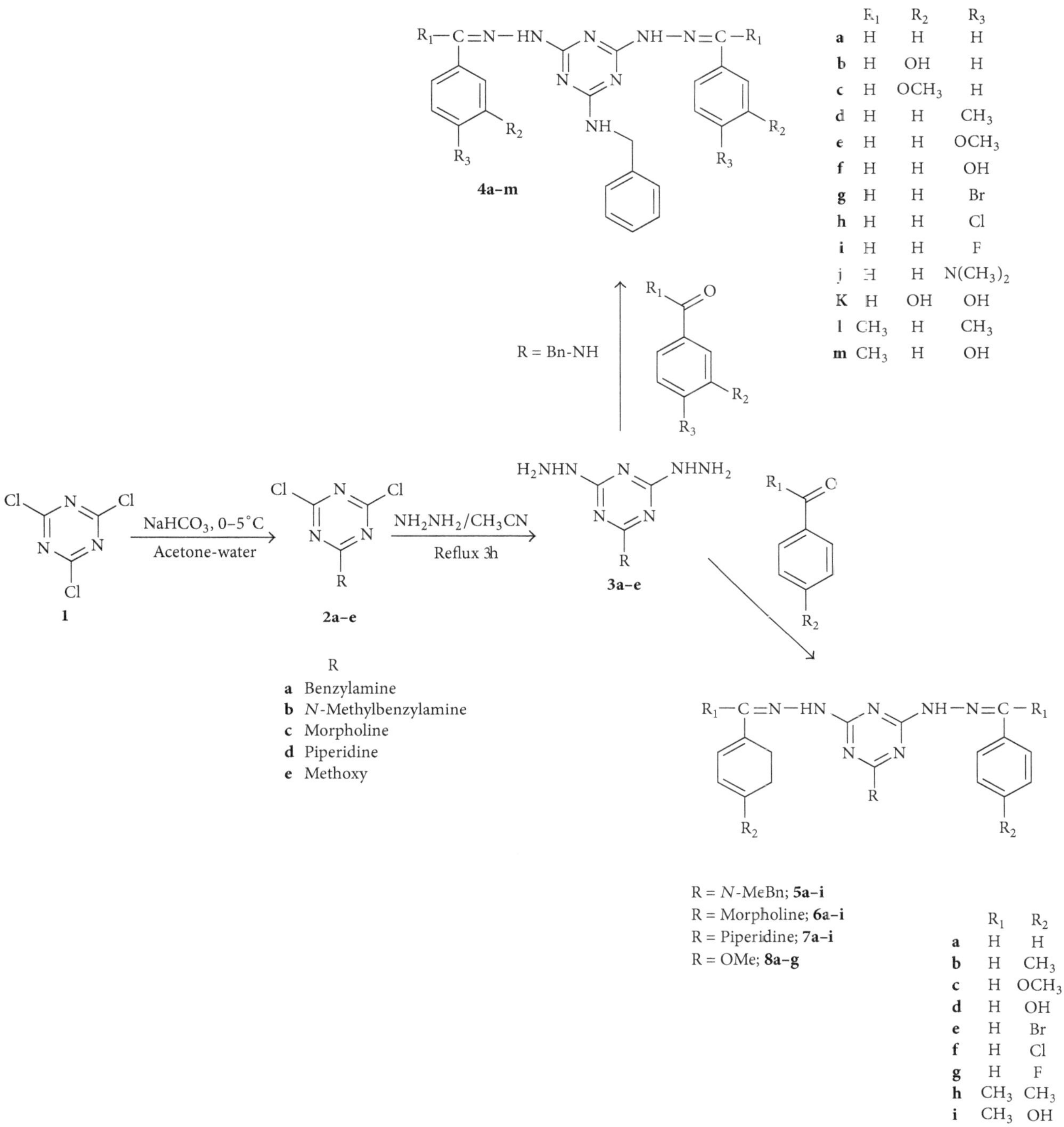

Scheme 1: Synthesis of 6-substituted-s-triazine-Schiff base derivative.

substituted acetophenone in ethanol as solvent and in the presence of drops of acetic acid (Scheme 1).

The ^{1}H NMR spectrum of **4b** (Figure 1) as a prototype for the **4a–m** series displayed two doublet peaks at δ 4.54 (J = 6.4 Hz) and 7.04 ppm (J = 8.0), attributed to resonances of protons of CH_2N and 2H-6 respectively, doublet of doublet at δ 6.77 ppm which integrated two protons for 2H-4, two triplets at δ 7.22 (J = 8.0 Hz) and 7.32 ppm (J = 7.6) for H-4$'$, 2H-5, and H-3$'$, H-5$'$, and four singlets at δ 7.08, 7.38, 8.06, and 9.54 ppm for 2H-2, H-2$'$ combined H-6$'$, CH, and OH, respectively, while the NH proton appeared as abroad singlet at δ 10.82 ppm. The ^{13}C NMR spectrum also reveals the symmetry of **4b**, as indicated by the appearance of fourteen distinct carbon peaks, among which are four peaks at δ 112.9, 116.8, 118.3, 126.9, 127.6, 128.6, 130.1, 136.8, 140.7, and 157.9 ppm related to the aromatic carbons, in addition to four peaks at δ 43.8, 143.0, 164.6, and 166.6 ppm belonging to (CH_2NH), and (C=N), respectively.

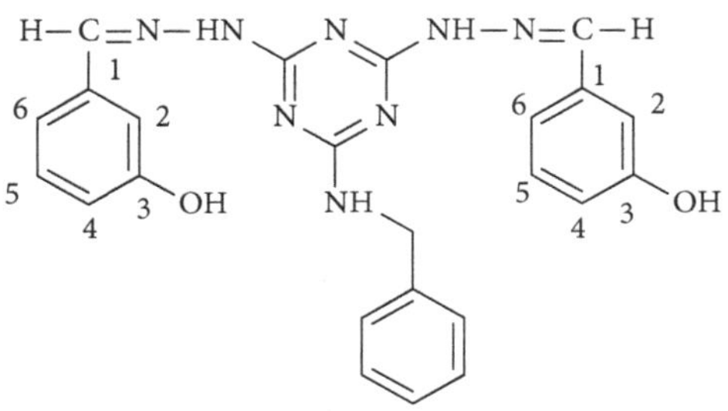

FIGURE 1: Structure of **4b**.

The ^{1}H NMR spectrum of **5b** as a prototype for the **5a–i** series exhibited three singlets at δ 2.33, 3.10, and 4.87 ppm, which were assigned to protons of $2CH_3$, CH_3N, and CH_2N, respectively. This spectrum showed three signals for aromatic rings at δ 7.21–7.26 (m, 4H, 2H-3, and 2H-5), 7.31–7.38 (m, 5H, H-2′, H-3′, H-4′, H-5′, and H-6′), and 7.53 (d, 4H, J = 8.0 Hz, 2H-2, and 2H-6) ppm. This spectrum also showed a sharp singlet at δ 8.11 ppm integrating one proton, which was assigned to CH, in addition the broad singlet at δ 11.90 ppm for NH. The ^{13}C NMR spectrum of **5b** showed fourteen individual carbon signals. The three signals in the sp^3 region appeared at δ 21.4, 34.1, and 51.1 ppm for ($2CH_3$), (CH_3N), and (CH_2N), respectively, in addition to signals at δ142.8, 164.7, and 165.9 ppm for (C=N). Assignments of resonances of six carbons of the benzyl were at δ127.4, 128.1, 128.8, and 138.7 ppm. The signals of the remaining aromatic carbons appeared at δ126.8, 129.8, 132.8, and 139.1 ppm.

The ^{1}H NMR spectrum of **6a** as a prototype for the **6a–i** series showed two sharp singlets at δ 3.66 and 3.78 ppm for 2 OCH_2- and 2 NCH_2- related to the morpholine, respectively, singlet at δ 8.14 ppm for CH, and a broad singlet at δ 11.0 for NH. Three signals for aromatic protons appeared at δ 7.42 (t, 4H, J = 7.2 Hz, 2H-3, and 2H-5) and 7.65 (d, 4H, J = 7.6 Hz, 2H-2, and 2H-6). The ^{13}C NMR spectrum also confirms the structure for compound **6a**, as indicated by the appearance of two signals at δ 43.8 and 66.5 ppm for (2 NCH_2-) and (2 OCH_2-), respectively, for the morpholino moiety, while the other three signals appeared at δ142.8, 164.8, and 165.3 ppm for (C=N), in addition to the observation of the aromatic ring carbons appearing at δ126.9, 129.2, 129.6, and 135.5 ppm, respectively. The LC-MS of compound **6a** (Supporting Information, Figure S24 in Supplementary Data) the exact molecular mass [M + H] 403 (m/z calcd. 402.45) with R_t 15.3 min, using buffer A: 0.1% formic acid in H_2O; and buffer B: 0.1% formic acid in CH_3CN in 30 min.

The ^{1}H NMR spectrum of **7i** as a prototype for **7a–i** series showed three singlet peaks at δ 1.55, 1.64, and 3.84 ppm for the $2CH_2$, CH_2, and $2NCH_2$ (piperidine moiety), while it showed a singlet at δ 2.31 ppm for the methyl group of the acetophenone moiety and two broad singlets at δ 9.90 and 11.0 ppm for OH and NH, respectively, beside two doublets forming AB system at δ 6.82 (J = 8.0 Hz) and 7.70 ppm (J = 7.2 Hz) for the aromatic proton H-3, H-5 and H-2, H-6, respectively. The ^{13}C NMR spectrum of **7i** exhibited absorption peak at δ 14.4 related to methyl group, three peaks for the piperidine residue at δ 24.6, 26.0, and 44.3 ppm related to (CH_2), ($2CH_2$), and ($2CH_2N$), respectively, and absorption peaks at δ 154.2 and 160.0 ppm related to (C=N), besides four peaks at δ115.6, 128.2, 131.1, and 159.1 ppm for the aromatic carbons. The LC-MS of compound **7i** (Supporting Information, Figure S42 in Supplementary Data) using buffer A: 0.1% formic acid in H_2O and buffer B: 0.1% formic acid in CH_3CN in 30 min showed one peak at R_t 14.08 min with the expected mass [M + 2H] 463.2 (m/z calcd. 460.53).

The ^{1}H NMR spectrum of **8b** as a prototype for the **8a–g** series revealed the resonances of protons methyl group at δ 2.33 ppm (s, $2CH_3$) and methoxy group at δ 3.89 ppm (s, OCH_3), in addition to the two doublets that appeared at δ 7.25 (d, 4H, J = 7.2 Hz) and 7.56 ppm (d, 4H, J = 7.2 Hz) representing of 2H-2, 2H-3, and 2H-5, 2H-6 (AB system) while singlet (integrated for one proton) appearing at δ 8.14 ppm was attributed to CH. Finally, the proton of NH group appeared as broad singlet at δ 11.28 ppm. The ^{13}C NMR spectrum of **8b** showed signal at δ 21.5 for the methyl carbon while methoxy carbon appeared at δ 54.4 ppm and the aromatic carbons signals at δ 127.1, 129.8, 132.4, and 139.6 ppm. The three signals at δ144.3, 165.5, and 166.2 ppm in the later spectrum were assigned to (C=N).

3.2. Biology

3.2.1. Bioactivity of Tested Compounds against the Pathogenic Bacteria and Fungus Candida albicans. In the present study, twenty-two compounds from the above series **4–8** were tested for bioactivities against the selected bacteria and fungi; only eight compounds have antiactivities of target pathogenic bacteria. The results in Table 1 showed that the tested compounds, namely, **4f, 4k, 4l, 4m, 5d, 5g, 6i,** and **6h,** have bioactivity against the target pathogenic bacteria as antimicrobial agents and do not have antagonistic effect against fungus *Candida albicans.* The present study also showed that compounds **4k** and **5g** have wide-range effect presently in Gram-positive and Gram-negative bacteria, while compounds **6i** and **6h** were specific in the effect against the Gram-negative and Gram-positive bacteria, respectively. Table 1 also showed that compounds **4f** and **4l** showed activity against *Streptococcus mutans,* while compounds **4m** and **5d** showed activity against *Staphylococcus aureus.*

A lowest concentration of chemical compound that inhibits the bacterial growth is the minimum inhibitory concentration (MIC, μg/mL). The MIC of compound **4k** that inhibited all the tested pathogenic bacteria was 31.25 μg/mL, while the MIC of compound **4f** that inhibited the *Streptococcus mutans* was 62.5 μg/mL, the MIC of compound **4l** that inhibited *Streptococcus mutans* was 100 μg/mL, and the MIC of compound **4m** that inhibited *Staphylococcus aureus* was 62.50 μg/mL. The MIC of compound **5d** that inhibited *Staphylococcus aureus* was 31.25 μg/mL, the MIC of compound **5g** that inhibited *Staphylococcus aureus* and *Salmonella typhimurium* was 100 μg/mL, the MIC of compound **6i** that inhibited *Escherichia coli and Salmonella typhimurium* was 62.50 μg/mL, and the MIC of compound **6h** that inhibited *Streptococcus mutans* and *Staphylococcus aureus* was 31.25 and 62.50 μg/mL, respectively (Table 2).

Table 1: Bioactivities of the synthetic organic compounds against some pathogenic bacteria and unicellular fungus *Candida albicans* at 100 μg/paper disc (6 mm).

Compd. number	*Streptococcus mutans*	*Staphylococcus aureus*	*Escherichia coli*	*Salmonella typhimurium*	*Candida albicans*
Control	−	−	−	−	−
4k	+	+	+	+	−
4j	−	−	−	−	−
4f	+	−	−	−	−
4e	−	−	−	−	−
4i	−	−	−	−	−
4l	+	−	−	−	−
4m	−	+	−	−	−
5c	−	−	−	−	−
5d	−	+	−	−	−
5e	−	−	−	−	−
5f	−	−	−	−	−
5g	−	+	−	+	−
5i	−	−	−	−	−
6d	−	−	−	−	−
6f	−	−	−	−	−
6g	−	−	−	−	−
6i	−	−	+	+	−
6h	+	+	−	−	−
8c	−	−	−	−	−
8d	−	−	−	−	−
8f	−	−	−	−	−
8g	−	−	−	−	−

Table 2: Minimum inhibitory concentration (MIC, μg/mL) of bioactive synthetic compounds against target pathogenic bacteria.

Compd. number	*Streptococcus mutans* wild strain	*Staphylococcus aureus*	*Escherichia coli*	*Salmonella typhimurium*
4k	31.25	31.25	31.25	31.25
4f	62.50	-	-	-
4l	100.00	-	-	-
4m	-	62.50	-	-
5d	-	31.25	-	-
5g	-	100.00	-	100.00
6i	-	-	62.50	62.50
6h	31.25	62.50	-	-

It is obvious from Table 1 that small structural variations in s-triazine ring may induce an effect on antibacterial activity, such as in series of benzyl derivatives (**4** series) and the *N*-methylbenzyl derivatives (**5** series); the benzyl derivatives showed more activity than the *N*-benzyl derivatives. The same was observed when the benzyl was substituted by methoxy (**8** series); the benzyl derivatives showed more activity. Also the substituent effect on the benzylidene moiety has a great effect on the antibacterial activity as shown in case of **4** series and **5** series, and it is interesting to note that in most of the cases the fluorine derivatives showed higher activity than the chlorine and bromine derivatives as in case of **5** series. On the other hand, not all the tested compounds showed any antifungal activity; this observation agreed with the reported data for s-triazine derivatives [48, 49].

4. Conclusion

The present work represents the synthesis, characterization, and biological activity of novel series of the s-triazine *bis*-Schiff base. Only twenty-two compounds from the above series **4a–m, 5a–i, 6a–i, 7a–i,** and **8a–g** were tested against Gram-positive and Gram-negative bacteria according to their solubility in aqueous DMSO. Only eight compounds **4k, 4f, 4l, 4m, 5d, 5g, 6i,** and **6h** showed bioactivity against the target pathogenic bacteria as antimicrobial agents and did not show antagonistic effect against fungus *Candida albicans*. Two compounds **4k** and **5g** have wide-range effect presently in Gram-positive and Gram-negative bacteria, while compounds **6i** and **6h** have specific effect against the Gram-negative and Gram-positive bacteria, respectively.

In addition, compounds **4f** and **4l** showed activity against *Streptococcus mutans* while compounds **4m** and **5d** showed activity against *Staphylococcus aureus*.

It is clear that the substituent in the benzylidene as well as the triazine ring may have great effect on the antimicrobial activity. Further investigation is being run in our lab to get a clear depiction on the mode of action and the relation between the biological activity and substituent effect on the s-triazine moiety.

Conflicts of Interest

The authors declare that they have no conflicts of interest.

Acknowledgments

The authors extend their appreciation to the Deanship of Scientific Research at King Saud University for funding this work through Research Group no. RGP-234, Saudi Arabia.

References

[1] A. Sousa-Pedrares, N. Camiña, J. Romero, M. L. Durán, J. A. García-Vázquez, and A. Sousa, "Electrochemical synthesis and crystal structure of cobalt(II), nickel(II), copper(II), zinc(II) and cadmium(II) complexes with 2-pyridinecarbaldehyde-(2′-aminosulfonylbenzoyl)hydrazone," *Polyhedron*, vol. 27, no. 16, pp. 3391–3397, 2008.

[2] M. Sönmez, M. Çelebi, and I. Berber, "Synthesis, spectroscopic and biological studies on the new symmetric Schiff base derived from 2,6-diformyl-4-methylphenol with N-aminopyrimidine," *European Journal of Medicinal Chemistry*, vol. 45, no. 5, pp. 1935–1940, 2010.

[3] S.-G. Teoh, G.-Y. Yeap, C.-C. Loh, L.-W. Foong, S.-B. Teo, and H.-K. Fun, "Inner coordination sphere tin(IV) complexes with some O,N,N-terdentate N-(2-hydroxybenzaldehyde)-1-amino-2-phenyleneimine and N-(2-hydroxy-1-naphthaldehyde)-1-amino-2-phenyleneimine and O,N,N,O-quadridentate N,N′-bis(2-hydroxybenzaldehyde)-1,2-phenylenediimine and N,N′-bis(2-hydroxy-1-naphthaldehyde)-1,2-phenylenediimine Schiff bases," *Polyhedron*, vol. 16, no. 13, pp. 2213–2221, 1997.

[4] G. P. Pokhariyal, B. K. Lal, and V. Rastogi, "Synthesis, spectral and thermal studies of some divalent metal complexes with 8-acetyl-7-hydroxy-4-methyl coumarrin oxime-hdrazone," *Asian Journal of Chemistry*, vol. 9, pp. 541–543, 1997.

[5] P. Barbazán, R. Carballo, B. Covelo, C. Lodeiro, J. C. Lima, and E. M. Vázquez-López, "Synthesis, characterization, and photophysical properties of 2-hydroxybenzaldehyde [(1E)-1-pyridin-2-ylethylidene]hydrazone and its rhenium(I) complexes," *European Journal of Inorganic Chemistry*, no. 17, pp. 2713–2720, 2008.

[6] S. Banerjee, S. Mondal, W. Chakraborty et al., "Interaction of boric acid with salicyl derivatives as an anchor group of boron-selective adsorbents," *European Journal of Organic Chemistry*, pp. 2785–2790, 2008.

[7] N. Ghavtadze, R. Fröhlich, and E.-U. Würthwein, "2H-pyrrole derivatives from an aza-nazarov reaction cascade involving indole as the neutral leaving group," *European Journal of Organic Chemistry*, no. 21, pp. 3656–3667, 2008.

[8] J. P. Cornelissen, J. H. Van Diemen, L. R. Groeneveld, J. G. Haasnoot, A. L. Spek, and J. Reedijk, "Synthesis and properties of isostructural transition-metal (copper, nickel, cobalt, and iron) compounds with 7,7′,8,8′-tetracyanoquinodimethanide(1-) in an unusual monodentate coordination mode. Crystal structure of bis(3,5-bis(pyridin-2-yl)-4-amino-1,2,4-triazole)bis(7,7′,8,8′- tetracyanoquinodimethanido)copper(II)," *Inorganic Chemistry*, vol. 31, no. 2, pp. 198–202, 1992.

[9] F. Kandil, M. K. Chebani, and W. Al Zoubi, "Synthesis of Macrocyclic Bis-Hydrazone and Their Use in Metal Cations Extraction," *ISRN Organic Chemistry*, vol. 2012, pp. 1–8, 2012.

[10] N. Nawar, M. A. Khattab, and N. M. Hosny, "Some metal(II) complexes of o-aminoacetophenone benzoylhydrazone (AABH): Their preparation, characterization and antimicrobial activity," *Synthesis and Reactivity in Inorganic, Metal-Organic, and Nano-Metal Chemistry*, vol. 29, no. 8, pp. 1365–1384, 1999.

[11] M.-D. Gui, R.-H. Zhang, H. Yan, H.-M. Wang, X.-B. Chen, and S.-J. Zhu, "Synthesis and magnetic and catalytic properties of manganous isonicotinoyl hydrazone complexes," *Synthesis and Reactivity in Inorganic, Metal-Organic, and Nano-Metal Chemistry*, vol. 28, no. 8, pp. 1381–1391, 1998.

[12] Z. H. Chohan and A. Rauf, "Some biologically active mixed ligand complexes of Co(II),Cu(II) and Ni(II) with ONO, NNO and SNO donor nicotinoylhydrazine-derived ligands," *Synthesis and Reactivity in Inorganic, Metal-Organic, and Nano-Metal Chemistry*, vol. 26, no. 4, pp. 591–604, 1996.

[13] S. K. Agarwal, J. Jain, and S. Chand, "Synthetic and antimicrobial studies of hexacoordinated ternary complexes of Mn(II) and Cu(II)," *Asian Journal of Chemistry*, vol. 14, no. 1, pp. 489–492, 2002.

[14] X. Zhu, C. Wang, W. Dong et al., "Synthesis, structure, and biological activities of the O,O′-diethylphosphorohydrazonothionate ligand and complexes with nickel(II), copper(II), and zinc(II)," *Synthesis and Reactivity in Inorganic, Metal-Organic, and Nano-Metal Chemistry*, vol. 32, no. 3, pp. 475–487, 2002.

[15] K. A. El-Manakhly, H. A. Bayoumi, M. M. Ezz Eldin, and H. A. Hammad, "Physicochemical and antimicrobial studies of some metal complexes," *Journal of the Indian Dental Association*, vol. 76, no. 2, pp. 63-64, 1999.

[16] K. P. Deepa, K. K. Aravindakshan, and F. Suhara, "Synthesis, characterization and antifungal studies of metal, complexes of benzoyl- and salicylylhydrazones of N-methylacetoacetanilide," *Asian Journal of Chemistry*, vol. 13, no. 2, pp. 513–523, 2001.

[17] A. K. Panda, D. C. Dash, P. Mishra, and H. Mohanty, "Metallo ligands: Part V - Homo and bimetallic trinuclear complexes of copper(II), nickel(II) and cobalt(II) with 1,8-di-(2′-hydroxyphenyl)-4,5-diphenyl-2,3,6,7-tetrazaoctan-1,3,5,7- tetraene," *Indian Journal of Chemistry - Section A Inorganic, Physical, Theoretical and Analytical Chemistry*, vol. 35, no. 10, pp. 848–851, 1996.

[18] M. F. Iskander, L. El-Sayed, N. M. H. Salem, W. Haase, H. J. Linder, and S. Foro, "Synthesis, characterization and magnetochemical studies of some copper(II) complexes derived from n-salicylidene-n-alkanoylhydrazins: X-ray crystal and molecular structure of bis [monochloro-(μ-n-salicylidenemyristoylhydrazine) ono(-1)] dicopper(II)," *Polyhedron*, vol. 23, no. 1, pp. 23–31, 2004.

[19] C. Li, L. Wang, X. Meng, and H. Zhao, "Synthesis, characterization and antitumor activity of benzaldehyde nitrogen mustard picolinoyl hydrazone complexes," *Transition Metal Chemistry*, vol. 24, no. 2, pp. 206–209, 1999.

[20] Q. Zhang, C. He, P. Yin, and J. M. Shreeve, "Insensitive nitrogen-rich materials incorporating the nitroguanidyl functionality," *Chemistry - An Asian Journal*, vol. 9, no. 1, pp. 212–217, 2014.

[21] A. El-Faham, S. M. Osman, H. A. Al-Lohedan, and G. A. El-Mahdy, "Hydrazino-methoxy-1,3,5-triazine derivatives' excellent corrosion organic inhibitors of steel in acidic chloride solution," *Molecules*, vol. 21, no. 6, article no. 714, 2016.

[22] G. A. El-Mahdy, H. H. Al-Rasheed, M. A. Alshaikh, H. A. Al-Lohedan, and A. El-Faham, "2,4-Dihydrazino-6-Morpholino-1,3,5-Triaizne (DHMT) and 2,4-Dihydrazino-6-Piperidino-1,3, 5-Triaizne (DHPT) as promising corrosion inhibitors of steel in acidic media," *International Journal of Electrochemical Science*, vol. 11, no. 7, pp. 5459–5472, 2016.

[23] P. Gamez and J. Reedijk, "1,3,5-Triazine-based synthons in supramolecular chemistry," *European Journal of Inorganic Chemistry*, no. 1, pp. 29–42, 2006.

[24] S. M. Soliman, A. El-Faham, and J. H. Albering, "Synthesis, X-ray crystal structure and DFT studies of two octahedral cobalt(II) complexes with N,N,N-tridentate triazine-type ligand," *Journal of Coordination Chemistry*, vol. 70, no. 13, pp. 2261–2279, 2017.

[25] S. Demeshko, S. Dechert, and F. Meyer, "Anion-π Interactions in a Carousel Copper(II)-Triazine Complex," *Journal of the American Chemical Society*, vol. 126, no. 14, pp. 4508-4509, 2004.

[26] J. A. Zerkowski, C. T. Seto, and G. M. Whitesides, "Solid-state structures of "rosette" and "crinkled tape" motifs derived from the cyanuric acid-melamine lattice," *Journal of the American Chemical Society*, vol. 114, no. 13, pp. 5473–5475, 1992.

[27] J. Ramírez, A.-M. Stadler, J. M. Harrowfield et al., "Coordination architectures of large heavy metal cations (Hg_2+ and Pb_2+) with bis-tridentate ligands: Solution and solid-state studies," *Zeitschrift für anorganische und allgemeine Chemie*, vol. 633, no. 13-14, pp. 2435–2444, 2007.

[28] J. Ramírez, A.-M. Stadler, L. Brelot, and J.-M. Lehn, "Coordinative, conformational and motional behaviour of triazine-based ligand strands on binding of Pb(II) cations," *Tetrahedron*, vol. 64, no. 36, pp. 8402–8410, 2008.

[29] A. Das, S. Demeshko, S. Dechert, and F. Meyer, "A new triazine-based tricompartmental ligand for stepwise assembly of mononuclear, dinuclear, and 1D-polymeric heptacoordinate manganese(II)/azido complexes," *European Journal of Inorganic Chemistry*, no. 8, pp. 1240–1248, 2011.

[30] F. Yaoting, L. Gang, L. Zifeng, H. Hongwei, and M. Hairong, "Synthesis, structure and third-order nonlinear optical properties of 1,3,5-triazine-based Zn(II) three-dimensional supramolecule," *Journal of Molecular Structure*, vol. 693, no. 1-3, pp. 217–224, 2004.

[31] L. Pavelek, V. Ladányi, M. Nečas, S. Vallová, and K. Wichterle, "Dioxouranium complexes with pentadentate s-triazine Schiff base ligands: Synthesis, crystal structure and optical properties," *Polyhedron*, vol. 107, pp. 89–96, 2016.

[32] L. Pavelek, V. Ladányi, M. Nečas, Z. Moravec, and K. Wichterle, "Synthesis and characterization of lanthanide complexes with a pentadentate triazine-based ligand," *Polyhedron*, vol. 119, pp. 134–141, 2016.

[33] L. Tei, Z. Baranyai, E. Brücher et al., "Dramatic increase of selectivity for heavy lanthanide(III) cations by tuning the flexibility of polydentate chelators," *Inorganic Chemistry*, vol. 49, no. 2, pp. 616–625, 2010.

[34] F. W. Lewis, L. M. Harwood, M. J. Hudson et al., "Highly efficient separation of actinides from lanthanides by a phenanthroline-derived bis-triazine ligand," *Journal of the American Chemical Society*, vol. 133, no. 33, pp. 13093–13102, 2011.

[35] M. J. Hudson, L. M. Harwood, D. M. Laventine, and F. W. Lewis, "Use of soft heterocyclic N-donor ligands to separate actinides and lanthanides," *Inorganic Chemistry*, vol. 52, no. 7, pp. 3414–3428, 2013.

[36] M. G. B. Drew, C. Hill, M. J. Hudson, P. B. Iveson, C. Madic, and T. G. A. Youngs, "Solvent extraction and lanthanide complexation studies with new terdentate ligands containing two 1,3,5-triazine moieties," *Dalton Transactions*, no. 2, pp. 244–251, 2004.

[37] T. Vilaivan, N. Saesaengseerung, D. Jarprung, S. Kamchonwongpaisan, W. Sirawaraporn, and Y. Yuthavong, "Synthesis of solution-phase combinatorial library of 4,6-diamino-1,2-dihydro-1,3,5-triazine and identification of new leads against A16V+S108T mutant dihydrofolate reductase of Plasmodium falciparum," *Bioorganic & Medicinal Chemistry*, vol. 11, no. 2, pp. 217–224, 2003.

[38] J. N. Sangshetti and D. B. Shinde, "One pot synthesis and SAR of some novel 3-substituted 5,5-diphenyl-1,2,4-triazines as antifungal agents," *Bioorganic & Medicinal Chemistry Letters*, vol. 20, no. 2, pp. 742–745, 2010.

[39] W. Lv, B. Banerjee, K. L. Molland et al., "Synthesis of 3-(3-aryl-pyrrolidin-1-yl)-5-aryl-1,2,4-triazines that have antibacterial activity and also inhibit inorganic pyrophosphatase," *Bioorganic & Medicinal Chemistry*, vol. 22, no. 1, pp. 406–418, 2014.

[40] K. K. Bansal, D. Kakde, U. Gupta, and N. K. Jain, "Development and characterization of triazine based dendrimers for delivery of antitumor agent," *Journal of Nanoscience and Nanotechnology*, vol. 10, no. 12, pp. 8395–8404, 2010.

[41] N. C. Desai, A. H. Makwana, and R. D. Senta, "Synthesis, characterization and antimicrobial activity of some novel 4-(4-(arylamino)-6-(piperidin-1-yl)-1,3,5-triazine-2-ylamino)-N-(pyrimidin-2-yl)benzenesulfonamides," *Journal of Saudi Chemical Society*, vol. 20, no. 6, pp. 686–694, 2016.

[42] G. Blotny, "Recent applications of 2,4,6-trichloro-1,3,5-triazine and its derivatives in organic synthesis," *Tetrahedron*, vol. 62, no. 41, pp. 9507–9522, 2006.

[43] C. A. M. Afonso, N. M. T. Lourenço, and A. A. de Rosatella, "Synthesis of 2,4,6-tri-substituted-1,3,5-triazines," *Molecules*, vol. 11, no. 1, pp. 81–102, 2006.

[44] V. V. Bakharev, A. A. Gidaspov, V. E. Parfenov et al., "Synthesis of 4-amino-6-chloro-1,3,5-triazin-2(1H)-ones," *Russian Chemical Bulletin*, vol. 61, no. 1, pp. 99–112, 2012.

[45] W. Huang, W. Zheng, D. J. Urban et al., "N4-Phenyl modifications of N2-(2-hydroxyl)ethyl-6-(pyrrolidin-1-yl)-1,3,5-triazine-2,4-d iamines enhance glucocerebrosidase inhibition by small molecules with potential as chemical chaperones for Gaucher disease," *Bioorganic & Medicinal Chemistry Letters*, vol. 17, no. 21, pp. 5783–5789, 2007.

[46] D. A. Vanden and A. J. Berghe, "Screening methods for antibacterial and antiviral agents from higher plants," in *Methods in Plant Biochemistry*, P. M. Dey and J. D. Harbone, Eds., pp. 47–69, Academic Press, London, 1991.

[47] S. T. Asundaria, S. A. Patel, K M. Mehta, and K. C. Patel, "Synthesis, characterization and antimicrobial studies of some novel 1,3,4-thiadiazolium-2-thiolate derivatives," *South African Journal of Chemistry*, vol. 63, pp. 141–144, 2015.

[48] K. Srinivas, U. Srinivas, K. Bhanuprakash, K. Harakishore, U. S. N. Murthy, and V. J. Rao, "Synthesis and antibacterial activity of various substituted s-triazines?" *European Journal of Medicinal Chemistry*, vol. 41, pp. 1240–1246, 2006.

[49] H. H. Al-Rasheed, M. Al Alshaikh, J. M. Khaled, N. S. Alharbi, and A. El-Faham, "Ultrasonic irradiation: synthesis, characterization, and preliminary antimicrobial activity of novel series of 4,6-disubstituted-1,3,5-triazine containing hydrazone derivatives," *Journal of Chemistry*, vol. 2016, Article ID 3464758, 9 pages, 2016.

25

Synthesis and Crystalline Structure of Zinc Complexes with Antihypertensive Drug Lisinopril

Márcia C. de Souza,[1] Luan F. Diniz,[1] Chris H. J. Franco,[1] and Renata Diniz[1,2]

[1]*Department of Chemistry, Federal University of Juiz de Fora, Juiz de Fora, Minas Gerais 36036-900, Brazil*
[2]*Department of Chemistry, Federal University of Minas Gerais, Belo Horizonte, Minas Gerais 31270-901, Brazil*

Correspondence should be addressed to Renata Diniz; dinizr@ufmg.br

Academic Editor: Henryk Kozlowski

The structural investigation of Zn^{2+} complexes with the ligand lisinopril (LIS), an inhibitor of angiotensin-converting enzyme (ACE), was performed. The main objective is to compare if Zn-LIS coordination *in vitro* is similar to that observed *in vivo*. Two zinc complexes were obtained from different synthetic routes. The synthesis of LISZn1 used stirring, while for LISZn2 involved solvothermal conditions, which favoured the full deprotonation of lisinopril ligand. In this sense, the different synthetic routes resulted in the formation of complexes with notorious chemical and structural differences. The crystal structure of LISZn2 showed that the ligand is coordinated to Zn^{2+} ion by oxygen and nitrogen atoms which is different from that observed *in vivo*. *In vitro*, the coordination of lisinopril occurs only by an oxygen atom of the central carboxylate group. LISZn2 forms a one-dimensional (1D) coordination polymer and presents disorder atoms in its unit cell.

1. Introduction

Hypertension is a disease that affects millions of people around the world. This disease is identified if the blood pressure levels maintain above the reference values [1]. Angiotensin-converting enzyme (ACE) inhibitors are considered the most effective drugs used to treat the arterial hypertension. ACE is a zinc-containing metallopeptidase that catalyzes the conversion of the angiotensin I-inactive decapeptide to angiotensin II-potent vasopressor octapeptide. This reaction promotes vasoconstriction and hydrolysis of the vasodilator bradykinin [2–4]. It plays an essential role in blood pressure regulation, and its interaction with an ACE inhibitor (ACEI) sets an important research topic for treatment of hypertension [2]. Some synthetic ACEIs are useful in various cardiovascular pathologies, acting in both reduction of blood pressure and also in the treatment for postmyocardial infarction [3, 5].

Lisinopril, (*S*)-1-[N^2-(1-carboxy-3-phenylpropyl)-L-lysyl]-L-proline, (Figure 1), is an angiotensin-converting enzyme (ACE) inhibitor, widely used in the treatment of hypertension, congestive heart failure, acute myocardial infarction, and diabetic nephropathy [3, 6–8]. The mechanism of action of the lisinopril is known to be related to its ability to interact with ACE forming a (*in vivo*) compound with the Zn^{2+} ion present in ACE. The Zn-ACEI occurs through the central carboxylate group of lisinopril molecule [9]. Electrostatic interactions, hydrogen bonds, and aromatic stacking appear to be sufficient to act as a backstop, positioning the substrate molecule [2, 9]. There are some studies investigating the nature of the interaction between lisinopril and ACE or metal ions by crystal structure analyses [9, 10]. A dihydrate phase of lisinopril is recognized as the most stable crystalline form, and it is extensively used within medicine [6, 11].

This work consists of structural characterization of zinc complexes with ACE inhibitor lisinopril. The LISZn structures are described to evaluate whether the interaction between lisinopril and Zn^{2+} ion *in vitro* resembles that proposed *in vivo*. The molecular and structural details reported in this study may be useful for the development of novel and more effective ACEI to combat high blood pressure.

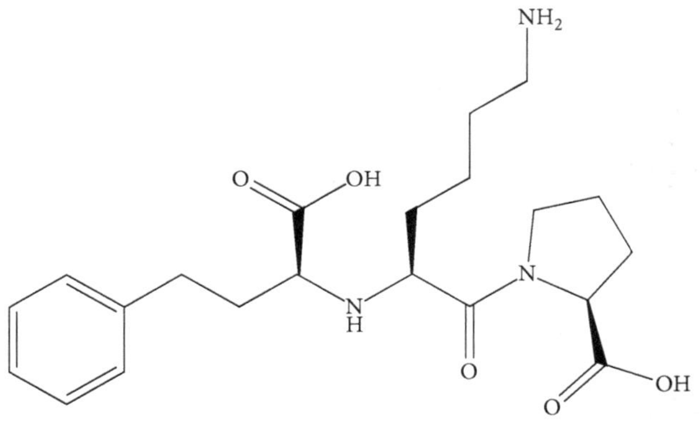

Figure 1: Molecular structure of the lisinopril.

2. Materials and Methods

All products used in the syntheses and analysis present analytical grade, and they were used without purifying. Elemental analyzes (C, H, and N) were performed on a Perkin-Elmer 2400 analyzer. Spectroscopic data in the infrared region were recorded on the Alpha Bruker FT-IR spectrophotometer in the 4000–400 cm^{-1} range with 64 scans and a 4 cm^{-1} resolution for samples using KBr pellets. The Raman spectra were obtained on a Bruker FRS 100 spectrometer using Nd3 +/YAG laser radiation (1064 nm) and a CCD detector. The measures were obtained with 256 scans, 4 cm^{-1} of spectral resolution, and 100 mW laser power.

2.1. Synthesis and Chemical Characterization of the LISZn1. A methanolic solution (2 mL) of $Zn(ClO_4)_2{\cdot}6H_2O$ (1.13 mmol) was added dropwise to a mixture of lisinopril dihydrate (1.13 mmol) and triethylamine (1.09 mmol) in methanol (15 mL) under stirring [10]. The resulting solution was maintained at room temperature. After few hours was observed the formation of a colourless material that visually looked to be single crystals with a yield of 52%. Elemental analysis points to $C_{27}H_{50}Cl_2N_4O_{15}Zn$: calcd. C 40.18, H 6.24, N 6.94; found C 41.03, H 6.13, N 6.88.

2.2. Synthesis and Chemical Characterization of the LISZn2. A methanolic solution of $Zn(ClO_4)_2{\cdot}6H_2O$ (0.89 mmol), 8 mL, was mixed with another aqueous solution, 10 mL, of lisinopril dihydrate (3.40 mmol) and triethylamine (3.26 mmol). This mixture was transferred to a 23 mL Teflon-line Parr acid digestion bomb. The reaction vessel was heated at 353 K for 38 h and, after this time, was slowly cooled to 298 K. At the end of this process, colourless crystals appropriate for single crystal X-ray diffraction analysis were obtained with a yield of 60%. Elemental analysis to $C_{21}H_{37}N_3O_9Zn$: calcd. C 46.63, H 6.89, N 7.77; found C 46.37, H 6.41, N 7.62.

Caution! Perchlorate salts with organic solvents are potentially explosive. Only small amounts of material should be prepared, and they should be handled with care.

2.3. Thermal Analyses (TGA/DTA). Thermal analyses (TGA/DTA) measurements were carried out on a Shimadzu TG-60 with a heating rate of 10 $K{\cdot}min^{-1}$ from room temperature to 874 K and in an air atmosphere (rate flow of 50 $mL{\cdot}min^{-1}$).

2.4. Single Crystal X-Ray Diffraction. Single crystal X-ray diffraction data were collected using a SuperNova diffraction system from Agilent/Rigaku Technologies with CuKα (λ = 1.54056 Å) at 120 K. Data collection, reduction, and cell refinement were performed with the CrysAlisPro [12]. A multiscan absorption correction was applied [13]. The structure was solved and refined by means of SHELXL-2014/7 program package [14]. Anisotropic displacement parameters were assigned to all nonhydrogen atoms. H atoms were located in different Fourier maps and subsequently geometrically optimized and allowed as riding atoms, with C-H = 0.93 Å for aromatic CH groups and 0.97 Å for secondary CH_2 groups with $U_{iso}(H) = 1.2\ U_{eq}$ (for secondary CH_2 groups). The positions of other H atoms were refined freely. The structures were drawn by the programs ORTEP-3 for Windows [15] and Mercury [16].

2.5. Powder Crystal X-Ray Diffraction. X-ray powder diffraction data were collected using a Bruker D8 Advance DaVinci diffractometer, equipped with CuKα (λ = 1.5418 Å) radiation, Ni-filter, LynxEye Detector, and Bragg–Brentano geometry. Data were collected between 5 and 50° in 2θ with a step size of 0.01° and the count time of 0.1 s per step for LISZn2 and with a step size of 0.02° and the count time of 0.05 s per step for LISZn1. Soller slit with a 0.02 mm divergent slit and a 0.6 mm sample slit were used. For all analyses, a voltage of 40 kV and current of 40 mA were applied to generate the incident radiation, and the measurements were performed at room temperature (298 K).

3. Results and Discussion

The combination of zinc perchlorate salt with the lisinopril drug has produced two new complexes (LISZn1 and LISZn2) with 1 : 1 metal:ligand stoichiometric ratio and lattice water, whereas the LISZn1 complex also contains lattice perchlorate ions and triethylamine, suggested by elemental analysis. The synthesis of LISZn2 involved solvothermal conditions, which favoured the full deprotonation of lisinopril ligand, while the synthesis of LISZn1 used only stirring. These two distinct synthetic methodologies employed led to the formation of compounds with chemical and structural differences that can be seen by their distinct solid habits (Figure 2).

3.1. Thermal Analyses (TGA/DTA). Thermogravimetric analysis (TGA) and differential thermal analysis (DTA) curves for the LISZn1 and LISZn2 complexes are depicted in Figures 3 and 4, respectively. It can be observed that the curves are quite different indicating that there are chemical differences between these compounds. Thermogravimetric

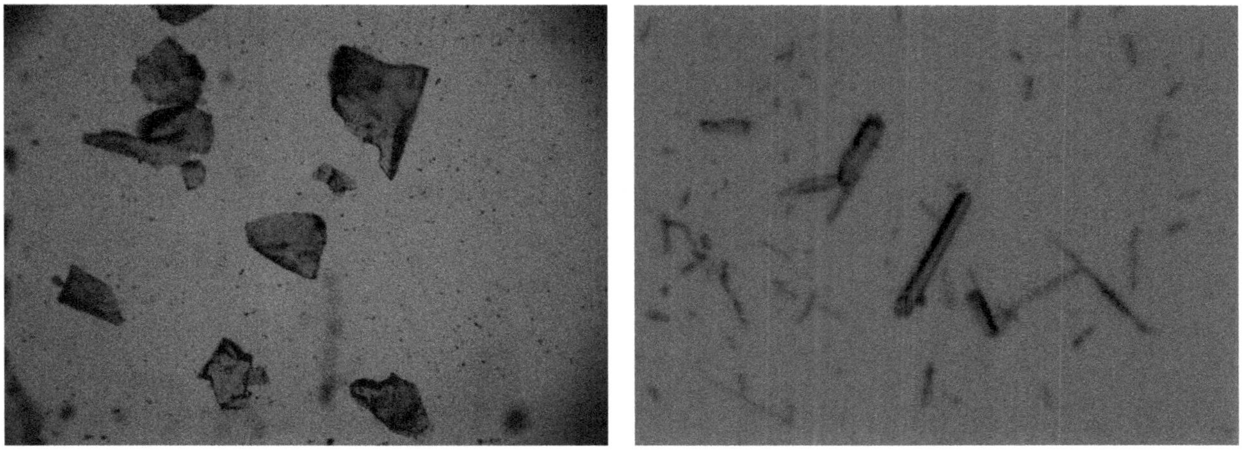

(a) (b)

FIGURE 2: Pictures taken at the polarized light microscope of (a) LISZn1 and (b) LISZn2 showing their distinct solid habits.

curve for LISZn1 displays four weight losses. The first one occurs between 298 and 377 K and corresponds to the weight loss of two moles of water that is in the crystalline lattice and 0.5 mol of triethylamine (obsd. 10.93%, calcd. 10.73%) per mol of the complex. Afterwards, three consecutive steps of weight losses above 500 K can be associated to the loss of 0.5 mol of the remained triethylamine, two moles of perchlorate ion and thermal decomposition of lisinopril drug (obsd. 74.34%, calcd. 74.71%). Finally, at 874 K, the residual percentage weight agrees with ZnO (obsd. 14.73%, calcd. 14.51%). For the LISZn2 complex, the curve shows three weight losses. The first one observed between 308 and 411 K is assigned to the loss of four moles of the lattice water (obsd. 12.10%, calcd. 13.33%) per mol of the complex. The second and third weight losses together refer to the thermal decomposition of lisinopril drug (obsd. 72.83%, calcd. 71.67%). The final residue at 874 K is in good agreement with one mol of ZnO (obsd. 15.07%, calcd. 15.00%). In the DTA curves of these complexes, one endothermic event at 334 K and 378 K for LISZn1 and LISZn2, respectively, can be noticed corresponding to the dehydration process. The other events in DTA curves for LISZn1 and LISZn2 are exothermic and can be associated with the thermal decomposition of both compounds. Thermal analysis results of LISZn2 complex are consistent with the X-ray crystallographic data.

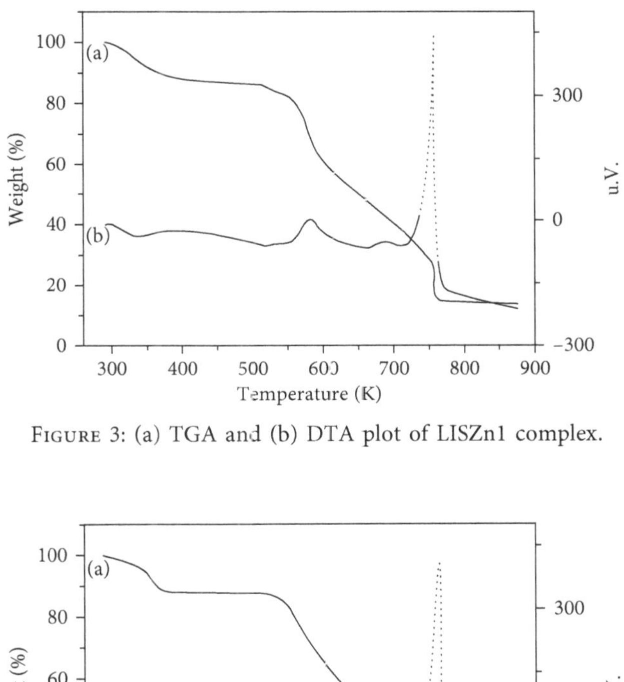

FIGURE 3: (a) TGA and (b) DTA plot of LISZn1 complex.

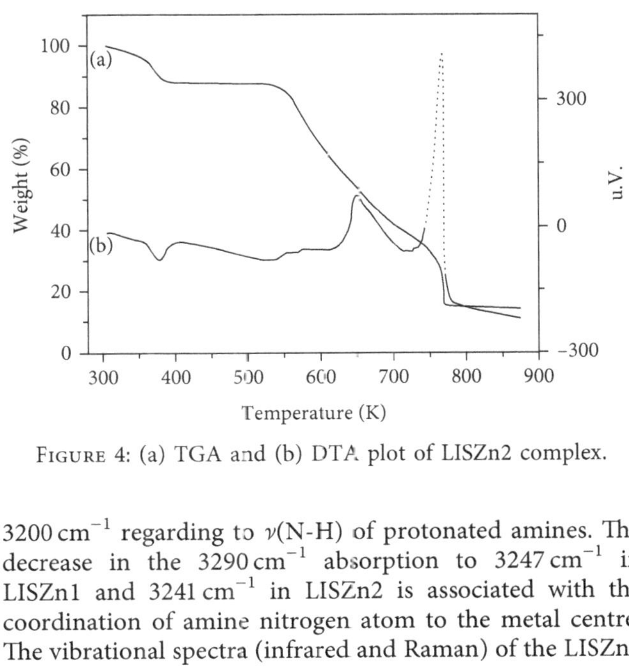

FIGURE 4: (a) TGA and (b) DTA plot of LISZn2 complex.

3.2. Spectroscopic Analyses. The spectroscopic data and vibrational assignments for lisinopril, LISZn1, and LISZn2 are shown in Table 1. The IR and Raman vibrational spectra are shown in the supplementary material (Figures S1 and S2, respectively). In the solid state, the main significant characteristic of the lisinopril structure is its double zwitterionic character. Both nitrogen atoms of amine groups are protonated, while the two carboxylic acid groups exist as carboxylate anions [6]. This behaviour directly influences the appearance of its vibrational spectra. The IR spectrum of the free lisinopril ligand shows one band with weak intensity at 3556 cm^{-1} related to the O-H stretching vibration of water molecules and some broad bands in the range of 3400 and 3200 cm^{-1} regarding to ν(N-H) of protonated amines. The decrease in the 3290 cm^{-1} absorption to 3247 cm^{-1} in LISZn1 and 3241 cm^{-1} in LISZn2 is associated with the coordination of amine nitrogen atom to the metal centre. The vibrational spectra (infrared and Raman) of the LISZn1

Table 1: Principal IR and Raman bands (cm^{-1}) for lisinopril, LISZn1, and LISZn2 compounds.

Lisinopril		LISZn1		LISZn2		Assignment[a]
IR[b]	Raman[b]	IR	Raman	IR	Raman	
3556w		3450m		3452m		$\nu(OH)$
3397w				3315vw		$\nu(NH)$
3290vw		3247vw		3241vw		$\nu(NH_2)$
3027vw	3058m	3024vw	3058m	3027vw	3056m	$\nu(CH)$
2966w	2982m	2952w	2979m	2955w	2983w	$\nu(CH)$, $\nu_a(CH_3)$
2925vw	2934s		2927s		2935w	$\nu_a(CH_2)$
2877vw	2878w	2874vw	2881m	2885vw	2865w	$\nu_s(CH_2)$, $\nu_s(CH_3)$
1657s	1654w	1602vs	1604m	1603vs	1604vs	$\nu(CO)_{amide}$
1578m	1585m		1585m		1585m	$\nu_a(COO)$
1391m	1398vw	1392m	1398vw	1396m	1400vw	$\nu_s(COO)$
		1088m	933vs			$\nu(ClO)$

[a]Tentative assignment according to [6, 18–20]. ν = stretching; a = antisymmetric; s = symmetric; d = deformation. [b]s = strong; m = medium; w = weak; v = very.

and LISZn2 display characteristic bands of aliphatic and aromatic moiety of the structure of lisinopril, such as $\nu(CH)$, $\nu_a(CH_3)$, $\nu_s(CH_3)$, $\nu_a(CH_2)$, and $\nu_s(CH_2)$. The band at 1657 cm^{-1} and 1654 cm^{-1} observed in the IR and Raman spectra of lisinopril attributed to ν(CO)amide shifted to 1602 cm^{-1} in IR and 1604 cm^{-1} in Raman for LISZn1 and 1603 cm^{-1} in IR and 1604 cm^{-1} in Raman for LISZn2. These results suggest that the oxygen atom of the amide group is also coordinated to Zn(II) ion.

The lisinopril Raman spectrum exhibits bands related to $\nu_a(COO)$ (at 1585 cm^{-1}) and $\nu_s(COO)$ (at 1398 cm^{-1}) that agree with the complete dissociation of the drug in the solid state. These bands are observed in spectra of LISZn1 and LISZn2 at 1585 cm^{-1} ($\nu_a(COO)$ for both complexes) and at 1398 and 1400 cm^{-1} ($\nu_s(COO)$ for LISZn1 and LISZn2, respectively). Some information about the coordination mode can be obtained by the difference between the wave numbers of these modes, thus indicating whether the compound is in its ionic or coordinate form or on the type of coordination of the ligand with the metal site [17]. Differences determined from the Raman spectra are 187 cm^{-1}, 185 cm^{-1}, and 187 cm^{-1} for LISZn1, LISZn2, and ionic compound, respectively [6] These values very close are inconclusive regarding the lisinopril's coordination mode to the metal site in the LISZn1 and LISZn2 complexes. Additionally, spectroscopic data of LISZn1 display a medium intensity band at 1088 cm^{-1} (in IR) and a strong intensity band at 933 cm^{-1} (in Raman) assigned to the presence of noncoordinated perchlorate ion. IR spectra of the LISZn1 and LISZn2 complexes show a broad band centred at 3450 cm^{-1} and 3452 cm^{-1}, respectively, that are attributed to the O-H stretching vibration of water molecules. Spectroscopic data suggest similar coordination mode for both complexes. The vibrational results are in agreement with the X-ray diffraction data for LISZn2.

3.3. Powder Crystal X-Ray Diffraction. The simulated diffractogram generated from the crystal structure of LISZn2 crystal structure is in excellent agreement with the experimental diffractogram on the bulk sample of LISZn2 (Figures 5(a) and 5(b)). Qualitative analysis of the experimental and simulated diffraction patterns show that the Bragg peaks are in the same position. This proved that all further experimental results obtained for the complex LISZn2 could be interpreted based on the crystallographic model established in this study, i.e., the crystal structure is representative of the whole sample. In order to analyse the crystallinity of the solid LISZn1, powder X-ray diffraction measurement was realized (Figure 5(c)). No crystallinity was found in the diffraction pattern confirming its amorphous nature.

3.4. Crystallographic Analysis. The LISZn2 complex asymmetric unit is shown in Figure 6 in which can be seen four lisinopril drug molecules, four crystallographically independent zinc (II) ions, and sixteen water molecules. The complex is neutralized through full deprotonation of lisinopril ligand molecules and crystallizes in a noncentrosymmetric space group, $P3_2$ (Table 2), building a coordination polymer. In Figure S3 (Supplementary Material), it is possible to see a section of the polymeric structure of LISZn2.

The refinement was not satisfactory using default methods. The first problem was to determine the space group. There were systematic absences for a 3_1 or a 3_2 axis. It was possible to obtain the crystal structure of LISZn2 in the space group $P3_1$ with relatively low *R* values (*R* and *wR* indices for $[F_{obs} > 4\sigma(F_{obs})]$ equal to 0.066 and 0.1731, respectively). However, in crystals that crystalize in a noncentrosymmetric space groups is important the determination of absolute structure analyzing the Flack parameter. The Flack parameter obtained for this complex was 0.71(2) [21], which means that the absolute structure is wrong and the space group $P3_2$ is the correct one instead of $P3_1$. The refinement using the space group $P3_2$ presented the Flack parameter equal to 0.023(11) indicating that $P3_2$ is the correct space group [22]. All necessary changes have been made, and the refinement became satisfactory (Table 2).

The refinement of the LISZn2 complex shows some atoms (derived from lisinopril ligand and water molecules) with disordered effect in the unit cell. The O31A, O35A, and O36A oxygen atoms of water molecules; the C47A, C48A, and C64A carbon atoms; and the N3A nitrogen atom of the lisinopril ligand exhibited these disorder effects (Figure 7).

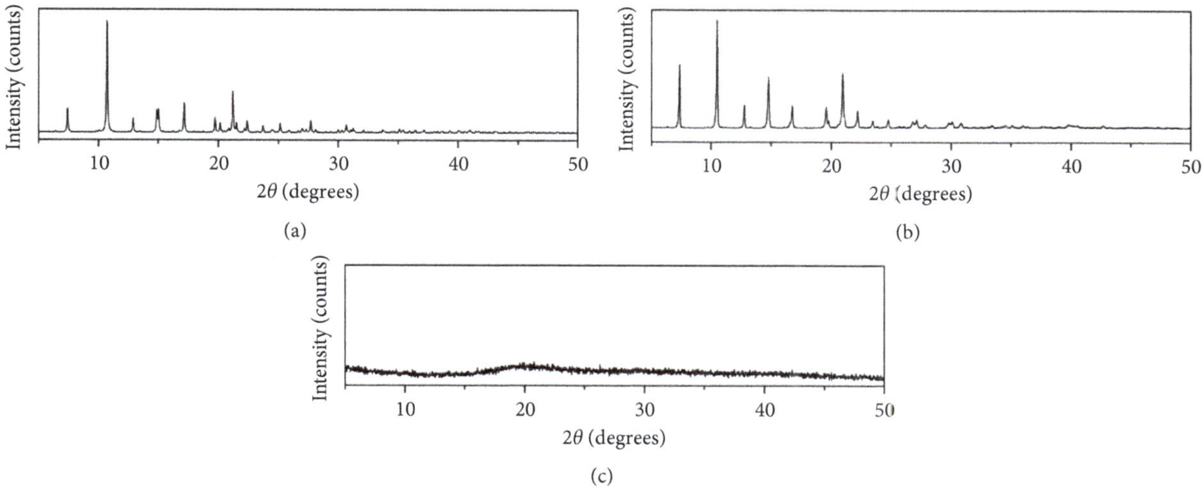

FIGURE 5: Diffractograms of the complexes: (a) simulated for LISZn2, (b) experimental for LISZn2, and (c) experimental for LISZn1.

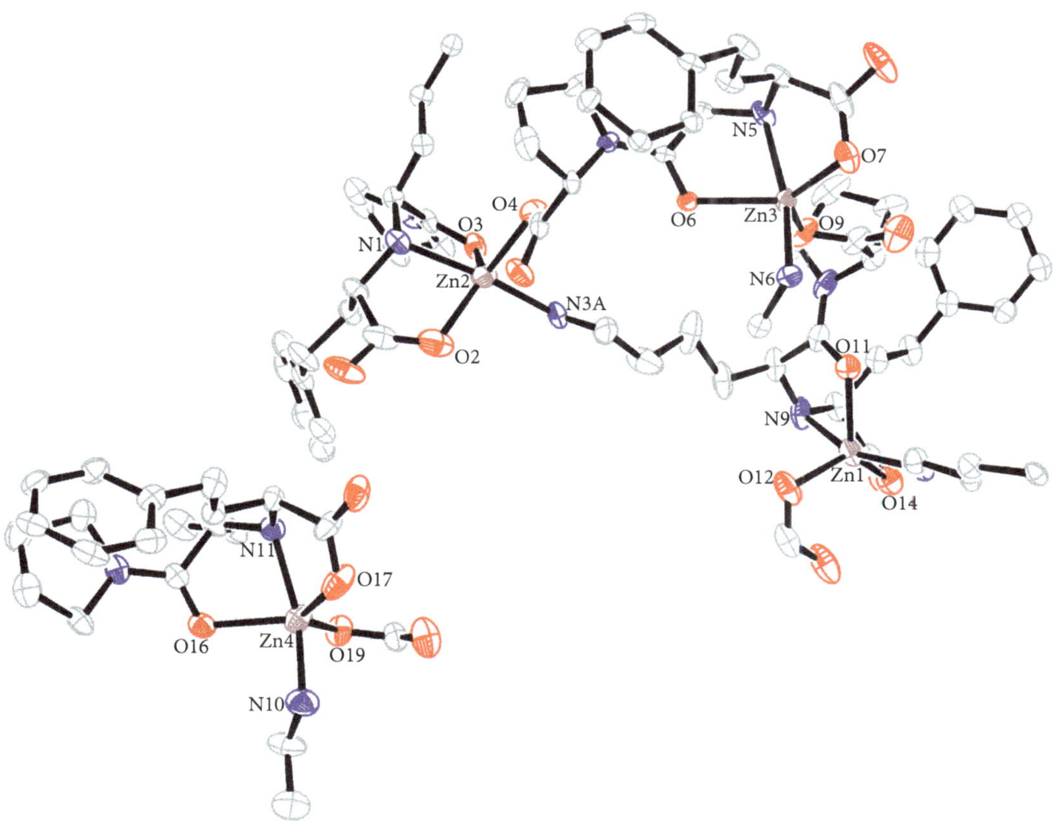

FIGURE 6: The asymmetric unit of the LISZn2 complex. Thermal ellipsoids are drawn at the 30% probability level. Some atoms are omitted for clearness.

The disorder refined from O31A, O35A, O36A/O31B, O35B, O36B, and C64A/C64B shows occupational values of 0.488(6):0.512(3) and 0.800(3):0.200(2), respectively. The amine chain atoms were refined together with the same occupation value. Each group is associated with one occupational value, and both contain the same atoms, however, with different crystallographic positions. The C47A, C48A, and N3A atoms are the first component of the disorder with an occupational distribution of 51%, and the C47B, C48B, and N3B atoms are the second component with an occupational distribution of 49%. The four crystallographically independent Zn^{2+} ions are pentacoordinated. In Zn1, Zn2, and Zn3 metal sites, a distorted square pyramidal geometry was observed and confirmed by the calculation of the t values, which presented values of 0.45, 0.33, and 0.48 respectively, similar to those observed in copper-lisinopril

Table 2: Crystal data for LISZn2 compound.

Compound	LISZn2
Chemical formula	$C_{84}H_{148}N_{12}O_{36}Zn_4$
Formula weight (g·mol^{-1})	2163.61
Crystal system	Trigonal
Space group	$P3_2$
a (Å)	27.4727(7)
b (Å)	27.4727(7)
c (Å)	11.4613(3)
α (°)	90
β (°)	90
γ (°)	120
Volume (Å^3)	7491.5(4)
Z	3
Temperature (K)	120.0(1)
$d_{calc.}$ (g·cm^{-3})	1.439
μ (KαCu) (mm^{-1})	1.835
Crystal size (mm)	0.12 × 0.11 × 0.09
Radiation	λ = 1.5418 Å (KαCu)
θ limits (°)	3.1980–72.5730
Refl. measured/independent	24164/15047
Refl. observed [$F_{obs} > 4\sigma_{(Fobs)}$]	11695
Parameters	1189
Flack parameter	0.023(11)
R_{int}	0.0356
R [$F_{obs} > 4\sigma_{(Fobs)}$]	0.0609
R indices for all data	0.0777
wR [$F_{obs} > 4\sigma_{(Fobs)}$]	0.1635
Goodness-of-fit (GOF)	1.06
Weighting scheme	$W = 1/[\sigma^2(Fo^2) + (0.1137P)^2 + 0.0000P]$ Where $P = (Fo^2 + 2Fc^2)/3$
$(\Delta\rho)_{min}$, $(\Delta\rho)_{max}$ (e·Å^{-3})	−0.71, 1.55

complex described in the literature [10]. Three positions in the basal plane are occupied by oxygen atoms of central carboxylate (Zn-O 2.101(2) Å) and carbonyl (Zn-O 2.164 (3) Å) groups as also the secondary amine nitrogen (Zn-N 2.122(2) Å) of one lisinopril ligand, while the fourth is occupied by primary amine nitrogen (Zn-N 2.035(2) Å) of a second lisinopril ligand. An oxygen atom of the prolyl carboxylate group of a third lisinopril ligand is located at the apex with an average Zn-O bond distances equal to 2.025(2) Å. The Zn4 metal site has a very much distorted trigonal bipyramidal geometry with t value equal to 0.58. The Zn4 metal site is coordinated to oxygen atoms of central and prolyl carboxylate (Zn-O 2.062(5) Å and Zn-O 1.999(4) Å, respectively) and carbonyl (Zn-O 2.171(5) Å) groups in the basal plane as well as primary and secondary amine nitrogen atoms (Zn-N 2.012(8) Å and Zn-N 2.126 (6) Å, respectively) in the axial positions. The average values found for the Zn-O and Zn-N bonds distances are 2.092(3) Å and 2.076(1) Å, respectively. The average Zn-O bond distance is similar to that observed in complex lis-enzyme described in the literature [9].

In Table 3, selected bond lengths (Å) and angles (°) of LISZn2 are listed. The interaction *in vitro* among lisinopril and Zn^{2+} ion is similar to which is suggested *in vivo*, once the lisinopril coordination occurs through the oxygen atom of the central carboxylate moiety [9]. In the structure of lis-enzyme, the Zn^{2+} ion also is coordinated to two histidine molecules by two nitrogen atoms of the imidazole ring and to one glutamic acid by the oxygen atom present in the environment. In the LISZn2 was observed that the coordination environment around zinc ion is formed by oxygen atoms from prolyl carboxylate and carbonyl groups, as also by nitrogen atoms of primary and secondary amines.

Furthermore, the crystal packing stability occurs through classical hydrogen bonds (OH ··· and NH ··· O) formed among water molecules and carboxylate and amine groups of the LIS ligand. Other hydrogen bonds among water molecules in the structure are also observed. The O···O distances from classical hydrogen bonds are in the 2.808 to 3.069 Å range (Table 4).

In this compound, we describe the structure considering the symmetry operation from the space group $P3_2$ and to facilitate the description; only the equivalent general positions were used providing an insight between structure and symmetry. The space group diagram within the structure is given in Figure 8(a). Note that there are four different Zn atom sites specified in the unit cell, numbered Zn1, Zn2, Zn3, and Zn4; the atomic coordinates of Zn atoms are shown in Table S1 (Supplementary Material). The 3_2 screw axis running parallel to the c-axis and all atoms must be placed in equivalent general positions making two similar one-dimensional chains. The first chain is localized in the vertex of the unit cell, and it is formed by Zn4–Zn4 coordinate bond motifs that propagate along the 3_2 screw axis, as is possible to see in Figure 8(b). This representation was performed by TOPOS program package [23]. The second chain built by Zn1–Zn2–Zn3 is replicated three times, by the application of the symmetry screw axis. Thus, the unit cell contains four chains that interact by hydrogen bonds with water molecules. The pyrrolidine ring of the ligand is represented with a solid line for clarity. In addition, each chain related by symmetry equivalence is illustrated with different colours in Figure 8(a).

Comparing the LISZn2 conformation with the LIS, conformation was realized [6]. The more important dihedral angles observed in dehydrated lisinopril are 175.7(4)° (C14-C15-C16-C17), 179.6(2)° (C9-C8-C7-C1), −5.2(3)° (N22-C26-C27-O28), −76.8(1)° (O21-C20-C14-N13), 0.26(1)° (O21-C20-N22-C26), −69.3(2)° (C16-C17-C18-N19), and −17.5(4)° (N13-C9-C10-O11). The equivalent angles in LISZn2 are 175.4(9)°, −179.2(8)°, −23.1(1)°, −37.7(14)°, 176.8 (2)°, 173.9(3)°, and −38.9(17)° (Figure S4). The first and second dihedral angles of LISZn2 are similar to the angles of the LIS structure. The most important difference was observed in the other dihedral angles, which could be related to the coordination of the LIS to the Zn^{2+} ion. Similar angles in the crystal structure of lis-enzyme have the values equal to 50.6(3)°, −30.7(2)°, −0.6(2)°, 175.3(3)°, and 125.4(2)° [9]. These angles also disagreed with values found in LISZn2 due to the difference in coordination modes of LIS ligand and by the protein environment that also has a very important influence on the conformation of this molecule. The coordination of LISZn2 occurs via the oxygen atoms (carbonyl and carboxylate groups) also as the nitrogen atoms of the

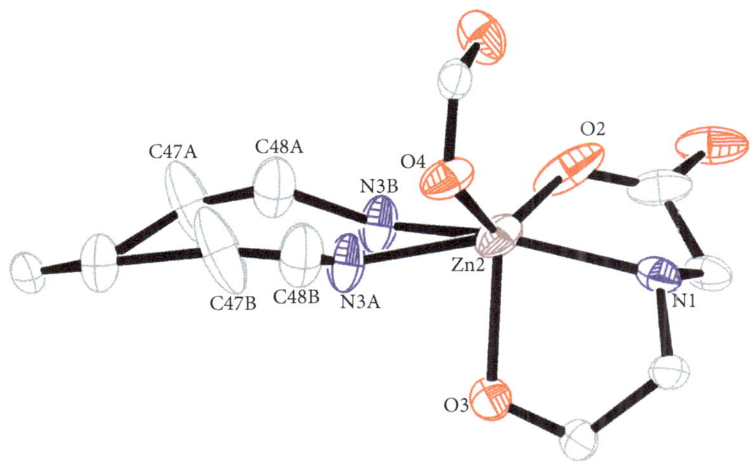

FIGURE 7: The carbon and nitrogen atoms with order and disorder effects in LISZn2. Anisotropic displacement ellipsoids are drawn at the 30% probability level. In order to clarify, some atoms are omitted.

TABLE 3: Selected bond lengths (Å) and angles (°) of the LISZn2 compound.

	Bond lengths (Å)		
Zn1-O12	2.036(7)	Zn3-O7	2.094(6)
Zn1-N8	2.022(7)	Zn3-N5	2.107(7)
Zn1-O14	2.116(8)	Zn3-O6	2.171(5)
Zn1-N9	2.134(8)	Zn4-O19	1.999(6)
Zn1-O11	2.158(7)	Zn4-N10	2.012(8)
Zn2-O4	2.033(6)	Zn4-O17	2.062(7)
Zn2-O2	2.092(9)	Zn4-N11	2.126(7)
Zn2-N1	2.125(7)	Zn4-O16	2.171(7)
Zn2-O3	2.164(6)	N7-C43	1.314(10)
Zn3-O9	2.005(7)	N7-C40	1.501(14)
	Bond angles (°)		
O12-Zn1-N8	99.6(3)	O9-Zn3-O7	127.2(3)
N8-Zn1-N9	157.7(3)	O7-Zn3-N5	79.7(2)
O14-Zn1-O11	125.9(3)	O19-Zn4-N10	100.5(4)
N3A-Zn2-O2	100.8(8)	N10-Zn4-N11	160.9(4)
O4-Zn2-O3	102.0(2)	O19-Zn4-O16	106.9(3)
N3A-Zn2-N1	154.3(6)		
	Torsion angles (°)		
C76-C77-C78-C79	175.2(9)	O16-C80-N12-C81	176.7(12)
C67-C68-C69-C70	−179.8(8)	C78-C79-C64A-N10	173.9(3)
N12-C82-C65-O19	−22.9(1)	N11-C67-C66-O17	−38.9(17)
O16-C80-C76-N11	−37.7(14)		

TABLE 4: Hydrogen-bonding geometry of LISZn2 compound.

D-H···A	D-H (Å)	H ··· A (Å)	D ··· A (Å)	D-H ··· A (°)
N1-H1A ··· O21	0.9800	2.0700	3.010(12)	160.00
N5-H5A ··· O26	0.9800	2.0800	2.999(18)	156.00
N11-H11A ··· O29	0.9800	1.8300	2.759(17)	158.00
O23-H23D ··· O25	0.8200	2.0200	2.808(15)	161.00
O24-H24C ··· O14	0.8400	2.4100	3.069(15)	135.00

amine groups, whereas that of the complex lis-enzyme occurs only by oxygen atom of the central carboxylate group of lisinopril molecule.

4. Conclusions

In this work, two new Zn^{2+} complexes with the antihypertensive drug lisinopril were obtained from different synthetic routes and structurally characterized by thermal, spectroscopy, and X-ray diffraction techniques. The synthesis of LISZn2 involved solvothermal conditions, while the synthesis of LISZn1 involved only using stirring. The synthetic differences resulted in the formation of two zinc complexes with notorious structural differences once LISZn1 is amorphous while LISZn2 is crystalline. Furthermore, chemical differences were also observed, wherein LISZn1 complex contains perchlorate ions indicating that in this complex, the lisinopril molecule is neutral. On the contrary, in LISZn2, the results show that the lisinopril species are in ionic form. Moreover,

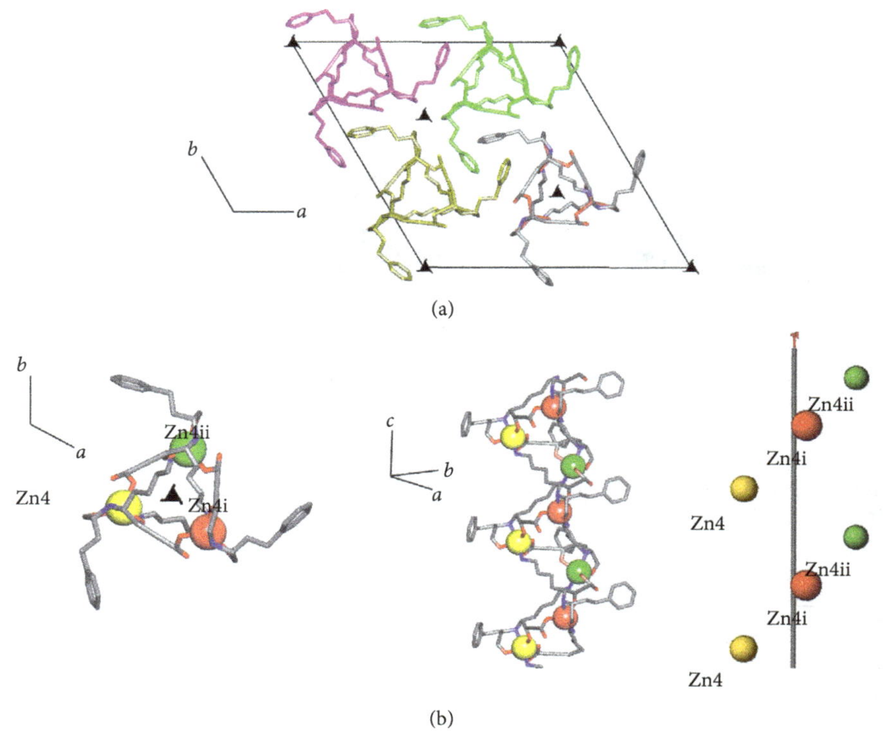

FIGURE 8: (a) Unit cell and space group diagram within four one-dimensional chains for structure. (b) Zn4–Zn4 coordinate bond motifs that propagate along a 3_2 screw axis parallel to the c-axis crystallographic. Symmetry codes: (i) $-y, x-y, z+2/3$ and (ii) $-x+y, -x, z+1/3$. Some atoms were omitted for more clearness.

the LISZn1 presents water and triethylamine molecules in the lattice, while in LISZn2, only water molecules were observed in the lattice. The X-ray diffraction data show that LISZn2 is 1D coordination polymer and exhibits the LIS coordinated to the metal site via oxygen and nitrogen atoms. The interaction between the lisinopril and the Zn^{2+} ion in LISZn2 is similar only in the Zn-O coordination bond. The spectroscopic data confirm that the coordination of lisinopril with zinc ion occurs by carboxylate, carbonyl, and amine groups, which agrees with the crystallographic analysis, and that LISZn1 coordination probably is similar to LISZn2.

Conflicts of Interest

The authors declare that there are no conflicts of interest regarding the publication of this paper.

Acknowledgments

The authors are thankful to Prof. Max Siebel from John Hopkins University (Baltimore, USA) for single crystal X-ray facilities and the Brazilian Agencies FINEP and CAPES for financial support and scholarship, respectively. The authors thank the Brazilian Agencies FAPEMIG (CEX-APQ-01919-17) and CNPq (307599/2017-5) for the research financial support.

References

[1] F. Nobre, "VI diretrizes brasileiras de hipertensão," *Revista Brasileira de Hipertensão*, vol. 17, no. 1, pp. 4–64, 2010.

[2] Z. Su, L. Chen, Y. Liu et al., "35MHz quartz crystal microbalance and surface plasmon resonance studies on the binding of angiotensin converting enzyme with lisinopril," *Biosensors and Bioelectronics*, vol. 26, no. 7, pp. 3240–3245, 2011.

[3] J. C. Joyner, L. Hocharoen, and J. A. Cowan, "Targeted catalytic inactivation of angiotensin converting enzyme by lisinopril-coupled transition-metal chelates," *Journal of the American Chemical Society*, vol. 134, no. 7, pp. 3396–3410, 2012.

[4] L. Hocharoen, J. C. Joyner, and J. A. Cowan, "N-versus C-domain selectivity of catalytic inactivation of human angiotensin converting enzyme by lisinopril-coupled transition metal chelates," *Journal of Medicinal Chemistry*, vol. 56, no. 24, pp. 9826–9836, 2913.

[5] R. Mukherjee, A. McCaddon, C. A. Smith et al., "Synthesis, synchrotron X-ray diffraction, and kinetic studies on the formation of a novel thiolatocobalamin of captopril: evidence for cis-trans isomerization in the β-axial ligand," *Inorganic Chemistry*, vol. 48, no. 19, pp. 9526–9534, 2009.

[6] M. Sorrenti, L. Catenacci, D. L. Cruickshank et al., "Lisinopril dihydrate: single-crystal X-ray structure and physicochemical characterization of derived solid forms," *Journal of Pharmaceutical Sciences*, vol. 102, no. 10, pp. 3596–3603, 2013.

[7] M. Miclaus, L.-G. Grosu, X. Filip et al., "Optimizing structure determination from powders of crystalline organic solids with

high molecular flexibility: the case of lisinopril dihydrate," *CrystEngComm*, vol. 16, no. 3, pp. 299–303, 2014.

[8] F. Qin, D. Wang, S. Yang et al., "Quantitative determination of lisinopril in human plasma by high performance liquid chromatography—tandem mass spectrometry and its application in a pharmacokinetic study," *Biomedical Chromatography*, vol. 26, no. 6, pp. 691–696, 2012.

[9] R. Natesh, S. L. Schwager, E. D. Sturrock et al., "Crystal structure of the human angiotensin-converting enzyme-lisinopril complex," *Nature*, vol. 421, no. 6922, pp. 551–554, 2003.

[10] E. B. Gonzalez, E. Farkas, A. A. Soudi et al., "Metal complexes of the angiotensin-converting enzyme inhibitor, lisinopril. Solution studies and the crystal and molecular structure of a dimeric copper(II)-lisinopril complex," *Journal of the Chemical Society, Dalton Transactions*, vol. 21, no. 13, pp. 2377–2379, 1997.

[11] K. Fujii, H. Uekusa, N. Itoda et al., "Mechanism of dehydration–hydration processes of lisinopril dihydrate investigated by ab initio powder X-ray diffraction analysis," *Crystal Growth and Design*, vol. 12, no. 12, pp. 6165–6172, 2012.

[12] Rigaku Corporation, *CrysAlisPro*, Rigaku Corporation, Tokyo, Japan, 2015.

[13] R. H. Blessing, "An empirical correction for absorption anisotropy," *Acta Crystallographica Section A Foundations of Crystallography*, vol. 51, no. 1, pp. 33–38, 1995.

[14] G. Sheldrick, "SHELXT-Integrated space-group and crystal-structure determination," *Acta Crystallographica Section A Foundations and Advances*, vol. 71, no. 1, pp. 3–8, 2015.

[15] L. J. Farrugia, "ORTEP-3 for Windows-a version of ORTEP-III with a Graphical User Interface (GUI)," *Journal of Applied Crystallography*, vol. 30, no. 5, p. 565, 1997.

[16] C. F. Macrae, P. R. Edgington, P. McCabe et al., "Mercury: visualization and analysis of crystal structures," *Journal of Applied Crystallography*, vol. 39, no. 3, pp. 453–457, 2006.

[17] G. B. Deacon and R. J. Phillips, "Relationships between the carbon-oxygen stretching frequencies of carboxylato complexes and the type of carboxylate coordination," *Coordination Chemistry Reviews*, vol. 33, no. 3, pp. 227–250, 1980.

[18] S.-L. Wang, S.-Y. Lin, and T.-F. Chen, "Thermal-dependent dehydration process and intramolecular cyclization of lisinopril dihydrate in the solid state," *Chemical Pharmaceutical Bulletin*, vol. 48, no. 12, pp. 1890–1893, 2000.

[19] S.-L. Wang, C.-H. Chuang, and S.-Y. Lin, "pH-dependent coordination of metal–lisinopril complex investigated by attenuated total reflection/fourier transform infrared spectroscopy," *Chemical Pharmaceutical Bulletin*, vol. 50, no. 1, pp. 78–82, 2002.

[20] J. Hinojosa-Torres, J. M. Aceves-Hernández, J. Hinojosa-Torres et al., "Degradation of lisinopril: a physico-chemical study," *Journal of Molecular Structure*, vol. 886, no. 1–3, pp. 51–58, 2008.

[21] H. D. Flack and G. Bernardinelli, "Absolute structure and absolute configuration," *Acta Crystallographica Section A Foundations of Crystallography*, vol. 55, no. 5, pp. 908–915, 1999.

[22] P. Muller, R. Herbst-Irmer, A. L. Spek et al., *Crystal Structure Refinement: A Crystallographer's Guide to SHELXL*, Oxford University Press, New York, NY, USA, 1st edition, 2006.

[23] V. A. Blatov, A. P. Shevchenko, and V. N. Serezhkin, "TOPOS3.2: a new version of the program package for multipurpose crystal-chemical analysis," *Journal of Applied Crystallography*, vol. 33, no. 4, p. 1193, 2000.

Permissions

All chapters in this book were first published in JC, by Hindawi Publishing Corporation; hereby published with permission under the Creative Commons Attribution License or equivalent. Every chapter published in this book has been scrutinized by our experts. Their significance has been extensively debated. The topics covered herein carry significant findings which will fuel the growth of the discipline. They may even be implemented as practical applications or may be referred to as a beginning point for another development.

The contributors of this book come from diverse backgrounds, making this book a truly international effort. This book will bring forth new frontiers with its revolutionizing research information and detailed analysis of the nascent developments around the world.

We would like to thank all the contributing authors for lending their expertise to make the book truly unique. They have played a crucial role in the development of this book. Without their invaluable contributions this book wouldn't have been possible. They have made vital efforts to compile up to date information on the varied aspects of this subject to make this book a valuable addition to the collection of many professionals and students.

This book was conceptualized with the vision of imparting up-to-date information and advanced data in this field. To ensure the same, a matchless editorial board was set up. Every individual on the board went through rigorous rounds of assessment to prove their worth. After which they invested a large part of their time researching and compiling the most relevant data for our readers.

The editorial board has been involved in producing this book since its inception. They have spent rigorous hours researching and exploring the diverse topics which have resulted in the successful publishing of this book. They have passed on their knowledge of decades through this book. To expedite this challenging task, the publisher supported the team at every step. A small team of assistant editors was also appointed to further simplify the editing procedure and attain best results for the readers.

Apart from the editorial board, the designing team has also invested a significant amount of their time in understanding the subject and creating the most relevant covers. They scrutinized every image to scout for the most suitable representation of the subject and create an appropriate cover for the book.

The publishing team has been an ardent support to the editorial, designing and production team. Their endless efforts to recruit the best for this project, has resulted in the accomplishment of this book. They are a veteran in the field of academics and their pool of knowledge is as vast as their experience in printing. Their expertise and guidance has proved useful at every step. Their uncompromising quality standards have made this book an exceptional effort. Their encouragement from time to time has been an inspiration for everyone.

The publisher and the editorial board hope that this book will prove to be a valuable piece of knowledge for researchers, students, practitioners and scholars across the globe.

List of Contributors

Carlos Martínez-Conde, Edgar Rosales-Ortega, Juan José Ramírez-Espinosa, Samuel Estrada-Soto and Gabriel Navarrete-Vázquez
Facultad de Farmacia, Universidad Autónoma del Estado de Morelos, 62209 Cuernavaca, MOR, Mexico

Erika Gutierrez-Lara
Facultad de Farmacia, Universidad Autónoma del Estado de Morelos, 62209 Cuernavaca, MOR, Mexico
Departamento de Farmacobiología, Cinvestav-Coapa, 14330 Ciudad de México, Mexico

Julio C. Rivera-Leyva
Facultad de Farmacia, Universidad Autónoma del Estado de Morelos, 62209 Cuernavaca, MOR, Mexico
Laboratorio de Nutrición Experimental, Instituto Nacional de Pediatría, 04150 Ciudad de México, Mexico

David Centurión
Departamento de Farmacobiología, Cinvestav-Coapa, 14330 Ciudad de México, Mexico

Karla Carvajal and Daniel Ortega-Cuellar
Laboratorio de Nutrición Experimental, Instituto Nacional de Pediatría, 04150 Ciudad de México, Mexico

Le-Ming Jiang, Shao-Ping Nie, Dan-Fei Huang, Zhi-Hong Fu and Ming-Yong Xie
State Key Laboratory of Food Science and Technology, Nanchang University, Nanchang 330047, China

Carlos A. González-González, Juan Javier Mejía Vega, Ricardo García Monroy, Davir González-Calderón, David Corona-Becerril, Aydeé Fuentes-Benítes and Carlos González-Romero
Departamento de Química Orgànica, Facultad de Química, Universidad Autónoma del Estado de Mèxico, Paseo Colón Esquina Paseo Tollocan s/n, 50120 Toluca, MEX, Mexico

Joaquín Tamariz Mascarúa
Departamento de Química Orgànica, Escuela Nacional de Ciencias Biológicas, Instituto Politècnico Nacional, Prol. Carpio y Plan de Ayala s/n, 11340 Ciudad de Mèxico, Mexico

Yonghai Yu
School of Bio-science and Food Engineering, Changchun University of Science and Technology, Changchun 130-600, China

Yujiao Tang
School of Bio-science and Food Engineering, Changchun University of Science and Technology, Changchun 130-600, China
Division of Food Bioscience, College of Biomedical and Health Sciences, Konkuk University, Chungju 27478, Republic of Korea

Meiqi Fan, Young-Jin Choi, Sang-Ho Moon and Eun-Kyung Kim
Division of Food Bioscience, College of Biomedical and Health Sciences, Konkuk University, Chungju 27478, Republic of Korea

Eun-Ju Choi
Department of Physical Education, College of Education, Daegu Catholic University, Gyeongsan 38430, Republic of Korea

Trishna Debnath
Department of Food Science and Biotechnology, Dongguk University, Goyang 10326, Republic of Korea

Il Nam Lee
R&D Department, Korea Mcnulty Co., Ltd., Cheonan 31009, Republic of Korea

Ellis Benjamin
Department of Chemistry, Stockton University, 101 Vera King Farris Drive, Galloway, NJ 08205-9441, USA

Yousef M. Hijji
Department of Chemistry and Earth Sciences, Qatar University, Doha, Qatar

Zheng Ruan
State Key Laboratory of Food Science and Technology, Nanchang University, Nanchang, Jiangxi 330047, China

Guiping Guan
State Key Laboratory of Food Science and Technology, Nanchang University, Nanchang, Jiangxi 330047, China
College of Bioscience and Biotechnology, Hunan Agricultural University, Changsha, Hunan 410128, China

Roisbel Aroche
State Key Laboratory of Food Science and Technology, Nanchang University, Nanchang, Jiangxi 330047, China
College of Bioscience and Biotechnology, Hunan Agricultural University, Changsha, Hunan 410128, China
Faculty of Veterinary Medicine, University of Granma, Bayamo, 85100 Granma, Cuba

Yordan Martínez
College of Bioscience and Biotechnology, Hunan Agricultural University, Changsha, Hunan 410128, China

Shile Lan
College of Bioscience and Biotechnology, Hunan Agricultural University, Changsha, Hunan 410128, China
Faculty of Veterinary Medicine, University of Granma, Bayamo, 85100 Granma, Cuba

Samuel Waititu
Swine and Poultry Nutritionist, Fortified Nutrition Limited, Winnipeg, MB, Canada R2J OK4

Charles M. Nyachoti
Animal Science Department, University of Manitoba, Winnipeg, MB, Canada R3T 2N2

Dairon Màs
Laboratory of Animal Nutrition, Faculty of Natural Sciences, Autonomous University of Queretaro, 76230 Santiago de Querètaro, QRO, Mexico

Pierre-Philipe Roy, Diene Faye, Sébastien Blanchard, Marc Cormier, Jérémie A. Doiron, Marc E. Surette and Mohamed Touaibia
Department of Chemistry and Biochemistry, Universitè de Moncton, Moncton, NB, Canada E1A 3E9

Hongbo Li, Shutao Lei, Yongqiong Tang, Liyuan Zou, Hongqin Hu and Suqin Wang
Key Laboratory of Functional Small Organic Molecule, Ministry of Education, College of Chemistry and Chemical Engineering, Jiangxi Normal University, Nanchang 330022, China

Jhonatan R. Guarín and Juan Carlos Moreno-Pirajan
Department of Chemistry, Sciences Faculty, Research Group in Porous Solids and Calorimetry, University of the Andes, Bogota 111711, Colombia

Liliana Giraldo
Departamento de Química, Facultad de Ciencias, Universidad Nacional de Colombia, Bogota 111321, Colombia

Stefany Fanelli, Eliane Gandolpho Totòli and Hèrida Regina Nunes Salgado
School of Pharmaceutical Sciences, Universidade Estadual Paulista, Araraquara, SP, Brazil

Alexander Zimmermann
Institute of Biochemistry and Biotechnology, Faculty of Natural Science I, Martin-Luther-University Halle-Wittenberg, Halle (Saale), Germany

Vu Dinh Tien, Vu Van Vu, Trinh Thi Nhung, Nguyen Thi Thanh, Vu Thu Thuy and Tran Khac Vu
School of Chemical Engineering, Hanoi University of Science and Technology, No. 1 Dai Co Viet, Hai Ba Trung, Hanoi, Vietnam

Luu Van Chinh
Institute of Natural Products Chemistry, Vietnam Academy of Science and Technology, 18 Hoang Quoc Viet, Cau Giay, Hanoi, Vietnam

Do Thi Thao
Institute of Biotechnology, Vietnam Academy of Science and Technology, 18 Hoang Quoc Viet, Cau Giay, Hanoi, Vietnam

Nguyen Hai Nam
Hanoi University of Pharmacy (HUP), 13-15 LeThanh Tong, Hanoi, Vietnam

Angela Koeckritz
Leibniz Institute for Catalysis at the University of Rostock, Albert-Einstein-Str. 29A, 18059 Rostock, Germany

Nur Rahimah Fitrah Mohd Sofyan, Syahrina Nur 'Ain Abdul Halim, Nur Adila Fatin Mohd Khir, Nor Fadilah Rajab and Rozie Sarip
Department of Chemistry, Faculty of Science, University of Malaya, Lembah Pantai, 50603 Kuala Lumpur, Malaysia

Fariza Juliana Nordin
Faculty of Health Sciences, Universiti Kebangsaan Malaysia, Kuala Lumpur Campus, Jalan Raja Muda Abdul Aziz, 50300 Kuala Lumpur, Malaysia

Mohd Ridzuan Mohd Abd Razak and Amirrudin Muhammad
Bioassay Unit, Herbal Medicine Research Centre, Institute for Medical Research, Jalan Pahang, 50588 Kuala Lumpur, Malaysia

Mohamed El-Naggar
Chemistry Department, Faculty of Sciences, University of Sharjah, Sharjah 27272, UAE

Abdel-Nasser El-Shorbagi
College of Pharmacy, University of Sharjah, Sharjah 27272, UAE
Sharjah Institute for Medical Research, University of Sharjah, Sharjah 27272, UAE

Dina H. Elnaggar
Applied Organic Chemistry Department, National Research Centre, Dokki, Giza 12622, Egypt

Mohamed A. Al-Omar
Drug Exploration & Development Chair (DEDC), Pharmaceutical Chemistry Department, College of Pharmacy, King Saud University, Riyadh 11451, Saudi Arabia

Abd El-Galil E. Amr
Drug Exploration & Development Chair (DEDC), Pharmaceutical Chemistry Department, College of Pharmacy, King Saud University, Riyadh 11451, Saudi Arabia
Department of Applied Organic Chemistry, Industrial Chemical Division, National Research Center, Cairo, Dokki 12622, Egypt

Elsayed A. Elsayed
Bioproducts Research Department, Zoology Department, Faculty of Science, King Saud University, Riyadh, Saudi Arabia
Chemistry of Natural and Microbial Products Department, National Research Centre, Dokki 12622, Cairo, Egypt

Carolynne Zie Wei Sie, Zainab Ngaini, Nurashikin Suhaili and Eswaran Madiahlagan
Faculty of Resource Science and Technology, Universiti Malaysia Sarawak, 94300 Kota Samarahan, Sarawak, Malaysia

Haroon ur Rashid
Department of Chemistry, Sarhad University of Science and Information Technology, Peshawar, Khyber Pakhtunkhwa, Pakistan

Sher Wali Khan, Noor Rehman and Muhammad Tariq
Department of Chemistry, Shaheed BB University, Sheringal, Upper Dir, Khyber Pakhtunkhwa, Pakistan

Momin Khan
Department of Biotechnology, Quaid-I-Azam University, Islamabad 45320, Pakistan
Department of Microbiology, Institute of Basic Medical Sciences, Khyber Medical University, Peshawar, Pakistan

Akhtar Nadhman
Institute of Integrative Biosciences (IIB), CECOS University of Information Technology and Emerging Sciences, Peshawar, Khyber Pakhtunkhwa, Pakistan

Sammer Yousuf
H. E. J. Research Institute of Chemistry, International Center for Chemical and Biological Sciences, University of Karachi, Karachi 75270, Pakistan

Sraa Abu-Melha
Department of Chemistry, Faculty of Science of Girls, King Khaled University, Abha, Saudi Arabia

Agnieszka Sobczak, Monika A. Lesniewska-Kowiel, Izabela Muszalska, Artur Firlej, Judyta Cielecka-Piontek, Szymon Tomczak and Anna Jelńiska
Department of Pharmaceutical Chemistry, Faculty of Pharmacy, Poznan University of Medical Sciences, Grunwaldzka 6, 60-780 Poznań, Poland

Bolesław Barszcz
Institute of Molecular Physics Polish Academy of Sciences, Smoluchowskiego 17, 60-179 Poznań, Poland

Irena Oszczapowicz
Department of Modified Antibiotics, Institute of Biotechnology and Antibiotics, Starościńska 5, 02-515 Warszawa, Poland

Bao-Hua Song and Chen Li
Inner Mongolia University for Nationalities, Tongliao, Inner Mongolia, China

Gui-Feng An
Inner Mongolia University for Nationalities, Tongliao, Inner Mongolia, China
The First Affiliated Hospital of Inner Mongolia University for Nationalities, Tongliao, Inner Mongolia, China

Kornelia Czaja, Jacek Kujawski, Elhbieta Jodłowska-Siewert, Paulina Szulc and Marek K. Bernard
Department of Organic Chemistry, Faculty of Pharmacy, Poznan University of Medical Sciences, Grunwaldzka 6 Street, 60-780 Poznań, Poland

Tomasz Ratajczak
Centre of New Technologies, University of Warsaw, Banacha 2C, 02-097 Warsaw, Poland

Dominika Krygier and Marcin K. Chmielewski
Institute of Bioorganic Chemistry, Polish Academy of Sciences, Z. Noskowskiego Str. 12/14, 61-704 Poznań, Poland

Nedhal A. Al-Douri, Ashok K. Shakya and Ghaleb A. Oriquat
Pharmacy Department, Faculty of Pharmacy and Medical Sciences, Al-Ahliyya Amman University, Al Salt Road, Amman 19328, Jordan

Randa N. Haddadin
Pharmaceutics and Pharmaceutical Technology Department, School of Pharmacy, The University of Jordan, Amman, Jordan

Shelley L. Cooke
Department of Materials Science and Engineering, Virginia Polytechnic Institute and State University, Blacksburg, VA 24061, USA

Abby R. Whittington
Department of Materials Science and Engineering, Virginia Polytechnic Institute and State University, Blacksburg, VA 24061, USA
Department of Chemical Engineering, Virginia Polytechnic Institute and State University, Blacksburg, VA 24061, USA

Siham Lahsasni, Dunya A. M. Al-Hemyari, Yahia Nasser Mabkhoot, Fadilah S. Aleanizy, Asma A. Alothman and Zainab M. Almarhoon
Department of Chemistry, King Saud University, Riyadh 11451, Saudi Arabia

Hazem A. Ghabbour
Department of Pharmaceutical Chemistry, College of Pharmacy, King Saud University, Riyadh 11451, Saudi Arabia
Department of Medicinal Chemistry, Faculty of Pharmacy, University of Mansoura, Mansoura 35516, Egypt

Sulan Liao
College of Tea and Food Science, Wuyi University, Wuyishan 354300, China

Qilin Weng
Information Technology & Laboratory Management Center, Wuyi University, Wuyishan 354300, China

Hessa H. Al-Rasheed, Monirah Al Alshaikh and Mohammed R. H. Siddiqui
Department of Chemistry, College of Science, King Saud University, Riyadh 11451, Saudi Arabia

Ayman El-Faham
Department of Chemistry, College of Science, King Saud University, Riyadh 11451, Saudi Arabia
Chemistry Department, Faculty of Science, Alexandria University, Ibrahimia, Alexandria 12321, Egypt

Essam N. Sholkamy and Ahmed S. Al-Obaidi
Department of Botany and Microbiology, College of Science, King Saud University, Riyadh 11451, Saudi Arabia

Màrcia C. de Souza, Luan F. Diniz and Chris H. J. Franco
Department of Chemistry, Federal University of Juiz de Fora, Juiz de Fora, Minas Gerais 36036-900, Brazil

Renata Diniz
Department of Chemistry, Federal University of Juiz de Fora, Juiz de Fora, Minas Gerais 36036-900, Brazil
Department of Chemistry, Federal University of Minas Gerais, Belo Horizonte, Minas Gerais 31270-901, Brazil

Index